DISCRETE MATHEMATICS

By
Vinay Kumar

FIRST EDITION 2002

REPRINTED 2011

Distributors:

MICRO BOOK CENTRE
2, City Centre, CG Road,
Near Swastic Char Rasta,
AHMEDABAD-380009 Phone: 26421611

COMPUTER BOOK CENTRE
12, Shrungar Shopping Centre, M.G. Road,
BANGALORE-560001 Phone: 5587923, 5584641

MICRO BOOKS
Shanti Niketan Building, 8, Camac Street,
KOLKATTA-700017 Phone: 2826518, 2826519

BUSINESS PROMOTION BUREAU
8/1, Ritchie Street, Mount Road,
CHENNAI-600002 Phone: 28534796, 28550491

DECCAN AGENCIES
4-3-329, Bank Street,
HYDERABAD-500195 Phone: 24756400, 24756967

MICRO MEDIA
Shop No. 5, Mahendra Chambers, 150 D.N. Road,
Next to Capital Cinema V.T. (C.S.T.) Station,
MUMBAI-400001
Ph.: 22078296, 22078297, 22002732

BPB PUBLICATIONS
B-14, Connaught Place, **NEW DELHI-110001**
Phone: 23325760, 23723393, 23737742

INFOTECH
G-2, Sidhartha Building, 96 Nehru Place,
NEW DELHI-110019
Phone: 26438245, 26415092, 26234208

INFOTECH
Shop No. 2, F-38, South Extension Part-1
NEW DELHI-110049
Phone: 24691288, 24641941

BPB BOOK CENTRE
376, Old Lajpat Rai Market,
DELHI-110006 PHONE: 23861747

LIMITS OF LIABILITY AND DISCLAIMER OF WARRANTY

The Author and Publisher of this book have tried their best to ensure that the programmes, procedures and functions described in the book are correct. However, the author and the publishers make no warranty of any kind, expressed or implied, with regard to these programmes or the documentation contained in the book. The author and publishers shall not be liable in any event of any damages, incidental or consequential, in connection with, or arising out of the furnishing, performance or use of these programmes, procedures and functions. Product name mentioned are used for identification purposes only and may be trademarks of their respective companies.

All trademarks referred to in the book are acknowledged as properties of their respective owners.

ISBN 81-7656-639-X

Published by Manish Jain for BPB Publications, B-14, Connaught Place, New Delhi-110 001 and Printed by him at Pressworks, New Delhi.

CONTENTS

Page

Chapter 1 Set and Relation....1
1.1 Sets and Its Representation....1
1.2 Types of Sets....4
1.3 Operations on Sets....8
1.4 Principles of Inclusion and Exclusion....13
1.5 Relations and Its Type....15
1.6 Equivalence Relation....21
1.7 Digraph and Matrix Representation of Relation....29
1.8 Closures of Relations....32
Exercise....37

Chapter 2 Function and Generating Function.... 47
2.1 Types of Functions....49
2.2 Order of Function....55
2.3 Sequences and Series....61
2.4 Generating Functions and Series....68
Exercise....75

Chapter 3 Solving Recurrence Relation....82
3.1 Classifications of Recurrence Relation....82
3.2 Backtracking and Forward Chaining Method....84
3.3 Characteristics Equation Method....89
3.4 Generating Function Method....109
Exercise....119

Chapter 4 Combinatorics....124
4.1 Methods of Proofs....124
4.2 Permutations....139
4.3 Combinations....149
4.4 Pigeon Hole Principles....159
Exercise....165

Chapter 5 Group....171
5.1 Definitions and Properties....171
5.2 Subgroup 185
5.3 Homomorphism of groups....195
5.4 Cyclic Group....201
5.5 Permutation Group....206
Exercise....220

Page

Chapter 6 Ordered Set....225
- 6.1 Poset....225
- 6.2 Lattice....240
- 6.3 Finite Boolean Algebra....250
- 6.4 Optimization of Boolean Expression....260
- 6.5 Propositional Calculus....276
- Exercise....285

Chapter 7 Graph Theory and Tree....292
- 7.1 Graphs Introduction....292
- 7.2 Different Types of Graphs....299
- 7.3 Path and Circuits....308
- 7.4 Shortest Path Algorithms....322
- 7.5 Coloring of Graphs....333
- 7.6 Definitions and Types of Tree....343
- 7.7 Tree Traversals....348
- 7.8 Minimum Spanning Tree....352
- Exercise....358

Chapter 8 Finite Automata....368
- 8.1 Grammar and languages....368
- 8.2 Regular Expressions....380
- 8.3 Finite State Machine....389
- 8.4 Minimization of Finite State Machine....400
- Exercise....411

Chapter 9 Derived Algebraic Structures....422
- 9.1 Matrix....422
- 9.2 Ring....429
- 9.3 Field....443
- 9.4 Vector Space....451
- Exercise....462

BIBLIOGRAPHY....467
INDEXES....477

PREFACE

What should be *"a course on discrete mathematics"*? No two scientists, involved in teaching this course, will ever arrive at the same list of topics. However, there are sure to be some common topics. I have decided the contents of this book based on the syllabus followed in institutions where computer science is taught as the main subject at the graduation level of computer engineering and master level of computer applications.

This book is a detailed set of lecture notes I prepared for the students pursuing ***B-level*** course of DOE at Mahan Computer Systems, New Delhi, India. I have taken care to simplify the subject to the maximum.

This book is intended to be a textbook for the students pursuing B.E./B.Tech in computer science or Master of Computer Application or equivalent courses. Topics included are self-contained. Sequence is maintained in such a way that no prerequisite is necessary. Though, it is assumed that this course is taught after a basic course on computer fundamental and data structure. Only necessary theorems have been included; and wherever required, I have tried to demonstrate their applicability using appropriate examples.

The book contains nine chapters. Each chapter is divided into number of sections. The **chapter one** describes the basic concepts of sets and relations. Since the principle of inclusion and exclusion is based on the cardinality of sets, this topic too is included in this chapter. The **chapter two** contains definitions of functions and sequences. The discussion relating to generating function in this chapter is restricted to binomial and exponential generating function only. Recurrence relation and recursive function is an important topic taught in this course. **Chapter three** describes how to solve a recursive equation.

Chapter four contains different methods of proof and way to solve counting problems using method of permutation, combination and pigeonhole principle. This chapter is named '*combinatorics*' because it deals with counting problems. **Chapter five** contains basic definitions and properties of a group. This chapter in included and placed in the middle of this book with twin purposes: first to make the reader aware about the concept of a mathematical structure and second to prepare them for the subsequent chapters which deals with different mathematical structures.

Chapter six describes the concept of partial ordered sets (Poset). The evolution of the notion of a finite Boolean algebra from the primitive concept of sets and relations is explained here. We have concluded this chapter with a section on propositional calculus to explain "how a mathematical theory when supplemented by an appropriate technology brings a revolutionary change in industry and day to day life of human being as a whole."

Chapter seven contains topics related to graphs and trees. Algorithms related to graph/tree traversal, minimum spanning tree, shortest path etc is mentioned in this chapter. Reader may realize that these concepts are very useful in determining the topology of a computer communication network. The last **Chapter eight** describes a finite automata, language and grammar. We have restricted our discussion in this book to the deterministic finite state machine. **Chapter nine** is included as an extension of the chapter five.

Each example is numbered in sequence. The sequence is maintained for a section. Wherever reference is made to an example say **n**, it means example **n** of the current section, unless otherwise stated. I have extensively used diagrams, as a diagrammatic representation of a fact is more apprehensive than its textual counterpart. A number of problems are provided at the end of each chapter as an exercise. Some facts and definitions may be found in the exercises.

It will be honest on my part to accept that it is not possible to include every thing in one book. Consequently, it may be possible that I may have omitted some *definitions* or *facts* related to the topics covered. I have consulted many books, research papers, websites and hand notes. A list of references under **Bibliography** is presented at the end of the book.

Many people contributed directly or indirectly to the completion of this book. Thanks are due to Captain R. K. Mittal, Director, MCS, who was able to convince me to write this book. Otherwise, I would not have even initiated this task. I am grateful to the students of MCS, who always encouraged me and many times thanked me for teaching them this course. Thanking for an otherwise thankless job is itself an encouragement. Special thanks to my student Miss Amita Jain, who made me, realize that I can indeed write a book on discrete mathematics. Among my colleagues in National Informatics Center, where presently I am working as scientist, some pulled me down, some encouraged me and some gave constructive suggestions. I am grateful to all of them. I would like to single out my boss, Dr. (Mrs) Vandana Sharma, Sr. Technical Director, for giving me very useful tips and guidance all along and Miss Bhavna Chopra, who contributed a lot by sparing her time in proofreading the work.

Last, but not the least, I specially thank to my wife Rakhi and daughter Rakvi who tolerated me all along while I devoted my time completing this book. I recall the word of Rakvi, "*Kaash main bhi book hoti, pappa ke hathon mein hameshe rahti*!" She uttered this line while she was trying to talk to me and I was ignoring her to complete a paragraph of this book.

I dedicate this book to the students of MCS.

Vinay Kumar

CHAPTER ONE
Sets and Relation

Whenever a word finds its usage in many areas, it becomes very difficult to give a universally accepted meaning to the word. Therefore, the word get. its contextual meaning. For examples: network, communication, inputs etc. ***Set*** is one of those words. While writing a program in a structured programming language, we declare a variable of a type before using it. Once we have declared a variable of a type say ***integer***, the variable can hold only an integer value. What I am trying to convey to you that we can always make collection of certain type of objects. We can give a name to it. Then, we can use the name to refer to this collection, thereafter.

1.1 Sets and Its Representation

A set can be defined as a "*well defined collection of objects without repetition*". A set is known by its elements. A set is said to be well defined if it can be determined that a given object is a member of the set. This can be made possible, if the nature of the objects of a set is known. For example, a set of characters, a set of digits, a set of positive integers, a set of male students of a class, a set of female students of a class etc. This implies that all objects of a set share a common property. A set can be described by the common property of elements of that set. If an object does not possess that property, it does not belong to the set.

Example 1 D = {0, 1, 2, 3, 4, 5, 6, 7, 8, 9}. This is a set. It is a collection of all digits. ♦

In the example 1, all objects are placed inside '{' and '}'. The collection is given a name D. Thus, D is a set of all digits. An object belonging to a set is called **member** of that set. Thus, 0 is a member of D. 5 is a member of D. 10 is not a member of D. 'a' is not a member of D. In brief, we say that any thing other than 0 to 9 is not a member of D. The set D can also be written as

$$D = \{x \mid x \text{ is a digit}\}$$

In this representation, we have used common property of the members. A set can also be represented in diagrammatic way. The set D of example 1 can be shown as a diagram of figure 1.1.1.

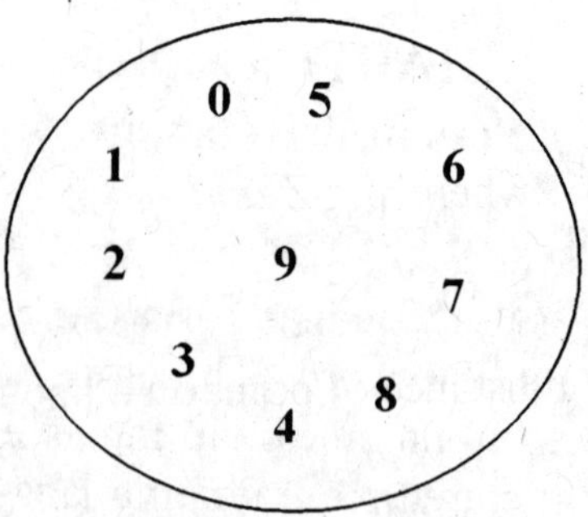

Figure 1.1.1

Membership of a Set

If an object x is a member of a set A, it is written as **x ∈ A**, and is read as 'x belongs to A'. The symbol ∈ stands for 'belongs to'. If an object x is not a member of A then, it is written as **x ∉ A,** and is read as 'x does not belong to A'. The symbol ∈ is also called 'membership operator'

Example 2 Let A = {a, e, i, o, u}. This is a set of lower case vowels of roman alphabets. Here, a ∈ A, i ∈ A, but b∉ A. ♦

Example 3 Let B = {x | x is a rational number}. Since 0 is a rational number, we can write 0 ∈ B. √2 is not a rational number. Therefore, √2 ∉ B. ♦

Some Standard Sets

In this book, we shall be using different sets. Some of the sets have standard notations. Wherever the notation appears, it represents the corresponding sets unless otherwise stated.

1. Set of Integers

$$Z = \{-\infty, \ldots, -3, -2, -1, 0, 1, 2, 3, \ldots, \infty\}$$

2. Set of Positive Integers

$$Z_+ = \{1, 2, 3, \ldots, \infty\}$$

3. Set of Negative Integers

$$Z_- = \{-1, -2, -3, \ldots, -\infty\}$$

4. Set of Natural Numbers

$$N = \{1, 2, 3, \ldots, \infty\}$$

5. Set of Whole Numbers

$$I = \{0, 1, 2, 3, \ldots, \infty\}$$

6. Set of Rational Numbers

$$Q = \{x \mid x = \frac{p}{q} \text{ where, } p \in Z \text{ and } q \in Z_+\}$$

7. Set of Real Numbers

R = {x | x is a distance of point on a line from origin}

8. Set of complex Numbers

C = {z = x + iy | z is a point on a 2-D plane and i = √-1}

9. Set of Alphabets

$\Sigma = \{a, b, c, \ldots, z\}$

The set of whole numbers is also a set of non-negative integers. Zero is neither a positive integer nor a negative integer There is always a controversy regarding inclusion of zero in the set of natural numbers. Since counting starts from **zero**, it must be included in N, but some authors argue otherwise. Since we have a set I which solves this purpose, we shall exclude zero from N.

Set Representation

A set is constituted by its elements. Elements of a set is denoted by lowercase alphabets like a, b, c etc and a set is denoted by uppercase alphabets like A, B, C etc. Any set can be represented in the following three ways:

- Enumerated way
- Symbolic way, and
- Diagrammatic way

Enumerated way: In this method, all members of a set are listed explicitly. For example,

A = {1, 2, 3, 4, 5} is a set of all positive integers less than or equal to 5.
B = {a, e, i, o, u} is a set of vowels of lowercase Roman alphabets.
B = {0, 1} is a set of binary symbols.

This method of representation is also called ***Tabulation method***.

Symbolic way: While representing a set in symbolic way, we use a general statement describing the common property of elements of the set. For example,

Z = {x | x is an integer} is a set of all integers
P = {x | x is a prime number} is a set of all prime numbers
C ={x | x is serial number of PCs manufactured by HCL} is a set of serial numbers of all personal computers manufactured by Hindustan Computer Limited.
A = $\{x \mid x^2 \leq 16$ and x is an integer$\}$ is another example of a set represented in symbolic way.

The symbolic representation of a set is also called '***set former method***' or '***descriptive phrase method***' or '***set theoretic method***'.

Diagrammatic way: This method of representation is suitable for a set having few elements. In this method, two geometrical entities: *circle and point* are used. A circle is used to represent a set and a point is used to represent an element of the set. The number of elements in a set determines the number of points to be used. The diagram is called **Venn diagram**. For example, a set A = {2, 3, 5, 7, 11} can be shown in diagrammatic way as below. More about this we shall study in the section 1.3.

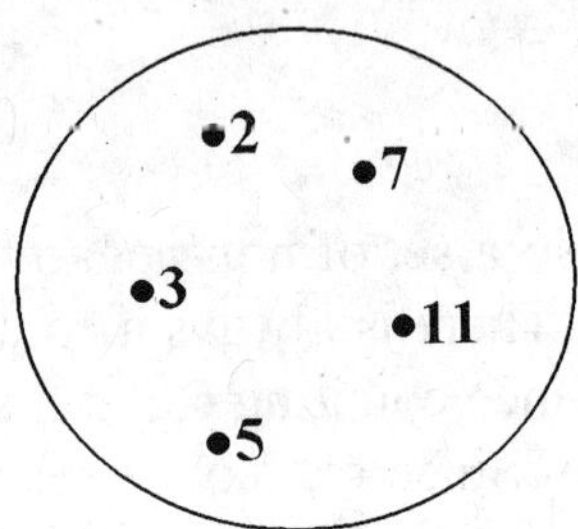

Figure 1.1.2

1.2 Types of Set

Nature of elements determines nature of a set e.g. set of students, set of courses etc. Number of elements in a set determines what type of set we are dealing with. The number of elements of a set A is called ***cardinality*** of the set A and is denoted by n(A) or |A|.

Finite Set

Any set A is set to be finite if A has finite number of elements i.e. elements of set A can be counted by a finite number. For example,

F = {a, b, x, z, 0, 18, p}; |F| = 7
S = {x | x is a floating point operation executed in a second}
C = {x | x is a name of TV channels broadcast over a cable}

Infinite Set

If a set has an infinite number of elements, it is said to be an infinite set i.e., elements of such a set cannot be counted by a finite number. For example,

S = {x | x is a natural number}
C = {x | x is a grain of sand on sea shore}

Point Set

This set is defined as a collection of points along a straight or curved line. A *point set* may be finite or infinite depending upon the context in which point is defined. For example, let P and L be point set defined as below.

P = {x | x is pixels along a line on VDU}
L = {x | x is a point on any geometric line}

Clearly, P is a finite set whereas L is an infinite set.

Null Set

A set containing no element is called a *null set*. A null set is denoted by ϕ (pronounced as 'phi'). Following sets are examples of a null set.

A = {x | x is a married bachelor}
B = {x | x is a register capable of storing infinite number}
C = {x | x is real and $x^2 = -1$}

Unit Set

Any set containing only one element is called a *unit set* or *singleton set*. For example,

A = {a}
B = {x | x is an integer that is neither negative nor positive}

Universal Set

A *universal set* is the totality of elements under consideration. It is the set of all possible elements relevant to a particular application. It is denoted by Ω (pronounced as 'omega'). Following are some examples of a universal set.

Ω = {Head, Tail}
Ω = {0, 1}
Ω = {x | x is an ASCII code}
Ω = {x | x is roll number of students of MCS}

Disjoint Set

Any two sets A and B are said to be disjoint if they have no elements in common. For example, let A = {x | x is odd integer} and B = {x | x is even integer}. Since sets A and B have no element in common, they are disjoint. Another example can be given as:

M = {x | x is a pointing device of a computer system}
C = {x | x is a keyboard}

Here, M and C are two sets having no elements in common. Hence, they are disjoint.

Overlapping Set

Any two sets A and B are overlapped if they have at least one element in common. For example, let A = {x | x is a name of student pursuing DM course} and B = {x | x is a name of student pursuing UNIX course}. It is possible that a few students may be pursuing both the courses in the same semester. Hence, sets A and B are overlapping sets. Another example can be given as:

A = {x | x is a character represented by 7 bits}
B = {y | y is a character represented by 8 bits}

Since characters, represented by 7 bits are also represented by 8 bits, sets A and B are overlapping sets.

Subset and Superset

Let A and B be any two sets. Set A is said to be a **subset** of B if all elements of A are in B. For example, let A = {a, b} and B = {a, b, c}. Since all elements of A are also in B, A is a subset of B. Whenever a set A is a subset of a set B, it is written as $A \subseteq B$ and we call that set A is *contained in* set B or B *contains* A. If set A is a subset of set B, B is called **superset** of A and is written as $B \supseteq A$. As for example, let Z_+ be the set of positive integer and Z be the set of all integer, then $Z_+ \subseteq Z$ and $Z \supseteq Z_+$.

Proper Subset

Let A and B be any two sets. Set A is said to be a **proper subset** of B if

- All elements of A are in B and
- There exists at least one element x in B such that x is not in A.

For example, let A = {a, b} and B = {a, b, c}. Since all elements of A are also in B, and $\exists$ $c \in B$ such that $c \notin A$, A is a proper subset of B. Whenever a set A is a proper subset of a set B, it is written as $A \subset B$ and we call that set A is *properly contained in* set B or B *properly contains* A.

It is important to note that a *null set is subset of every set* and *any set is subset of itself*.

Equal Set

Let A and B be two sets. Set A is set to be ***equal*** to set B if and only if

1. $A \subseteq B$ and
2. $B \subseteq A$

That means every element of one set is also an element of another set. Equal sets are also called ***identical*** sets. Equal sets have exactly the same element, possibly in different order. As for example, let A = {x | x is letter of word "equal"} and B = {e, q, u, a, l}. Here, A = B. See also the following examples.

Example 1 Let A = {x | x is a digit} and B = {x | x is a non-negative integer and $x < 10$}. Integers that are non-negative and less than 10, are digits 0, 1, 2, …, 9. Thus, A = B. ♦

Example 2 Let A = {a, b, 2, 3} and B = {2, 3, a, b}. Here, though the elements in A and B are in different order, both have the same elements. Hence, A = B. ♦

Equivalent Set

Two sets A and B are said to be equivalent if both have same number of elements i.e. $|A| = |B|$. Which all elements are in A and which all are in B is not important. For example, let A = {1, 2, 3} and B = {x, y, z}, then both A and B are equivalent sets. Notice that they are not identical sets. Let us consider another example as below.

Example 3 Let A = {x | x is an even integer} and B = {y | y is an odd integer}. Obviously, both A and B are equivalent sets. ♦

It is important to note here that two identical sets are always equivalent but the reverse is not true.

Power Set

Sometimes, a set may contain another set as its element. If all the elements of a set are sets, we call the set as '***set of sets***'. If some elements of a set are primitive element and other are sets; we call the set as a '***mixed set***'. For example, let A = {{1}, {1, 2}, {3}}. Since all elements of A are sets, A is set of sets.

Example 4 Let A = {1, 2, {3}, {2}, {1}}. Since some of the elements of A are primitive element and some are sets; A is a mixed set. ♦

Let A = {1, 2, 3}. Consider all possible distinct subsets of A. Since ϕ is a subset of every set, $\phi \subseteq A$. Also, every set is a subset of itself thus, $A \subseteq A$. Other distinct subsets are: {1}, {2}, {3}, {1, 2}, {1, 3} and {2, 3}. The set of these subsets is {ϕ, {1}, {2}, {3}, {1, 2}, {1, 3}, {2, 3}, {1, 2, 3}}. This set is a set of all possible subsets of the set A. We call the set as **Power Set** of A and is represented as either P(A) or 2^A. Therefore, a ***power set of*** any set is defined as a collection of all possible distinct subsets of the set. Therefore,

$$P(A) = \{\phi, \{1\}, \{2\}, \{3\}, \{1, 2\}, \{1, 3\}, \{2, 3\}, \{1, 2, 3\}\}$$

Example 5 Let A = {a, b}. Then, P(A) = {{ϕ, {a}, {b}, {a, b}}. ♦

Example 6 Let B = {x}. Then P(B) = {{ϕ, {x}}. ♦

From above examples, it is obvious that if $|A| = n$ then, $|P(A)| = 2^n$. This implies that if number of elements in A is 1 then, its power set will have $2^1 = 2$ elements as shown in example 6 above. Similarly, if number of elements in A is 3 then, its power set contains $2^3 = 8$ elements. Now, let us co-relate the concept of power set with that of bit combinations that we have studied in our computer fundamental courses. Let A be any set. Let us mark the position of elements in A, if it is non-empty, as first, second, third etc. We can represent any subset of A by a bit-string indicating absence of an element by '0' and presence by '1'. We can have a bit representation for every possible subset of A. The length of each bit–string is equal to $|A|$. For illustration see the following example.

***Example** 7* Let A = {1, 2, 3}. All possible subsets of A, theirs bit representation and remarks are given the following table 1.2.1.

Sr. No.	Subsets	Bits Representation	Remark
1.	ϕ	000	None of the 1, 2, 3 is present
2.	{1}	100	Only '1' is present
3.	{2}	010	Only '2' is present
4.	{3}	001	Only '3' is present
5.	{1, 2}	110	'3' is absent
6.	{1, 3}	101	'2' is absent
7.	{2, 3}	011	'1' is absent
8.	{1, 2, 3}	111	All are present

Table 1.2.1

We know that we can have 2^3 possible bit combinations from 3 bits and 2^3 possible subsets of a set having 3 elements. In general, if we have a set of n elements its power set will have 2^n elements.

Multi Set

We have studied that a set is a collection of distinct objects. In real life, we may encounter many situations when collections are not distinct. For example, consider the collection of names of students in a class. Possibly, we may have two or more students having same name. A set in which an element may appear more than once is called a ***multi set***.

Example 8 Let A = {a, a, a, b, b, d}. Since 'a' appears three times and 'b' appears two times in A, we call A as a multi set. ♦

Frequency of appearance of an element x, in a multi set A, is called ***multiplicity*** of x in A. In the example 8, multiplicity of a is 3, that of b is 2 and of d is 1. *Cardinality* of a multi set is given by the count of distinct element in the multi set. Thus, cardinality of multi set A of example A is 3. It is important to note that *a set is a special case of multi set in which multiplicity of every element is one.*

1.3 Operations on Sets

It is required, most of the time, to know combined, common or relative properties of sets. For this we use set operators. Binary set operators are Union ($\cup$), Intersection ($\cap$), Set difference (–), Symmetric difference ($\oplus$) and Cartesian Product (X). The unary set operator is the set complement (A^C) where, A is any set. Let us see them one by one to know what they yield when applied on sets.

Union $A \cup B$

The union of two sets A and B is the set of all elements belonging to set A or to set B i.e. $A \cup B = \{x \mid x \in A \text{ or } x \in B\}$.

Example 1 Let $A = \{x \mid x \text{ is a rational number}\}$ and $B = \{x \mid x \text{ is an irrational number}\}$. Then $A \cup B = \{x \mid x \text{ is a real number}\}$. ♦

Example 2 Let $A = \{x \mid x \text{ is an instruction to be executed when if condition is true}\}$ and $B = \{x \mid x \text{ is an instruction to be executed when if condition is false}\}$. Then,

$$A \cup B = \{x \mid x \text{ is an instruction in if-else body}\}$$ ♦

Example 3 Let $A = \{1, 2, 3\}$ and $B = \{2, 3, 4\}$. Then, $A \cup B = \{1, 2, 3, 4\}$. ♦

Intersection $A \cap B$

The intersection of two sets A and B is defined as the collection of all elements belonging to both A and B i.e. $A \cap B = \{x \mid x \in A \text{ and } x \in B\}$.

Example 4 Consider the sets A and B of example 3. Then, $A \cap B = \{2, 3\}$. ♦

Example 5 Consider the sets of example 1. Since a number is either a rational or irrational (it can never be both), we have $A \cap B = \phi$. ♦

Set Difference A – B

The set difference A – B is defined as the collection of all the elements belonging to A which does not belong to B i.e. $A - B = \{x \mid x \in A \text{ and } x \notin B\}$. Similarly, set difference B – A is defined as the collection of all the elements belonging to B which does not belong to set A i.e. $B - A = \{x \mid x \in B \text{ and } x \notin A\}$.

Example 6 Let $A = \{a, b, x, y\}$ and $B = \{c, d, x, y\}$. Then $A - B = \{a, b\}$ and $B - A = \{c, d\}$. ♦

Symmetric Difference

The symmetric difference of two sets A and B is defined as the collection of element belonging to either A or B but not to both i.e.,

$$A \oplus B = \{x \mid (x \in A \text{ and } x \notin B) \text{ or } (x \notin A \text{ and } x \in B)\} = (A - B) \cup (B - A)$$

Example 7 Consider the sets of example 6. Then, $A \oplus B = \{a, b, c, d\}$. ♦

Example 8 Consider the sets of example 1. Here, $A - B = A$ and $B - A = B$. Thus, $A \oplus B = A \cup B$ ♦

Example 9 Let $A = \{a, b, c, d\}$ and $B = \{a, c, e, f, g\}$. Then $A \oplus B = \{b, d, e, f, g\}$. ♦

Complement of Set

Let A be a set and Ω be its universal set. The complement of A, denoted by A^C (A' or $\bar{A}$) is defines as the collection of all elements of Ω which are not in A i.e. $A^C = \{x \mid x \notin A\}$.

Example 10 Let Ω = {x | x is an integer} and A = {x | x is an even integer}. Then, A^C = {x | x is an odd integer}. ♦

Example 11 Let A = {1, 2, 3} and Ω = {x | x is a digit}. Then, A^C = {0, 4, 5, 6, 7, 8, 9}. ♦

Example 12 Let A = {0} and Ω = {0, 1}. Then A^C = {1}. Similarly, if B = {1} then B^C = {0}. This property of complement conveys that if we have only two possibilities: OFF(0) and ON(1), then complement of 1 is 0 and that of 0 is 1. ♦

Results of the different set operations discussed above can be shown diagrammatically using Venn Diagram. The figure 1.3.1(a) shows the set operation of union of two sets. The operation of intersection of two sets is shown in the figure 1.3.1(b). The set difference operations of A – B and B – A are shown in the figures 1.3.1(c) and 1.3.1(d) respectively. The figure in 1.3.1(e) represents the symmetric difference of two sets A and B. The last figure 1.3.1(f) stands for set operation of complement.

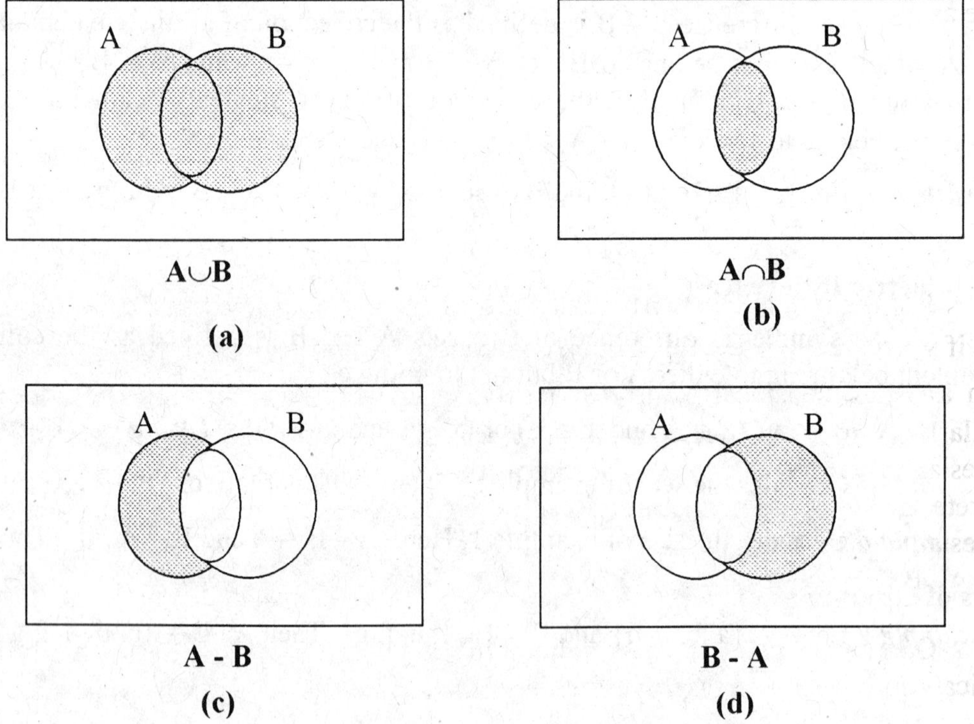

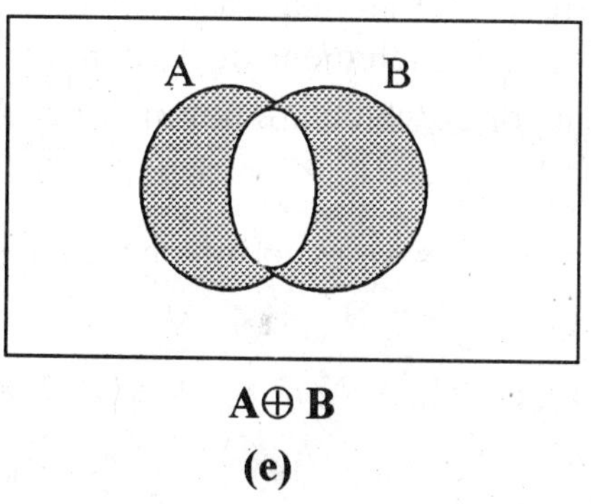

A⊕ B

(e)

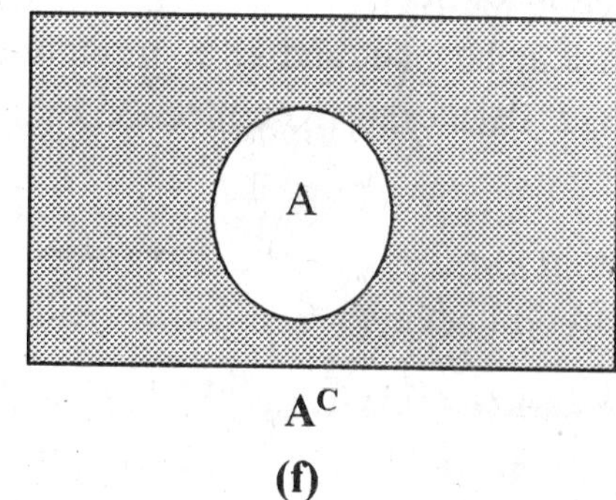

A^C

(f)

Cartesian Product

Let A = {a, b} and B = {x, y, z}. Let us construct pairs (x, y) in such a way that first element x of the pair comes from set A and the second element y comes from set B. This results into following collection of pairs.

$$\{(a, x), (a, y), (a, z), (b, x), (b, y), (b, z)\}$$

This collection of pairs is product set of sets A and B. We call this product set as **Cartesian Product** of sets A and B. It is written as A x B and defined as collection of all **ordered pairs** (x, y) such that x ∈ A and y ∈ B. In symbolic form of notation, we can write as

$$A \times B = \{(x, y) \mid x \in A \text{ and } y \in B\}$$

Similarly,

$$B \times A = \{(x, y) \mid x \in B \text{ and } y \in A\}$$

If A and B are defined as above, we have

$$B \times A = \{(x, a), (y, a), (z, a), (x, b), (y, b), (z, b)\}$$

It is important to note here that pairs (x_1, y_1) and (x_2, y_2) are said to be equal if and only if $x_1 = x_2$ and $y_1 = y_2$. This can be better illustrated by taking two points (1, 3) and (3, 1) in a two dimensional plane. Obviously, these two points are not the same points. Similarly, (x, a) is not equal to (a, x) if x ≠ a. This implies that, in general, A x B ≠ B x A. Cartesian product is the basis of **Relation** on sets. Relation is a very important topic in discrete mathematical structure. Readers are advised to understand the concept of Cartesian product thoroughly.

Laws of Operations on Sets

Operations on sets have some properties that are used in various real life applications to prove or disprove theorems and statements. These rules are listed below.

Proof of some of the laws listed below is given. In the similar way other laws can be proved. Readers are advised to prove them.

A. *Idempotent Laws* or *Laws of Inclusion* or *Tautology*

1. $A \cup A = A$ 2. $A \cap A = A$

B. *Commutative Laws*

3. $A \cup B = B \cup A$ 4. $A \cap B = B \cap A$

C. *Associative Laws*

5. $A \cup (B \cup C) = (A \cup B) \cup C$ 6. $A \cap (B \cap C) = (A \cap B) \cap C$

D. *Distributive Laws*

7. $A \cup (B \cap C) = (A \cup B) \cap (A \cup C)$ 8. $A \cap (B \cup C) = (A \cap B) \cup (A \cap C)$

E. *Identity Laws*

9. $A \cup \phi = A$ 10. $A \cap \phi = \phi$
11. $A \cup \Omega = \Omega$ 12. $A \cap \Omega = A$

F. *Laws of Complement*

13. $A \cup A^C = \Omega$ 14. $A \cap A^C = \phi$
15. $(A^C)^C = A$ 16. $\phi^C = \Omega$
17. $\Omega^C = \phi$

G. *De Morgan's Laws*

18. $(A \cup B)^C = A^C \cap B^C$ 19. $(A \cap B)^C = A^C \cup B^C$

***Proof of Law** 7:* The general principle used in proving the equality of two sets A and B is to show that $A \subseteq B$ and $B \subseteq A$. To prove this law, we have to prove that

$$A \cup (B \cap C) \subseteq (A \cup B) \cap (A \cup C) \quad \text{———} \quad (1)$$
$$(A \cup B) \cap (A \cup C) \subseteq A \cup (B \cap C) \quad \text{———} \quad (2)$$

Let us first prove the validity of expression (1). Let $x \in A \cup (B \cap C)$. Then,

$$\begin{aligned} x \in A \cup (B \cap C) &\Rightarrow x \in A \text{ or } x \in B \cap C \\ &\Rightarrow x \in A \text{ or } (x \in B \text{ and } x \in C) \\ &\Rightarrow (x \in A \text{ or } x \in B) \text{ and } (x \in A \text{ or } x \in C) \\ &\Rightarrow (x \in A \cup B) \text{ and } (x \in A \cup C) \\ &\Rightarrow x \in (A \cup B) \cap (A \cup C) \end{aligned}$$

Therefore, $A \cup (B \cap C) \subseteq (A \cup B) \cap (A \cup C)$. Similarly, we can prove the validity of expression (2). Combining (1) and (2), we get

$$A \cup (B \cap C) = (A \cup B) \cap (A \cup C)$$

Proved.

Proof of Law 18: Let x be any element of $(A \cup B)^C$. Then, we have

$$\begin{aligned} x \in (A \cup B)^C &\Rightarrow x \notin (A \cup B) \\ &\Rightarrow x \notin A \text{ and } x \notin B \\ &\Rightarrow x \in A^C \text{ and } x \in B^C \\ &\Rightarrow x \in A^C \cap B^C \end{aligned}$$

Therefore, $(A \cup B)^C \subseteq A^C \cap B^C$. Similarly, we can prove the validity of expression $A^C \cap B^C \subseteq (A \cup B)^C$. Combining these two results, we get

$$(A \cup B)^C = A^C \cap B^C$$

Proved.

1.4 Principle of Inclusion and Exclusion

This principle is based on the cardinality of finite sets. Let us begin with an example of finding all integers between 1 and 100 that are divisible by 3 or 7. Let A and B be two sets defined as:

$$A = \{x \mid x \text{ is divisible by 3 and } 1 \le x \le 100\} \text{ and}$$
$$B = \{x \mid x \text{ is divisible by 7 and } 1 \le x \le 100\}$$

Here, problem is to find the number of elements in $A \cup B$ i.e. count of those numbers which are divisible either by 3 or 7. Also, there are numbers that are divisible by both 3 and 7 i.e. $A \cap B$ is not a null set. This implies that while counting the numbers, the common elements of A and B have been ***included*** twice, so once it has to be ***excluded***. Therefore, we subtract the count of numbers in $A \cap B$. Hence, we can write

$$|A \cup B| = |A| + |B| - |A \cap B|$$

In the above equation, if the count of any three are known, the other can be found. In the given context, $|A|$, $|B|$ and $|A \cap B|$ can be computed as below:

$$|A| = 33 \qquad \left[n = \frac{l-a}{d} + 1 = \frac{99-3}{3} + 1 = 33\right]$$

$$|B| = 14 \qquad \left[n = \frac{l-a}{d} + 1 = \frac{98-7}{7} + 1 = 14\right]$$

$$|A \cap B| = 4 \qquad \left[n = \frac{l-a}{d} + 1 = \frac{84-21}{21} + 1 = 4\right]$$

Therefore,

$$|A \cup B| = 33 + 14 - 4 = 43.$$

Ans.

Now, we are in a position to know formal statements related to *Principle of Inclusion and Exclusion.* These are listed below.

1. Let A and B be two sets then

$$|A \cup B| = |A| + |B| - |A \cap B|$$

If A and B are disjoint sets, then

$$|A \cup B| = |A| + |B|$$

2. Let A, B and C be three sets then

$$|A \cup B \cup C| = |A| + |B| + |C| - |A \cap B| - |A \cap C| - |B \cap C| + |A \cap B \cap C|$$

If A, B and C are mutually disjoint sets, then

$$|A \cup B \cup C| = |A| + |B| + |C|$$

3. In general, if $A_1, A_2, A_3, \ldots, A_n$ be n sets, then

$$\left|\bigcup_{i=1}^{n} A_i\right| = |A_1| + |A_2| + |A_3| + \cdots + |A_n| - |A_1 \cap A_2| - |A_1 \cap A_3| - \cdots - |A_1 \cap A_n| - \cdots$$

$$- |A_{n-1} \cap A_n| + |A_1 \cap A_2 \cap A_3| + |A_1 \cap A_2 \cap A_4| + \cdots + (-1)^{n-1}\left|\bigcap_{i=1}^{n} A_i\right|$$

If all A_i's are mutually disjoint sets, then

$$\left|\bigcup_{i=1}^{n} A_i\right| = |A_1| + |A_2| + |A_3| + \cdots + |A_n| = \sum_{i=1}^{n} |A_i|$$

4. Let A and B be any two sets, then

$$|A \cap B| \le \min(|A|, |B|)$$
$$|A \oplus B| = |A| + |B| - 2|A \cap B|$$
$$|A - B| = |A| - |A \cap B|$$
$$|A - B| \ge |A| - |B|$$
$$|A \times B| = |A| * |B|$$

See the following example to learn how to apply these rules to solve problems. Many counting problems have been solved using this set of rules. Readers are suggested to solve exercises given at the end of this chapter to get themselves accustomed with such type of problems and theirs solutions.

Example 1 At a university 60% of the teacher play tennis, 50% of them play bridge, 70% jog, 20% play tennis and bridge, 30% play tennis and jog and 40% play bridge and jog. What is the percentage of teachers who jog, play tennis and play bridge?

Solution: Let us assume that there are 100 teachers in the university. Let T, B and J be the set of teachers who play tennis, who play bridge and who jog respectively. Therefore, we have

$$|T| = 60 \qquad |B| = 50 \qquad |J| = 70$$
$$|T \cap B| = 20 \qquad |T \cap J| = 30 \qquad |B \cap J| = 40$$
$$|T \cup B \cup J| = 100 \qquad |T \cap B \cap J| = ?$$

From the formula

$$|A \cup B \cup C| = |A| + |B| + |C| - |A \cap B| - |A \cap C| - |B \cap C| + |A \cap B \cap C|$$

We have

$$100 = 60 + 50 + 70 - 20 - 30 - 40 + |T \cap B \cap J|$$

Or $\quad |T \cap B \cap J| = 100 - 90 = 10$

Therefore, there are 10% of the teachers who jog, play tennis and play bridge.

Ans.

1.5 Relation and Its Type

A *relation* is an association between two or more things. When a relation suggest a correspondence or an association between the elements of two sets, it is a *binary* or *dyadic* relation. For three elements it is *ternary* or *triadic* and so on. In most of the part of this book, we shall be dealing with binary relation only. We have studied Cartesian product of the two sets. We know what is an ordered pair. Now let us get down to the job of defining a ***relation.***

Let A and B be any two non-empty sets. Then A x B is defined as:

$$A \times B = \{(x, y) \mid x \in A \text{ and } y \in B\}$$

Let R be any subset of A x B. Then R may contain zero or more (at the most all) ordered pairs of A x B. This set R is called a relation from a set A to set B. *In general, we define a relation R from a set a to another set B as any subset of A x B.* ***Whenever an ordered pair (x, y) $\in$ R, we say that xRy and it is read as "x is R related to y". Similarly, if (x, y) $\notin$ R, we say that x~Ry and is read as "x is not related to y ".*** If R is a subset of A x A i.e., R is a relation from set A to itself, then R is said to be a relation on A.

Example 1 Let A = {1, 2, 3} and B = {x, y, z}. Then,

A x B = {(1, x), (1, y), (1, z), (2, x), (2, y), (2, z), (3, x), (3, y), (3, z)}

Since a relation from A to B is defined as any possible subset of A x B, {(1, x)}, {(1, x), (2, x)} etc are relation from set A to set B. ♦

In the previous example, in fact, we can have $2^9 = 512$ possible relations from set A to set B. Reader may verify this. *In general, if* $|A| = n$ *and* $|B| = m$, *then* $|A \times B| = mn$ *and so number of possible subsets of A x B =* 2^{mn}. Depending upon the *nature* of ordered pairs a relation R contains from the set A x B, we get different types of relations. The following sections deals with different types of relations.

Types of Relation

Let R be a relation from set A to set B. The product set A x B is the exhaustive list of all possible ordered pairs that R may contain. This set A x B is used to define different types of relation R. Here, set A and B have been assumed to be non-empty.

Universal Relation

A relation R from set A to set B is said to be a ***universal*** relation if

R = A x B.

Example 2 Let A = {1, 2} and B = {a, b}. Here, R = {(1, a), (1, b), (2, a), (2, b)} is a universal relation from set A to set B. ♦

Example 3 Let A = {x, y}. Here R = {(x, x), (x, y), (y, x), (y, y)} is a universal relation on the set A. ♦ .

Complement of a Relation

A relation R from a set A to set B is said to be complement of another relation S from set A to set B if

R = A x B – S **or** S = A x B – R

R and S are called complement of each other. It can be easily observed that if R and S are two relations from set A to set B and if they are complement of each other, then

$R \cup S = A \times B$ and $R \cap S = \phi$

Example 4 Let A = {1, 2} and B = {3, 4}. Let R be a relation from set A to set B defined as {(1, 3), (2, 4)}. Let S be its complement. Then S is given by

A x B – R = {(1, 4), (2, 3)} ♦

The complement of a relation R is denoted as $\bar{R}$ or R^C or R′. In this text we shall use the notation R^C. If R is a relation on a set A then its complement R^C is given by AxA – R.

Null, Void or Empty Relation

Any subset of A x B is a relation from set A to set B. A null set ϕ is a subset of A x B. Thus any relation R is called an *empty relation* from set A to set B if $R = \phi$.

***Example* 5** Let A = {x | x is name of males of a town} and B = {y | y is name of females of that town}. Let R be a relation from A to B defined as:

xRy if and only if "x *is sister of y*".

Since no male can be sister of any other person, $R = \phi$. ♦

If R is a relation on A and if $R = \phi$, R is called an empty relation on A.

Inverse Relation

Let R be a relation from set A = {0, 1} to B = {F, T} and is given as R = {(0, T), (1, F)}. The inverse of a relation R, denoted as R^{-1}, is given by {(T, 0), (F, 1)}. It can be easily observed that R^{-1} is obtained by just reversing the position of elements of ordered pairs of R i.e. (0, T) to (T, 0) and (1, F) to (F, 1). Further, it can also be noticed that $R^{-1} \subseteq B \times A$ i.e. R^{-1} is a relation from B to A whenever R is a relation from A to B. Therefore, we can define inverse of a relation as follow:

"*If* $R = \{(x, y) \mid x \in A \text{ and } y \in B\}$ *be a relation from A to B then* R^{-1} *is a relation from B to A and is given as* $\{(y, x) \mid y \in B \text{ and } x \in A\}$."

In the same way, if R is a relation on A and $R = \{(a, b) \mid a, b \in A\}$ then its inverse, R^{-1}, is given by the set $\{(b, a) \mid (a, b) \in R\}$.

There are some types of relation that is always defined on a set. In the remaining part of this section of the chapter, we shall be discussing relation defined on a set. In some books the following types of relations are placed under properties of relation, but we shall continue to treat them as types of relation.

Reflexive Relation

A relation R defined on a set A is said to be a ***reflexive*** relation if $\forall x \in A$, x R x i.e., $\forall x \in A$, $(x, x) \in R$. (Read as for all x belonging to A (x, x) is in R).

Example 6 Let A = {1, 2, 3}. Let R_1 = {(1, 1), (2, 2), (3, 3)} be a relation on A. Then R_1 is a reflexive relation on A. ♦

Example 7 Let A be a set of example 6 and R_2 be a relation on A defined as R_2 = {(1, 1), (1, 2), (2, 2), (3, 1), (3, 3)}. Since $\forall x \in A$, $(x, x) \in R_2$, R_2 is a reflexive relation on A. ♦

Example 8 Let R_3 be a relation on set A of example 6, defined as R_3 = {(1, 1), (2, 3), (3, 3)}. Since for $2 \in A$, $(2, 2) \notin R_3$, R_3 is not a reflexive relation on A. ♦

Example 9 Let R_4 be a relation on set A of example 6, defined as $R_4 = \{(1, 1), (1, 2), (1, 3)\}$. Since for 2 and 3 $\in$A, we have neither $(2, 2) \in R_4$ nor $(3, 3) \in R_4$, R_4 is not a reflexive relation on A. ♦

Example 10 Let R_5 be a relation on set A of example 6, defined as $R_5 = \{(1, 2), (1, 3), (2, 3)\}$. Since for 1, 2 and 3 $\in$A, we have none of the (1, 1) or (2, 2) or (3, 3) is in R_5, R_5 is not a reflexive relation on A. ♦

Irreflexive Relation

A relation R on a set A is said to be an ***irreflexive*** relation if $\forall x \in A$, x~Rx i.e., $\forall x \in A$, $(x, x) \notin R$. (Read as for all x belonging to A (x, x) is **not** in R). The relation R_5 defined in the example 10 above is an example of irreflexive relation.

Non-Reflexive Relation

A relation R on a set A is said to be a **non-*reflexive*** relation if R is neither *reflexive* nor *irreflexive* i.e. for some x$\in$A, xRx and for some other x$\in$A, x ~R x. The relation R_3 defined in the example 8 and R_4 defined in the example 9 above are examples of non-reflexive relation.

Symmetric Relation

A relation R defined on a set A is said to be a ***symmetric*** relation if for any x, y $\in$ A, if $(x, y) \in R$ then $(y, x) \in R$ i.e. $x R y \Rightarrow y R x$. In the previous example 6, relation R_1 is a symmetric relation, by default. Since $\exists$ no (x, y) in R_1 such that (y, x) is not in R_1. Relation R_2 of the example 7 is not a symmetric relation because, $(1, 2) \in R_2$ but $(2, 1) \notin R_2$. Similarly, relations R_3, R_4 and R_5 of examples 8, 9 and 10 respectively, are not symmetric relations. Readers may verify this. Another examples of a symmetric relation are given below.

Example 11 Let A be the same set of example 6 and R_6 be a relation on A defined as $\{(1, 2), (2, 1), (3, 1), (1, 3)\}$. Here, R_6 is a symmetric relation. ♦

Example 12 Let A be the same set of example 6 and R_7 be a relation on A defined as $\{(1, 2), (2, 1), (1, 1)\}$. Here, R_7 is a symmetric relation. ♦

Asymmetric Relation

A relation R on a set A is called an ***asymmetric*** relation

$$\text{if } (x, y) \in R \Rightarrow (y, x) \notin R \text{ for } x \neq y$$

i.e. presence of pair (x, y) in R excludes the possibility of presence of (y, x) in R. Amongst relations R_1 to R_7 given in examples 6 to 12 respectively, R_1, R_2, R_3, R_4, R_6 and R_7 are not an asymmetric relation whereas R_5 is an asymmetric relation. If x = y then the relation should be irreflexive to be asymmetric. In fact a relation that is irreflexive and anti-symmetric is called an asymmetric relation.

Anti-Symmetric Relation

A relation R on a set A is called an ***anti-symmetric*** relation if for x, y∈A

$$(x, y) \text{ and } (y, x) \in R \Leftrightarrow x = y$$

i.e. $x \neq y \Rightarrow$ either x ~R y or y ~R x or both. In the relations R_1 to R_7 given in examples 6 to 12 respectively, R_1, R_2, R_3, R_4 and R_5 are examples of anti-symmetric relation. Observe that a relation may be symmetric as well as anti-symmetric at the same time (not always). Here, R_1 is both symmetric and anti-symmetric.

Transitive Relation

A relation R on a set A is called a ***transitive*** relation if for x, y, z∈A

$$(x, y) \text{ and } (y, z) \in R \Rightarrow (x, z) \in R$$

i.e., if xRy and yRz then xRz.. In the examples 6 through 12, relation R_1 is transitive by default. Relation R_2 is also transitive. Verify that R_3 is also transitive. What can you say about R_4 and R_5?

Intransitive Relation

A relation R defined on a set A is said to be ***intransitive*** if presence of pairs (x, y) and (y, z) together in R excludes the possibilities of presence of (x, z) in R i.e.

$$(x, y) \text{ and } (y, z) \in R \Rightarrow (x, z) \notin R$$

In the previous examples 6 through 12, R_1 is intransitive relation by default. Relation R_6 is also an example of intransitive relation. Verify about the other relations given in example 7, 8, 9, 10 and 12.

Non-transitive Relation

A relation R on a set A is said to be a **non-*transitive*** relation if R is neither *transitive* nor *intransitive*. Does a relation of this type exist?

Now let us discuss a few examples to learn the process of showing that a given relation is of a particular type or not?

Example 13 Let Z is the set of all integers and R be a relation defined on Z as "equal to" What types of relation R is on Z?

Solution: Given that R is a relation on Z thus, $R \subseteq Z \times Z$. Here, R contains only those ordered pairs (x, y) of Z x Z in which x = y (x is equal to y) i.e.

$$R = \{(x, x) \mid x \in Z\}$$

(i) R is not a universal relation since $R \neq Z \times Z$.

(ii) R is not a NULL relation since R contains ordered pairs like (1, 1), (2, 2) etc.

(iii) R is inverse of itself i.e. $R = R^{-1}$.

(iv) Complement of R i.e. $R^C = \{(x, y) \mid x, y \in Z \text{ and } x \neq y\}$.

(v) R is a reflexive relation since every integer is equal to itself.

(vi) R is not irreflexive and also not non-reflexive.
(vii) R is symmetric, not asymmetric and anti-symmetric by default.
(viii) R is transitive and intransitive by default.
(ix) R is not non-transitive.

Ans.

Example 14 Let H is the set of all human beings on earth and R be a relation defined on H as "is brother of". What types of relation R defines on Z?

Solution: According to the definition of a relation, $R \subseteq H \times H$. Also, R contains only those pairs (x, y) such that x and y are humans and "x is brother of y". *Note that there may be many humans who are not brother of any humans (females).*

(i) Since $R \neq H \times H$, R is not a universal relation.
(ii) Since a male is a brother of himself, R is not empty and hence not a null relation.
(iii) Next, if x is brother of y then it is not necessary that y is also a brother of x (y may be female). Thus $R^{-1} = \{(x, y) \mid y \text{ is brother of } x\}$.
(iv) $R^C = \{(x, y) \mid x \text{ is not a brother of } y\}$.
(v) R is not reflexive because $\exists$ x (*female*) $\in$ H such that $(x, x) \notin R$. Similarly, R is not irreflexive as $\exists$ x(*male*) $\in$ H such that $(x, x) \in R$. Thus, R is a non-reflexive relation.
(vi) R is not a symmetric relation because whenever $(x, y) \in R$ i.e. x is brother of y, it is not necessary that y is also a brother of x. *y may be sister of x.* So (y, x) may not be in R. R is not an anti-symmetric relation as presence of both (x, y) and (y, x) in R does not imply that x = y. R is not an asymmetric relation also. As presence of (x, y) in R does not rule out the possibility of (y, x) in R. *The relation R on set H, here, is a very good example of a relation that is **non-symmetric**.*
(vii) R is a transitive relation and hence it is not intransitive and not non-transitive.

Ans.

Example 15 Let Z_+ be a set of positive integers and R be a relation defined on Z_+ as xRy iff x = 2y. Find hat type of relation it is?

Solution: A few members of R can be given as (1, 2), (2, 4), (3, 6) etc. It is obvious that R is neither universal nor a null relation. $R^{-1} = \{(x, y) \mid 2x = y\}$. $R^C = \{(x, y) \mid x \neq 2y\}$. Verify that R is not reflexive, it is irreflexive and not non-reflexive. R is not a symmetric relation. However, it is asymmetric and anti-symmetric relation. R is an intransitive relation, however it is neither transitive nor non-transitive.

Ans.

1.6 Equivalence Relation

The concept of relation, specially equivalence relation and the order relation (to be discussed in the chapter 6), is very important so far as the understanding of different topics of discrete mathematical structure are concerned. Therefore, readers are advised to read this topic well.

"Any relation R defined on a non-empty set A is said to be an equivalence relation if

- R is reflexive i.e. $\forall x \in A$. $(x, x) \in R$
- R is symmetric i.e. whenever $(x, y) \in R$, there is $(y, x) \in R$
- R is transitive i.e. If (x, y) & $(y, z) \in R$ then $(x, z) \in R$".

Example 1 Let A be the set of all triangles in a plane and let R be the relation on A defined as aRb if and only if "a is congruent to b" for a, b $\in$ A. Show that R is an equivalence relation.

Solution: In order to show that a relation R is an equivalence relation, we have to prove that R is reflexive, symmetric and transitive.

Reflexivity: Since every triangle $x \in A$ is congruent to itself, we can write $(x, x) \in R$ $\forall x \in A$. Thus, R is a reflexive relation.

Symmetry: Let x and y be any two triangles of A such that $(x, y) \in R$. This implies that x is congruent to y and hence y is also congruent to x. Therefore, $(y, x) \in R$. This shows that R is a symmetric relation.

Transitivity: Let x, y and z be any three triangles of set A such that (x, y) and $(y, z) \in R$. This implies that 'x is congruent to y' and 'y is congruent to z'. Therefore, x is also congruent to z and hence $(x, z) \in R$. Thus, R is a transitive relation.

Therefore, the given relation R on the set A is an equivalence relation.

Proved.

Example 2 Let Z be the set of all integers and R be a relation defined on Z such that for any a, b $\in$ Z, aRb if and only if $ab \geq 0$. Determine whether R is an equivalence relation or not.

Solution: *Reflexivity:* Let $x \in Z$, then either $x < 0$ or $x > 0$ or $x = 0$. In all these cases, $x.x \geq 0$ i.e. $\forall x \in Z$, $(x, x) \in R$. Thus, R is reflexive.

Symmetry: Let x and y be any two integers and $(x, y) \in R$. This shows that $x.y \geq 0$ and hence $y.x \geq 0$. Thus, $(y, x) \in R$. Hence R is a symmetric relation also.

Transitivity: Let x, y and z be any three elements of Z such that (x, y) and (y, z) $\in$ R. Thus, we have $xy \geq 0$ and $yz \geq 0$. This implies that

Either $x = y = 0$ then $z = 0$ or $z < 0$ or $z > 0$

Or $x < 0$ and $y < 0$ then $z = 0$ or $z < 0$ [*z cannot be greater than 0*]

Or $x > 0$ and $y > 0$ then $z = 0$ or $z > 0$ [*z cannot be less than 0*]

In all the three cases listed above $xz \geq 0$ i.e. (x, z) $\in$ R. But, what about the situation when $x > 0$, $y = 0$ and $z < 0$? We have both $xy \geq 0$ and $yz \geq 0$, but $xz < 0$. It shows that R is not a transitive relation. Therefore, R is not an equivalence relation.

Proved.

Example 3 Let Z be the set of integers and R be a relation defined on Z as aRb if and only if $a \equiv_3 b$. Prove that R is an equivalence relation on Z.

Solution: Here $\equiv_3$ is read as '*congruence modulo 3*'. Any integer a is said to 'congruence modulo 3' another integer b, if both a and b yield the same remainder when divided by 3. For example, 1 is 'congruence modulo 3' to 4. Similarly, $2 \equiv_3 5$, $2 \equiv_3 8$, $2 \equiv_3 11$ etc.

Now let us consider what remainder we get when a negative integer is divided by a positive integer. Reader may ask, "why am I giving emphasis on this?" Most of the students, whom I have taught, do not know this very fact *that remainder can never be negative* i.e. if a number is divided by m, then remainder must be between 0 and m–1. Thus, when –5 is divided by 6, the remainder will be 1 and quotient will be –1. Similarly, when –1 is divided by 3, the remainder is 2 and quotient is –1. Thus $-1 \equiv_3 2$ (and not $-1 \equiv_3 1$ as commonly perceived). Now let us solve the actual problem of proving that $\equiv_3$ is an equivalence relation on Z.

Reflexivity: Let x $\in$ Z be any integer, then $x \equiv_3 x$ since both yield the same remainder when divided by 3. Thus, (x, x) $\in$ R $\forall$ x $\in$ Z. This proves that R is a reflexive relation.

Symmetry: Let x and y be any two integers and (x, y) $\in$ R. This shows that $x \equiv_3 y$ and hence $y \equiv_3 x$. Thus, (y, x) $\in$ R. Hence R is a symmetric relation also.

Transitivity: Let x, y and z be any three elements of Z such that (x, y) and (y, z) $\in$ R. Thus, we have $x \equiv_3 y$ and $y \equiv_3 z$. It implies that (x–y) and (y–z) are divisible by 3. Therefore, $(x - y) + (y - z) = (x - z)$ is also divisible by 3 i.e. $x \equiv_3 z$. Hence, (x, y) and (y, z) $\in$ R $\Rightarrow$ (x, z) $\in$ R. So R is a transitive relation.

Therefore, R is an equivalence relation.

Proved.

R-relative sets

Let R be a relation from a set A to set B. The *R-relative* set of an element $x \in A$, denoted as R(x), is defined as the set containing those elements of B that are related to x i.e.

$$R(x) = \{y \mid y \in B \text{ and } xRy\}$$

Similarly, R-relative set of a subset C of A is defined as the set containing those elements of B that are associated with one or more elements of C i.e.

$$R(C) = \{y \mid y \in B \text{ and } xRy \text{ for } x \in C\}$$

Example 4 Let A = {1, 2, 3, 4, 5, 6} and B = {x, y, z, w}. Let R be a relation from A to B and is defined as R = {(1, x), (5, y), (3, z), (1, w), (2, x)}. Thus,

$$R(1) = \{x, w\}$$

Now, let C = {1, 2, 4, 5} be a subset of A. Then R-relative set of C is given as

$$R(C) = \{w, x, y\}$$ ♦

Example 5 Let A = {a, b, c, d} and R be any relation defined on A such that

$$R = \{(a, a), (b, b), (a, b), (b, c), (c, a), (c, b), (d, c)\}$$

Find R (a), R (b). If $A_1 = \{c, d\}$ then find R (A_1).

Solution: Readers are advised to try on their own as explained in the previous example 4. ♦

Domain and Range of a Relation

Let R be a relation from a set A to another set B. The *Domain* of R is the set of all first elements of ordered pairs of R and *Range* of R is the set of all second elements of ordered pairs of R. Thus,

Dom(R) = $\{a \mid (a, b) \in R \text{ for some } b \in B\}$ and
Rang(R) = $\{b \mid (a, b) \in R \text{ for some } a \in A\}$

Example 6 Let A be a set of all men living in a certain locality of Delhi and B the set of all women in the same locality. Let R be a relation from set A to set B defined as aRb iff 'a is the husband of b". Here Dom(R) is the set of all husbands and the Rang(R) is the set of all wives. ♦

Example 7 In the previous example 5, Dom(R) = {a, b, c, d} and Rang(R) = {a, b, c}. ♦

Example 8 In the previous example 4, Dom(R) = {1, 2, 3, 5} and Rang(R) = {w, x, y, z}. ♦

Theorem 1 Let R and S be two relations from set A to set B. If R(x) = S(x) $\forall x \in A$ then prove that R = S.

Proof: Here, we have to prove that every ordered pairs (x, y) of R is also in S and vice versa.

Let (x, y) be any element of R then $xRy \Rightarrow y \in R(x)$

$\Rightarrow y \in S(x)$ $[R(x) = S(x)]$

$\Rightarrow xSy$

$\Rightarrow (x, y) \in S$

Therefore, $R \subseteq S$ ____________________(1)

Similarly, let $(x, y) \in S(x)$ then $xSy \Rightarrow y \in S(x)$

$\Rightarrow y \in R(x)$ $[R(x) = S(x)]$

$\Rightarrow xRy$

$\Rightarrow (x, y) \in R$

Therefore, $S \subseteq R$ ____________________(2)

From inequality (1) and (2), we have R = S.

Proved.

Theorem 2 Let R be a relation from set A to set B and let A_1 and A_2 be subsets of A then prove that

(i) $A_1 \subseteq A_2 \Rightarrow R(A_1) \subseteq R(A_2)$

(ii) $R(A_1 \cup A_2) = R(A_1) \cup R(A_2)$

Proof: (i) Let y be any element of $R(A_1)$. Then $\exists\, x \in A_1$ such that $(x, y) \in R$. Since $A_1 \subseteq A_2$, we have $x \in A_2$ also. This implies that $y \in R(A_2)$. Therefore, $R(A_1) \subseteq R(A_2)$. ♦

(ii) Since $A_1 \subseteq A_1 \cup A_2$ and $A_2 \subseteq A_1 \cup A_2$, from part (i), we can conclude that

$R(A_1) \subseteq R(A_1 \cup A_2)$

and $R(A_2) \subseteq R(A_1 \cup A_2)$

Combining these two results, we get

$R(A_1) \cup R(A_2) \subseteq R(A_1 \cup A_2)$ ________________(1)

Now to show the equality, we have to show that $R(A_1 \cup A_2) \subseteq R(A_1) \cup R(A_2)$. Let y be any element of $R(A_1 \cup A_2)$, then $\exists\, x \in A_1 \cup A_2$ such that $(x, y) \in R$.

$\Rightarrow x \in A_1$ or $x \in A_2$

$\Rightarrow y \in R(A_1)$ or $y \in R(A_2)$

$\Rightarrow y \in R(A_1) \cup R(A_2)$

Therefore, $R(A_1 \cup A_2) \subseteq R(A_1) \cup R(A_2)$ ________________(2)

Combining inequality (1) and (2), we have the result.

Proved.

Theorem 3 Let R be a relation from set A to set B then for all subsets A_1 and A_2 of A prove that

$$R(A_1 \cap A_2) = R(A_1) \cap R(A_2) \text{ iff } R(a) \cap R(b) = \phi \text{ for any } a, b \in A \text{ and } a \neq b.$$

Proof: Let y be any element of $R(A_1 \cap A_2)$. Then

$$\begin{aligned} y \in R(A_1 \cap A_2) &\Rightarrow \exists\, x \in A_1 \cap A_2 \text{ such that } xRy \\ &\Rightarrow \exists\, x \in A_1 \text{ and } x \in A_2 \text{ such that } xRy \\ &\Rightarrow y \in R(A_1) \text{ and } y \in R(A_2) \\ &\Rightarrow y \in R(A_1) \cap R(A_2) \end{aligned}$$

Therefore, $R(A_1 \cap A_2) \subseteq R(A_1) \cap R(A_2)$ ________________(1)

Now, let y be any element of $R(A_1) \cap R(A_2)$. Then

$$\begin{aligned} y \in R(A_1) \cap R(A_2) &\Rightarrow y \in R(A_1) \text{ and } y \in R(A_2) \\ &\Rightarrow \exists\, x \in A_1 \text{ such that } xRy \text{ and } \exists\, z \in A_2 \text{ such that } zRy \\ &\Rightarrow x \text{ must be equal to } z \text{ since } R(x) \cap R(z) = \phi \text{ for } x \neq z \\ &\Rightarrow \exists\, x \in A_1 \text{ such that } xRy \text{ and } \exists\, x \in A_2 \text{ such that } xRy \\ &\Rightarrow \exists\, x \in A_1 \cap A_2 \text{ such that } xRy \\ &\Rightarrow y \in R(A_1 \cap A_2) \end{aligned}$$

Therefore, $R(A_1) \cap R(A_2) \subseteq R(A_1 \cap A_2)$ ________________(2)

Combining inequality (1) and (2), we have the result.

Proved.

Note: For any two element a, b of A such that $a \neq b$ and $R(a) \cap R(b) \neq \phi$, we have

$$R(A_1 \cap A_2) \subseteq R(A_1) \cap R(A_2)$$

Example 9 Let A = {1, 2, 3}, B = {x, y, z, w, p, q} and R from A to B is given as

$$R = \{(1, x), (1, z), (2, w), (2, p), (2, q), (3, y)\}$$

For all possible subsets A_1 and A_2 of A, prove or disprove that

$$R(A_1 \cap A_2) = R(A_1) \cap R(A_2)$$

Solution: Readers are advised to try on their own. ♦

Partition of a set

A *partition* of a non-empty set A is defined as a collection of subsets A_i's of A such that

- Union of all such subsets A_i is equal to A i.e. $\cup A_i = A$
- All such subsets are mutually disjoint i.e. $A_i \cap A_j = \phi$.for $i \neq j$ and
- All subsets A_i's are non-empty i.e. $| A_i | \neq \phi$.

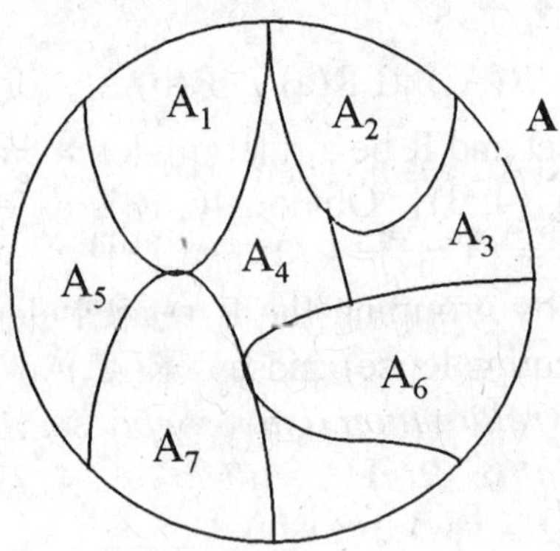

Figure 1.6.1

See the figure 1.6.1 shown above. Let A be a non-empty set and A_1, A_2, A_3, A_4, A_5, A_6 and A_7 be subsets of A. It is obvious from the figure 1.6.1 that

(i) $\bigcup_{i=1}^{7} A_i = A$

(ii) $A_i \cap A_j = \Phi$ for $i \neq j$

(iii) $| A_i | \neq 0$

A *partition* is represented by P. Each subset of A in P is called ***block*** or ***cell.*** A partition of a set is also called ***quotient Set.***

Example 10 Let A = {1, 2, 3}, list all partitions of A.

Solution: Let us first list all possible subsets A. These are: ϕ, {1}, {2}, {3}, {1, 2}, {1, 3}, {2, 3}, {1, 2, 3}. Since, we have to consider only non-empty subsets, we are not considering ϕ at all. Various possible partitions of A are:

(i) {{1}, {2}, {3}} (ii) {{1, 2, 3}} (iii) {{1}, {2, 3}}
(iv) {{2}, {1, 3}} (v) {{3}, {1, 2}}

Verify that there exists no other partition for A. ♦

Example 11 Let Z be a set of all integers, A be the set of all positive integers and B be the set of all negative integers including **zero.** Thus, P = {A, B} is a partition of Z. ♦

Example 12 Let A = {a, b, c, d, e, f, g, h} be a set. Let A_1, A_2, A_3, A_4 and A_5 are subsets of A given as: A_1 = {a, b, c, d}, A_2 = {b, d}, A_3 = {a, c, e, f, g, h}, A_4 = {a, c, e, g} and A_5 = {f, h}. The set {A_2, A_4, A_5} is a partition of A however, {A_1, A_2, A_3} is not a partition of set A as $A_1 \cap A_3 \neq \phi$.♦

Equivalence Partition

Let A = {1, 2, 3, 4} be a set and R be a relation defined on A as R = {(1, 1), (2, 2), (3, 3), (4, 4), (1, 2), (2, 1), (3, 4), (4, 3)}. Obviously, R is an equivalence relation on A. Also, R(1) = {1, 2} = R(2), R(3) = {3, 4} = R(4) and {{1, 2}, {3, 4}} is a partition of A. This partition has been obtained by grouping the R-related elements of A in one subset. Subsets {1, 2} and {3, 4} are equivalence classes of A. An equivalence class can be defined as "*If R is an equivalence relation on a non-empty set A then equivalence class of any element a ∈ A, denoted by* ***[a] or R(a)***, *is collection of all elements x ∈ A such that aRx i.e. [a] = {x | x ∈ A and aRx}*"

Let A be a non-empty set and R be an equivalence relation defined on A. An ***equivalence partition*** of set A is defined as the collection of *equivalence classes* A_1, A_2, A_3, ...A_n such that

(i) $\bigcup_{i=1}^{n} A_i = A$

(ii) $A_i \cap A_j = \Phi \quad \text{for } i \neq j$

(iii) $|A_i| \neq 0 \quad \text{for } 1 \leq i \leq n$

(iv) *If* $x, y \in A_i$ *then xRy and*

(v) *If* $x \in A_i$ *and* $y \in A_j$ *for* $i \neq j$ *then* $x \sim Ry$

An equivalence partition of a set A by an equivalence relation R is denoted by A/R and is also called a ***quotient set*** of A by R.

Example 13 Let Z be a set of all integers and $\equiv_m$ be a relation defined on Z. Find the equivalence partition of Z by $\equiv_m$.

Solution: It is known that $\equiv_m$ stands for '*congruence modulo m*'. The possible remainders when an integer is divided by m are 0, 1, 2, 3, ..., m–1. Let us group all the elements of Z which yield remainder 0 in [0], which yield remainder 1 in [1] and so on up to [m–1]. After that, we have the following equivalent classes of z by $\equiv_m$.

[0] = {......, –3m, –2m, –m, 0, m, 2m, 3m, ...}

$[1] = \{......, -3m+1, -2m+1, -m+1, 1, m+1, 2m+1, 3m+1, ...\}$
$[2] = \{......, -3m+2, -2m+2, -m+2, 2, m+2, 2m+2, 3m+2, ...\}$

.

.

$[10] = \{......, -3m+10, -2m+10, -m+10, 10, m+10, 2m+10, 3m+10, ...\}$

.

.

$[m-1] = \{......, -2m-1, -m-1, -1, m-1, 2m-1, 3m-1, ...\}$

Here, P = {[0], [1], [2], ..., [m–1]} is the equivalence partition of Z by $\equiv_m$.

Ans.

Example 14 Show that aRb iff R(a) = R(b), where R is an equivalence relation on set A and a, b are any two elements of set A.

Solution: If part i.e. aRb if R(a) = R(b). Let R(a) = R(b). Since R is reflexive, we have b $\in$ R(b). Now

$$b \in R(b) \Rightarrow b \in R(a) \qquad \text{[Since } R(a) = R(b)]$$
$$\Rightarrow aRb$$

Therefore, aRb **if** R(a) = R(b) ________________(1)

Only if Part i.e. aRb only if R(a) = R(b) i.e. if aRb then R(a) = R(b). Let aRb i.e. a is related to b. This shows that b $\in$ R(a). Since R is an equivalence relation, both a and b belong to the same equivalence class which is either R(a) or R(b).

Therefore, If aRb **then** R(a) = R(b) ________________(2)

Combining (1) and (2), we have

aRb **iff** R(a) = R(b)

Proved.

Theorem 4 Let R be an equivalence relation defined on a non-empty set A then show that R always induces an equivalence partition of A. Conversely, if P be a partition of set A and R is defined on set A as xRy iff 'x and y both belong to the same block of P' then R is an equivalence relation on set A.

Proof: Let us first prove the first part. Let R be any equivalence relation on set A. Let us make equivalence classes of elements of A as follows:

Step 1: Choose any element a of A and find R(a).

Step 2: If R(a) = A, its over, otherwise choose another element b $\in$ A such that b $\notin$ R(a) and find R(b).

Step 3: If A is not equal to union of the already obtained equivalence classes, then choose an element x∈ A such that x is left out and not included in any of the equivalence classes already determined. Find R(x).

Step 4: Repeat the step 3 until every element of set A has been included in some classes.

The collection of these equivalence classes is a ***partition*** of set A induced by the equivalence relation R.

To prove the next part, let P be a partition of set A. Since an element a of set A can find place in only one block of P, aRa ∀ a∈A. Thus R is reflexive. Let a and b be any two elements of A such that aRb. By the definition of R, a and b both belong to the same block and hence bRa. This shows that R is symmetric also. At last, let a, b and c be any three elements of set A such that aRb and bRc i.e. a, b and c all belong to same block of P. Since a and c both belong to the same block of P, aRc. Hence R is transitive also. Therefore, R is an equivalence relation.

Proved.

1.7 Digraph and Matrix Representation of a Relation

A graph contains a set of points (also called nodes or vertices) and a set of arcs (also called edges) connecting a pair of nodes. Every edge has a *from-node* and a *to-node*. We can represent every edge by an ordered pair of *from-node* and a *to-node* like (*from-node, to-node*). A binary relation too contains ordered pairs. Let R be a relation on a set A. Let us treat set A as a set of nodes and R as a set of edges, then R can very well be represented by a digraph. More about graph, we shall learn in the chapter 7. In computer science such representation has importance for storing a relation in memory for its processing. Digraph and matrix are some of the available data structures, which we use to store a relation while processing.

Since there is no way to deal with infinite amount of information in computer, the digraph and matrix form of representation is used for only those relations that are defined on finite set. Let us see some examples to learn the method to represent relation in digraph form and matrix form.

Example 1 Let A = {1, 2, 3, 4} and R be a relation on set A given as R = {(1, 1), (1, 2), (2, 1), (2, 2), (2, 3), (2, 4), (3, 4), (4, 1)}. Represent this relation in digraph form.

Solution: Consider set A as a set of nodes and R as a set of directed arcs. This means that (1, 2) ∈ R ⇒ there is an arc from node 1 to node 2. Thus, digraph for R can be shown as in the figure 1.7.1.

Ans.

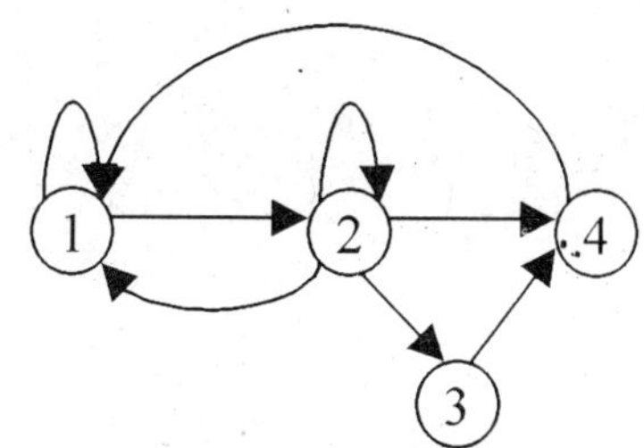

Example 2 Let A = {a, b, c, d} and B ={1, 2, 3}. Let R be relation defined from set A to set B and is given as R = {(a, 1), (a, 2), (b, 1), (c, 2), (d, 1)}. Draw a digraph for R.

Solution: Here, we have two sets of nodes and R is a set of arcs from one set of nodes to another set of nodes. Thus, digraph for R is given in the figure 1.7.2.

Ans.

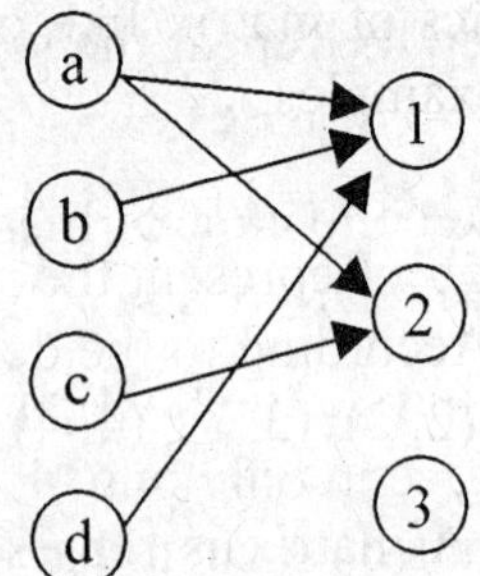

Figure 1.7.2

Example 3 Find the set A and relation R defined on set A from the following digraph shown in the figure 1.7.3.

Solution: Here A is set of nodes i.e. A = {1, 2, 3, 4} and R is set of arcs i.e. R = {(1, 1), (1, 3), (2, 3), (3, 2), (3, 3), (4, 3)}.

Ans.

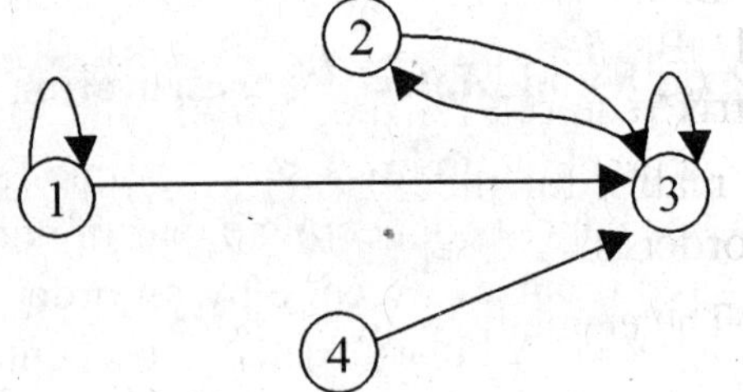

Figure 1.7.3

Example 4 1 Let A = {1, 2, 3, 4} and B = {1, 4, 6, 8, 9}. Let R be a relation from set A to set B and defined as aRb iff $b = a^2$. Represent this relation in digraph form.

Solution: The relation R can be given as {(1, 1), (2, 4), (3, 9)}. Thus, the required digraph is shown in the figure 1.7.4.

Ans.

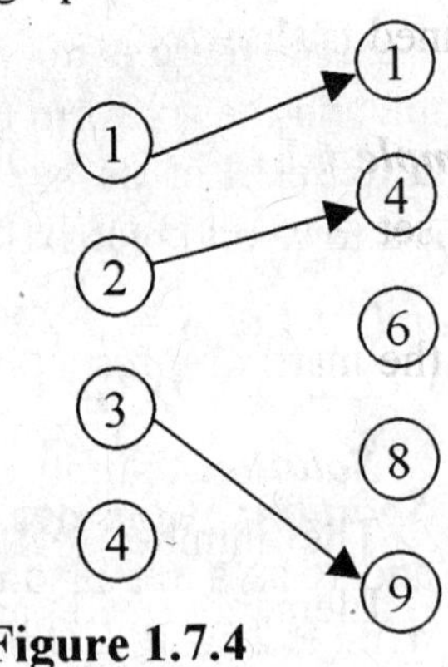

Figure 1.7.4

Now let us discuss the matrix representation of a relation. A matrix is a rectangular arrangement of set of elements in columns and rows form. Each element of a matrix is referred as a_{ij} of ith row and jth column. It is suggested to the readers to brush up the basics of matrix like operations on matrices and different types of matrix. Let us see some examples.

Example 5 Let a = {1, 2, 3, 4, 6} be a set and R be a relation on A defined as aRb iff 'a is multiple of b'. Represent the relation in matrix form.

Solution: According to the definition of R, its ordered pairs can be enumerated as R={(1, 1), (2, 2), (3, 3), (4, 4), (6, 6), (2, 1), (3, 1), (4, 1), (6, 1), (4,2), (6,3), (6,2)}

A matrix is represented by number of rows and columns it has. It is important to determine the order of matrix needed to represent the given relation. Since a relation is defined from a set A to another set B, |A| determined rows of the matrix and |B| determines the column of the matrix required for the relation. Let M_R be the matrix for the relation R: A →B, then its order is |A|x|B|.

$$\begin{array}{c} \\ \\ M_R = \end{array}\begin{array}{c} \\ 1 \\ 2 \\ 3 \\ 4 \\ 6 \end{array}\begin{array}{c} \begin{array}{ccccc} 1 & 2 & 3 & 4 & 6 \end{array} \\ \begin{bmatrix} 1 & 0 & 0 & 0 & 0 \\ 1 & 1 & 0 & 0 & 0 \\ 1 & 0 & 1 & 0 & 0 \\ 1 & 1 & 0 & 1 & 0 \\ 1 & 1 & 1 & 0 & 1 \end{bmatrix}_{5\times 5} \end{array}$$

Figure 1.7.5

The elements a_{ij} of M_R are given according to the following rule:

$$a_{ij} = \begin{cases} 0 & if \quad (i,j) \notin R \\ 1 & if \quad (i,j) \in R \end{cases}$$

Here, the elements of set A have been numbered from 1 to |A| and is referred as i. The elements of set B has been numbered from 1 to |B| and is referred as j. The matrix so obtained is shown in the figure 1.7.5 above.

Ans.

Example 6 Let A = {a_1, a_2, a_3} and B = {b_1, b_2, b_3, b_4} be two sets and R be a relation from set A to set B given by

R = {(a_1, b_1), (a_1, b_4), (a_2, b_2), (a_2, b_3), (a_3, b_1), (a_3, b_3)}

Find the matrix representation of R i.e. M_R.

Solution: Number of rows of M_R = |A| = 3, and The number of columns of M_R = |B| = 4. Elements of M_R are determined according to the ordered pairs in R. Thus, the matrix M_R is shown in the figure 1.7.6.

$$\begin{array}{c} \\ \\ M_R = \end{array}\begin{array}{c} \\ a_1 \\ a_2 \\ a_3 \end{array}\begin{array}{c} \begin{array}{cccc} b_1 & b_2 & b_3 & b_4 \end{array} \\ \begin{bmatrix} 1 & 0 & 0 & 1 \\ 0 & 1 & 1 & 0 \\ 1 & 0 & 1 & 0 \end{bmatrix}_{3\times 4} \end{array}$$

Ans.

Figure 1.7.6

***Example** 7* Let A = {1, 2, 3} be a set and R be a relation on A and given as

R = {(1, 1), (2, 2), (3, 3), (1, 2), (2, 1)}

Find the matrix representation of R i.e. M_R.

Solution: Number of rows of M_R = |A| = 3, and The number of columns of M_R = |A| = 3. Elements of M_R are determined according to the ordered pairs in R. Thus, the matrix M_R is shown in the figure 1.7.7.

Ans.

$$M_R = \begin{array}{c} \\ 1 \\ 2 \\ 3 \end{array}\begin{array}{c} \begin{array}{ccc} 1 & 2 & 3 \end{array} \\ \begin{bmatrix} 1 & 1 & 0 \\ 1 & 1 & 0 \\ 0 & 0 & 1 \end{bmatrix}_{3\times 3} \end{array}$$

Figure 1.7.7

Note: The following observations may be very useful in determining whether a relation is reflexive, symmetric or transitive on a given set. Let M_R be the matrix of a relation R defined on a set A, then

1. *R is reflexive iff all the diagonal elements of M_R are 1.*
2. *R is symmetric iff $M_R = M_R^T$*
3. *R is transitive iff $M_R + M_R^2 = M_R$*

1.8 Closure of Relation

A relation R defined on a set A may attain a property if some ordered pairs are added to it. For example, a relation R = {(1, 1), (1, 2), (2, 2), (1, 3)} is not a reflexive relation on set A = {1, 2, 3}. If we add the ordered pair (3, 3) to R, it becomes a reflexive relation on A. Let us call this relation as S. Here,

$$S = R \cup \{(3, 3)\}$$

It is obvious that S is reflexive and it contains R. If we have any other relation say T on A and if T is reflexive and contains R, it must contain S also. Thus, we say that S is closest to the relation R. Similar examples can be given for R to attain any other properties. In this illustration, S is the reflexive closure of relation R. Let us define closure of a relation in formal way. In this text, we shall restrict ourselves to reflexive, symmetric and transitive closure only.

Reflexive Closure

"Reflexive closure of a relation R is the smallest reflexive relation S that contains R."

A diagonal (or identity) relation Δ on a non-empty set A is defined as

$$\Delta = \{(x, x) \mid x \in A\}$$

Let R be any relation on A. If R is reflexive then $R \cup \Delta = R$. If R is not a reflexive relation then $\exists\ x \in A$, such that $(x, x) \notin R$ i.e. some (x, x) are needed to make R a reflexive relation. Let $S = R \cup \Delta$. Then S is the smallest reflexive relation containing R. Thus we say that S is the reflexive closure of R. In fact, for any given relation R, $R \cup \Delta$ is the reflexive closure of R.

Example 1 Let A = {x, y, z} and a relation R on set A is given as

$$R = \{(x, x), (x, y), (y, z), (x, z)\}$$

Here, to make R a reflexive relation ordered pairs (y, y) and (z, z) are to be included in R. The reflexive closure S of $R = R \cup \Delta = R \cup \{(x, x), (y, y), (z, z)\}$

$$\text{Or } S = \{(x, x), (x, y), (y, z), (x, z), (y, y), (z, z)\}$$ ♦

Symmetric Closure

"Symmetric closure of a relation R is the smallest symmetric relation S that contains R."

Let R be a relation defined on a non-empty set A. The symmetric closure S of R is given by $R \cup R^{-1}$.

Example 2 In the example 1, relation R is not symmetric on A. To make it a symmetric relation, ordered pairs (z, x), (y, x) and (z, y) are to be added to it. We have R^{-1} = {(x, x), (z, x), (y, x) (z, y)}. If we define $S = R \cup R^{-1}$, then S is a symmetric relation on A and it contains R. Also any other symmetric relation on A that contains R, must be containing S also. Thus S is a symmetric closure of R. And; S = {(x, x), (x, y), (y, z), (x, z), (z, x), (y, x) (z, y)}. ♦

Transitive Closure

"Transitive closure of a relation R is the smallest transitive relation S that contains R."

The method of finding transitive closure of a relation is not as straightforward as we have seen in the case of reflexive and symmetric closure. The matrix multiplication method is used for this purpose. The matrix M_R corresponding to the given relation R is treated as a Boolean matrix and hence Boolean 'AND' and Boolean 'OR" are used for multiplication and addition of the elements of M_R respectively. The steps involved are illustrated by the following example.

Example 3 Let A = {1, 2, 3, 4} and R be a relation defined on A as R = {(1, 2), (2, 3), (3, 4), (2, 1)}. Find the transitive closure of R.

Solution: We use the following algorithm to find the transitive closure of the relation.
Step 1: First convert the given relation in matrix form and call it M_R Thus,

$$M_R = \begin{bmatrix} 0 & 1 & 0 & 0 \\ 1 & 0 & 1 & 0 \\ 0 & 0 & 0 & 1 \\ 0 & 0 & 0 & 0 \end{bmatrix} \qquad \text{Matrix to start with}$$

Step 2: Compute different powers of M_R using Boolean matrix multiplication. Stop the computation when M_R^n is equal to any of M_R, M_R^2, M_R^3, ..., M_R^{n-1}. Here M_R^n is the nth power of M_R. *Readers should note that in Boolean matrix multiplication,* 0+0 = 0.0 = 1.0 = 0; 1.1 = 1+1 = 1+0 = 1.

$$M_R^2 = \begin{bmatrix} 0 & 1 & 0 & 0 \\ 1 & 0 & 1 & 0 \\ 0 & 0 & 0 & 1 \\ 0 & 0 & 0 & 0 \end{bmatrix} \begin{bmatrix} 0 & 1 & 0 & 0 \\ 1 & 0 & 1 & 0 \\ 0 & 0 & 0 & 1 \\ 0 & 0 & 0 & 0 \end{bmatrix} = \begin{bmatrix} 1 & 0 & 1 & 0 \\ 0 & 1 & 0 & 1 \\ 0 & 0 & 0 & 0 \\ 0 & 0 & 0 & 0 \end{bmatrix}$$

$$M_R^3 = M_R^2 M_R = \begin{bmatrix} 1 & 0 & 1 & 0 \\ 0 & 1 & 0 & 1 \\ 0 & 0 & 0 & 0 \\ 0 & 0 & 0 & 0 \end{bmatrix} \begin{bmatrix} 0 & 1 & 0 & 0 \\ 1 & 0 & 1 & 0 \\ 0 & 0 & 0 & 1 \\ 0 & 0 & 0 & 0 \end{bmatrix} = \begin{bmatrix} 0 & 1 & 0 & 1 \\ 1 & 0 & 1 & 0 \\ 0 & 0 & 0 & 0 \\ 0 & 0 & 0 & 0 \end{bmatrix}$$

$$M_R^4 = M_R^3 M_R = \begin{bmatrix} 0 & 1 & 0 & 1 \\ 1 & 0 & 1 & 0 \\ 0 & 0 & 0 & 0 \\ 0 & 0 & 0 & 0 \end{bmatrix} \begin{bmatrix} 0 & 1 & 0 & 0 \\ 1 & 0 & 1 & 0 \\ 0 & 0 & 0 & 1 \\ 0 & 0 & 0 & 0 \end{bmatrix} = \begin{bmatrix} 1 & 0 & 1 & 0 \\ 0 & 1 & 0 & 1 \\ 0 & 0 & 0 & 0 \\ 0 & 0 & 0 & 0 \end{bmatrix}$$

Similarly, we have $M_R^5 = M_R^4 M_R = \begin{bmatrix} 0 & 1 & 0 & 1 \\ 1 & 0 & 1 & 0 \\ 0 & 0 & 0 & 0 \\ 0 & 0 & 0 & 0 \end{bmatrix}$ *[Veryfy it]*

It is obvious from above, that $M_R^5 = M_R^3$ and $M_R^4 = M_R^2$. This pattern will be repeated and, in general, we have $M_R^n = M_R^2$ when **n is even** and $M_R^n = M_R^3$ when **n is odd**.

Step 3: Select distinct matrices M_R^n from the computed list of matrices M_R, M_R^2, M_R^3 ... determined in the step 2. In this case, we have three distinct matrices M_R, M_R^2 and M_R^3.

Step 4: Add the matrices selected in the step 3. Use Boolean OR on the corresponding elements of matrices while adding them together. This matrix corresponds to the transitive closure of R. **A transitive closure of R is denoted by R^{∞} and the corresponding matrix by M_R^{∞}**. Therefore,

$$M_R^{\infty} = M_R + M_R^2 + M_R^3 = \begin{bmatrix} 0 & 1 & 0 & 0 \\ 1 & 0 & 1 & 0 \\ 0 & 0 & 0 & 1 \\ 0 & 0 & 0 & 0 \end{bmatrix} + \begin{bmatrix} 1 & 0 & 1 & 0 \\ 0 & 1 & 0 & 1 \\ 0 & 0 & 0 & 0 \\ 0 & 0 & 0 & 0 \end{bmatrix} + \begin{bmatrix} 0 & 1 & 0 & 1 \\ 1 & 0 & 1 & 0 \\ 0 & 0 & 0 & 0 \\ 0 & 0 & 0 & 0 \end{bmatrix}$$

$$= \begin{bmatrix} 1 & 1 & 1 & 1 \\ 1 & 1 & 1 & 1 \\ 0 & 0 & 0 & 1 \\ 0 & 0 & 0 & 0 \end{bmatrix} \quad \text{Final Matrix}$$

Step 5: Convert the M_R^{∞} into R^{∞} i.e. transitive closure. Therefore, we have

$$R^{\infty} = \{(1, 1), (1, 2), (1, 3), (1, 4), (2, 1), (2, 2), (2, 3), (2, 4), (3, 4)\}$$

Ans.

The five steps algorithm explained in the previous example is used to compute the reachability matrix in a graph. The same algorithm is used to determine whether a graph is connected or not. We shall learn all these and more about application of transitive closure in graph theory, in the chapter 7. For now, let us learn a more efficient algorithm due to Warshall to find transitive closure of a given relation. This algorithm is also used to find all pair shortest path in a weighted graph. Reader is therefore, advised to understand the algorithm to acquaint themselves before actually using this in the chapter 7.

Warshall 's Algorithm

To explain this algorithm, we shall continue with the same relation as given in the previous example.

Step 1: Convert the given relation into matrix form M_R and call this matrix as initial Warshall matrix –denoted by W_0.

$$W_0 = \begin{bmatrix} 0 & 1 & 0 & 0 \\ 1 & 0 & 1 & 0 \\ 0 & 0 & 0 & 1 \\ 0 & 0 & 0 & 0 \end{bmatrix} \quad \text{Initial Warshall 's matrix}$$

Step 2: For K = 1 to |A|, A is the set on which relation R is defined, compute W_k –called Warshall's matrix at k^{th} stage. To compute W_k from W_{k-1}, we proceed as follows:

Step 2.1: Transfer all 1's from W_{k-1} to W_k,

Step 2.2: List the rows in column K of W_{k-1} where, the entry in W_{k-1} is 1. Say these rows $p_1, p_2, p_3, \ldots$;

Similarly, list the columns in row K of W_{k-1} where, the entry in W_{k-1} is 1. Say these columns $q_1, q_2, q_3, \ldots$;

Step 2.3: Place 1 at the locations (p_i, q_j) in W_k if 1 is not already there.

Now, let us apply this on W_0 of the step 1. Here we have to find W_1, W_2, W_3, and W_4 as |A| = 4.

For K = 1. We have to find W_1.

Step 2.1: Transfer all 1's from W_0 to W_1.

Step 2.2: Here K = 1. Thus in W_0, we have '1' in column 1 at row = 2 and '1' in row 1 at column = 2. Therefore, W_1 will have an additional '1' at (2, 2). The Warshall 's matrix W_1 is given in the right side.

$$W_1 = \begin{bmatrix} 0 & 1 & 0 & 0 \\ 1 & 1 & 1 & 0 \\ 0 & 0 & 0 & 1 \\ 0 & 0 & 0 & 0 \end{bmatrix}$$

For K = 2. We have to find W_2.

Step 2.1: Transfer all 1's from W_1 to W_2.

Step 2.2: Here K = 2. Thus in W_1, we have '1' in column 2 at row = 1, 2 and '1' in row 2 at column = 1, 2, 3. Therefore, W_2 will have additional '1' at (1, 1), (1, 2), (1, 3), (2, 1), (2, 2) and (2, 3) if '1' is not already there.

Warshall 's matrix W_2 is given in the right side.

$$W_2 = \begin{bmatrix} 1 & 1 & 1 & 0 \\ 1 & 1 & 1 & 0 \\ 0 & 0 & 0 & 1 \\ 0 & 0 & 0 & 0 \end{bmatrix}$$

For K = 3. We have to find W_3.

Step 2.1: Transfer all 1's from W_2 to W_3.

Step 2.2: Here K = 3. Thus in W_2, we have '1' in column 3 at row = 1, 2 and '1' in row 3 at column = 4. Therefore, W_3 will have additional '1' at (1,4) and (2,4). The W_3 is given in the right side.

$$W_3 = \begin{bmatrix} 1 & 1 & 1 & 1 \\ 1 & 1 & 1 & 1 \\ 0 & 0 & 0 & 1 \\ 0 & 0 & 0 & 0 \end{bmatrix}$$

For K = 4. We have to find W_4.

Step 2.1: Transfer all 1's from W_3 to W_4.

Step 2.2: Here K = 4. Thus in W_3, we have '1' in column 4 at row = 1, 2, 3 and '1' in row 4 at nowhere. Therefore, W_4 will have no additional '1'. The W_4 is equal to W_3 as shown in the right side matrix.

$$W_4 = \begin{bmatrix} 1 & 1 & 1 & 1 \\ 1 & 1 & 1 & 1 \\ 0 & 0 & 0 & 1 \\ 0 & 0 & 0 & 0 \end{bmatrix}$$

Therefore, W_4 is the matrix for the transitive closure of the given relation. Hence,

$$R^{\infty} = \{(1, 1), (1, 2), (1, 3), (1, 4), (2, 1), (2, 2), (2, 3), (2, 4), (3, 4)\}.$$

A conscious reader may ask that why to use this complicated method once we have a simple method of matrix multiplication to do the same job? The answer lies in the efficiency of the Warshall 's algorithm as outlined below:

1. *In Warshall's algorithm, it is not required to keep a copy of matrices of every stage.*
2. *Once we have computed* $W_{|A|}$, *we have the result available. No Boolean addition needed on the previously computed matrices.*

Sometimes in Warshall 's algorithm, for a relation R defined on a set A having |A| = n, we may have $W_k = W_{k+1}$ for k, k+1 < n. It does not mean that we should stop the procedure there. The procedure must be proceeded up to |A|.

Example 4 Let A = {1, 2, 3, 4, 5} and R is a relation on A given as

R = {(1, 1), (1, 2), (2, 1), (2, 2), (3, 3), (4, 3), (4, 4), (4, 5), (5, 4), (5, 5)}

Find transitive closure of R using Warshall 's algorithm.

Solution: Readers should try to do that and verify that above remark.

The concept of *enumeration* of items has given the notion of set. The enumeration is used extensively in computer programming languages. The concept of Boolean algebra and formal logic theory has evolved over the period from the basic concept of set and relation. It is the concept of relation that has simplified the representation of graphs and hence helped in finding solution to many problems related to graph and tree. It is suggested to the reader to read this chapter and solve the problems given in the exercise below to have a first hand idea about the things to follow in the subsequent chapters.

Exercise

1. Determine whether each of the following statements is true or false. Justify your answer.
   (a) $\phi \subseteq \phi$ (b) $\phi \in \phi$ (c) $\phi \in \{\phi\}$ (d) $\{\phi\} \subseteq \phi$ (e) $\{\phi\} \subseteq \{\phi\}$
   (f) $\{\phi\} \in \phi$ (g) $\{\phi\} \in \{\phi\}$ (h) $\{a, b\} \in \{a, b, c, \{a, b, c\}\}$ (i) $\{a, \phi\} \in \{a, \{a, \phi\}\}$

2. For $A = \{a, b, \{a, c\}, \phi\}$, determine the following sets:
   (a) $A - \{a\}$ (b) $A - \phi$ (c) $A - \{a, b\}$ (d) $A - \{\phi\}$ (e) $A - \{a, c\}$
   (f) $A - \{\{a, b\}\}$ (g) $A - \{\{a, c\}\}$ (h) $\{a\} - A$ (i) $\phi - A$ (j) $\{\phi\} - A$
   (k) $\{a, c\} - A$ (l) $\{\{a, c\}\} - A$ (m) $\{a\} - \{A\}$

3. Let $A = \{\phi\}$ and $B = P(P(A))$. Determine whether each of the following statements is true or false. Justify your answer.
   (a) $\phi \subseteq B$ (b) $\phi \in B$ (c) $\{\{\phi\}\} \in B$
   (d) $\{\{\phi\}\} \subseteq B$ (e) $\{\phi\} \subseteq B$ (f) $\{\phi\} \in B$

4. Let $A = \{a, \{a\}\}$. Determine whether each of the following statement is true or false. Wherever required explain your answer.
   (a) $\phi \subseteq P(A)$ (b) $\phi \in P(A)$ (c) $\{\{a\}\} \in P(A)$
   (d) $\{\{a\}\} \subseteq P(A)$ (e) $\{a\} \subseteq P(A)$ (f) $\{a\} \in P(A)$
   (g) $\{\{\{a\}\}\} \subseteq P(A)$ (h) $\{\{\{a\}\}\} \in P(A)$ (i) $\{a, \{a\}\} \subseteq P(A)$
   (j) $\{a, \{a\}\} \in P(A)$ (k) $A \cup P(A) = P(A)$ (l) $A \cap P(A) = P(A)$
   (m) $\{A\} \cup P(A) = P(A)$ (n) $\{A\} \cap P(A) = P(A)$ (o) $A - P(A) = A$
   (p) $P(A) - A = P(A)$

5. Determine set A if the Ω is a set of (i) positive integers, (ii) the integer, (iii) the real number and (iv) the complex number.
   (a) $A = \{x \mid x \text{ is of the form } 3y + 1 \text{ for } y \in N\}$
   (b) $A = \{x \mid x \text{ is of the form } y - 4 \text{ for } y \in N\}$
   (c) $A = \{x \mid x^2 + x - 2 = 0\}$
   (d) $A = \{x \mid x^2 - x - 1 \leq 0\}$
   (e) $A = \{x \mid x^3 - 2x^2 - 2x - 3 = 0\}$

6. For any three arbitrary sets A, B and C show that
   (a) $(A - B) - C = A - (B \cup C)$
   (b) $(A - B) - C = (A - C) - B$
   (c) $(A - B) - C = (A - C) - (B - C)$

7. For any three sets A, B and C draw Venn diagram for the following conditions:
   (a) $(A \cup B) \subseteq B$ and $B \subseteq A$
   (b) $A \subseteq B$, $A \subseteq C$, $(B \cap C) \subseteq A$ and $A \subseteq (B \cap C)$
   (c) $(A \cap B \cap C) = \phi$, $(A \cap B) \neq \phi$, $(A \cap C) \neq \phi$ and $(B \cap C) \neq \phi$

8. For any three arbitrary sets A, B and C, answer the following questions giving the proper justification.
   (a) For $(A \cap B) = (A \cap C)$ and $(\bar{A} \cap B) = (\bar{A} \cap C)$, is it necessary that $B = C$?
   (b) For $(A \cup B) = (A \cup C)$, is it necessary that $B = C$?
   (c) For $(A \cap B) = (A \cap C)$, is it necessary that $B = C$?
   (d) For $(A \oplus B) = (A \oplus C)$, is it necessary that $B = C$?

(e) For $A \subseteq B$ and $C \subseteq D$, is it necessary that $(A \cup C) \subseteq (B \cup D)$?
(f) For $A \subseteq B$ and $C \subseteq D$, is it necessary that $(A \cap C) \subseteq (B \cap D)$?

9. What can you say about the sets P and Q if
(a) $P \cap Q = P$
(b) $P \cup Q = P$
(c) $P \oplus Q = P$
(d) $P \cap Q = P \cup Q$
(e) $P - Q = Q$
(f) $P - Q = Q - P$

10. (a) Give an example of sets A, B and C such that $A \in B$, $B \in C$ and $A \notin C$.
(b) Given that $(A \cup C) \subseteq (B \cup C)$ and $(A \cup \bar{C}) \subseteq (B \cup \bar{C})$, prove that $A \subseteq B$.

11. Determine whether each of the following statements is true for arbitrary sets A, B and C. Justify your answer.
(a) If $A \in B$ and $B \subseteq C$ then $A \in C$.
(b) If $A \in B$ and $B \subseteq C$ then $A \subseteq C$
(c) If $A \subseteq B$ and $B \in C$ then $A \in C$.
(d) If $A \subseteq B$ and $B \in C$ then $A \subseteq C$

12. Let A, B and C be any three arbitrary sets. Specify the conditions under which the following statements are true..
(a) $(A - B) \cup (A - C) = A$
(b) $(A - B) \cup (A - C) = \phi$
(c) $(A - B) \cap (A - C) = \phi$
(d) $(A - B) \oplus (A - C) = \phi$

13. A survey was conducted among 1000 people. 595 of them are democrats, 595 wear glasses and 550 like ice cream. 395 democrats wear glasses. 350 democrats like ice cream. 400 of the people wear glasses and like ice cream. 250 democrats wear glasses and like ice cream. Answer the following.
(a) How many people are not democrats who do not wear glasses and do not like ice cream?
(b) How many people are democrats who do not wear glasses and do not like ice cream?

14. The 60,000 fans who attended the homecoming football game bought up all the paraphernalia for their cars. Altogether, 20,000 bumper stickers, 36,000 window decals and 12,000 key rings were sold. We know that 52,000 fans bought at least one item and no one bought more than one of a given item. Also, 6000 fans bought both decals and key rings, 9000 bought both decals and bumper stickers and 5000 bought both key rings and bumper stickers. Answer the following.
(a) How many fans bought all the three items?
(b) How many fans bought exactly one item?
(c) Someone questioned the accuracy of the total number of purchasers: 52,000 (Given that all the other numbers have been confirmed to be correct). This person claimed the total number of purchasers to be either 60,000 or 44,000. How do you dispel the claim?

15. Out of a total of 130 students, 60 are wearing hats to class, 51 are wearing scarves and 30 are wearing both hats and scarves. Of the 54 students who are wearing sweaters, 26 are wearing hats, 21 are wearing scarves and 12 are wearing both hats and scarves. Everyone wearing neither a hat nor a scarf is wearing gloves. Find the number of students:
(a) who are wearing gloves,
(b) who are not wearing a sweater , wearing hat but not wearing scarves.
(c) who are not wearing sweater and wearing neither a hat nor a scarf.

16. Among 100 students, 32 study mathematics, 20 study physics, 45 study biology, 15 study mathematics and biology, 7 study mathematics and physics, 10 study physics and biology and 30 do not study any of the three subjects. Find the number of students
(a) studying all the three subjects
(b) studying exactly one of the three subject.

17. Seventy-five children went to an amusement park where they can ride on the merry-go-round, collar coaster and ferris wheels. It is known that 20 of them have taken all the three rides and 55 of them have taken at least two of the three rides. Each ride costs 50p and the total receipt of the amusement park was Rs. 70. Determine the number of children who did not try any of the rides.

18. Write a program fragment (any language) to perform the following task
    (a) To test whether x is an element of a set A where, A is any finite set with $|A| = n$.
    (b) To find $\bar{A}$ where, $\Omega = \{a, b, \ldots, z\}$ and $A \subseteq \Omega$.
    (c) To test whether $A \subseteq B$ where, A and B are any set defined on $\Omega = \{1, 2, 3, \ldots, 100\}$.
    (d) To find all elements x such that $\{x, x+1\} \subseteq A$ where, A is defined on $\Omega = \{1, 2, 3, \ldots, 100\}$.

19. In each part, find x or y so that the following statements are true.
    (a) $(x, 3) = (4, 3)$ (b) $(a, 3y) = (a, 9)$ (c) $(3x+1, 2) = (7, 2)$
    (d) (C++, PASCAL) = (y, x) (e) $(4x, 6) = (16, y)$ (f) $(2x - 3, 3y - 1) = (5, 5)$
    (g) $(x^2, 25) = (49, y^2)$ (h) $(x, y) = (x^2, y^2)$

20. Let $A = \{a, b\}$ and $B = \{4, 5, 6\}$. List the elements in
    (a) A x B (b) B x A (c) A x A (d) B x B

21. A genetics experiment classifies fruit flies according to the following two criteria:

    Gender: male (m), female (f)

    Wing span short (s), medium (m), long (l)

    (a) How many categories are in this classification scheme?
    (b) List all the categories in this classification scheme.

22. A car manufacturer makes three different types of car frames: sedan(s), coupe(c), van(v) and two types of engines: gas(g), diesel(d). List all possible models of cars.

23. Prove the following:
    (a) If A has three elements and B has $n \geq 1$ elements, then $|A \times B| = 3n$.
    (b) If $|A_1| = n_1$, $|A_2| = n_2$ and $|A_3| = n_3$ then $|A_1 \times A_2 \times A_3| = n_1 n_2 n_3$.

24. Sketch A x B and B x A for sets A and B given as below:
    (a) If A = {a | a is real number} and B = {1, 2, 3}.
    (b) If A = {a | a is real number and $-2 \leq a \leq 3$} and B = { b | b is real number and $1 \leq b \leq 5$}.

25. For any set A, B, C and D prove the followings:
    (a) If $A \subseteq C$ and $B \subseteq D$ then $A \times B \subseteq C \times D$.
    (b) $A \times (B \cup C) = (A \times B) \cup (A \times C)$
    (c) Given that $A \times B \subseteq C \times D$, does it necessarily follow that $A \subseteq C$ and $B \subseteq D$.
    (d) $(A \cap B) \times (C \cap D) = (A \times C) \cap (B \times D)$.
    (e) Use the set A = (1, 2, 4}, B = {2, 5, 7} and C = {1, 3, 7} to investigate whether

$$A \times (B \cap C) = (A \times B) \cap (A \times C).$$

26. For any sets A, B, C and D confirm or disprove the followings:
    (a) $(A \cup B) \times (C \cup D) = (A \times C) \cup (B \times D)$
    (b) $(A - B) \times (C - D) = (A \times C) - (B \times D)$
    (c) $(A \oplus B) \times (C \oplus D) = (A \times C) \oplus (B \times D)$
    (d) $(A - B) \times C = (A \times C) - (B \times C)$
    (e) $(A \oplus B) \times C = (A \times C) \oplus (B \times C)$

27. For any sets A and B answer the following:
   (a) Is the set A x ϕ well defined?
   (b) If A x B = ϕ, what can you say about the sets A and B ?
   (c) Is it possible that A x A = ϕ for some set A?

28. List all the partitions of sets
   (a) A = {1, 2, 3} (b) B = {a, b, c, d}

29. Let A = {1, 2, 3, 4, 5, 6, 7, 8, 9, 10} and
   A_1 = {1, 2, 3, 4} A_2 = {5, 6, 7} A_3 = {4, 5, 7, 9}
   A_4 = {4, 8, 10} A_5 = {8, 9, 10} A_6 = {1, 2, 3, 6, 8, 10}
   Which of the following are partitions of A?
   (a) {A_1, A_2, A_5} (b) {A_1, A_3, A_5} (c) {A_3, A_6} (d) {A_2, A_3, A_4}

30. Let A_1 be the set of positive integers and A_2 be the set of all negative integers. Is {A_1, A_2} a partition of Z? Here Z is the set of all integers. Explain your conclusion.

31. Let B = {0, 3, 6, 9, ...}. Give a partition B containing
   (a) Two infinite subsets (b) Three infinite subsets

32. In the following examples, determine whether the relation R on the set A is reflexive, irreflexive, symmetric, asymmetric, anti-symmetric or transitive.
   (a) A is set of integers and aRb iff $a \leq b+1$.
   (b) A is set of positive integers and aRb iff $|a - b| \leq 2$.
   (c) A is set of integers and aRb iff a + b is even.
   (d) A is set of positive integers and aRb iff GCD(a, b) = 1. In this case we say that a and b are relatively prime.
   (e) A is set of positive integers and aRb iff $a = b^k$ for some positive integer k.
   (f) A is set of positive integers and aRb iff a - b is an odd positive integer.
   (g) A is the set of all people and aRb iff "a is brother of b".
   (h) A is the set of all binary strings and aRb iff a and b have equal number of 0's.
   (i) A is the set of books and aRb iff "a costs more **and** a contains more pages than that in b".
   (j) A is the set of books and aRb iff "a costs more **or** a contains more pages than that in b".

33. Determine whether the relation R on the set A is an equivalence relation.
   (a) A = {a, b, c, d} and R = {(a, a), (b, a), (b, b), (c, c), (d, d), (d, c)}.
   (b) A = {1, 2, 3, 4, 5} and R = {(1, 1), (1, 2), (1, 3), (2, 1), (2, 2), (3, 1), (2, 3), (3, 3), (4, 4), (5, 2), (5, 5)}.
   (c) A is set of all members of the software-of-the month Club and aRb iff a and b buy the same programs.
   (d) A is set of all members of the software-of-the month Club and aRb iff a and b buy the same number of programs.
   (e) A is set of all people in the Indian electoral list database and aRb iff a and b have the same last name.
   (f) A is set of all triangles in the plane and aRb iff a is similar to b.
   (g) $A = Z^+ \times Z^+$ and (a, b)R(c, d) iff b = d.

34. Let A = {6:00, 6:30, 7:00, ..., 9:30, 10:00} denote the set of nine half-hour periods in the evening. Let B = {3, 12, 15, 17} denote the set of the four local television channels. Let R and S be two binary

relations from A to B. What possible interpretations can be given to the binary relations R, S, $R \cup S$, $R \cap S$, $R \oplus S$ and $R - S$?

35. Let I be the set of all integers
   (a) Is there a natural way to interpret the ordered pairs in I x I as geometric points in the plane?
   (b) Let R be a binary relation on I x I such that the ordered pair of ordered pairs ((a, b), (c, d)) is in R iff a -c = b- d. What is the geometric interpretation of the binary relation R?
   (c) Let S be a binary relation on I x I such that the ordered pair of ordered pairs ((a, b), (c, d)) is in S iff **sqrt (($(a - c)^2 + (b - d)^2$))** less than equal to 10. What is a geometric interpretation of S?
   (d) Give the geometric interpretation of $R \cup S$, $R \cap S$, $R \oplus S$ and $R - S$?

36. Let S = {1, 2, 3, 4, 5} and let A = S x S. A relation R is defined on set A in such a way that

   (a, b) R (a', b') iff ab' = a'b

   (a) Show that R is an equivalence relation.
   (b) Compute A /R.

37. If {{a, c, e}, {b, d, f}} is a partition of the set A = {a, b, c, d, e, f}, determine the corresponding equivalence relation R.

38. Find complement and inverse of the following relations defined from A to B.
   (a) A = {a, b, c, d}, B = {1, 2, 3} and R = {(a, 1), (a, 2), (b, 1), (c, 2), (d, 1)}.
   (b) A ={IBM, COMPAQ, Dell, Gateway, Zenith}, B = {750C, PS60, 450SV, 4/33S, 525SX, 466V, 486SL} and R = {(IBM, 750C), (Dell, 466V), (COMPAQ, 450SV), (Gateway, PS60)}.
   (c) A ={1, 2, 3, 4}, B = {1, 4, 5, 8, 9} and R = $\{(a, b) | b = a^2\}$.
   (d) A ={1, 2, 3, 4, 8}, B = {1, 4, 6, 9} and R = {(a, b)| a divides b}.
   (e) A ={1, 2, 3, 4, 8} = B and R = $\{(a, b) | a + b \leq 9\}$.

39. Let A and B be any two sets such that |A| = n and |B| = m. Answer the following:
   (a) How many different relations are there from A to B?
   (b) How many different binary relations on A are there?
   (c) How many of relations of (b) are reflexive?
   (d) How many of relations of (b) are symmetric?

40. Use the fact that "better than" is a transitive binary relation to prove the claim "A ham sand-witch is better than eternal happiness".
   ***Hint**: "Nothing is better than eternal happiness.*

41. Let A = {1, 2, 3, 4, 5, 6, 7} and R = {(1, 2), (1, 4), (2, 3), (2, 5), (3, 6), (4, 7)}. Compute the restriction of R to the following subset B of A.
   (a) B = {1, 2, 4, 5}   (b) B = {2, 3, 4, 6}

42. Let A is set of positive integers and R be a relation defined on A as aRb iff $a = b^k$, for some positive integer k. Which of the following belongs to R?
   (a) (4, 16)   (b) (1, 7)   (c) (8, 2)   (d) (3, 3)   (e) (2, 32)

43. Let A be a set of real number and R be a relation defined on A as below. Find Dom(R) and Rang(R).
   (a) For any two elements a and b of A, aRb iff 2a + 3b = 6.
   (b) For a, b $\in$ A, aRb iff $a^2 + b^2 = 25$.

44. Let A = R. Give a description of the relation specified by the shaded region shown in the figure ex44.

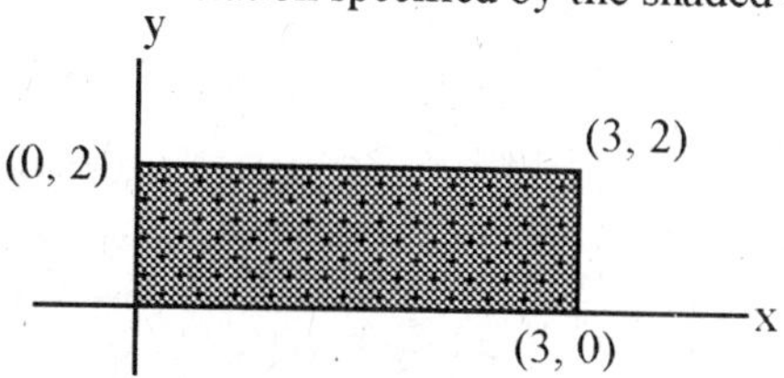

Figure ex44

45. Show the following relations, given in the form of matrix, diagrammatically and determine whether the relation is an equivalence relation.

(a) Let $A = \{1, 2, 3, 4\}$ and $M_R = \begin{bmatrix} 1 & 1 & 0 & 1 \\ 0 & 1 & 1 & 0 \\ 0 & 0 & 1 & 1 \\ 1 & 0 & 0 & 0 \end{bmatrix}$ (b) Let $A = \{a, b, c\}$ and $M_R = \begin{bmatrix} 1 & 0 & 0 \\ 0 & 1 & 1 \\ 0 & 1 & 1 \end{bmatrix}$

(c) Let $A = \{a, b, c, d, e\}$ and $M_R = \begin{bmatrix} 1 & 1 & 0 & 0 & 0 \\ 0 & 0 & 1 & 1 & 0 \\ 0 & 0 & 0 & 1 & 1 \\ 0 & 1 & 1 & 0 & 0 \\ 1 & 0 & 0 & 0 & 0 \end{bmatrix}$ (d) Let $A = \{a, b, c\}$ and $M_R = \begin{bmatrix} 1 & 0 & 0 \\ 0 & 1 & 0 \\ 0 & 0 & 1 \end{bmatrix}$

Figure ex45

46. Find the matrix for the following relations, shown diagrammatically and determine whether the relation is an equivalence relation.

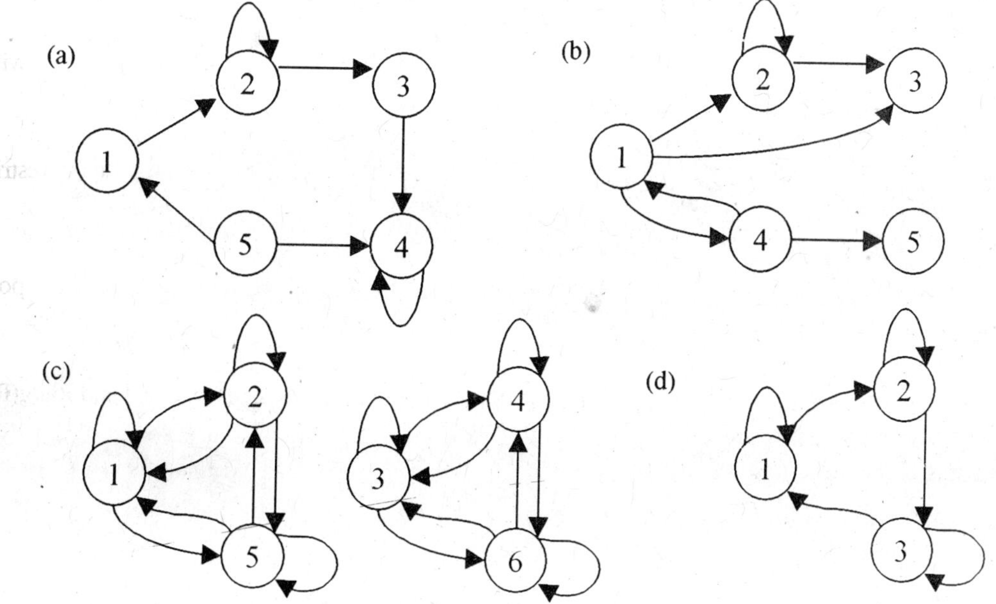

Figure ex46

47. Determine whether the following relations, shown in diagrammatic way, is reflexive, irreflexive, symmetric, asymmetric, anti-symmetric or transitive.

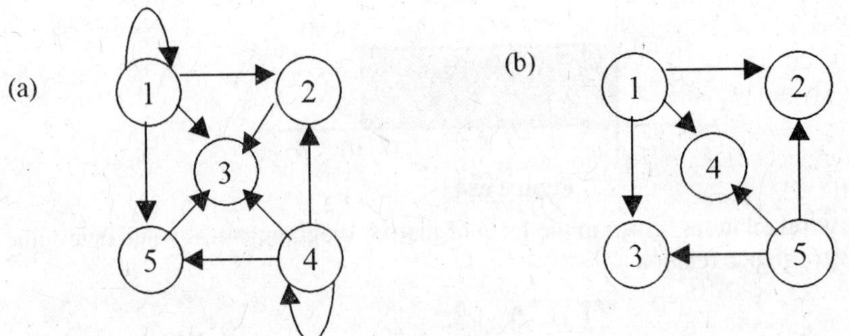

Figure ex47

48. Show the following relations R defined on set A, as given below, in diagrammatic way.
   (a) A = {1, 2, 3, 4, 5} and R = {(1, 2), (2, 1), (3, 4), (4, 3), (3, 5), (5, 3), (4, 5), (5, 4), (5, 5)}
   (b) A = {a, b, c, d} and R = {(a, b), (b, a), (a, c), (c, a), (a, d), (d, a)}

49. Compute $\overline{R}$, $\overline{S}$, $R \cap S$, $R \cup S$, R^{-1} and S^{-1} for relations R and S given by the following digraph/ matrix.

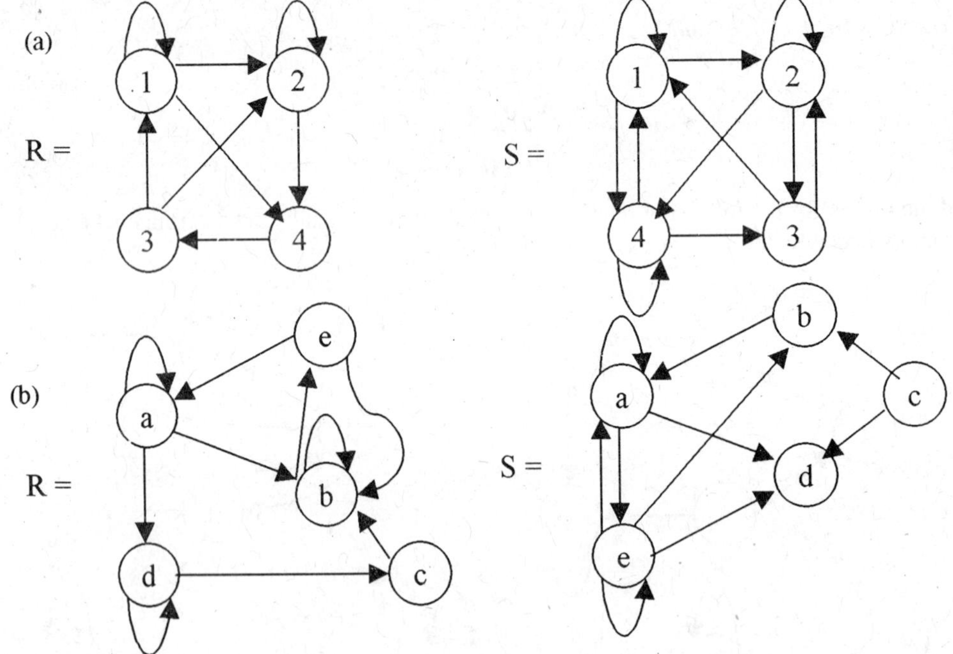

Figure ex49a

(c) $M_R = \begin{bmatrix} 1 & 1 & 0 & 1 \\ 0 & 0 & 0 & 1 \\ 1 & 1 & 1 & 0 \end{bmatrix}$ $M_s = \begin{bmatrix} 0 & 1 & 1 & 0 \\ 1 & 0 & 0 & 1 \\ 1 & 1 & 0 & 0 \end{bmatrix}$

(d) $M_R = \begin{bmatrix} 1 & 0 & 1 & 0 \\ 0 & 0 & 0 & 1 \\ 1 & 1 & 1 & 0 \end{bmatrix}$ $M_s = \begin{bmatrix} 1 & 1 & 1 & 1 \\ 0 & 0 & 0 & 1 \\ 0 & 1 & 0 & 1 \end{bmatrix}$

Figure ex49b

50. Prove the followings:
   (a) If a relation on a set A is transitive and irreflexive then it is asymmetric.
   (b) If a relation R is symmetric then R^2 is also symmetric.
   (c) If R is symmetric and transitive then R is not irreflexive.
   (d) A relation is reflexive and circular iff it is an equivalence relation. A relation R on a set A is called ***circular*** if aRb and bRc imply cRa.
   (e) If R and S are equivalence relations on A then $R \cap S$ is an equivalence relation on A.
   (f) Let R be a symmetric and transitive relation on a set A. Show that if for every a in A there exists b in A such that (a, b) is in R, then R is an equivalence relation.
   (g) Let R be a transitive and reflexive relation on A. Let T be a relation on A such that (a, b) is in T iff both (a, b) and (b, a) are in R. Show that T is an equivalence relation.
   (h) Let R be a binary relation and S = {(a, b) | (a, c) and (c, b) ∈ R for some c}. Show that if R is an equivalence relation then S is also an equivalence relation.
   (i) Let R be a reflexive relation on a set A. Show that R is an equivalence relation iff

$$(a, b) \text{ and } (a, c) \in R \Rightarrow (b, c) \in R$$

51. A binary relation on a set that is *reflexive* and *symmetric* is called a ***compatible relation.*** Now answer the following:
   (a) Let A be a set of people and R be a binary relation on set A such that (a, b) is in R iff "a is a friend of b". Show that R is a compatible relation.
   (b) Let A be a set of English words and R be a binary relation on set A such that any two words are R-related iff they have one or more letters in common. Show that R is a compatible relation.
   (c) Let R and S be two compatible relations on A then determine whether $R \cap S$ and $R \cup S$ are compatible relations.
   (d) Let A be a set. A ***cover*** of A is a set of nonempty subsets of A, say $\{A_1, A_2, A_3, \ldots A_n\}$, such that union of A_i's is equal to A. Suggest a way to define a compatible relation on A from a ***cover*** of A.
   (e) Give two more examples of compatible relation.

52. (a) Show that the transitive closure of a symmetric relation is symmetric.
   (b) Is the transitive closure of an anti-symmetric relation always anti-symmetric?
   (c) Show that the transitive closure of a compatible relation is an equivalence relation.

53. (a) Let R and S be binary relations from A to B. Is it true that $(R \cup S)^{-1} = R^{-1} \cup S^{-1}$?
   (b) Let R be a binary relation on A. Show that if R is an equivalence relation then R^{-1} is also an equivalence relation.

54. (a) Let A = {1, 2, 3} and R = {(1, 1), (1, 2), (2, 3), (1, 3), (3, 1), (3, 2)}. Compute the matrix $M_{R^{\infty}}$.
   (b) List the relation R^{∞} whose matrix was computed in part (a).

55. Let A = {a, b, c, d, e} and R be a relation onset A such that M_R is given as below.

$$M_R = \begin{bmatrix} 1 & 0 & 0 & 1 & 0 \\ 0 & 1 & 0 & 0 & 0 \\ 0 & 0 & 0 & 1 & 1 \\ 1 & 0 & 0 & 0 & 0 \\ 0 & 1 & 0 & 0 & 1 \end{bmatrix}$$

Figure ex55

Find the transitive closure of R using Warshall's algorithm.

56. Find the transitive closure of the following relations given in matrix form as below.

$$(a)\ M_R = \begin{bmatrix} 1 & 0 & 0 & 1 \\ 1 & 1 & 0 & 0 \\ 0 & 0 & 1 & 0 \\ 0 & 0 & 0 & 1 \end{bmatrix} \quad (b)\ M_R = \begin{bmatrix} 1 & 1 & 0 & 0 \\ 1 & 0 & 0 & 0 \\ 0 & 0 & 0 & 0 \\ 0 & 0 & 1 & 0 \end{bmatrix}$$

$$(c)\ M_R = \begin{bmatrix} 1 & 0 & 0 & 1 \\ 0 & 1 & 1 & 0 \\ 0 & 1 & 1 & 0 \\ 1 & 0 & 0 & 1 \end{bmatrix} \quad (d)\ M_R = \begin{bmatrix} 0 & 0 & 0 & 1 \\ 1 & 0 & 0 & 1 \\ 0 & 1 & 0 & 1 \\ 0 & 0 & 1 & 0 \end{bmatrix}$$

Figure ex56

57. Prove that if R is reflexive and transitive, then $R^n = R$ for all n.

♦♦♦

CHAPTER TWO
Function and Generating Function

A function is defined in terms of some variables. If x is a variable then f(x) is said to be a function of one variable. Similarly, if x and y are two variables then f(x, y) is called a function of two variables. A function of n variables can be denoted by $f(x_1, x_2, x_3, x_4, \ldots, x_n)$. Here, $x_1, x_2, x_3, x_4, \ldots, x_n$ are called arguments or parameters of function f. Each of these parameters has a defined type say- integer, float etc. The value returned by the function is again of some type. A function, simply, maps a set of values to another set of values. In this chapter, we shall learn this and other related concepts. Sequence is another important topics in the discrete mathematical structure. We shall learn this and about a function, called Generating Function that generates sequences.

Function

A function is defined from a set A to a set B. Let A = {a, b, c, d, e} and B = {1,2,3,4,5,6}, and f: A→B be a function defined from set A to set B as shown below.

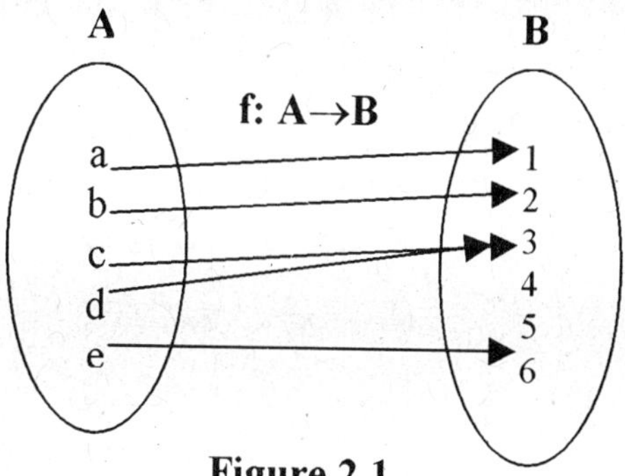

Figure 2.1

The diagram shows that element **a** of A is associated to element **1** of B, **b** of A to **2** of B, **c** of A to **3** of B, **d** of A to **3** of B and **e** of A to **6** of B. We say that, **1** is a **f-image** of **a** (or **image** of **a** under f). **a** is called **pre-image** of **1 under f**. If we arrange **a** and its image **1** as ordered pair **(a,1)**, then f can be represented as **f = {(a,1),(b,2),(c,3),(d,3),(e,6)}**, a set of ordered pairs of all the associations shown in the above diagram. Here, $f \subseteq A \times B$. We know that any subset of A×B is a relation from set A to set B, f is also a relation from A to B. Every function is a relation but the reverse is not true. A function f: A→B said to be defined from set A to set B if and only if f is **everywhere defined** on set A, i.e.

(1) Every element of A is associated with some element of B, and
(2) There exists no element in set A such that it is associated with more than one element of B i.e. for any two elements **a** and **b** of A if **a = b** then **f(a) = f(b)**.

The **figure 2.1,** is an example of a valid function from a nonempty set A to a nonempty set B. Whereas, the diagram shown in **figure 2.2** does not correspond to a valid function f: A→B, since –2 ∈ A and 2 ∈ A have no image in B under f.

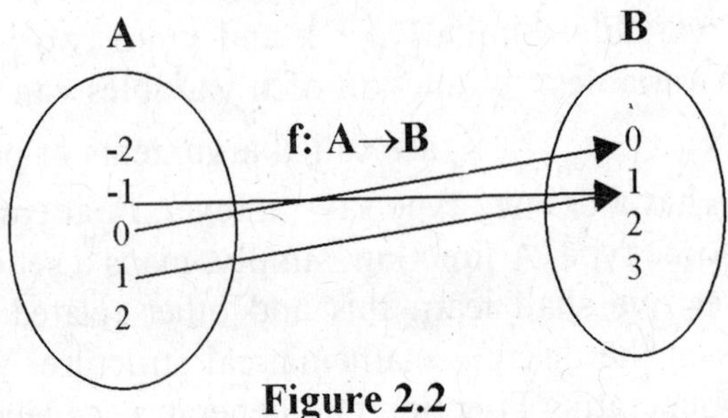

Figure 2.2

Another example of an invalid function can be given by diagram shown in the **figure 2.3**. Where, an element 4 ∈ A has two images –2 and 2 in B under function f. The function f: A→B is defined as f(x) = √x. This function is not defined on the set of integers, on the set of rational numbers, on the set of real numbers and many more such sets containing at least one negative number.

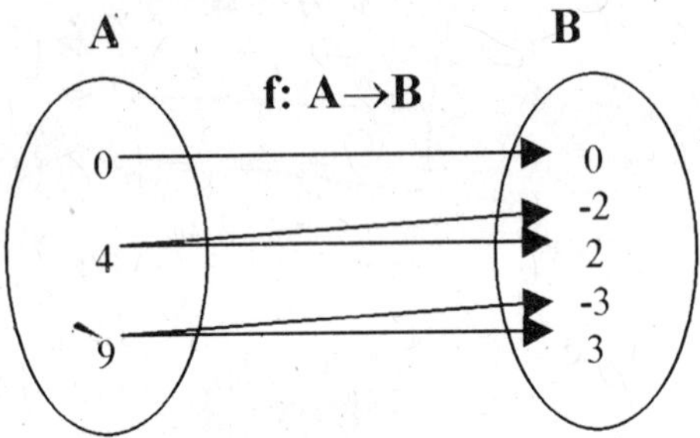

Figure 2.3

In function **f: A→B** set A is called **Domain** set of f and set B is called **Range** set of **f**. Thus any relation **R: A→B** is a function if **Domain(R) = A** and **R-relative** set of every element of A is a **Singleton** set. A function is also called **Mapping** or **Transformation**.

***Example* 1** Let A be a set of all even integers and B be a set of all odd integers and let **f: A→B** be a function defined as f(x) = x+1. Show that **f** is a function from set A to set B.

Solution: Let x be any element of set A. Since x is even, x+1 is odd. Therefore f(x)∈B. For every even integer x ∈ A, ∃ x+1 = y, an odd integer in B such that f(x) = y. And for any two elements x_1 and x_2 of A if $x_1 = x_2$ then $f(x_1) = f(x_2)$ i.e. no element of A has more than one image in B. Hence f is a function from set A to set B.

***Example* 2** Let A be a set of all integers and B be a set of all **non negative** integers and let **f: A→B** be a function defined as $f(x) = \sqrt{x}$. Show that **f** is not a function from set A to set B.

***Solution*:** Since there exists no real square root for a negative integer, there are many elements x of A which have no image in B, i.e. f is not everywhere defined. Hence f is not a function.

2.1 Types of Functions

In a function **f: A→B** one element of **A** cannot be associated with more than one elements of **B**, but more than one elements of **A** may get associated with one element of **B**. This gives two types of functions. One is **One to One** function and other is **Many to One** function. Further, there may be some elements left in set B which is (are) not associated with any of the element of A. Accordingly, we can have **Into** and **Onto** function. Let A be a set {'**One to One**', '**Many to One**'} and B be another set {**Onto**, **Into**}. Then all possible functions can be given by the elements of A×B. These are **One to One Onto, One to One Into, Many to One Onto,** and **Many to One Into** functions. Before discussing these functions, we shall discuss **Identity** function because it is required for understanding some important properties of a function. Let **I** be a function defined from set A to itself, called $\mathbf{I_A}$, such that $\mathbf{I_A}(x) = x, \forall x \in A$. This function is called **Identity** function on set A.

Let **f: A→B** and **g: B→C** be two functions. The **g 0 f,** called **composition** of functions f and g, is a function from set A to set C. The diagram shown in **figure 2.1.1** demonstrates a composition of functions f and g.

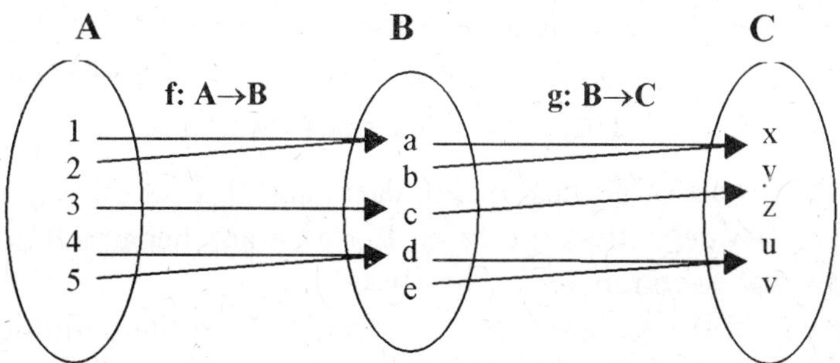

Figure 2.1.1

In the above diagram, f = {(1,a), (2,a), (3,c), (4,d), (5,d)} and g = {(a, x), (b, x), (c, y), (d, u), (e, u)}. The composition of these two functions **g 0 f** is given by the set of ordered pairs {(1,x), (2,x), (3,y), (4,u), (5,u)}. Clearly, **g 0 f** is a function from set A to set C. The method to find **g 0 f** can be illustrated as below.

The elements of f can be arranged in two lines: the first line contains elements from set A and second line contains images of corresponding elements under f. Similarly, we can have arrangement for the elements of g. The arrangement is shown in **figure 2.1.2** on the next page.

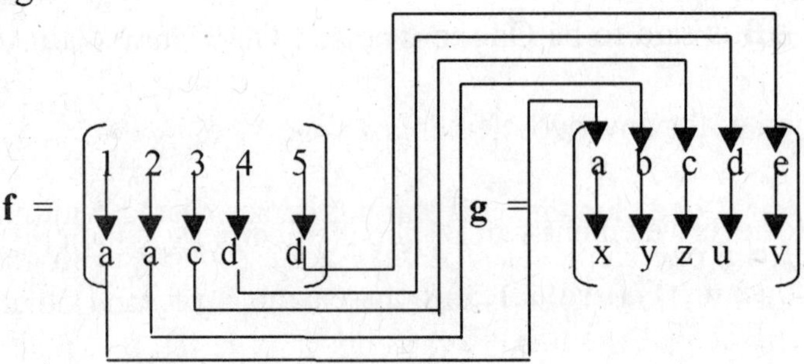

Figure 2.1.2

The **g o f**(x) is also written as g(f(x)). Now, g(f(1)) = g(a) = x, g(f(2)) = g(a) = x, g(f(3)) = g(c) = y, g(f(4)) = g(d) = u, and g(f(5)) = g(d) = u. This is the way to find the composition of two functions. In general, if f_1, f_2, f_3, ..., f_n are n functions then theirs composition is given as $f_n \circ f_{n-1} \circ \ldots \circ f_3 \circ f_2 \circ f_1$.

It is important to note that whenever **g 0 f** is possible, it is not necessary that **f 0 g** is also possible. And, if it is possible, **g 0 f = f 0 g** is not always true. That is to say that a function composition is not a **commutative** operation. But it is an associative operation i.e. if **f**, **g** and **h** are any three functions then **h 0 (g 0 f) = (h 0 g) 0 f**, if the composition is possible.

A function is said to be **invertible** if its inverse exists. Let **f: A→B** be any function then a function **g: B→A** is called inverse of **f** if **g o f = f o g = I_A**. Inverse of a function f is denoted as f^{-1}. Therefore, $f^{-1} \circ f = f \circ f^{-1} = I_A$.

Example 3 Let A is set of all negative integers and B a set of all positive integers. Let f be function **f: A→B** defined as f(x) = -x. Let g be another function **g: B→A** defined as g(y) = -y. Show that g is an inverse function of f.

Solution: Here **g o f**(x) = g(f(x)) = g(-x) = -(-x)) = x , by the definition of functions f and g. Similarly, **f o g**(y) = f(g(y)) = f(-y) = -(-y) = y, again by the definition of functions f and g. Also we know that if I_A is an identity function on A then $I_A(x) = x$ for every element x of A. Thus, we have **g o f = f o g = I_A**. Therefore f and g are inverse of each other i.e. $\mathbf{f^{-1} = g}$ and $\mathbf{g^{-1} = f}$.

Note: 1. Inverse of inverse of a function is the function itself i.e. $\mathbf{(f^{-1})^{-1} = f}$.

2. $\mathbf{(g \circ f)^{-1} = f^{-1} \circ g^{-1}}$ and $\mathbf{(f \circ g)^{-1} = g^{-1} \circ f^{-1}}$.

3. In general, if $f_1, f_2, f_3, \ldots, f_n$ are n functions then
$$(f_n \circ f_{n-1} \circ \ldots \circ f_3 \circ f_2 \circ f_1)^{-1} = f_1^{-1} \circ f_2^{-1} \circ f_3^{-1} \circ \ldots \circ f_{n-1}^{-1} \circ f_n^{-1}.$$

One to One Onto function

A function **f: A→B** is said to be One to One and Onto from a set A to a set B if and only if

1. **f** is One to One i.e. for any two elements $\mathbf{x_1}$ and $\mathbf{x_2}$ of A if $\mathbf{x_1 \neq x_2}$ then $\mathbf{f(x_1) \neq f(x_2)}$, and
2. **f** is **onto**, i.e. for every element **y** in B ∃ an element **x** in A such that **f(x) = y**.

The diagram shown in **Figure 2.1.3** is an example of an One to One Onto function from set A to set B defined as f(x) = x + 3.

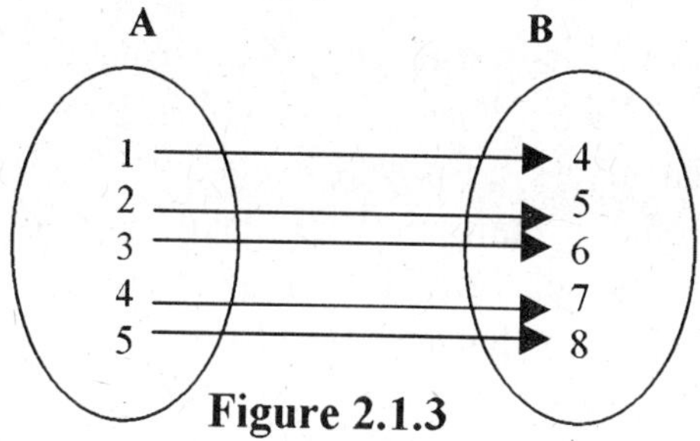

Figure 2.1.3

When a function is One to One it is called **injection** and if it is One to One Onto it is called **bijection.** The following theorem shows that the inverse of a function f exists if and only if f is One to One and Onto.

Theorem 1 Let f be a function **f: A→B** such that f is **bijection,** then show that $\mathbf{f^{-1}}$ is also a function from B to A, and it is a **bijection**.

Proof: Let **f: A→B** be a one to one and onto function from A to B. Then Dom(f) = A and Range(f) = B. Let $\mathbf{f^{-1}}$ be a function from B to A. Then we have to show that $\mathbf{f^{-1}}$: **B→ A** is a function and it is one to one and onto.

Since Range(f) = B, $\mathbf{f^{-1}}$ is everywhere defined on B. Also f is one to one, so no two or more elements of A have same image in B. Therefore, no two or more elements of B can have same pre-image in A. Hence, $\mathbf{f^{-1}}$ is a function from B to A.

Let y_1 and y_2 be any two elements of B then ∃ x_1 and x_2 in A such that $f(x_1) = y_1$ and $f(x_2) = y_2$. Now, $f(x_1) = y_1$ and $f(x_2) = y_2 \Rightarrow x_1 = \mathbf{f^{-1}}(y_1)$ and $x_2 = \mathbf{f^{-1}}(y_2)$. Since f is one to one, we have $y_1 \neq y_2$ for $x_1 \neq x_2$, i.e., for $y_1 \neq y_2$ we have $\mathbf{f^{-1}}(y_1) \neq \mathbf{f^{-1}}(y_2)$. Therefore $\mathbf{f^{-1}}$ is one to one function. Since f is a function from A to B, every element of A is pre-image of some element of B, i.e. every element of A is image of some element of B under $\mathbf{f^{-1}}$. Hence $\mathbf{f^{-1}}$ is onto.

Proved.

Theorem 2 Let f be a function **f: A→B,** defined as f(x) = y and that f is a **bijection.** Show that a function **g: B→A** defined as g(y) = x is inverse of f and it is also a **bijection**.
Proof: Let x be any element of A. Then there exists a y in B such that f(x) = y by definition of f. Now, (**g o f**)(x) = g(f(x)) = g(y) = x. Similarly, (**f o g**)(y) = f(g(y)) = f(x) = y. Therefore, we have **g o f = f o g = I_A**, i.e. g = **f** $^{-1}$. From ***theorem 1***, we can show that g is one to one and onto.

Proved.

Theorem 3 Let A and B be two finite sets of same cardinality, and let **f: A→B** is a function then

(i) If f is one to one then it is onto.
(ii) If f is onto then it is one to one.

Proof: (i) Let |A| be n and since f is one to one, B must have at least n elements for f to be defined from A to B. Since |B| = |A|, ∃ no element in B which is not an image of any element of A. This proves that function f is onto.
(ii) Let f is not one to one then f in many to one. Since |A| = |B|, there must exist at least one element in B which is not associated with any element of A. This leads to conclusion that f is not onto. This is contrary to that fact that f is onto. Thus our assumption that f is not one to one is wrong. Hence f is one to one.

Proved.

***Example* 4** Let A is a set of all positive real numbers and B is a set of all real numbers. Let f be a function **f: A→B** defined as f(x) = $\log_e$x. Show that f is one to one and onto function.
Solution: Let x_1 and x_2 be any two elements of A such that
Now,

$$\log_e x_1 \neq \log_e x_2$$

$$\log_e x_1 \neq \log_e x_2 \Leftrightarrow e^{x_1} \neq e^{x_2} \Leftrightarrow x_1 \neq x_2$$

Therefore, f is one to one. Now, let y be any element of B, then e^y is a positive real number in A, such that $\log_e e^y$ = y, i.e. for every element y in B ∃ e^y in A, such that $\log_e e^y$ = y. Hence f is onto also. ♦

One to One Into function

A function **f: A→B** is said to be One to One and Into from a set A to a set B if and only if

1. **f is One to One** i.e. for any two elements $\mathbf{x_1}$ and $\mathbf{x_2}$ of A if $\mathbf{x_1 \neq x_2}$ then $\mathbf{f(x_1) \neq f(x_2)}$, and

2. f is **into** i.e. ∃ at least one element **y** in B such that it is **not** an image of any element x of A.

Example 5 Let A is the set of positive integers and B is a set of even integers. Let f be a function **f: A→B** defined as f(x) = 2x. Show that f is one to one and into function.

Solution: Let x_1 and x_2 be any two distinct elements of A then $2\mathbf{x_1} \neq 2\mathbf{x_2}$, i.e. f is one to one. Since there are many (**infinite**) negative even integers in B which are not associated with any of the element of A (twice of a positive integer is positive integer only). **f** is an into function.

A diagrammatic example of an One to One Into function can be given as in **Figure 2.1.4.** The function **f** from set A to set B is defined as f(x) = x + 3.

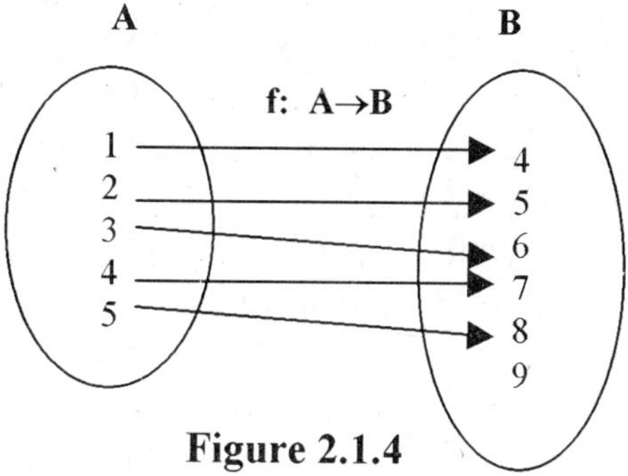

Figure 2.1.4

Many to One onto function

A function **f: A→B** is said to be Many to One and Onto from a set A to a set B if and only if

1. **f** is **Many to One** i.e. ∃ at least one pair of distinct elements $\mathbf{x_1}$ and $\mathbf{x_2}$ in A such that $\mathbf{f(x_1) = f(x_2)}$, and
2. **f** is **onto**, i.e. for every element **y** in B ∃ an element **x** in A such that **f(x) = y**.

In the following examples, shown in **Figure 2.1.5,** f is a function from A to B. Distinct elements pair 1 and 2 have same image 4 in B. F is many to one function. Further, ∃ no element in B that is not an image of any element of A. Thus f is onto.

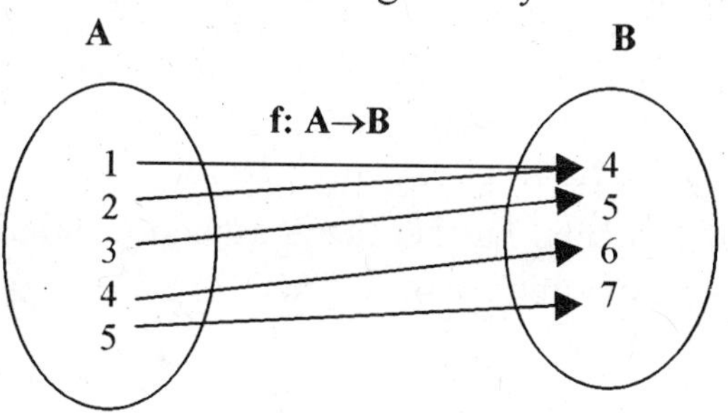

Figure 2.1.5

***Example* 6** Let A is the set of integers and B is the set of all non-negative integers. Let f be a function **f: A→B** defined as f(x) = |x|. Show that f is many to one and onto function.
Solution: Let x be any integer then –x is also an integer and | ±x | is magnitude of x i.e. |x|. Thus for every two members of A (except zero) we have one image in B. **f** is many to one. Next, for any element y in B we have an integer y in A such that |y| = y. **f** is onto. ♦

Many to One into function

A function **f: A→B** is said to be Many to One and Into from a set A to a set B if and only if

1. **f** is **Many to One** i.e. ∃ at least one pair of distinct elements $\mathbf{x_1}$ and $\mathbf{x_2}$ in A such that $\mathbf{f(x_1) = f(x_2)}$, and
2. **f** is **into** i.e. ∃ at least one element **y** in B such that it is **not** an image of any element x of A.

***Example* 7** Let A and B be set of real numbers. Let f be a function **f: A→B** defined as $f(x) = x^2$. Show that f is many to one and into function.
Solution: Let x be any real number then –x is also a real number and x^2 is a positive real number. Thus for every two members of A (except zero) we have one image in B. **f** is many to one. Next, ∃ real numbers in B (all negative) which are not square of any real numbers. Therefore, **f** is into function.

In **Figure 2.1.6,** f is a function from A to B. Distinct elements pair 1 and 2 have same images 4 in B. f is many to one function. Further, ∃ an element 8 in B which is not an image of any element of A, thus f is into.

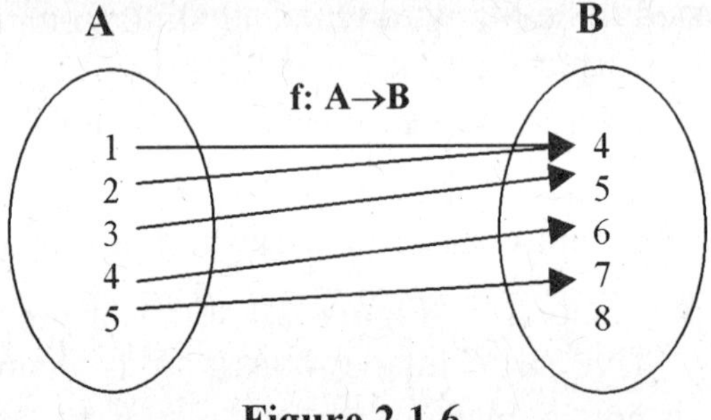

Figure 2.1.6

***Example* 8** Let A ={-π,π} and B be set of real numbers. Let f be a function **f: A→B** defined as f(x) = sin(x). Show that f is many to one and into function.
Solution: Since, sin (-π) = sin (π) = 0, i.e. for two distinct elements of A, we have same image in B, f is many to one function. Further there are many elements (in fact, all except zero) in B which are not image of either -π or π. Therefore, f is into function. ♦

It is important to mention, here, two more types of functions, which are frequently used. One is **Constant function** and other is **real valued function.** A function is said to be real valued if its range set is the set of real numbers. Examples 4, 7 and 8 mentioned in this section are examples of a **real valued function**. A function **f: A→B** is called a **Constant function** if for every $x \in A$, there is only one image say k in B. For example f(x) = 5 is a constant function from R to R. A diagrammatic example is given in **figure 2.1.7**.

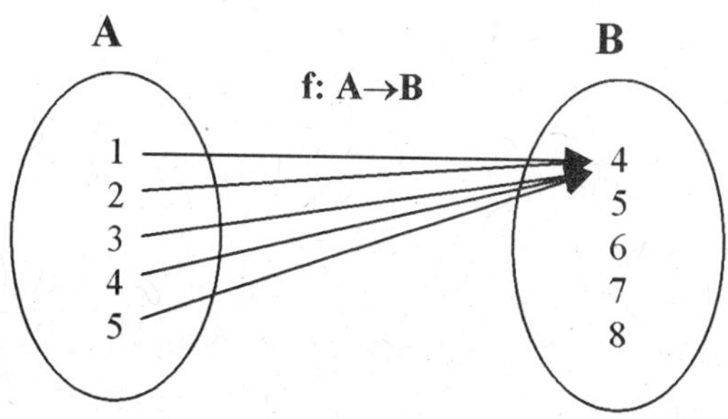

Figure 2.1.7

In this example $f(x) = 4 \ \forall x \in A$. Thus f is a constant function. **Can we say that f is many to one into also in this example?**

2.2 Order of Function

We know about algorithm. There are criteria for deciding whether an algorithm is good or relatively poor. To state simply that an algorithm is a method for solving a problem is inadequate as many assumptions are hidden e.g. tools and techniques required, storage requirements, time needed to find a solution etc. An algorithm is then defined as a discrete, deterministic method for solving a problem that terminates in a finite number of steps, whatever be the input data.

This definition of an algorithm does not provide criteria for the goodness of an algorithm. Several concepts are involved in such a judgement. Some of these are: correctness, tolerant to erroneous data, number of steps needed to find solution, main memory requirement, elegance, specificity, generality etc. Some of these are measurable criteria whereas others are subjective in nature. **Time**, number of steps needed to find the solution, and **storage**, main memory requirement can be measured but criteria like elegance, specificity etc, are not measurable. Order of a function is used to measure time and storage requirement (complexity) of an algorithm. *Time complexity of a problem is defined as time complexity of a best possible algorithm for solving that problem.* Let us consider the following algorithm of sorting an array of n items in ascending order.

```
BUB_SORT(list, no_of_item)
{
    for (int i =0; i < no_of_item  ; i++)
        for (int j = i; j < no_of_item -1 ; j++)
            if (list[j] > list[j+1])
            {
                int tmp = list[j];
                list[j] = list[j+1];
                list[j+1] = tmp;
            }
}
```

Algorithm 2.2.1

Let number of items is n. Size of every item in the list of this algorithm, is fixed. Thus storage complexity is a function of n, and it is directly proportional to n, i.e. as n increases the storage requirement increases. Let f and g are functions whose domains are subsets of Z^+. We say f is O(g) if there exists constants c and k such that

$$|f(n)| \leq c.\ |g(n)| \quad \forall\ n \geq k.$$

If f is O (g), then f increases no faster than g. In the above example, let f be storage complexity function then f(n) = n + 3; n is number of items, two counter variable i & j and one tmp variable used for swapping.

$$f(n) = n+3 \ \leq\ n+n \ \leq\ 2n = 2.g(n)$$

Where, g(n) = n. Thus f(n) is O(g) i.e. f(n) is O(n). Let us consider, now, the time complexity of the same algorithm. To find the required time, we take into account the number of comparisons needed to perform the above task. There are n–1 comparison in first pass, n – 2 in second pass, n – 3 in third pass & so on, and 1 comparison in last pass. Therefore, there are in total

$$1+2+3+\cdots+n-1=\frac{n(n-1)}{2} \quad \textit{comparisons.}$$

Let f(n) be the function of time complexity of the above algorithm, then

$$f(n) = \frac{1}{2}(n^2 - n)$$

$$or, \qquad \leq \frac{1}{2}(n^2 + n)$$

$$or, \qquad \leq \frac{1}{2}(n^2 + n^2)$$

$$or, \qquad \leq n^2$$

$$\therefore \quad time \quad complexity \quad is\ of\ O(n^2)$$

Now, let us consider another algorithm of finding whether a relation R defined on a set A is transitive or not. Let $|A| = n$ and $|R| = p$. Let M_R be the matrix of R and A = {1, 2, 3, …, N}. R is stored in a two-dimensional array, say MAT. If ordered pair $(i, j) \in R$, we have MAT(i, j) ← 1.

```
TRANS (MAT, n)
{
    RESULT ← T
    for (int i = 0; i < n; i++)
      for (int j = 0; j < n; j++)
        if (MAT(i, j) == 1)
        {
           for (int k = 0; k < n; k++)
           if ((MAT(j, k) == 1) and (MAT(i, k) == 0))
            RESULT ← F
        }
}
```

Algorithm 2.2.2

Amount of storage required, for matrix MAT, is n^2. Thus if f is the measure of storage complexity of this algorithm then f(n) is $O(n^2)$. We shall now find time complexity of the algorithm. Observe that outer two loops corresponding to i & j run n times each. If (i,j) is not in R , the only test "if (MAT(i, j) == 1)" is performed. If the test is false, the rest of statements are not executed. Since only p ordered pairs are in R, "if (MAT (i, j) == 1)" will be true for p times only. That means n^2 - p times, only up to "if (MAT (i, j) == 1)" is executed and the rest of the instructions are not executed. The case when condition "if (MAT (i, j) == 1)" is true, an additional loop corresponding to k is executed n times. This means the test "if ((MAT (j, k) == 1) and (MAT (i, k) == 0))" is executed np times. Thus, total number of execution required by the algorithm 2.2.2 is given as $n^2 - p + np$. If f be the time complexity function of n, then f(n) is given as

$$f(n) = n^2 - p + np$$

Since p must be between 0 and n^2, let $p = k\,n^2$, where $0 \le k \le 1$. Therefore,

$$f(n) = n^2 - k\,n^2 + k\,n^3 = k\,n^3 + (1 - k)n^2$$

So f(n) is $O(n^3)$. Having discussed this, we shall now discuss a few examples before taking up the rules to determine order of a function.

Example 1 Show that

$$f(n) = \frac{n^3}{2} + \frac{n^2}{2}$$

is O(g) for $g(n) = n^3$

Solution: To see this we proceed as below:

$$f(n) = \frac{n^3}{2} + \frac{n^2}{2} \le \frac{n^3}{2} + \frac{n^3}{2} = 1 * n^3$$

$$or, \quad f(n) \le 1 * n^3$$

If c = 1 and k = 1, we have $|f(n)| \le c.|g(n)|\ \forall\ k \ge n$ and so f is O(g). ♦

Example 2 Show that $f(n) = 3n^4 - 5n^2$ and $g(n) = n^4$ are of same order.

Solution:
$$\begin{aligned} f(n) &= 3n^4 - 5n^2 \\ &\le 3n^4 - 5n^2 \\ &\le 3n^4 + 5n^2 \\ &\le 3n^4 + 5n^4 \\ &= 8n^4 \end{aligned}$$

Now, let c = 8 and k = 1, then $|f(n)| \le c.\ |g(n)|$ for all $n \ge k$. Thus f is O (g). And,

$$\begin{aligned} g(n) &= n^4 \\ &= 3n^4 - 2n^4 \\ &\le 3n^4 - 5n^2 \text{ for } n \ge 2 \end{aligned}$$

If we put c = 1 and k = 2, then $|g(n)| \le c.\ |f(n)|$ for all $n \ge k$. Therefore g is O (f). ♦

Example 3 Show that every logarithmic function f(n) = $\log_b n$ has the same order as g(n) = $\log_2 n$ . .

Solution: We know the logarithmic change of base identity $\mathbf{\log_a x = \log_a b * \log_b x}$. Let a = 2, then for any positive integer b ≥ 3, $\mathbf{\log_a b} \geq 1$ and is a constant. Also 1/ $\mathbf{\log_a b}$ is a constant. So for any base b ≥ 3, we have,

$$\log_2 n = \log_2 b * \log_b n$$

$$or, \ |\log_2 n| = \log_2 b * |\log_b n|$$

$$or, \ |g(n)| = \log_2 b * |f(n)|$$

i.e., g is O(f). Similarly, we can write

$$\log_b n = \frac{1}{\log_2 b} * \log_2 n$$

$$or, \ |\log_b n| = \frac{1}{\log_2 b} * |\log_2 n|$$

$$or, \ |f(n)| = \frac{1}{\log_2 b} * |g(n)|$$

Or, f is O (g). This shows that every logarithmic function is of same order as that of $\log_2 n$. That means while mentioning the complexity of an algorithm and if it is of logarithmic order, base of log is not important. ♦

Let F be the set of all functions whose domains are subsets of set of positive integer. Clearly an element of F can be used to determine complexity of an algorithm. Let Θ (big theta) be a relation on F defined as f Θ g if and only if both f and g are of the same order for any function f and g belonging to F.

Theorem 1 Relation Θ, defined as above on F is an equivalence relation.

Proof: Since every function f of F is of order itself, the relation Θ is reflexive. Let f and g be any two elements of F and if f Θ g, then both f and g are of same order, so g Θ f, i.e., Θ is symmetric. Next, let f, g and h be any three elements of F such that f Θ g and g Θ h. This implies that f and g are of the same order & g and h are of same order. This shows that f and h are also of same order, i.e., f Θ h and so Θ is transitive. Therefore Θ is an equivalence relation.

Proved.

We know that every equivalence relation on a set induces an equivalence partition. The relation Θ partitions F into equivalence classes. Each equivalence class contains functions that are of same order. As usual, we use any simple function of an equivalence class to represent that class and hence to represent the order of all functions in that class. One Θ class [f] is said to be **lower** than another Θ class [g] if a representative function f from [f] is of lower order than that of any g from [g]. This implies that a function from class [f] grows more slowly than a function from class [g] as magnitude of n, value from domain of function or number of items to be processed by the algorithm, increases. It is the Θ class of a function that gives the information we need for analysis of algorithm.

Rules for Determining the Θ Class of a Function

1. A constant function has zero growth as n increases. This class of function is represented as $\Theta(1)$ class. In the hierarchy of time complexity, this class of function corresponds to the lowest class. All **hash functions** belong to this class.

2. Any logarithmic function grows more slowly than any power function with positive exponent, i.e., $\Theta(\log n)$ is lower than $\Theta(n^k)$ for $k > 0$.

3. Any exponential function with base greater than 1 grows more rapidly than any power function, i.e., $\Theta(n^k)$ is lower than $\Theta(a^n)$ for any k and $a > 1$.

4. $\Theta(n^a)$ is lower than $\Theta(n^b)$ if and only if $a < b$, i.e. in case of two power functions the hierarchy is decided on the basis of theirs exponents.

5. Among the exponential functions the hierarchy is determined on the basis of the base of the function, i.e., $\Theta(a^n)$ is lower than $\Theta(b^n)$ if and only if $a < b$.

6. If a function is multiplied by a non zero constant, its order remains unchanged, i.e., for any non zero k $\Theta(kf) = \Theta(f)$ for any f.

7. While determining the hierarchy of order of composition of functions, the hierarchy of constituent functions is the determining factor. If h is a non zero function and $\Theta(f)$ is lower than $\Theta(g)$, then $\Theta(fh)$ is lower than $\Theta(gh)$.

8. Finally, if $\Theta(f)$ is lower than $\Theta(g)$, then $\Theta(f+g) = \Theta(g)$.

In computer science while analyzing algorithm, most frequently used Θ-classes of functions are constant Θ(1), logarithmic, power $\Theta(n^k)$, exponential $\Theta(a^n)$ and composition of these functions only. The above set of rules is sufficient to determine the complexity of any algorithm and of any problem.

Example 4: Determine the Θ class of the following functions.
(i) $f(n) = 4n^4 - 6n^7 + 25n^3$ (ii) $g(n) = \log n - 3n$
(iii) $h(n) = (1.001)^n + n^{15}$

Solution:

(i) Here, $f(n) = 4n^4 - 6n^7 + 25n^3$. From rule of determining order of power function, $\Theta(f) = \Theta(n^7)$.

(ii) From rule 2, we know that any logarithmic function grows more slowly than any power function, so (log n -3n) grows more slowly than n, thus Θ(g) = Θ(n).

(ii) Rule 3 above states that any exponential function with base greater than 1 grows more rapidly than any power function, so n^{15} grows more slowly than $(1.001)^n$, thus $\Theta(h) = \Theta((1.001)^n)$.

2.3 Sequences and Series

Let us consider the following collection of numbers.

(i) 28, -2.25, 27, 32, 7, ...
(ii) 2, 7, 11, 19, 31, 51, ...
(iii) 1, 2, 3, 4, 5, ...
(iv) 20, 18, 16, 14, 12, ...
(v) 4, 8, 16, 32, 64, ...

In the collection (i), it is not possible to know which number will follow 7 as no definite rule is followed by the elements in this collection. In the **second** collection no rule is noticed except that all are positive and are in ascending order. In collection **three**, if one asks " what is the nth element?" Prompt reply comes from some corner that it is **n.** This is possible because every element is arranged according to some rule. The rule here is, succeeding element is one greater than its predecessor. Similarly, we can predict that nth term in (iv) is **20 – 2n (for n = 0, 1, 2, ..., n)** and that in (v) it is $\mathbf{4 \times 2^n}$ **(n = 0, 1, 2, ..., n).** Therefore collections in (iii), (iv) and (v) are **sequence** but that in (i) and (ii) are not. We can, then, define a sequence as:

"An ordered collection of numbers $a_0, a_1, a_2, a_3, \ldots, a_n, \ldots$; is a sequence if according to some definite rule or law, there is a definite value of a_n, called the term or elements of the sequence corresponding to any value of the natural number n."

Clearly, t_0 is the initial term of the sequence, t_1 is the first term, t_2 is the second term, …, t_n is the nth term. In the nth term, t_n, by successively substituting 0, 1, 2, 3, …, for n we get t_0, t_1, t_2, t_3, and so on. It is obvious now that nth term of a sequence is a function of natural number n. A sequence is denoted by $\{t_n\}$ or by (t_n) or simply by **t**. in which t_n is the nth term of sequence. The nth term t_n is called **general term** of the sequence.

Example 1: Let **t** be a sequence in which general term t_n is given by = 2n + 3, then by substituting n = 0, 1, 2, ..., we get t_0, t_1, t_2 and so on respectively as given below.

When $n = 0, t_0 = 2 * 0 + 3 = 3$
$n = 1, t_1 = 2 * 1 + 3 = 5$
$n = 2, t_2 = 2 * 2 + 3 = 7$
$n = 3, t_3 = 2 * 3 + 3 = 9$
Similarly, when $n = 10, t_{10} = 2 * 10 + 3 = 23$ and so on.

The above example shows that there is correspondence of one to one between the set of natural numbers and the elements of a sequence. If f is function from N, the set of natural number, to R, the set of real number then depending upon the f(N) ⊂ R, we have different sequences. A function **f: N→R** gives a sequence of real numbers. There could be infinite many such functions, and so are infinite many possible sequences of real numbers. We call these functions as **discrete numeric functions** or simply **numeric functions**. We can say f(0) is $\mathbf{a_0}$, f(1) is $\mathbf{a_1}$, f(2) is $\mathbf{a_2}$ and so on. The sequence **a,** is called **numeric function**. We shall discuss how to find a **generating function** for a given sequence i.e. for a given **numeric function** and how to find a **numeric function** from a given **generating function**. Before that, let us learn about different types of sequences and way to represent a sequence.

Depending upon the number of terms in a sequence, we may have a **finite** or **infinite** sequence. A sequence is said to be finite if number of terms is finite i.e. there is a last term in the sequence. For example, the sequence of dates in the month of August 2001, which are Mondays, i.e.

6, 13, 20,27

This is, obviously, a finite sequence. A sequence is said to be **infinite** if there are infinite many terms in that sequence i.e. each term is followed by another term. In such a sequence there is no last term. For examples,

(i) Sequence of positive odd integers

1, 3, 5, 7,...

(ii) Sequence of positive even integers

2, 4, 6, 8, ...

(iii) Sequence of prime numbers

2, 3, 5, 7, 11, 13, 17, 19, 23...

Further, if the absolute value of any terms of a sequence do not exceed a definite value, then the sequence is called a **bounded sequence**. Thus, a sequence **t** is said to be bounded if there exists a positive number K such that

$$|t_n| < K \ \forall n.$$

If there exists no such numbers K satisfying the condition mentioned above, the sequence is said to be **unbounded**. All finite sequences are bounded sequence. The following sequences are infinite and bounded.

$$(i)\ 1, \frac{1}{2}, \frac{1}{3}, \frac{1}{4}, \cdots\cdots, \frac{1}{n}, \cdots\cdots$$

$$(ii)\ 1, \frac{1}{2}, \frac{2}{3}, \frac{3}{4}, \cdots\cdots, \frac{n}{n+1}, \cdots\cdots$$

$$(iii)\ 1, -1, 1, -1, 1, \ \cdots\cdots, (-1)^n, \cdots\cdots$$

However, the following sequences are unbounded

(i) 1, 2, 3, 4, ...,(n+1), ...

(ii) 1, 2, 4, 8, 16, ...,2^n, ...

(iii) 1, 3, 5, 7, ..., 2n+1, ...

A sequence **t** in which a term t_{n+1} is never less than any of its previous terms i.e. $t_{n+1} \geq t_n$ for every value of n, then **t** is called **monotonic increasing** sequence. Similarly, a sequence **t** in which a term t_{n+1} is never greater than any of its previous terms i.e. $t_{n+1} \leq t_n$ for every value of n, then **t** is called **monotonic decreasing** sequence. A monotonic increasing or monotonic decreasing sequence is called **monotonic sequence**. The following sequences are monotonic increasing.

(i) 1, 2, 3, 4, ...,(n+1), ...
(ii) 1, 2, 4, 8, 16, ...,2^n, ...
(iii) 1, 3, 5, 7, ..., 2n+1, ...

However, the sequences given below are monotonic decreasing.

$$1, \frac{1}{2}, \frac{1}{3}, \frac{1}{4}, \ldots\ldots$$

The following is an example of non-monotonic sequence.

$$1-\frac{1}{3!}, 1-\frac{1}{3!}+\frac{1}{5!}, 1-\frac{1}{3!}+\frac{1}{5!}-\frac{1}{7!}, \ldots\ldots$$

In principle, a sequence can be specified by exhaustively listing the values of terms (0th, 1st, 2nd etc. elements) of that sequence. But, in practice, it is not possible to enlist all the elements of an infinitely long sequence. Thus we need a representation of sequence which is not infinitely long. A sequence **t** may be expressed by an algebraic formula describing its general term t_n. This formula for t_n may be in terms of only its position number n in the sequence or in terms of previous sequence terms t_{n-1}, t_{n-2}, t_{n-3}, ... etc. The former is called **explicit formula** and the latter is called **recurrence formula** or **recurrence equation**.

Example 2 The sequence 5, 10, 20, 40,..., can be represented by an explicit formula

$$a_n = 5 * 2^n \quad for \quad n \geq 0$$

And, by an recurrence formula as

$$a_n = 2a_{n-1} \quad for \quad n \geq 1 \; with\; initial\; condition\; a_0 = 5$$

Example 3 The sequence 3, 7, 11, 15,..., can be represented by an explicit formula

$$a_n = 4n + 3 \quad for \quad n \geq 0$$

And, by an recurrence formula as

$$a_n = a_{n-1} + 4 \quad for \quad n \geq 1 \; with\; initial\; condition\; a_0 = 3$$

Example 4 The sequence 87, 82, 77, 72, 67, ..., can be represented by an explicit formula

$$a_n = 87 - 5n \quad for \quad n \geq 0$$

And, by an recurrence formula as

$$a_n = a_{n-1} - 5 \quad for \quad n \geq 1 \; with \; initial \; condition \; a_0 = 87$$

Example 5 The sequence -4, 16, -64, 256, 1024, ..., can be represented by an explicit formula

$$a_n = (-4)^{n+1} \quad for \quad n \geq 0$$

And, by an recurrence formula as

$$a_n = (-4) * a_{n-1} \quad for \quad n \geq 1 \; with \; initial \; condition \; a_0 = -4$$

We have noticed that in recurrence formula an **initial condition**, also called the **terminating condition**, is used. As in a recursive function, a recursive formula must have a terminating condition. The number of terminating conditions depends upon the order of a recurrence relation. We shall learn about this in the next chapter.

Example 6 Find the different terms of numeric function

$$\mathbf{a} = a_n = \begin{cases} 2r & 0 \leq r \leq 11 \\ 3^r - 1 & r > 11 \end{cases}$$

Solution: The sequence terms for 0 to 11 is given by formula 2r and then it is given by the formula 3^r-1. Putting r = 0, 1, 2,..., 11, 12, 13, ..., we get the sequence as below.

a = 1, 2, 4, 6, ...,22, 531440, 1594322, ...

A set contains only distinct element, as we have learnt in the first chapter. The order of appearance of an element in a set is not important. Whereas, in a sequence order of term is important but an element may be repeated any number of times. **A set corresponding to a sequence** is collection of all **distinct elements** forming that sequence. For example, set corresponding to sequence 1, 2, 3, 1, 2, 3, 1, 2, 3, ... is {1, 2, 3}. Similarly, a set corresponding to sequence 1, 2, 3, 4, ..., is {1, 2, 3, 4, ..}, the set of positive integers. On the other hand, if a set is given, we can form a sequence from the elements of that set. For example, let A = {0, 1}. Then we can have sequences like

1, 0, 1, 0, ...
1, 1, 1, ...
0, 0, 0, ...
And so on.

All such sequences arising from set A. In fact, we can have infinite many sequences arising from the set A. Some of the sequences will be finite and some infinite.

The indicated sum $t_0 + t_1 + t_2 + t_3 + \ldots + t_n + \ldots$, of the terms of a sequence $\mathbf{t} = (t_0, t_1, t_2, t_3, \ldots)$ is called **series**. For example,

$$1 + 2 + 3 + 4 + \ldots$$
$$1 - 2 + 4 - 8 + 16 - \ldots + (-2)^n + \ldots$$

are series. A series may be finite or infinite according to the sequence it corresponds to. If $S_n = t_0 + t_1 + t_2 + t_3 + \ldots + t_n$, then S_n is called the sum to n terms of **t, or nth partial sum** of the series

$$\sum_{i=0}^{\infty} t_n$$

The sequence in example 6 has been represented using **characteristic function** notation. Many sequences are represented in that way. Particularly, when nature of algebraic formula differs from one subset of natural numbers to another subset of natural number for a sequence. This gives the concept of selection of definition (**block of statements**) according to situation. If there are two definitions and one is to be selected, the selection is called **Binary Branching**, or simply **Branching** (If Then Else). If one selection is to be made out of many (but finite), the selection is called **Multiple Branching** (Switch case). This concept has come from the definition of the **characteristic function**. A function **f: A→{0, 1}** from a nonempty set A to set {0, 1} defined as

$$f(x) = \begin{cases} 0 & if \quad x \notin A \\ 1 & if \quad x \in A \end{cases}$$

is called characteristic function of set.

Theorem 1 Characteristic functions of subsets satisfy the following properties:

(*i*) $f_{A \cap B} = f_A f_B$

(*ii*) $f_{A \cup B} = f_A + f_B - f_A f_B$

(*iii*) $f_{A \oplus B} = f_A + f_B - 2 f_A f_B$

Proof: (i) Let x is any element of $A \cap B$. Then, $x \in A$ and $x \in B$. Therefore,

$$f_{A\cap B}(x) = 1 \Leftrightarrow f_A(x) = 1 \text{ and } f_B(x) = 1$$
$$\Leftrightarrow f_A(x) f_B(x) = 1$$

Similarly, if $x \notin A \cap B$ *then,*

$$f_{A\cap B}(x) = 0 \Leftrightarrow f_A(x) = 0 \text{ or } f_B(x) = 0$$
$$\Leftrightarrow f_A(x) f_B(x) = 0$$

Therefore, $f_{A\cap B} = f_A f_B$ Proved.

(ii) Let x is any element of $A \cup B$. Then $x \in A$ or $x \in B$. There are three possibilities only. (a) $x \in A$ and $x \in B$, (b) $x \notin A$ and $x \in B$ and (c) $x \in A$ and $x \notin B$. Therefore,

$$f_{A\cup B}(x) = 1 \Leftrightarrow f_A(x) = 1 \text{ and } f_B(x) = 1$$
$$\text{or, } f_A(x) = 1 \text{ and } f_B(x) = 0$$
$$\text{or, } f_A(x) = 0 \text{ and } f_B(x) = 1$$

Similarly, if $x \notin A \cap B$ *then,*

$$f_{A\cup B}(x) = 0 \Leftrightarrow f_A(x) = 0 \text{ and } f_B(x) = 0$$
$$\text{and } f_A(x) f_B(x) = 0$$

Therefore, $f_{A\cup B} = f_A + f_B - f_A f_B$ Proved.

(iii) Let x is any element of $A \oplus B$. Then, there are two possibilities only. (a) $x \in A$ and $x \notin B$ and (b) $x \notin A$ and $x \in B$. Further, if x is not a member of $A \oplus B$, then either x is in both A and B or not in A and B. Therefore,

$$f_{A\oplus B}(x) = 1 \Leftrightarrow f_A(x) = 1 \text{ and } f_B(x) = 0$$
$$\text{or, } f_A(x) = 0 \text{ and } f_B(x) = 1$$

Similarly, if $x \notin A \cap B$ *then,*

$$f_{A\oplus B}(x) = 0 \Leftrightarrow f_A(x) = 0 \text{ and } f_B(x) = 0$$
$$\text{or, } f_A(x) = 1 \text{ and } f_B(x) = 1$$

Therefore, $f_{A\oplus B} = f_A + f_B - 2 f_A f_B$ Proved.

Proved.

2.4 Generating Function

An important idea in mathematics is to establish connection between two fields in order to apply knowledge in one field to the other field, or at least take a problem of one field and transform it to a problem in the other field. This idea motivates the idea of a **generating function,** which establishes a connection between functions of a real variable and sequences of numbers.

A **generating function** can be defined as "Given a numeric sequence **a** = a_0, a_1, a_2, ... a_n, ... the series

$$f(x) = a_0 + a_1x + a_2x^2 + \cdots\cdots + a_nx^n + \cdots\cdots$$

is called the **generating function** of the sequence."

Suppose **h** = h_0, h_1, h_2, ... h_n, ... is a sequence of numbers. We write it as an infinite sequence, but we mean to include finite sequences also as well. **If h_0, h_1, h_2, ..., h_n is finite sequence, we make it of infinite length, simply, by setting $h_r = 0$ for r > n.** A generating function g(x) for the sequence **h** is an infinite series such that

$$g(x) = h_0 + h_1x + h_2x^2 + \cdots + h_nx^n + \cdots = \sum_{n=0}^{\infty} h_nx^n$$

Thus, the function generates the sequence as its sequence of coefficients. If the sequence is finite then there is an **m** for which **h_r = 0 for r > m.** In this case g(x) is an ordinary polynomial in x of degree m. The inspiration for this idea is the **Binomial Theorem**. The function g(x) = $(1+x)^m$ generates the binomial coefficients **h_r = C(m, r).** Therefore a generating function in which coefficients of x^n are sequence terms of the sequence **h**, is called **Binomial Generating function** of the sequence **h**. If we recall our basic mathematical knowledge of school days, the binomial coefficient **C(m, r)** gives us the total number of **combinations** of **r** selections from **m** objects. Thus we can infer from here that the concept of binomial generating function can be used to solve problems of combinations. How to do it, we will learn in the **chapter 4**.

Permutation is a sister term (one may use brother to avoid gender bias) while discussing combination. One comes to mind whenever other is referred. We know that **P(m, r) = r! * C(m, r)**. This means if **C(m, r)** is coefficient of x^r then **P(m, r)** is coefficient of **x^r /r!**. Now, let **h(x)** is a **generating function** for a sequence **a** given in series form as below.

$$h(x) = a_0 + a_1 x + a_2 \frac{x^2}{2!} + a_2 \frac{x^3}{3!} + \cdots = \sum_{n=0}^{\infty} a_n \frac{x^n}{n!}$$

This series is exponential. The generating function h(x) defined as above for a sequence **a** = a_0, a_1, a_2, ... a_n, ... is called **Exponential Generating** function. Clearly coefficients of exponential generating function gives the value of permutation when **r** selections are made from n objects. This is only tip of the iceberg. The answers to many-many counting problems are just coefficient of certain polynomial generating functions. Actually, **De Moivre** introduced the notion of generating function in 1730 AD to solve recurrence equations. It took almost 530 years to find a closed form formula for a recurrence equation defined by **Fibonacci** in 1200 AD. In this book we shall discuss **Binomial** and **Exponential** generating function only.

To find the generating function for a sequence means to find a closed form formula for f(x), one that has no ellipses(**...**). In essence, a generating function is that function whose coefficients are really the values that we seek. That is, we do not really care about evaluating the function but only about examining its coefficients.

Example 1 Find the generating function for the sequence **a** =1, 1, 1, 1, **...**

Solution: In this sequence the general term a_n = 1. Let f(x) be the binomial generating function for this sequence. Then, f(x) can be written as

$$f(x) = \sum_{n=0}^{\infty} a_n x^n = \sum_{n=0}^{\infty} 1 * x^n \qquad [\because a_n = 1]$$

$$= 1 + x + x^2 + x^3 + \cdots\cdots$$

$$= \frac{1}{1-x} \qquad [\because \textit{The series is geometric with x as common ratio}]$$

Therefore, $f(x) = \dfrac{1}{1-x}$ *is the binomial generating function for the sequence* **a** = 1, 1, 1, ...

Next, let g(x) be the exponential generating function for the same sequence then g(x) can be written as

$$g(x) = \sum_{n=0}^{\infty} a_n \frac{x^n}{n!} = \sum_{n=0}^{\infty} 1 * \frac{x^n}{n!} \qquad [\because a_n = 1]$$

$$= 1 + x + \frac{x^2}{2!} + \frac{x^3}{3!} + \cdots\cdots$$

$$= e^x \qquad [\because \text{The series is exponential with } x \text{ as base}]$$

Therefore, $f(x) = e^x$ *is the exponential generating function for the sequence* **a** = 1, 1, 1, ······

Example 2 Find generating function for the sequence **b** = 1, 3, 9, ..., 3^n, ...

Solution: In this sequence the general term $b_n = 3^n$. Let f(x) be the binomial generating function for this sequence. Then, f(x) can be written as

$$f(x) = \sum_{n=0}^{\infty} b_n x^n = \sum_{n=0}^{\infty} 3^n * x^n \qquad [\because b_n = 3^n]$$

$$= 1 + 3x + (3x)^2 + (3x)^3 + \cdots\cdots$$

$$= \frac{1}{1-3x} \qquad [\because \text{The series is geometric with } 3x \text{ as common ratio}]$$

Therefore, $f(x) = \frac{1}{1-3x}$ *is the binomial generating function for the sequence* **b** = 1, **3**, $\mathbf{3^2}$, $\mathbf{3^3}$ ······

Let g(x) be the exponential generating function for the same sequence. Then, g(x) can be written as

$$g(x) = \sum_{n=0}^{\infty} b_n \frac{x^n}{n!} = \sum_{n=0}^{\infty} 3^n \frac{x^n}{n!} \qquad [\because b_n = 3^n]$$

$$= 1 + 3x + \frac{(3x)^2}{2!} + \frac{(3x)^3}{3!} + \cdots\cdots$$

$$= e^{3x} \qquad [\because \textit{The series is exponential with } 3x \textit{ as base}]$$

Therefore, $f(x) = e^{3x}$ *is the exponential generating function for the sequence* $\mathbf{b} = 1, 3, 3^2, 3^3, \cdots\cdots$

Example 3 Find exponential generating function for sequence

$$\mathbf{t} = {}^nP_0, {}^nP_1, {}^nP_2, \cdots\cdots, {}^nP_n$$

Solution: The given sequence is finite. All terms in this sequence for m > n are zero. Let h(x) be the exponential generating function for the given sequence then h(x) can be written as

$$h(x) = \sum_{i=0}^{\infty} a_i \frac{x^i}{i!} = \sum_{i=0}^{\infty} {}^nP_i \frac{x^i}{i!} \qquad [\because a_i = {}^nP_i]$$

$$= {}^nP_0 + {}^nP_1 x + {}^nP_2 \frac{x^2}{2!} + \cdots\cdots + {}^nP_n \frac{x^n}{n!}$$

$$= {}^nC_0 + {}^nC_1 x + {}^nC_2 x^2 + \cdots\cdots + {}^nC_n x^n \qquad [\because {}^nP_r = r! * {}^nC_r]$$

$$= (1+x)^n$$

Therefore, the exponential generating function for the sequence $\mathbf{t} = {}^nP_0, {}^nP_1, {}^nP_2, \cdots\cdots, {}^nP_n$ *is given by*

$$h(x) = (1+x)^n \quad \textit{Ans.}$$

Example 4 Find the binomial generating function for the sequence

a =1, 2, 3, ….,r, …

Solution: In this sequence the general term $a_r = r+1$. Let f(x) be the binomial generating function of the given sequence, then it can be written as

$$f(x) = \sum_{r=0}^{\infty} a_r x^r = \sum_{r=0}^{\infty} (r+1) * x^r \qquad [\because a_r = r+1]$$

$$or, \quad f(x) = 1 + 2x + 3x^2 + 4x^3 + \cdots\cdots \qquad (1)$$

The equation (1) is arithmetic-geometric in which first factor is following arithmetic progression and second following geometric. We find the sum of the series by multiplying equation (1) with x, and then by subtracting the result from equation (1). Thus, we get

$$(1-x)f(x) = 1 + x + x^2 + x^3 + \cdots\cdots$$

$$or, \quad f(x) = \frac{1}{(1-x)^2} \quad \textit{This is the required binomial generating function.}$$

We have seen some examples to find a generating function for a given sequence (discrete numeric function). Similarly, if a generating function f(x) is given, we can find the sequence as coefficient of different powers of **x.** If f(x) is a binomial generating function we take coefficient of $\mathbf{x^n}$ for all n≥0. If, on the other hand, f(x) is an exponential generating function we collect the coefficient of $\mathbf{x^n/n!}$ for all n≥0.

Example 5 Find the discrete numeric function (sequence) whose exponential generating function is given by $\mathbf{2e^x}$.

Solution: The given generating function can be written in series form as

$$2e^x = 2\left[1 + x + \frac{x^2}{2!} + \frac{x^3}{3!} + \frac{x^4}{4!} + \cdots\cdots\cdots\right]$$

Clearly, the coefficient of $\mathbf{x^n/n!}$ for n≥0, are 2, 2, 2, ...Therefore, if the discrete numeric function is **a**, then **a** = 2, 2, 2, 2, ...

Ans.

Sometimes two or more sequences are considered together to know the combined behavior of these sequences. For example, if **a** is a sequence for the monthly income of husband and **b** is a sequence corresponding to monthly income of the wife, then sequence **a + b** is the sequence representing joint monthly income of the couple. Similarly, if **c** is the sequence representing monthly balance in a savings account and **d** is the sequence corresponding to the monthly interest rate, which fluctuates from month to month, then the terms of sequence **cd** represents interest earned in each month.

Example 6 For the following two sequences **a** and **b**, whose general terms are given find **a + b** and **ab.**

$$a_r = \begin{cases} 0 & \textit{for } 0 \le r \le 2 \\ 2^{-r} + 5 & \textit{for } r \ge 3 \end{cases}$$

and

$$b_r = \begin{cases} 3 - 2^r & \textit{for } 0 \le r \le 1 \\ r + 2 & \textit{for } r \ge 2 \end{cases}$$

Solution: Let **c** be the sequence for **a** + **b.** At r = 2, the general definitions of **a** and **b** cannot be simply added. For r ≥ 3 and 0 ≤ r ≤ 1, bottom and top definitions will be added, but at r = 2 the value will be determined and will be placed in the definition. Next let **d** = **ab**. In this case, we shall find the sequence formula for **d** by multiplying the definitions of **a** and **b** and setting the range for sequence position r accordingly. The result is shown below.

$$c_r = a_r + b_r = \begin{cases} 3-2^r & \text{for } 0 \le r \le 1 \\ 4 & \text{for } r = 2 \\ 2^{-r} + r + 7 & \text{for } r \ge 3 \end{cases}$$

and

$$c_r = a_r * b_r = \begin{cases} 0 & \text{for } 0 \le r \le 2 \\ r2^{-r} + 2^{-r+1} + 5r + 10 & \text{for } r \ge 3 \end{cases}$$

Ans.

Theorem 1 Let f(x) is the generating function for **a** and g(x) is the generating function for **b**. Show that f(x)+g(x) is the generating function for **a** + **b.**

Proof: Let

$$f(x) = \sum_{n=0}^{\infty} a_n x^n \quad \text{and} \quad g(x) = \sum_{n=0}^{\infty} b_n x^n$$

be the generating functions for **a** and **b**. Then,

$$f(x) + g(x) = \sum_{n=0}^{\infty} a_n x^n + \sum_{n=0}^{\infty} b_n x^n = \sum_{n=0}^{\infty} (a_n + b_n) x^n$$

Therefore, the sequence generated by $f(x) + g(x)$ *is*

$(a_0 + b_0), (a_1 + b_1), (a_2 + b_2), \cdots\cdots$

These terms are precisely the sum of the corresponding terms from **a** and **b.** Thus, f(x)+g(x) is generating function for the sum of the sequence **a** + **b**.

Proved.

In general, the above theorem can be applied to any finite number of sequences. The same composition cannot be applied for the product of two or more sequence, i.e., if *f(x) and g(x) are generating functions for sequences a and b respectively then f(x) *g(x) is not the generating function for **a *b**.*

***Theorem** 2* Let f(x) and g(x) are generating functions for finite sequences **a** and **b** respectively. The **convolution** of two sequences **a** and **b** is the sequence **c** defined as

$$c_n = \sum_{r=0}^{n} a_r b_{n-r}$$

Show that f(x)g(x) is the generating function for **convolution** of **a** and **b.**

Proof: Left as an exercise to the readers.

The generating function method can be used to solve many of the counting problems. We shall be using this method to solve problems related to permutation and combination in the chapter 4. In the chapter 3, we shall see how to use this method to solve some recurrence equations as well. But, it is important to discuss one problem, here, before leaving this topic. For k > 0 the function g(x) = $1/(1\text{-}x)^k$ generates the sequence a = {C(n + k -1, n) | n = 0, 1, 2, 3, ...}. Thus the nth coefficient is the number of ways to select **n objects** of **k types**. Obviously,

$$\frac{1}{(1-x)^k} = \left[\frac{1}{(1-x)}\right]^k = (1+x+x^2+x^3+\cdots\cdots)^k$$

If we carry out this multiplication k times, then x^n appear as many times as there are nonnegative integer solutions to the equation

$$x_1 + x_2 + x_3 + \cdots\cdots + x_k = n$$

And, we know that this number is C(n + k –1, n). We shall use this concept in the next example.

***Example** 7* Find the number of positive integral solution to the equation

x + y + z = 10

Solution: Here x, y and z are positive integers satisfying the given equations. Our task is to find the number of possible solutions to the above equation. Here x belongs to {1, 2, 3, 4, ...} and so are y and z. If different possible positive integers for x is represented by the powers of x then series for x can be written as

$$x + x^2 + x^3 + \cdots\cdots$$

This problem is reduced to selection of 10 objects of 3 kinds. The generating function f(x) for this problem is then written as

$$f(x) = (x + x^2 + x^3 + \cdots\cdots)^3$$
$$= \frac{x^3}{(1-x)^3}$$

Then, the required result is given by the coefficient of $\mathbf{x}^{10}$ in f(x). This is equal to the coefficient of $\mathbf{x}^7$ in $(1\text{-}x)^{-3}$. Which is

$$^{7+3-1}C_7 \quad [\because n = 7, k = 3 \text{ in the above formula}]$$
$$= {}^9C_7 = \frac{9*8}{2} = 36 \qquad \textbf{Ans.}$$

The same approach can be used to find number of solutions to various such equations in which values allowed for variables may be different. For example, a variable may take only digits as its value whereas, in the same equation another variable may be allowed to assume any positive integers. The generating function is found accordingly before calculating the coefficient. Readers are suggested to try problems given in the exercises below.

Exercise

1. Let A = {a, b, c, d} and B = {1, 2, 3}. Determine whether the relation R from A to B given below is a function. If it is a function, give its range.
   (a) R = {(a,1),(b,2),(c,1),(d,2)}
   (b) R = {(a,1),(b,2),(a,2),(c,1),(d,2)}
   (c) R = {(a,3),(b,2),(c,1) }
   (d) R = {(a,1),(b,1),(c,1),(d,1)}

2. Determine whether the relation R from A to B is a function.
   (a) A = the set of all holders of motor driving license in India, B = {x | x is a seven-digit number}, aRb if b is a's driving license number.

   (b) A = the set of citizen of India, B = {x | x is a seven-digit number}, aRb if b is a's passport number.

3. Let A = B = C = R, the set of real number, and let f: A→B and g: B→C be defined by f(a) = a -1 and g(b) = b^2. Find
   (a) (fog)(2) (b) (gof)(2) (c) (fog)(x)
   (d) (gof)(x) (e) (fof)(y) (f) (gog)(y)

4. Let A = B = C = R, the set of real number, and let f: A→B and g: B→C be defined by f(a) = a + 1 and g(b) = b^2 + 2. Find
   (a) (fog)(-2) (b) (gof)(-2) (c) (fog)(x)
   (d) (gof)(x) (e) (fof)(y) (f) (gog)(y)

5. If A has n elements then how many functions are there from A to A? How many bijections are there from A to A?

6. If A has m elements and B has n elements, how many functions are there from A to B?

7. Let f: A→B be a function with finite domain and range. Suppose that |Dom(f)| = n and |Ran(f)| =m. Prove that:
(a) If f is one to one, then m = n, and
(b) If f is not one to one, then m < n.

8. Let |A| = |B| = n and let f: A→B is function. Prove that the following three statements are equivalent.
(a) f is one to one,
(b) f is onto, and
(c) f is a bijection.

9. Let A = R, set of real number and B = Z, set of integers. Let f: A→B be a function defined as f(a) = the greatest integer less than equal to a. Verify that f is a function.

10. Prove that if f: A→B and g: B→C are bijection then gof is a bijection.

11. Let f: A→B and g: B→C are function. Show that if gof is one to one then f is one to one and when gof is onto then g is onto.

12. A bijection is also called a set bijection or set isomorphism. Show that a function from set of integer to the set of even integer defined as f(x) =2x is a set isomorphism.

13. Let C denote the set of complex numbers and let R denote the set of real numbers. Prove that the mapping f: C→R given by f(x + iy) = x, where x and y are real, is an onto mapping.

14. Let C denote the set of complex numbers and let R denote the set of real numbers. Prove that the mapping f: C→R given by f(x+iy) = |x+iy|, where x and y are real, is neither one to one nor onto.

15. Let R be the set of real numbers. Using the fact that every cubic equation with real coefficients has real root, show that $f(x) = x^3 - x$ defines a function from R to R.

In exercise 16 through 20 analyze the operation performed by the given pseudo code and give a function describing the number of steps required to carry out the job. Give the order of the function.

16.
```
A ← 1
B ← 1
UNTIL (B > 100)
a. B ← 2A – 2
b. A ← A + 3
```

17.
```
A ← 1
B ← 100
WHILE (A < B)
a. A ← A + 2
b. B ← B / 2
```

18.
```
X ← 1
Y ← 0
READ N
WHILE (X ≤ N)
```

19.
```
SUM ← 0
B ← 1
READ N
FOR(I = 0; I < 2N -1; I +=2)
```

a. Y ← Y + 1 SUM ← SUM +
b. X ← X + 1

20. Let A be an array of size N and X be an item being searched in the array A. The pseudo code is given below.

```
FUNCTION(A, X)
  FOUND ← FALSE
  K ← 1
  WHILE ((NOT FOUND) AND ( K < N))
  IF (A[K] == X) FOUND ← FALSE
  ELSE K ← K + 1
  RETURN
```

21. Show that $g(n) = n!$ is of $O(n^n)$.

22. Show that $g(n) = 1 + 2 + 3 + ... + n$ is of $O(n^2)$.

23. Show that $f(n) = n^2(7n - 2)$ is of $O(n^3)$.

24. Show that f and g have the same order for
 (a) $f(n) = 5n^2 + 4n + 3$ and $g(n) = n^2 + 100n$.
 (b) $f(n) = \log(n^3)$ and $g(n) = \log_5(6n)$.

25. Show that $f(n) = n\log(n)$ is of $O(g)$ for $g(n) = n^2$, but that g is not of $O(f)$.

26. Show that $f(n) = n^{100}$ is of $O(g)$ for $g(n) = 2^n$, but that g is not of $O(f)$.

27. We study the problem of computing x^n for given x and n, assuming that all intermediate results of computation are available to us. For a given integer, let $\mathbf{b_k b_{k-1} b_{k-2} \dots b_2\ b_1\ b_0}$ denote its binary number representation. For each b_i, replace b_i by SX if $b_i = 1$ and by S if $b_i = 0$. For example the binary number representation for 29 is is 11101, we get the sequence **SXSXSXSSX**. Removing leading SX, we are left with **SXSXSSX**. Now, if we interpret each S by 'square' and each X by 'multiply by x', we get the sequence of steps as {square, multiply by x, square, multiply by x, square, square, multiply by x} , where square means to compute the square of current result and multiply by x means to multiply the current result by x. Show that starting with x as the current result, for n=29, this sequence of steps indeed will compute x^{29}.
 (a) Given x, can you compute x^{16} in four multiplications?
 (b) Given x, in how many multiplications can you compute x^{15}?
 (c) Show how we can compute x^{73} using the above algorithm.
 (d) Find the time complexity of the algorithm.

28. Given an integer n, an addition chain for n is a sequence of integers $\mathbf{b_0 b_1 b_2 \dots b_m}$ such that $b_0 = 1$, $b_m = n$, and $b_i = b_j + b_k$ for $k \le j < i$. For example 1,2,3,6,7,14,28,29 is an addition chains for 29. What is the relationship between addition chains and evaluation of powers of x? Determine addition chains for 19, 33, 46, 79 and 87. In how many multiplications can you evaluate x^{19}, x^{33}, x^{46}, x^{79} and x^{87}?

29. Give the set corresponding to the following sequence.

(i) 2, 1, 2, 1, 2, 1, 2, 1
(ii) 0, 2, 4, 6, 8, 10, ...
(iii) aabbccddee...zz

30. Give three different sequences that can arise from the following set
(i) {x, y, z} (ii) {1, 2, 3, ...}

31. Write down the first ten terms of the sequence whose general term is given as below.
(i) $a_n = 5^n$ (ii) $b_n = 3n^2 + 2n - 6$
(iii) $c_n = c_{n-1} + 1.5$ and $c_1 = 2.5$ (iv) $d_n = -2d_{n-1}$ and $d_1 = -3$
(v) $f_{n+2} = 2f_n + f_{n+1}$ and $f_0 = 0, f_1 = 1$ (iv) $g_{n+2} = g_n^{\ 2} + g_{n+1}$ and $f_0 = 1, f_1 = 2$

32. Write the formula for the nth term of the sequence given below. Identify your formula as recursive (difference) or explicit (closed).
(i) 1, 3, 5, 7, ... (ii) 0, 3, 8, 15, 24, 35, ...
(iii) 1, -1, 1, -1, 1, -1, ... (iv) 0, 2, 0, 2, 0, ...
(v) 1, 4, 7, 10, 13, ... (vi) 2, 5, 8, 11, 14, 17, ...
(vi) 2, 5, 7, 12, 19, ...

33. Let A = {x | x is real and 0 < x < 1},
B = {x | x is real and x^2 +1= 0},
C = {x|x = 4m, m∈Z},
D = {(x, 3)| x is an English word of length 3}, and
E = {x|x∈Z and $x^2 \leq 100$}.
Identify each set as finite, countable, or uncountable.

34. Using characteristic functions, prove that $(A \oplus B) \oplus C = A \oplus (B \oplus C)$.

35. We define T-numbers recursively as follows:
a. 0 is a T-number
b. If X is a T -number, X+3 is a T-number.
Write a description of the set of T -numbers.

36. We define S-numbers recursively as follows:
a. 8 is a S-number
b. If X is a S -number and Y is a multiple of X then Y is a S-number,
c. If X is a S -number and X is a multiple of Y then Y is a S-number.
Write a description of the set of S -numbers.

37. Find the sum to infinite number of terms of the following series.
(i) $1 + 2r + 3r^2 + 4r^3 + \ldots (r < 1)$ (ii) $1.1 + 2.3x + 4.5x^2 + \ldots (x < 1)$
(iii) $1 + 3x + 5x^2 + 7x^3 + \ldots (x < 1)$ (iv) $1.2 + 2.3x + 3.4x^2 + \ldots (x < 1)$

(v) $1 + \frac{2}{3} + \frac{3}{9} + \frac{4}{27} + \cdots\cdots$

(vi) $1 + \left(1 + \frac{1}{2}\right)\frac{1}{3} + \left(1 + \frac{1}{2} + \frac{1}{2^2}\right)\frac{1}{3^2} + \cdots\cdots$

38. Sum up to n terms the following series
(i) 7 + 77 + 777 + ...
(ii) .4 + .44 + .444 + ...
(iii) .3 + .33 + .333 + ...
(iv) 1.2.3 + 2.3.5 +3.4.7 + ...
(v) $(x^2 + \frac{1}{x^2} + 2) + (x^4 + \frac{1}{x^4} + 5) + (x^6 + \frac{1}{x^6} + 8) + \cdots\cdots$

39. The natural numbers have been grouped as (1), (2,3),(4,5,6), ... Prove that the sum of the nth such group is $n(n^2 +1)/2$. Using this approach prove that the sum of numbers in any of the group given as (1), (1,3,5),(1,3,5,7,9), ... is the square of an odd number.

40. A ping pong ball is dropped to the floor from a height of 20 m. Suppose that the ball always rebounds to reach half of the height from which it falls. If a_r be height of a ball, that it reaches in the rth rebound, find the numeric function **a**.

41. Let **a** be a numeric function such that a_r is equal to the remainder when the integer r is divided by 17. Let **b** be a numeric function such that b_r is equal to 0 if the integer r is divisible by 3 and is equal to 1 otherwise.
(a) Let $c_r = a_r + b_r$. For what value of r is $c_r = 0$? For what values of r is $c_r = 1$?
(b) Let $d_r = a_r * b_r$. For what value of r is $d_r = 0$? For what values of r is $d_r = 1$?

42. Every particle inside a nuclear reactor splits into two particles in each second. Suppose one particle is injected into the reactor every second beginning at t = 0. How many particles are there in the reactor at the nth second? Find the numeric function and then a generating function for the numeric function so found.

43. After a careful analysis, a student concluded that a certain numeric function **a** is O(nlog n). Upon hearing that her mate reminded her that she had forgotten to specify the base of the logarithm. Had she forgotten something important?

44. Find generating function for each of the following discrete numeric functions:
(i) 1, -2, 3, -4, 5, -6, ...
(ii) 1, 2/3, 3/9, 4/27, ...
(iii) 1, 1, 2, 2, 3, 3, 4, 4, ...
(iv) 0 * 1, 1 * 2, 2 * 3, 3 *4, ...
(v) $0 * 5^0, 1 * 5^1, 2 * 5^2, 3 * 5^3, \ldots$

45. If x, y and z are digits, then find the number of possible solutions to the following equations:
(i) x + y + z = 10
(ii) x - y + z = 17
(iii) x + 2y + 3z = 8
(iv) x - 2y +2z = 37

46. If x, y and z are non-negative integers, then find the number of possible solutions to the following equations:
(i) x + y + z = 100
(ii) x - y + z = 27
(iii) x + 2y + 3z = 10
(iv) x - 2y +2z = 71

47. If x, y and z are positive integers, then find the number of possible solutions to the following equations:
(i) x + y + z = 16
(ii) x - y + z = -7
(iii) x + 2y + 3z = 30
(iv) x - 2y +2z = 0

48. Sum up to the 16 terms

$$\frac{1^3}{1}+\frac{1^3+2^3}{1+3}+\frac{1^3+2^3+3^3}{1+3+5}+\cdots\cdots$$

49. Sum up to infinity the following series

(i) $$\frac{1}{2*4}+\frac{1}{4*6}+\frac{1}{6*8}+\frac{1}{8*10}+\cdots\cdots\cdots$$

(ii) $$1+\frac{1}{1+2}+\frac{1}{1+2+3}+\cdots\cdots\cdots$$

50. Let f be a function from A to B. Determine whether each function f is one to one and whether it is onto.
(a) A = R, B= {x| x is real and x ≥ 0}; f(a) = |a|
(b) A = R × R, B = R; f((a, b)) =a
(c) Let S = {1, 2, 3}, T = {a, b}. Let A = B = S × T and let f be defined by f(n, a) = (n, b), n = 1, 2, 3, and f(n, b) = (1, a), n = 1, 2, 3.
(d) A = B = R × R ; f(a, b) = (a + b, a – b)
(e) A = R, B = {x | x is real and x ≥ 0}; f(a) = a^2

51. Let f: A→B and g: B → A. Verify that g = f^{-1}.

$(a) A = B = R; f(a) = \dfrac{a+1}{2}, g(b) = 2b - 1$

$(b) A = \{x \mid x \text{ is real and } x \geq 0\}; B = \{y \mid y \text{ is real and } y \geq -1\};$

$f(a) = a^2 - 1, g(b) = \sqrt{b+1}$

$(c) A = B = P(S), \text{where } S \text{ is a set. If } X \in P(S), \text{ let } f(X) = X = g(X).$

$(d) A = B = \{1,2,3,4\}; f = \{(1,4),(2,1),(3,2),(4,3)\}; g = \{(1,2),(2,3),(3,4),(4,1)\}$

52. Let f be a function from A to B. Find f^{-1}.

$(a) A = \{x \mid x \text{ is real and } x \geq -1\}; B = \{y \mid y \text{ is real and } y \geq 0\}; f(a) = \sqrt{a+1}$

$(b) A = B = R, f(a) = a^3 + 1$

$(c) A = B = R; f(a) = \dfrac{2a-1}{3}$

$(d) A = B = \{1,2,3,4,5\}; f = \{(1,3),(2,2),(3,4),(4,5),(5,1)\}$

53. Determine the sum

(a) $^nC_1 + 2\,^nC_2 + 3\,^nC_3 + \cdots + i\,^nC_i + \cdots n\,^nC_n$

(b) $^nC_0\,^mC_k + ^nC_1\,^mC_{k-1} + ^nC_2\,^mC_{k-2} + \cdots + ^nC_k\,^mC_0$ where $k \leq n$ and $k \leq m$

(c) $^{n}C_0 + 2*^{n}C_1 + 2^2*^{n}C_2 + \cdots + 2^n*^{n}C_n$

(d) $^{2n}C_n + ^{2n-1}C_{n-1} + ^{2n-2}C_{n-2} + \cdots + ^{n}C_0$

54. Let

$$A(z) = (1+z)^{2n} + z(1+z)^{2n-1} + \cdots + z^i(1+z)^{2n-i} + \cdots + z^n(1+z)^n$$

Show that the numeric function of this generating function is given by

$$a_r = \begin{cases} ^{2n+1}C_r & 0 \le r \le n \\ ^{2n+1}C_r - ^{n}C_{r-n-1} & n+1 \le r \le 2n+1 \\ 0 & otherwise \end{cases}$$

55. Show that

$$(^{n}C_0)^2 + (^{n}C_1)^2 + (^{n}C_2)^2 + \cdots + (^{n}C_n)^2 = ^{2n}C_n$$

56. Show that he generating function for **a,** where

$$a_n = ^{2n}C_n$$

is

$$\frac{1}{\sqrt{1-4z}}$$

57. Determine the generating function of the numeric function **a**, where the general term of **a** is given as

$$a_n = \begin{cases} 2^r & if\ r\ is\ even \\ -2^r & if\ r\ is\ odd \end{cases}$$

58. Determine the discrete numeric function corresponding to each of the following generating functions:

(i) $f(x) = \dfrac{1}{1-x^2}$

(ii) $f(x) = \dfrac{(1+x)^2}{(1-x)^4}$

(iii) $g(y) = (1+y)^n + (1-y)^n$

(iv) $h(z) = \dfrac{1}{5-6z+z^2}$

(v) $h(z) = \dfrac{1}{(1-z)(1-z^2)(1-z^3)}$

CHAPTER THREE
Solving Recurrence Relation

While learning programming languages, we have come across the term 'recursive function'. In our early school days we have learnt about 'recurring number'. Now we are facing the term 'recurrence relation' or '**recurrence equation**'. All these contains a common term 'recurrence', i.e. there is something that occurs again and again and repeats itself until it meets some terminating conditions.

In previous chapter, we have studied that a sequence of terms $a_0, a_1, a_2, a_3, \ldots, a_n, \ldots$, can be expressed in explicit way (also called closed form) or in recurrence form. In recurrence form, we express sequence term a_n in terms of previous sequence terms $a_{n-1}, a_{n-2}, a_{n-3}$ etc and terminating (also called initial) conditions. For example,

$$a_n = a_{n-1} + a_{n-2}; \text{ with } a_0 = 0 \text{ and } a_1 = 1$$

In this chapter, we shall learn how to solve a recurrence equation i.e. how to convert a recurrence form of representation of a sequence into its closed form of representation. There are only few types of recurrence relations, which can be solved, in closed form. It means any term in the sequence can be evaluated by plugging numbers into an equation instead of having to calculate the entire sequence. Classifying the recurrence relation helps to decide which, if any, techniques can be used to solve it.

3.1 Classification of Recurrence Relation

Here is a generic recurrence equation consisting of a sequence of terms a_n,

$$a_n + 3a_{n-1} - 2\sqrt{n}\, a_{n-2} + 2\sqrt{n}\, a_{n-2}a_{n-3} = f(n) \quad \text{———————} \quad (1)$$

In the above equation, the right side, f(n) is some function of n -the place at which a_n is in the sequence. In the standard form, the right side of a recurrence equation can contain any function of n but must not contain any of the a_ns. There are four characteristics of any recurrence equation which are needed to be understood to classify a given recurrence equation.

Homogeneous or Non-homogeneous

A recurrence equation is homogeneous if the right side of the standard form of the recurrence equation is zero, i.e. f(n) = 0. The above recurrence equation (1) is non-homogeneous but the equations (2) and (3) -given below, are homogeneous.

$$a_n + 3a_{n-1} - 2\sqrt{n}\, a_{n-2} + 2\sqrt{n}\, a_{n-2}a_{n-3} = 0 \quad (2)$$
$$a_n + 3a_{n-1} - 2\sqrt{n}\, a_{n-2} + 2\sqrt{n}\, a_{n-3} = 0 \quad (3)$$

Linear or Non-linear

A recurrence equation is linear if there are no products or powers of the sequence terms. The equation (1) and (2) above are non-linear whereas, equation (3) is a linear recurrence equation. The following equations are linear recurrence equation.

$$a_n + 3a_{n-1} - 2a_{n-2} = 0 \quad (4)$$
$$b_n + 4b_{n-1} = \sqrt{n} + 17 \quad (5)$$
$$c_n + \sqrt{n}c_{n-1} = n^n \quad (6)$$
$$d_n - 2d_{n/2} - 7c_n + \sqrt{n}\, c_{n-1} = n^n \quad (7)$$
$$e_n - \sqrt{n}\, e_{\sqrt{n}} = n^n \quad (8)$$

The recurrence equations listed below are examples of non-linear recurrence equations.

$$a_n^2 + 2a_na_{n-1} + 2a_{n-1}^2 = 0 \quad (9)$$
$$b_n \times 2b_{n-1} \times 3b_{n-2} \times \ldots \times nb_1 = \sqrt{n} + 17 \quad (10)$$
$$c_n + c_{n-1}c_{n-1} = n^n \quad (11)$$

Order of the Recurrence Equation

The order of a recurrence equation is the number of steps along the sequence from the first to the last member of the sequence in the equation. Some examples are given below to help you understand the order of a recurrence equation.

Sr. No.	Recurrence Equation	Order
1.	$a_n + 4a_{n-1} = n^2$	First order
2.	$b_n - 2b_{n-1} + b_{n-2} = 0$	Second order
3.	$c_n - c_{n-2} = 2^n$	Second order
4.	$d_n - 3d_{n-2} + d_{n-4} = 1$	Fourth order
5.	$c_n - c_{n/2} = 2^n$	Huh?

Table 3.1

The last recurrence equation does not have a constant order. That complicates the solution. The number of initial conditions required to express a sequence in recurrence form is equal to order of the recurrence equation if the equation is of constant order.

Constant verses Non-constant Coefficients

We make a distinction between recurrence equation in which the coefficient which multiply the sequence terms are constant, such as these:

$$a_n + 4a_{n-1} + a_{n-2} = n^2 \quad \text{———} \quad (12)$$

$$b_n - b_{n-1} - b_{n-2} = n \quad \text{———} \quad (13)$$

and those recurrence equations in which the coefficients are non constant - they vary with n, such as these :

$$a_n + 4na_{n-1} + a_{n-2} = n^2 \quad \text{———} \quad (14)$$

$$nb_n - (n-1)b_{n-1} - (n-2)b_{n-2} = n \quad \text{———} \quad (15)$$

If we form three sets, H, L and C of characteristics of a recurrence equations and let H = {Homogeneous, Non-homogeneous}, L = {Linear, Non-linear} and C = {Constant-coefficient, Non-constant-coefficient}. Then a recurrence equation can be of any type of ordered triplet of H x L x C, and each equation can be of either constant or variable order. So what? There are some classes of recurrence equations, which are always solvable. So it is important to recognize them.

1. First-order homogeneous or non-homogeneous, linear recurrence equations in which the coefficients are never zero; are always solvable.

2. Any constant order linear recurrence equations with constant coefficients which are homogeneous or whose right sides can be expressed as the product of polynomials in n and constants to the nth power; are also always solvable.

The techniques for solving these two classes of recurrence equations are discussed in detail in this chapter. Other recurrence equations are generally harder to solve although a few special cases have been solved.

3.2 Backtracking and Forward chaining Method

This is a very useful method to solve a first order linear recurrence equation with constant coefficient. The core of this technique is to identify a pattern emerging out from the arrangements of terms. **Once you have got the pattern, you have got the solution.** The solution so obtained is the closed form formula for the given recurrence equation. We shall learn this method through the following examples.

Example 1 Solve the recurrence equation $a_n = a_{n-1} + 3$ with $a_1 = 2$.
Solution: We shall use the backtracking method to solve this problem. In this method we shall start from a_n, and move backward towards a_1 to find a pattern, if any, to solve the problem. To backtrack, we keep on substituting the definition of a_n, a_{n-1}, a_{n-2} and so on until a recognizable pattern appears. The method is illustrated below.

$a_n = a_{n-1} + 3$
$= a_{n-2} + 3 + 3$ [since $a_{n-1} = a_{n-2} + 3$]
$= a_{n-2} + 2x3$
$= a_{n-3} + 3 + 2x3$ [since $a_{n-2} = a_{n-3} + 3$]
$= a_{n-3} + 3x3$
$= a_{n-4} + 3 + 3x3$ [since $a_{n-3} = a_{n-4} + 3$]
$= a_{n-4} + 4x3$

.
.
.

$= a_{n-(n-1)} + (n-1)x3$ **[note the pattern]**
$= a_1 + 3(n-1)$
$= 2 + 3(n-1)$ [since $a_1 = 2$ is the terminating condition]

Therefore, $a_n = 2 + 3(n-1)$.

Ans.

Now, to illustrate the forward chaining method, we shall continue with the same example. In this method, we begin from initial (terminating) condition and keep on moving to-wards the nth term until we get a clear pattern. The method is illustrated below.

$a_1 = 2$
$a_2 = a_1 + 3$
$a_3 = a_2 + 3$
$= a_1 + 2x3$
$a_4 = a_3 + 3$
$= a_1 + 2x3 + 3$
$= a_1 + 3x3$
$= a_1 + (4-1)x3$
$a_5 = a_1 + (5-1)x3$

.
.

$a_n = a_1 + (n-1)x3$

Therefore, $a_n = 2 + 3(n-1)$.

Ans.

The next and very useful method is **summation method** to solve a first order linear recurrence relation with constant coefficient. This is a form of backtracking method only, but more convenient to use. In this method we arrange the given equation in the following form:

$$a_n - ka_{n-1} = f(n)$$

and then backtrack till terminating condition. In the process, we get a number of equations. Add these equations in such a way that all intermediate terms get cancelled. Finally, we get the required solution. We illustrate the method on the same example under consideration. The given equation can be rearranged as

$$a_n - a_{n-1} = 3$$
$$a_{n-1} - a_{n-2} = 3$$
$$a_{n-2} - a_{n-3} = 3$$
$$\vdots$$
$$a_3 - a_2 = 3$$
$$a_2 - a_1 = 3 \text{ [we stop here, since } a_1 = 2 \text{ is given]}$$

Adding all, we get $a_n - a_1 = 3 + 3 + 3 + \ldots + (n-1)$ times

or, $a_n - a_1 = 3(n-1)$

Therefore, $a_n = 2 + 3(n-1)$.

Ans.

Example 2 Solve the recurrence equation $t_n = 2t_{n/2} + n$ with $t_1 = 1$.

Solution: We shall solve this recurrence equation first by **backtracking method** and then by **summation method.** Readers should try to solve it by **forward chaining method.**

Backtracking method: We have $t_n = 2t_{n/2} + n$ ——————— (1)

By repeated substitution of the definition of $t_{n/2}$, $t_{n/4}$, $t_{n/8}$ etc., we get the following results:

$$t_n = 2[2t_{n/4} + n/2] + n$$
$$= 2^2 t_{n/4} + 2 * n/2 + n$$
$$= 2^2 t_{n/4} + n + n \quad \text{———————— (2)}$$

$$t_n = 2^2[2t_{n/8} + n/4] + n + n$$
$$= 2^3 t_{n/8} + 2^2 * n/4 + n + n$$
$$= 2^3 t_{n/8} + n + n + n \quad \text{———————— (3)}$$

By closely observing equations (1), (2) and (3) we can write them as:

$$t_n = 2^{\log_2 2} t_{n/2} + n\log_2 2 \ldots\ldots\ldots\ldots from(1)$$
$$t_n = 2^{\log_2 4} t_{n/4} + n\log_2 4 \ldots\ldots\ldots\ldots from(2)$$
$$t_n = 2^{\log_2 8} t_{n/8} + n\log_2 8 \ldots\ldots\ldots\ldots from(3)$$

Similarly we can write,

$$t_n = 2^{\log_2 n} t_{n/n} + n\log_2 n$$
$$t_n = 2^{\log_2 n} t_1 + n\log_2 n$$
$$t_n = 2^{\log_2 n} + n\log_2 n, \ldots\ldots\ldots [\because t_1 = 1]$$

$$Therefore,\quad t_n = 2^{\log_2 2} + n\log_2 n \quad Ans.$$

Summation method: Arranging the given recurrence equation we get

$$t_n - 2t_{n/2} = n$$
$$t_{n/2} - 2t_{n/4} = \frac{n}{2}$$
$$t_{n/4} - 2t_{n/8} = \frac{n}{4}$$

...

$$t_4 - 2t_2 = 4$$
$$t_2 - 2t_1 = 2$$

The number of equations in this set of equations is $\log_2 n$. Now if we multiply the above equations by

$$1, 2, 4, \ldots, 2^{\log_2 n}$$

respectively from top to bottom, and then by adding them all together we get

$$t_n - 2^{\log_2 n} = n + n + n + \ldots \log_2 n \quad times$$
$$or,\quad t_n - 2^{\log_2 n} = n\log_2 n$$
$$\therefore\quad t_n = 2^{\log_2 n} + n\log_2 n \quad Ans.$$

Example 3: Solve the recurrence equation

$$t(n) = t(\sqrt{n}) + c\log_2 n \quad with \quad t(1) = 1.$$

Solution: Sometimes a sequence term t_n is also written as a function of its place in sequence, i.e., t_n may also be written as t(n). We solve this problem, here, by **forward chaining method.** The same problem will be solved by **characteristic equation method** in the next section of this chapter.

$$\begin{aligned} t(2) &= t(2^{1/2}) + c\log_2 2 \\ &= t(2^{1/4}) + c\log_2 2^{1/2} + c\log_2 2 \\ &= t(2^{1/8}) + c\log_2 2^{1/4} + c\log_2 2^{1/2} + c\log_2 2 \\ &= t(2^{1/2n}) + c\log_2 2^{1/n} + \cdots + c\log_2 2^{1/4} + c\log_2 2^{1/2} + c\log_2 2 \end{aligned}$$

Since t is defined for $n = 1$ *only, we have to let* $t(2^{1/2n}) \to t(1)$

This is possible when $n \to \infty$

i.e., when $n \to \infty \Rightarrow \dfrac{1}{2n} \to 0 \Rightarrow 2^{\frac{1}{2n}} \to 1$

Therefore, $t(2) = t(1) + [c\log_2 2 + c\log_2 2^{1/2} + c\log_2 2^{1/4} + \cdots\cdots]$

$$= t(1) + c\log_2 2\left[1 + \frac{1}{2} + \frac{1}{4} + \frac{1}{8} + \cdots\cdots\right] = t(1) + c\log_2 2\left[\frac{1}{1 - \frac{1}{2}}\right]$$

$$= t(1) + 2c\log_2 2$$

i.e., $t(2) = t(1) + 2c\log_2 2$

$\|ry$ $t(3) = t(1) + 2c\log_2 3$

$t(4) = t(1) + 2c\log_2 4$

$\vdots$

$t(n) = t(1) + 2c\log_2 n$

Therefore $t(n) = t(1) + 2c\log_2 n.$

3.3 Characteristic Equation Method

In principle, this method can be used to solve **any** constant order linear recurrence equation with constant coefficient. This recurrence relation may be homogeneous or non-homogeneous. Before attempting to solve any such problem, let us first, understand what is characteristic equation for a given recurrence equation and how to find it.

A given recurrence equation of the mentioned type can be arranged in **standard form** as:

$A_n + c_1A_{n-1} + c_2A_{n-2} + c_3A_{n-3} = \mathbf{RHS}$ ———————— (1)

Where c_1, c_2, c_3 are constant coefficients and **RHS** has one of the following forms:

Form	Examples
Homogeneous	0
A constant to the nth power	$2^n, \pi^{-s}, 2^{-n}, \sqrt{2}^n$
A polynomial in n	$3, n^2, n^2 - n, n^3+2n-1$
A product of a constant to the nth power and a polynomial in n	$2^n (n^2 + 2n - 1), (n-1)n^6, n6^n$
A linear combination of any of the above	$(2^n + 3^{n/2})(n^2 + 2n - 1) + 5$

Table 3.2

In the recurrence equation (1) , assigning RHS = 0, we get

$A_n + c_1A_{n-1} + c_2A_{n-2} + c_3A_{n-3} = \mathbf{0}$ ———————— (2)

This equation (2) gives the homogeneous part of the given recurrence equation. **Every recurrence equation has a homogeneous part.** If the recurrence relation is homogeneous then it has only homogeneous part and solving such equation is **one step process** (to be discussed in this section). On the other hand, if the given recurrence equation is non-homogeneous then its homogeneous part is obtained by assigning **RHS equal to zero**. **A characteristic equation corresponds to homogeneous part of the given recurrence relation**. The characteristic equation of (2) is given as:

$x^3 + c_1x^2 + c_2x + c_3 = 0$ ———————— (3)

This has been obtained by the following procedure:

- Find the order of the recurrence equation. Here it is 3.
- Take any variable (say x) and substitute A_n, A_{n-1}, A_{n-2}, by x^3, x^2, x respectively in the homogeneous part of the recurrence equation.

Equation so obtained is called **characteristic equation** of the given recurrence equation.

Example 1: The characteristic equation of the recurrence equation $c_n = 3\,c_{n-1} - 2\,c_{n-2}$; is given by

$$x^2 - 3x + 2 = 0$$

Example 2: The characteristic equation of the recurrence equation $f_n = f_{n-1} + f_{n-2}$; is given by

$$x^2 - x - 1 = 0$$

Example 3: The characteristic equation of the recurrence equation $A_n - 5A_{n-1} + 6A_{n-2} = 2^n + n$; is given by

$$x^2 - 5x + 6 = 0$$

Now, we shall discuss two theorems, which are the basis for finding the solution for the homogeneous part of any linear recurrence equation.

Theorem 1: If the characteristic equation $x^2 - r_1x - r_2 = 0$ of the recurrence equation $a_n = r_1a_{n-1} + r_2a_{n-2}$ has two distinct roots s_1 and s_2 then

$$a_n = us_1^{\,n} + vs_2^{\,n}$$

is the closed form formula for the sequence where u and v depend on the initial condition.

Proof: Since s_1 and s_2 are roots of

$$x^2 - r_1x - r_2 = 0 \qquad \text{(1)}$$

We have,

$$s_1^{\,2} - r_1s_1 - r_2 = 0 \qquad \text{(2)}$$

$$s_2^{\,2} - r_1 s_2 - r_2 = 0 \qquad \underline{\hspace{5cm}}(3)$$

Since u and v are dependent on the initial conditions, we have

$$a_1 = us_1 + vs_2 \qquad \& \qquad a_2 = us_1^{\,2} + vs_2^{\,2}$$

Now,

$$\begin{aligned}
a_n &= us_1^{\,n} + vs_2^{\,n} \\
&= us_1^{\,n-2} s_1^{\,2} + vs_2^{\,n-2} s_2^{\,2} \\
&= us_1^{\,n-2}[r_1 s_1 + r_2] + vs_2^{\,n-2}[r_1 s_2 + r_2] \qquad [\textit{from equation } (2) \,\&\, (3)] \\
&= r_1 us_1^{\,n-1} + r_2 us_1^{\,n-2} + r_1 vs_2^{\,n-1} + r_2 vs_2^{\,n-2} \\
&= r_1\left[us_1^{\,n-1} + vs_2^{\,n-1}\right] + r_2\left[us_1^{\,n-2} + vs_2^{\,n-2}\right] \\
&= r_1 a_{n-1} + r_2 a_{n-2}
\end{aligned}$$

i.e. $a_n = us_1^{\,n} + vs_2^{\,n}$ *is an explicit formula for the given recurrence equation.*

Proved.

Theorem 2: If the characteristic equation $x^2 - r_1 x - r_2 = 0$ of the recurrence equation $a_n = r_1 a_{n-1} + r_2 a_{n-2}$ has a single root s then

$$a_n = us^n + vns^n$$

is the closed form formula for the sequence where u and v depend on the initial condition.

Proof: Since s is roots of

$$x^2 - r_1 x - r_2 = 0 \qquad \underline{\hspace{5cm}}(1)$$

We have,

$$s^2 - r_1 s - r_2 = 0 \qquad \underline{\hspace{5cm}}(2)$$

Now,

$$\begin{aligned}
a_n &= us^n + vns^n = us^n + v(n-1)s^n + vs^n \\
&= us^{n-2}s^2 + v(n-1)s^{n-2}s^2 + vs^n \\
&= \left[us^{n-2} + v(n-1)s^{n-2}\right]s^2 + vs^n \\
&= \left[us^{n-2} + v(n-1)s^{n-2}\right](r_1 s + r_2) + vs^n \\
&= r_1\left[us^{n-1} + v(n-1)s^{n-1}\right] + r_2\left[us^{n-2} + v(n-1)s^{n-2}\right] + vs^n \\
&= r_1\left[us^{n-1} + v(n-1)s^{n-1}\right] + r_2\left[us^{n-2} + v(n-2)s^{n-2}\right] + r_2 vs^{n-2} + vs^n \\
&= r_1 a_{n-1} + r_2 a_{n-2} + vs^{n-2}\left[r_2 + s^2\right] \qquad \text{__________(3)}
\end{aligned}$$

From equation (2) we have

$$s = \frac{r_1 \pm \sqrt{r_1^2 + 4r_2}}{2} \qquad \left[\because \quad x = \frac{-b \pm \sqrt{b^2 - 4ac}}{2a}\right]$$

$\because$ *s is an equal root we have,*

$$\sqrt{r_1^2 + 4r_2} = 0 \quad or, \quad s = \frac{r_1}{2} \quad or, \quad r_1 = 2s$$

Now substituting $r_1 = 2s$ *in equation*(2), *we get*

$$s^2 - 2s^2 - r_2 = 0$$

$$or, \qquad -s^2 - r_2 = 0$$

$$or, \qquad s^2 + r_2 = 0 \qquad \text{\_\_\_\_\_\_\_\_\_\_(4)}$$

Using the result from equation (4) in equation (3), we have

$$a_n = r_1 a_{n-1} + r_2 a_{n-2}$$

Therefore the explicit formula for the recurrence equation $a_n = r_1a_{n-1} + r_2a_{n-2}$ is

$$a_n = us^n + vns^n$$

With initial conditions

$$a_1 = us + vs \qquad and \qquad a_2 = us^2 + 2vs^2$$

Proved.

Note:

1. **In general**, if $s_1, s_2, s_3, \ldots, s_r$ are r distinct roots of the characteristic equation of a rth order recurrence equation then its explicit formula is given by

$$a_n = u_1 s_1^{\ n} + u_2 s_2^{\ n} + u_3 s_3^{\ n} + \cdots + u_r s_r^{\ n};$$

where, $u_1, u_2, u_3, \ldots, u_r$ depend on initial conditions.

2. If s is r times equal root of the characteristic equation of a rth order recurrence equation then its explicit formula is given by

$$a_n = u_1 s^n + u_2 n s^n + u_3 n^2 s^n + \cdots + u_r n^{r-1} s^n;$$

where, $u_1, u_2, u_3, \ldots, u_r$ depend on initial conditions.

3. A combination of 1 and 2 is also possible, i.e. some roots of a characteristic equation are distinct and some are equal. In that case, for distinct roots we use method outlined in 1 and for repeated roots we use method mentioned in 2.

Having discussed the theorem, we shall now learn the way to solve a given recurrence equation using the characteristic equation method. This is based on the method of characteristic equations with undetermined constants used to solve the related continuous function in differential calculus. It relies on the **principle of uniqueness,** i.e. **any answer which works is the correct answer since there can be only one correct answer**.

There are two parts to the total solution. The **homogeneous** part of the solution depends only on what is on the left of the recurrence equation when expressed in the standard form. The **particular part** of the total solution depends on what is on the RHS and has the same form as the RHS. We will calculate the two parts separately and add them to form the total solution. There are four steps in the process as listed below.

Step 1: Find the homogeneous solution to the homogeneous equation. This results when you set the RHS to zero. **If it is already zero**, skip the next two steps and go directly to the **step 4.** Your answer will contain one or more "undetermined coefficients" whose values cannot be determined until **step 4**.

Step 2: Find the particular solution by **guessing a form similar to the RHS**. This step does not produce any additional undetermined coefficients, nor does it eliminate those from the **step 1**.

Step 3: Combine the homogeneous and particular solutions.

Step 4: Use boundary or initial conditions to eliminate the undetermined constants from the **step 1**.

This method is straightforward and very effective. Its major *limitation* is that it only works with a specific subset of recurrence equations. However, by using the change of variable method, a wider range of recurrence equations can be solved than is immediately apparent. In fact, the method presented here is useful for solving many recurrence equations that occur in the analysis of algorithm. Let us now demonstrate the procedure by solving some problems.

Example 4 Solve the recurrence equation

$$a_r - 7a_{r-1} + 10a_{r-2} = 2^r \quad \textit{with initial condition} \quad a_0 = 0 \quad \textit{and} \quad a_1 = 6.$$

Solution: The general solution, also called homogeneous solution, to the problem is given by homogeneous part of the given recurrence equation. The homogeneous part of the equation

$$a_r - 7a_{r-1} + 10a_{r-2} = 2^r \qquad \text{\_\_\_\_\_\_\_\_(1)}$$

is

$$a_r - 7a_{r-1} + 10a_{r-2} = 0 \qquad \text{\_\_\_\_\_\_\_\_(2)}$$

The characteristic equation of (2) is given as

$$x^2 - 7x + 10 = 0$$

$$or, \quad (x-2)(x-5) = 0$$

$$or, \quad x = 2 \quad and \quad x = 5.$$

Since the two roots 2 & 5 of characteristic equation are distinct, the homogeneous solution is given by

$$a_r = A2^r + B5^r \qquad \text{\_\_\_\_\_\_\_\_(3)}$$

Now, particular solution is given by

$$a_r = rC2^r$$

The reader must be wondering how this guess has been made. Please wait till the final solution of this problem. Thereafter we shall discuss the way, how to deal with particular solution before proceeding to the next example.

Substituting the value of a_r in equation (1), we get

$$C\left[r2^r - 7(r-1)2^{r-1} + 10(r-2)2^{r-2}\right] = 2^r$$

$$or,\quad C\left[4r - 7(r-1)2 + 10(r-2)\right]2^{r-2} = 2^r$$

$$or,\quad C\left[4r - 14(r-1) + 10r - 20\right] = 4$$

$$or,\quad C\left[4r - 14r + 14 + 10r - 20\right] = 4$$

$$or,\quad -6C = 4 \qquad or,\quad C = -\frac{2}{3}$$

$$\therefore\quad \textit{Particular solution is}\quad a_r = -\frac{2}{3}r2^r$$

The **complete,** also called **total**, solution is obtained by combining the homogeneous and particular solutions. This is given as

$$a_r = A2^r + B5^r - \frac{2}{3}r2^r \qquad \text{\_\_\_\_\_\_\_\_\_\_\_\_(4)}$$

The equation (4) contains two undetermined coefficients, A & B, which are to be determined. To find this, we use the given initial conditions for r = 0 and r = 1. Since values of a_0 and a_1 are given, putting r = 0 and r = 1 in equation (4), we get the following equations (5) and (6), respectively.

$$a_0 = A2^0 + B5^0 - \frac{2}{3} * 0 * 2^0$$

$$or,\quad 0 = A + B \qquad \text{\_\_\_\_\_\_\_\_\_\_\_\_(5)} \quad [\because a_0 = 0 \textit{ is given}]$$

And,

$$a_1 = A2^1 + B5^1 - \frac{2}{3} * 1 * 2^1$$

$$or, \quad 6 = 2A + 5B - \frac{4}{3} \qquad [\because a_1 = 6 \text{ is given}]$$

$$or, \quad 2A + 5B = \frac{22}{3} \qquad \text{\_\_\_\_\_\_\_\_\_\_\_\_\_\_\_\_\_\_\_\_\_\_\_\_\_\_\_\_\_\_\_\_\_\_(6)}$$

Solving equation (5) and (6), we get

$$A = -\frac{22}{9}, \quad and \quad B = \frac{22}{9}$$

Replacing A and B in equation (4) by its respective values, we get the closed form formula for the given recurrence equation. Thus,

$$a_r = -\frac{22}{9}2^r + \frac{22}{9}5^r - \frac{2}{3}r2^r$$

$$or, \quad a_r = \frac{22}{9}\left[5^r - 2^r\right] - \frac{2}{3}r2^r \qquad Ans.$$

As assured before, we shall now discuss the way to find a particular solution before proceeding further. **The form of a particular solution has nothing to do with the order of the recurrence relation**. It only **depends** on the form of the RHS of recurrence equation expressed in standard form. Guess a solution of the same form but with undetermined coefficients, which have to be calculated. We find their values by substituting the **guessed** particular solution into the recurrence equation. You may find the following table useful for guessing a particular solution.

RHS	Guessed Particular solution
17(constant)	C(constant)
π^n (Constant to the nth Power)	$C\pi^n$ (Constant to the nth power)
$2^n + 5^n + 3$ (Linear combination)	$D2^n + E5^n + F$ (Same linear combination)
$5n^3$ (Polynomial in n)	$An^3 + Bn^2 + Cn + D$ (Decreasing Polynomial)
$5n^3 - 1$(Polynomial in n)	$An^3 + Bn^2 + Cn + D$ (Decreasing Polynomial)
$3n^2 5^n$ (Linear combination)	$5^n (Bn^2 + Cn + D)$ (Linear combination)

Table 3.3

Notice how a polynomial in n produces a decreasing polynomial with all orders; down to, and including the constant term. Without those extra terms, usually the coefficients cannot be determined successfully.

When the RHS is of the form of homogeneous solution there is slight complication. You must have noticed that in example 4, the RHS was of the form 2^r so we should have guessed $C2^r$, instead we guessed

$$a_r = rC2^r$$

We cannot try a particular solution of the form:

$$a_r = C2^r$$

because that is already a homogeneous solution. So the left side will sum to zero. This is demonstrated by substituting

$$a_r = C2^r$$

in equation (1) of example 4. Thus, we have

$$C2^r - 7C2^{r-1} + 10C2^{r-2} = 2^r$$

$$or, \quad [4C - 14C + 10C]2^{r-2} = 2^r$$

$$or, \quad 0 = 4 \qquad \textit{which is impossible.}$$

This implies that coefficient C could not have been determined if we would have guessed the above way.

Example 5 Solve the recurrence equation

$$A_n - A_{n-1} - A_{n-2} = 2n \qquad with \quad A_0 = 0 \quad and \quad A_1 = 1.$$

Solution: The homogeneous part of the equation

$$A_n - A_{n-1} - A_{n-2} = 2n \qquad \text{\_\_\_\_\_\_\_\_\_\_(1)}$$

is

$$A_n - A_{n-1} - A_{n-2} = 0 \qquad \text{\_\_\_\_\_\_\_\_\_\_(2)}$$

The characteristic equation of (2) is given as

$$x^2 - x - 1 = 0$$

$$\therefore x = \frac{1 \pm \sqrt{1+4}}{2} \quad or, \quad x = \frac{1+\sqrt{5}}{2}, \quad and \quad \frac{1-\sqrt{5}}{2}$$

Thus, two roots of the characteristic equation are distinct. So, the homogeneous solution is given by

$$A_n = A\left[\frac{1+\sqrt{5}}{2}\right]^n + B\left[\frac{1-\sqrt{5}}{2}\right]^n \quad \text{\_\_\_\_\_\_\_\_\_\_(3)}$$

The particular solution is given by the guess

$$A_n = Cn + D$$

Substituting this value in equation (1), we get

$$Cn + D - C(n-1) - D - C(n-2) - D = 2n$$

or, $(Cn - Cn - Cn) + D + C - D + 2C - D = 2n$

or, $-Cn + 3C - D = 2n$

or, $-C = 2, \quad 3C - D = 0$

or, $C = -2, \quad D = -6$

$\therefore$ $A_n = -2n - 6$ *is the particular solution.*

Combining homogeneous and particular solutions, we get total solution as

$$A_n = A\left[\frac{1+\sqrt{5}}{2}\right]^n + B\left[\frac{1-\sqrt{5}}{2}\right]^n - 2n - 6 \quad \text{\_\_\_\_\_\_\_\_\_\_(4)}$$

The equation (4) contains two undetermined coefficients A & B, which are to be determined. To find this we use the given initial conditions for n = 0 and n = 1, since values of A_0 and A_1 are given. Putting n = 0 and n = 1 in equation (4), we get the following equations (5) and (6), respectively.

$$A_0 = A\left[\frac{1+\sqrt{5}}{2}\right]^0 + B\left[\frac{1-\sqrt{5}}{2}\right]^0 - 2*0-6$$

or, $0 = A + B - 6$ ________________(5) $\left[\because A_0 = 0 \text{ is given}\right]$

and

$$A_1 = A\left[\frac{1+\sqrt{5}}{2}\right]^1 + B\left[\frac{1-\sqrt{5}}{2}\right]^1 - 2*1-6$$

or, $1 = A\left[\frac{1+\sqrt{5}}{2}\right] + B\left[\frac{1-\sqrt{5}}{2}\right] - 8$ ______(6) $\left[\because A_1 = 1 \text{ is given}\right]$

Solving equation (5) and (6) we get,

$$B - \frac{1-\sqrt{5}}{1+\sqrt{5}}B = 6 - \frac{18}{1+\sqrt{5}}, \quad or, \quad B\frac{1+\sqrt{5}-1+\sqrt{5}}{1+\sqrt{5}} = \frac{6+6\sqrt{5}-18}{1+\sqrt{5}}$$

$$or, B = \frac{3(\sqrt{5}-2)}{\sqrt{5}}, \quad and \quad A = \frac{3(\sqrt{5}+2)}{\sqrt{5}}$$

Replacing A and B in equation (4) by its respective values, we get the closed form formula for the given recurrence equation. Therefore, the final solution is:

$$A_n = \frac{3(\sqrt{5}+2)}{\sqrt{5}}\left[\frac{1+\sqrt{5}}{2}\right]^n + \frac{3(\sqrt{5}-2)}{\sqrt{5}}\left[\frac{1-\sqrt{5}}{2}\right]^n - 2n - 6 \qquad Ans.$$

Example 6 Solve the recurrence equation

$$a_n - 8a_{n-1} + 16a_{n-2} = 0 \quad with \quad a_2 = 16 \quad and \quad a_3 = 80.$$

Solution: The given equation is homogeneous, so only the homogeneous part of the solution is required. In this case, we do not have to try for the particular solution and hence no combining of the particular part with the homogeneous part is needed. Here, we will go to the step 4 from the step 1, skipping the steps 2 & 3.

The characteristic equation of the given recurrence equation is

$$x^2 - 8x + 16 = 0$$

$$or, \quad (x-4)^2 = 0 \;\Rightarrow\; x = 4, \;\; 4$$

Since the two roots of characteristic equation are equal, the homogeneous solution is given by

$$a_n = (u + vn)4^n \qquad \text{\_\_\_\_\_\_\_\_\_\_(1)}$$

The equation (1) contains two undetermined coefficients, u & v, which are to be determined. To find this, we use the given initial conditions for n = 2 and n = 3, since values of a_2 and a_3 are given. Putting n = 2 and n = 3 in equation (1), we get the following equations (2) and (3), respectively.

When $n = 2,$

$$a_2 = (u + 2v)4^2, \quad or, \quad 16 = 16(u + 2v)$$

$$or, \quad u + 2v = 1 \qquad \text{\_\_\_\_\_\_\_\_\_\_(2)}$$

And,

When $n = 3,$

$$a_3 = (u + 3v)4^3, \quad or, \quad 80 = 64(u + 3v)$$

$$or, \quad u + 3v = \frac{5}{4} \qquad \text{\_\_\_\_\_\_\_\_\_\_(3)}$$

Solving equation (2) and (3) we have

$$u = \frac{1}{2}, \qquad and \qquad v = \frac{1}{4}$$

Replacing u and v by theirs respective values in (1), we have the solution in the form of closed formula. Therefore,

$$a_n = \left[\frac{1}{2} + \frac{n}{4}\right]4^n \qquad Ans.$$

Example 7 Solve the recurrence equation

$$a_n = a_{n-1} + 2(n-1) \qquad \textit{with} \quad a_0 = 1.$$

Solution: The given equation is non-homogeneous of order **one**. This can also be solved **either** by **backtracking,** or by **summation method**. Here, we shall solve it by characteristic equation method. The characteristic equation of the given recurrence equation is

$$x - 1 = 0, \qquad or, \quad x = 1$$

The characteristic equation has only one root, so the homogeneous solution is given by

$$a_n = A * 1^n \qquad \text{\_\_\_\_\_\_\_\_\_\_\_\_\_\_\_\_\_(1)}$$

Now to find particular solution, we guess

$$a_n = Bn + D$$

Now substituting this guess for a_n and a_{n-1} in the given recurrence equation, we get

$$Bn + D - B(n-1) - D = 2n - 2$$

$$or, \quad B = 2n - 2 \;\Rightarrow\; 0 = 2, \quad \textit{This is impossible.}$$

Notice that coefficient of n *in LHS is zero and in RHS it is* 2

This has happened because of our assumption (guess). Now we will modify our guess and take **a polynomial in n of a degree one higher than what we took earlier** as a guess to the particular solution. Let

$$a_n = Bn^2 + Cn + D$$

Now substituting this guess for a_n and a_{n-1} in the given recurrence equation, we get

$$Bn^2 + Cn + D - B(n-1)^2 - C(n-1) - D = 2n - 2$$

$$or, \quad 2Bn + (C - B) = 2n - 2, \;\Rightarrow\; 2B = 2, \quad and \quad C - B = -2$$

$$\Rightarrow\; B = 1, \quad and \quad C = -1 \quad and \quad D = 0$$

$$\therefore\; a_n = n^2 - n, \qquad \textit{is the particular solution.}$$

Here, we have assigned D =0. Readers may try assigning any finite value to D and may test the end result. The end result shall remain the same.

The complete solution is then given by combining homogeneous and particular solution as

$$a_n = A + n^2 - n \qquad \text{\_\_\_\_\_\_\_\_\_\_\_\_\_\_\_\_\_(2)}$$

To determine the value of A, we use the initial condition $a_0 = 1$. We have from equation (2), $a_0 = A$ or $A = 1$ when $n = 0$. Therefore,

$$a_n = n^2 - n + 1 \qquad Ans.$$

In example 7, initially, we guessed a particular solution that failed and then we guessed another that succeeded. Reader may ask, then **where to begin and where to stop**. Some people may find it easier to solve polynomial particular solutions by beginning at the top (at the order of RHS, or above if the RHS is of the form of homogeneous solution) and then by eliminating term until a solution is found. On the other hand, some may proceed in reverse direction and begin at the bottom (try a constant) and then work on first power of n, second power of n and so on until a solution is found. Pick whichever may work best for you. In any case, if you are able to find a set of coefficients which work then that is the solution. If you cannot find a consistent set of coefficients, your guess was poor. There will be no undetermined coefficients in the particular solution. If you have any, then what you have found is not a particular solution.

Example 8: Solve the recurrence equation

$$a_n - 5a_{n-1} + 6a_{n-2} = 2^n + n \quad \textit{with initial condition} \quad a_1 = 0 \quad \textit{and} \quad a_2 = 10.$$

Solution: The general to the problem is given by the homogeneous part of the given recurrence equation. The homogeneous part of the equation

$$a_n - 5a_{n-1} + 6a_{n-2} = 2^n + n \qquad \text{\_\_\_\_\_\_\_\_\_\_(1)}$$

is

$$a_n - 5a_{n-1} + 6a_{n-2} = 0 \qquad \text{\_\_\_\_\_\_\_\_\_\_(2)}$$

The characteristic equation of (2) is given as

$$x^2 - 5x + 6 = 0$$
$$or, \quad (x-2)(x-3) = 0$$
$$or, \quad x = 2 \quad and \quad x = 3$$

The two roots, 2 & 3, of characteristic equation are distinct. So the homogeneous solution is given by

$$a_n = A2^n + B3^n \qquad \text{\_\_\_\_\_\_\_\_\_\_(3)}$$

Let particular solution be

$$a_n = Cn2^n + Dn + E$$

Substituting this particular solution in equation (1) for a_n, a_{n-1} and a_{n-2}, we get

$$Cn2^n + Dn + E - 5\left[C(n-1)2^{n-1} + D(n-1) + E\right]$$
$$+6\left[C(n-2)2^{n-2} + D(n-2) + E\right] = 2^n + n$$

$$or,\quad C\left[n2^n - 5(n-1)2^{n-1} + 6(n-2)2^{n-2}\right] + D\left[n - 5(n-1) + 6(n-2)\right]$$
$$+E\left[1 - 5 + 6\right] = 2^n + n$$

$$or,\quad C\left[4n - 10(n-1) + 6(n-2)\right]2^{n-2} + D\left[n - 5n + 5 + 6n - 12\right] + 2E = 2^n + n$$

$$or,\quad C\left[4n - 10n + 10 + 6n - 12\right]2^{n-2} + D\left[2n - 7\right] + 2E = 2^n + n$$

$$or,\quad C\left[-2\right]2^{n-2} + D\left[2n - 7\right] + 2E = 2^n + n$$

$$or,\quad -2C = 4,\qquad 2D = 1 \qquad and \qquad -7D + 2E = 0$$

$$or,\quad C = -2,\qquad D = \frac{1}{2} \qquad and \qquad E = \frac{7}{4}$$

Therefore, particular solution is

$$a_n = -2n2^n + \frac{n}{2} + \frac{7}{4}$$

Total solution is given by

$$a_n = A2^n + B3^n - 2n2^n + \frac{n}{2} + \frac{7}{4} \qquad \text{\_\_\_\_\_\_\_\_\_\_\_\_(4)}$$

Using initial conditions for n = 1 and n = 2, we get the following two equations (5) and (6), respectively.

$$a_1 = A2^1 + B3^1 - 2*1*2^1 + \frac{1}{2} + \frac{7}{4}$$

$$or,\quad 0 = 2A + 3B - 4 + \frac{1}{2} + \frac{7}{4} \qquad [\because\ a_1 = 0]$$

$$or,\quad 2A + 3B = \frac{7}{4} \qquad \text{\_\_\_\_\_\_\_\_\_\_\_\_(5)}$$

And,

$$a_2 = A2^2 + B3^2 - 2*2*2^2 + \frac{2}{2} + \frac{7}{4}$$

$$or,\quad 10 = 4A + 9B - 16 + 1 + \frac{7}{4} \qquad [\because\ a_2 = 10]$$

$$or,\quad 4A + 9B = \frac{93}{4} \qquad \text{\_\_\_\_\_\_\_\_\_\_\_\_\_\_\_(6)}$$

Solving equation (5) and (6), we get

$$A = -9, \qquad and \qquad B = \frac{79}{12}$$

Replacing the values of A and B in equation (4), we get the closed form formula for the given recurrence equation.

$$a_n = -9*2^n + \frac{79}{12}3^n - 2n2^n + \frac{n}{2} + \frac{7}{4} \qquad Ans.$$

Example 9 Solve the recurrence equation

$$c_n = 3c_{n-1} - 2c_{n-2} \quad for\ n \geq 3 \quad with \quad c_1 = 5 \quad and \quad c_2 = 3$$

Solution: The given equation is homogeneous so only homogeneous part of the solution is required.

The characteristic equation of the given recurrence equation is

$$x^2 - 3x + 2 = 0$$

$$or,\quad (x-1)(x-2) = 0 \;\Rightarrow\; x = 1, \quad and \quad x = 2$$

The two roots of the characteristic equation are distinct. So the homogeneous solution is given by

$$c_n = u*1^n + v*2^n \qquad \text{\_\_\_\_\_\_\_\_\_\_\_\_\_\_\_(1)}$$

The equation (1) contains two undetermined coefficients, u & v, which are to be determined. To find this, we use the given initial conditions for n = 1 and n = 2, since the values of c_1 and c_2 are given. Putting n = 1 and n = 2 in equation (1), we get the following equations (2) and (3), respectively.

$$When \quad n = 1,$$

$$c_1 = u * 1 + v * 2 \quad or, \quad 5 = u + 2v$$

$$or, \quad u + 2v = 5 \qquad \text{\_\_\_\_\_\_\_\_\_\_\_\_(2)}$$

And,

$$When \quad n = 2,$$

$$c_2 = u * 1^2 + v * 2^2 \quad or, \quad 3 = u + 4v$$

$$or, \quad u + 4v = 3 \qquad \text{\_\_\_\_\_\_\_\_\_\_\_\_(3)}$$

By solving equation (2) and (3), we have

$$u = 7, \qquad and \qquad v = -1$$

Replacing u and v by theirs respective values in (1), we have the solution in the form of closed form formula. Therefore,

$$c_n = 7 - 2^n \qquad Ans.$$

Example 10 Solve the recurrence equation

$$T_n = T_{\sqrt{n}} + C\log_2 n \qquad with \quad T_1 = 1$$

by characteristic equation method.

Solution: The given equation can be written in standard form as

$$T_n - T_{\sqrt{n}} = C\log_2 n \qquad \text{\_\_\_\_\_\_\_\_\_\_\_\_(1)}$$

The homogeneous solution of (1) is given by the equation

$$T_n - T_{\sqrt{n}} = 0 \qquad \text{\_\_\_\_\_\_\_\_\_\_\_\_(2)}$$

The characteristic equation of (2) is

$$x - \sqrt{x} = 0 \qquad or, \quad x^2 - x = 0 \qquad or, \quad x(x-1) = 0$$

$$\therefore \quad x = 0, \quad and \quad x = 1$$

The homogeneous solution is then given by

$$T_n = A * 0^n + B * 1^n \qquad \text{\_\_\_\_\_\_\_\_\_\_\_\_(3)}$$

Guess the particular solution as

$$T_n = ZC\log_2 n$$

Using this particular solution in equation (1), we get

$$ZC\log_2 n - ZC\log_2 n^{\frac{1}{2}} = C\log_2 n$$

$$or, \quad \left[Z - \frac{Z}{2}\right] C\log_2 n = C\log_2 n$$

$$Z - \frac{Z}{2} = 1 \qquad or, \quad Z = 2$$

Therefore, particular solution is $T_n = 2C\log_2 n$.

Combining homogeneous and particular solution, we get the complete solution, which is:

$$T_n = A * 0^n + B * 1^n + 2C\log_2 n$$

$$or, \quad T_n = B + 2C\log_2 n$$

For n=1 the initial condition is given. Putting n = 1 in the above equation, we can find the value of undetermined coefficient B as below.

$$When \quad n = 1, \qquad T_1 = B + 2C\log_2 1,$$

$$or, \quad 1 = B + 2C * 0 \qquad [\because \; T_1 = 1 \;\; and \;\; \log_2 1 = 0]$$

$$or, \quad B = 1$$

Replacing B by 1 in the complete solution, we have

$$T_n = 1 + 2C\log_2 n \qquad Ans.$$

Before winding up this section, I would like to introduce –**the technique of change of variables.** This is very useful and effective for solving a range of recurrence equations, which are otherwise perceived to be unsolvable.

Example 11 Solve the recurrence equation

$$a_r = \sqrt{a_{r-1} + \sqrt{a_{r-2} + \sqrt{a_{r-3} + \sqrt{\cdots}}}} \qquad with \quad a_0 = 4.$$

Solution: **Wow**! Is it solvable by any of the above methods? At first, it seems no. But it is solvable. By squaring the given equation we have

$$a_r^{\ 2} = a_{r-1} + \sqrt{a_{r-2} + \sqrt{a_{r-3} + \sqrt{a_{r-4} + \sqrt{\cdots}}}}$$

$$or,. \quad a_r^{\ 2} = a_{r-1} + a_{r-1} \qquad \left[\because a_{r-1} = \sqrt{a_{r-2} + \sqrt{a_{r-3} + \sqrt{a_{r-4} + \sqrt{\cdots}}}}\right]$$

$$or, \quad a_r^{\ 2} = 2a_{r-1} \quad \Rightarrow \quad a_r^{\ 2} - 2a_{r-1} = 0 \qquad \text{\_\_\_\_\_\_\_\_(1)}$$

Now, what about equation (1)? Is it solvable? Again it seems no. Let us replace a_r with a new sequence term. Let

$$b_r = \log_2 a_r \quad \Rightarrow a_r = 2^{b_r}$$

Substituting the value of a_r in terms of the new sequence variable b_r in equation (1), we get

$$\left(2^{b_r}\right)^2 - 2 * 2^{b_{r-1}} = 0$$

$$or, \quad 2^{2b_r} - 2^{b_{r-1}+1} = 0$$

This is possible only if

$$2b_r - (b_{r-1} + 1) = 0 \qquad \text{\_\_\_\_\_\_\_\_(2)}$$

with initial condition $b_0 = \log_2 a_0 = \log_2 4 \quad \Rightarrow b_0 = 2$

Now, our job has been reduced to solve the equation (2), which is of **first order, linear and non-homogeneous.** The characteristic equation of (2) is

$$2x - 1 = 0 \qquad \Rightarrow \quad x = \frac{1}{2}$$

Therefore, homogeneous solution is

$$b_r = A\left(\frac{1}{2}\right)^r \qquad \text{\_\_\_\_\_\_\_\_(3)}$$

For particular solution, we guess $b_r = B$. Substituting this value for sequence term b_r and b_{r-1}, in equation (2), we have

$$2B - B = 1 \Rightarrow B = 1$$

$$\therefore \ particular \ solution \ is \quad b_r = 1$$

Then, the total solution is

$$b_r = A\left(\frac{1}{2}\right)^r + 1$$

Using the initial condition $b_0 = 2$, we get A = 1. Therefore the solution to the changed equation is

$$b_r = \left(\frac{1}{2}\right)^r + 1,$$

$$Since \ b_r = \log_2 a_r, \quad a_r = 2^{b_r}$$

Therefore, final solution is given by

$$a_r = 2^{\left(\frac{1}{2}\right)^r + 1} \qquad Ans.$$

Skill for solving problem by substitution method can be improved by rigorous practice only. There is no short cut to this. Readers are advised to work out the problems given in the exercises.

Example12 Solve the recurrence equation

$$a_n + 9a_{n-2} = 0 \quad with \ initial \ condition \quad a_0 = 0 \quad and \quad a_1 = 1.$$

Solution: The characteristic equation for the given recurrence equation is given by

$$x^2 + 9 = 0 \Rightarrow x^2 = -9 \Rightarrow x = \pm 3i$$

Therefore the homogeneous solution is

$$a_n = A(-3i)^n + B(3i)^n$$

Don't worry if the roots are imaginary or complex. When the values of the coefficients are determined at the very end, the resulting sequence values a_n will be real. Using the initial condition for n = 0 and n = 1, we get the following two equations.

$$0 = A(-3i)^0 + B(3i)^0$$

$$or, \quad A + B = 0 \qquad \text{\_\_\_\_\_\_\_\_\_\_(1)}$$

and,

$$1 = A(-3i)^1 + B(3i)^1$$

$$or, \quad -3iA + 3iB = 1 \qquad \text{\_\_\_\_\_\_\_\_\_\_(2)}$$

Now solving equation (1) and (2) we get values of A and B which are

$$A = \frac{-1}{6i} = \frac{i}{6} \quad and \quad B = \frac{-i}{6}$$

Substituting the value of A and B in homogeneous solution we get

$$a_n = \frac{i}{6}(-3i)^n + \frac{-i}{6}(3i)^n$$

You can verify that **all sequence terms will be real**.

3.4 Generating Function Method

One of the uses of generating function **(GF)** method is to find the closed form formula for a recurrence equation. Before using this method, **ensure that the given recurrence equation is in linear form.** A non-linear recurrence equation cannot be solved by the **GF** method. Use **substitution of variable** technique to convert a non-linear recurrence equation into linear. Solving a recurrence equation using **GF** method involves two steps process:

Step 1: Find generating function for the sequence for which the general term is given by recurrence equation.

Step 2: Find coefficient of x^n or $x^n/n!$ depending upon whether the GF is binomial or exponential to get a_n, the general term of the sequence.

The value so obtained will be an algebraic formula for a_n, expressed in terms of n which is the position of a_n in sequence. Our goal is to find the same while solving a recurrence equation. The process of doing this is illustrated with the following examples.

Example 1: Solve the following recurrence equation using generating function.

$$a_n = a_{n-1} + 2(n-1); \quad a_0 = 1$$

Solution: We will solve this using binomial generating function. Let f(x) be the binomial generating function for the sequence **a** of which general term is given by the given recurrence equation. Therefore, we have

$$f(x) = \sum_{n=0}^{\infty} a_n x^n = a_0 + \sum_{n=1}^{\infty} a_n x^n$$

$$or,\ f(x) = a_0 + \sum_{n=1}^{\infty} [a_{n-1} + 2(n-1)]x^n$$

$$= a_0 + \sum_{n=1}^{\infty} a_{n-1} x^n + 2\sum_{n=1}^{\infty} (n-1)x^n$$

$$= a_0 + x\sum_{n=1}^{\infty} a_{n-1} x^{n-1} + 2x\sum_{n=1}^{\infty} (n-1)x^{n-1}$$

$$= a_0 + x\sum_{m=0}^{\infty} a_m x^m + 2x\left[0 * x^0 + 1 * x^1 + 2 * x^2 + \cdots\right] \quad [n-1=m]$$

$$= a_0 + xf(x) + 2x^2\left[1 + 2x + 3x^2 + \cdots\right] \quad \left[\because f(x) = \sum_{m=0}^{\infty} a_m x^m\right]$$

$$or,\ (1-x)f(x) = a_0 + \frac{2x^2}{(1-x)^2} = 1 + \frac{2x^2}{(1-x)^2} \quad [\because a_0 = 1]$$

$$or,\ f(x) = \frac{1-2x+3x^2}{(1-x)^3}$$

$$or,\ f(x) = \frac{1}{(1-x)^3} - \frac{2x}{(1-x)^3} + \frac{3x^2}{(1-x)^3}$$ *is the GF of the equation.*

The value of a_n is given by the coefficient of x^n in f(x). Therefore,

$$Coeff\ of\ x^n = Coeff\ of\ x^n\ in\ \frac{1}{(1-x)^3} - 2x * Coeff\ of\ x^{n-1}\ in\ \frac{1}{(1-x)^3}$$

$$+3x^2 * Coeff\ of\ x^{n-2}\ in\ \frac{1}{(1-x)^3}$$

$$= {}^{3+n-1}C_n - 2 * {}^{3+(n-1)-1}C_{n-1} + 3 * {}^{3+(n-2)-1}C_{n-2}$$

$$or,\quad a_n = {}^{n+2}C_n - 2 * {}^{n+1}C_{n-1} + 3 * {}^{n}C_{n-2}$$

$$or,\quad a_n = \frac{n^2+3n+2}{2} - (n^2+n) + \frac{3(n^2-n)}{2} = \frac{n^2+3n+2-2n^2-2n+3n^2-3n}{2}$$

$$or,\quad a_n = \frac{2n^2-2n+2}{2} = n^2 - n + 1 \quad i.e.\ a_n = n^2 - n + 1 \quad Ans.$$

Example 2: Solve the following recurrence equation using generating function.

$$f_n = f_{n-1} + f_{n-2} \quad for\ n \geq 2 \quad and\ f_0 = f_1 = 1$$

Solution: We will solve this using binomial generating function. Let f(x) be the binomial generating function for the sequence **f** of which general term is given by the above recurrence equation. Then, we have

$$f(x) = \sum_{n=0}^{\infty} f_n x^n = f_0 + f_1 x + \sum_{n=2}^{\infty} f_n x^n$$

$$or,\quad f(x) = f_0 + f_1 x + \sum_{n=2}^{\infty} [f_{n-1} + f_{n-2}] x^n$$

$$= f_0 + f_1 x + \sum_{n=2}^{\infty} f_{n-1} x^n + \sum_{n=2}^{\infty} f_{n-2} x^n$$

$$= f_0 + f_0 x + \sum_{n=2}^{\infty} f_{n-1} x^n + \sum_{n=2}^{\infty} f_{n-2} x^n \quad [\because f_0 = f_1]$$

$$= f_0 + x \sum_{n=1}^{\infty} f_{n-1} x^{n-1} + x^2 \sum_{n=2}^{\infty} f_{n-2} x^{n-2}$$

$$= f_0 + x \sum_{n-1=0}^{\infty} f_{n-1} x^{n-1} + x^2 \sum_{n-2=0}^{\infty} f_{n-2} x^{n-2}$$

or, $f(x) = f_0 + xf(x) + x^2 f(x) = 1 + xf(x) + x^2 f(x)$

or, $f(x) = \dfrac{1}{(1 - x - x^2)}$ *is the GF of the equation.*

Using partial fraction, we can write f(x) as

$$f(x) = \frac{1}{\sqrt{5}}\left[\frac{A}{1 - x_1 x} - \frac{B}{1 - x_2 x}\right]$$

where, $x_1 = \dfrac{1+\sqrt{5}}{2}$, *and* $x_2 = \dfrac{1-\sqrt{5}}{2}$

where, $A = \dfrac{1+\sqrt{5}}{2}$, *and* $B = \dfrac{1-\sqrt{5}}{2}$

Therefore, coefficient of $x^n = \dfrac{1}{\sqrt{5}}\left[\left(\dfrac{1+\sqrt{5}}{2}\right)^{n+1} - \left(\dfrac{1-\sqrt{5}}{2}\right)^{n+1}\right]$

i.e., $f_n = \dfrac{1}{\sqrt{5}}\left[\left(\dfrac{1+\sqrt{5}}{2}\right)^{n+1} - \left(\dfrac{1-\sqrt{5}}{2}\right)^{n+1}\right]$ *Ans.*

By this time, you must have noticed the hidden rule. It is to arrange the Σ so that it conforms to the definition of the general term i.e., if a_n is defined for say $n \geq 4$, then before replacing a_n by its definition, we must arrange Σ so that its starting point begins from n = 4. Any early replacement will result in wrong result.

Example 3: Solve the following recurrence equation using generating function.

$$a_n - 8a_{n-1} + 16a_{n-2} = 0 \quad for\ n \geq 4 \quad and\ a_2 = 16, a_3 = 80.$$

Solution: We will use binomial generating function to solve this problem. Let f(x) be the binomial generating function for the sequence **a** of which general term is given by the above recurrence equation. One thing to be noticed here is that the sequence term starts from n =2, so summation will be used accordingly. f(x) is given as:

$$f(x) = \sum_{n=2}^{\infty} a_n x^n = a_2 x^2 + a_3 x^3 + \sum_{n=4}^{\infty} a_n x^n$$

$$or,\quad f(x) = a_2 x^2 + a_3 x^3 + \sum_{n=4}^{\infty} [8a_{n-1} - 16a_{n-2}]x^n$$

$$= a_2 x^2 + a_3 x^3 + 8\sum_{n=4}^{\infty} a_{n-1} x^n - 16\sum_{n=4}^{\infty} a_{n-2} x^n$$

$$= a_2 x^2 + a_3 x^3 + 8[a_3 x^4 + a_4 x^5 + \cdots] - 16[a_2 x^4 + a_3 x^5 + \cdots]$$

$$= a_2 x^2 + a_3 x^3 + 8[a_2 x^2 + a_3 x^3 + \cdots]x - 8a_2 x^3 - 16[a_2 x^2 + a_3 x^3 + \cdots]x^2$$

$$= a_2 x^2 + a_3 x^3 + 8xf(x) - 8a_2 x^3 - 16x^2 f(x)$$

$$or,\quad (1 - 8x + 16x^2)f(x) = a_2 x^2 + a_3 x^3 - 8a_2 x^3$$

Now, substituting the value of a_2 and a_3 in the above equation, we get

$$(1 - 8x + 16x^2)f(x) = 16x^2 + 80x^3 - 8 * 16x^3$$

$$or,\quad f(x) = \frac{16x^2 - 48x^3}{1 - 8x + 16x^2} = 16x^2 \frac{1 - 3x}{(1 - 4x)^2}$$

$$Therefore,\quad f(x) = 16x^2 \frac{1 - 3x}{(1 - 4x)^2}$$

To find the general term a_n, we will have to find the coefficient of x^n from f(x). Using Partial fraction method, we can write

$$\frac{1-3x}{(1-4x)^2} = \frac{A}{(1-4x)} + \frac{Bx}{(1-4x)^2}$$

Solving the above equation for A and B, we get

$A = 1, \quad B = 1$ *Therefore, the GF can be written as*

$$f(x) = 16x^2\left[\frac{1}{1-4x} + \frac{x}{(1-4x)^2}\right]$$

Hence, coefficient of x^n in f(x) is given as 16 *(coefficient of x^{n-2} in first term + coefficient of x^{n-3} in second term). This is equal to

$$16 * (4^{n-2} + (n-2)4^{n-3})$$

$$or, \quad a_n = 16 * 4^{n-2}\left[1 + \frac{n-2}{4}\right]$$

$$or, \quad a_n = 4^n\left[\frac{1}{2} + \frac{n}{4}\right] \qquad Ans.$$

Example 4: Solve the following recurrence equation using generating function.

$$c_n = 3c_{n-1} - 2c_{n-2} \quad for\ n \geq 3 \quad and\ c_1 = 5, c_2 = 3.$$

Solution: We will again use binomial generating function to solve this problem. Let f(x) be the binomial generating function for the sequence **c** of which general term is given by the above recurrence equation. Since the sequence term starts from n =1, so the summation will be used accordingly. f(x) is given by;

$$f(x) = \sum_{n=1}^{\infty} c_n x^n = c_1 x + c_2 x^2 + \sum_{n=3}^{\infty} c_n x^n$$

$$or,\quad f(x) = c_1x + c_2x^2 + \sum_{n=3}^{\infty}(3c_{n-1} - 2c_{n-2})x^n \qquad [by\ def^n\ of\ c_n]$$

$$= c_1x + c_2x^2 + 3\sum_{n=3}^{\infty}c_{n-1}x^n - 2\sum_{n=3}^{\infty}c_{n-2}x^n$$

$$= c_1x + c_2x^2 + 3[c_2x^3 + c_3x^4 + c_4x^5 + \cdots] - 2[c_1x^3 + c_2x^4 + c_3x^5 + \cdots]$$

$$or,\quad f(x) = c_1x + c_2x^2 + 3[c_1x + c_2x^2 + c_3x^3 + \cdots]x - 3c_1x^2$$
$$- 2[c_1x + c_2x^2 + c_3x^3 + \cdots]x^2$$

$$or,\quad f(x) = c_1x + c_2x^2 + 3xf(x) - 3c_1x^2 - 2x^2f(x)$$

$$or,\quad (1 - 3x + 2x^2)f(x) = c_1x + c_2x^2 - 3c_1x^2 = 5x + 3x^2 - 3*5x^2$$

$$Therefore,\quad f(x) = \frac{5x - 12x^2}{(1-x)(1-2x)} = x\frac{5 - 12x}{(1-x)(1-2x)}$$

To find the general term a_n, we shall have to find the coefficient of x^n from f(x). Using Partial fraction method, we can write

$$\frac{5 - 12x}{(1-x)(1-2x)} = \frac{A}{(1-x)} + \frac{B}{(1-2x)}$$

Solving the above equation for A and B, we get

$A = 7,\ B = -2$ *Therefore, the GF can be written as*

$$f(x) = x\left[\frac{7}{1-x} - \frac{2}{(1-2x)}\right]$$

Hence, coefficient of x^n in f(x) is given as (coefficient of x^{n-1} in first term + coefficient of x^{n-1} in second term). This is equal to

$$7 - 2*2^{n-1}$$

$$or,\quad a_n = 7 - 2^n \qquad Ans.$$

Now let us solve another problem by generating function method in which RHS is an exponential function. This problem has been solved using the binomial generating function.

Example 5: Solve the following recurrence equation using generating function.

$$a_n - 7a_{n-1} + 10a_{n-2} = 2^n \quad for\ n \ge 2 \quad and\ a_0 = 0, a_1 = 6.$$

Solution: Let f(x) be the binomial generating function for the sequence **a** of which general term is given by the above recurrence equation. Then, f(x) is given by

$$f(x) = \sum_{n=0}^{\infty} a_n x^n = a_0 + a_1 x + \sum_{n=2}^{\infty} a_n x^n$$

$$or,\ f(x) = a_0 + a_1 x + \sum_{n=2}^{\infty} (7a_{n-1} - 10a_{n-2} + 2^n) x^n \qquad [by\ def^n\ of\ a_n]$$

$$= a_0 + a_1 x + 7 \sum_{n=2}^{\infty} a_{n-1} x^n - 10 \sum_{n=2}^{\infty} a_{n-2} x^n + \sum_{n=2}^{\infty} 2^n x^n$$

$$= a_0 + a_1 x + 7x \sum_{n=1}^{\infty} a_{n-1} x^{n-1} - 10x^2 \sum_{n=2}^{\infty} a_{n-2} x^{n-2} + \sum_{n=2}^{\infty} 2^n x^n$$

$$or,\ f(x) = a_0 + a_1 x + 7x f(x) - 10x^2 f(x) + \frac{4x^2}{1-2x} \qquad \left[\because \sum_{n=2}^{\infty} 2^n x^n = \frac{4x^2}{1-2x}\right]$$

$$or,\ (1 - 7x + 10x^2) f(x) = a_0 + a_1 x + \frac{4x^2}{1-2x} = 6x + \frac{4x^2}{1-2x}$$

$$Therefore,\ f(x) = \frac{6x - 8x^2}{(1-2x)^2(1-5x)} = x\frac{6-8x}{(1-2x)^2(1-5x)}$$

As in the previous examples, we will have to find the coefficient of x^n from f(x). Using Partial fraction method, we can write

$$\frac{6x-8x^2}{(1-2x)^2(1-5x)}=\frac{A}{(1-5x)}+\frac{B}{(1-2x)}+\frac{Cx}{(1-2x)^2}$$

Solving the above equation for A, B and C, we get

$A=\frac{22}{9},\quad B=-\frac{22}{9},\quad$ *and* $C=-\frac{4}{3}$ *Therefore, the GF can be written as* :

$$f(x)=\frac{22}{9}\left[\frac{1}{1-5x}\right]-\frac{22}{9}\left[\frac{1}{1-2x}\right]-\frac{4}{3}\left[\frac{x}{(1-2x)^2}\right]$$

Hence, coefficient of x^n in f(x) is given as (coefficient of x^n in first term + coefficient of x^n in second term + coefficient of x^n in third term). This is equal to

$$\frac{22}{9}5^n-\frac{22}{9}2^n-\frac{4}{3}n2^{n-1}$$

$$or,\quad a_n=\frac{22}{9}\left[5^n-2^n\right]-\frac{2}{3}n2^n \qquad Ans.$$

A given recurrence equation can also be solved by using exponential generating function. If the problem is related to combination, we always use binomial generating function, whereas, if a problem is related to permutation, we use exponential generating function. The following problem is not solvable by using binomial generating function as it is related to permutation of numbers. Try to solve the following using binomial GF.

Example *6*: Solve the following recurrence equation using exponential generating function.

$$d_n=(n-1)(d_{n-1}+d_{n-2})\quad for\ n\geq 3\quad and\ d_1=0, d_2=1.$$

Solution: Let f(x) be the exponential generating function of the above recurrence relation, then f(x) is given as

$$f(x)=\sum_{n=0}^{\infty} d_n\frac{x^n}{n!} \qquad \text{\_\_\_\_\_\_\_\_\_\_(1)}$$

Differentiating (1) with respect to x, we get,

$$f'(x) = \sum_{n=0}^{\infty} d_n \frac{x^{n-1}}{(n-1)!} \qquad \text{\_\_\_\_\_\_\_\_\_\_}(2)$$

Replacing d_n by its definition, in the equation (2), we get

$$f'(x) = \sum_{n=0}^{\infty} (n-1)(d_{n-1} + d_{n-2}) \frac{x^{n-1}}{(n-1)!}$$

$$or, \quad = \sum_{n=0}^{\infty} (n-1)d_{n-1} \frac{x^{n-1}}{(n-1)!} + \sum_{n=0}^{\infty} (n-1)d_{n-2} \frac{x^{n-1}}{(n-1)!}$$

$$or, \quad f'(x) = \sum_{n=0}^{\infty} d_{n-1} \frac{x^{n-1}}{(n-2)!} + \sum_{n=0}^{\infty} d_{n-2} \frac{x^{n-1}}{(n-2)!}$$

$$= x \sum_{n=0}^{\infty} d_{n-1} \frac{x^{n-2}}{(n-2)!} + x \sum_{n=0}^{\infty} d_{n-2} \frac{x^{n-2}}{(n-2)!}$$

$$= x\left[d_0 . \frac{x^{-1}}{(-1)!} + d_1 . \frac{x^0}{(0)!} + d_2 . \frac{x^1}{1!} + d_3 . \frac{x^2}{2!} + \cdots\right]$$

$$+ x\left[d_0 + d_1 . x + d_2 . \frac{x^2}{2!} + d_3 . \frac{x^3}{3!} + \cdots\right]$$

$$or, \quad f'(x) = xf'(x) + xf(x)$$

$$or, \quad (1-x)f'(x) = xf(x) \Rightarrow \frac{f'(x)}{f(x)} = \frac{x}{1-x} \Rightarrow \frac{f'(x)}{f(x)} = \frac{1}{1-x} - 1$$

$$or, \quad \frac{f'(x)}{f(x)} = \frac{1}{1-x} - 1 \qquad \text{\_\_\_\_\_\_\_\_\_\_}(3)$$

Integrating both sides of equation (3) we get,

$$\log f(x) = -\log(1-x) - x$$

or, $\log f(x).(1-x) = -x$

or, $f(x).(1-x) = e^{-x}$

or, $f(x) = \frac{1}{1-x}e^{-x}$ *This is the generating function for the given recurrence equation.*

To find the general term from this generating function we will have to find the coefficient of $x^n/n!$. Which is given as

$$d_n = n!\left(1 - \frac{1}{1!} + \frac{1}{2!} - \cdots\cdots(-1)^n \frac{1}{n!}\right) \qquad Ans.$$

Readers are suggested to give emphasis on understanding the technique to solve a recurrence equation rather than simply following the examples given in the text. If there is only one method to solve a particular problem, one has no option but to follow it. This has the advantage of skipping the selection of method process. However, when there are multiple methods to solve the same problem, selection of choice always gives an edge to those who selects better method. A better method for one problem may be equally bad for the other. The selection of choice is perfected by experience.

Exercise

1 Solve the following recurrence equations:

(a) $a_r - 7a_{r-1} + 10a_{r-2} = 0$ *with* $a_0 = 0$ *and* $a_1 = 3$.

(b) $a_r - 4a_{r-1} + 4a_{r-2} = 0$ *with* $a_0 = 1$ *and* $a_1 = 6$.

(c) $a_r - 7a_{r-1} + 10a_{r-2} = 3^r$ *with* $a_0 = 0$ *and* $a_1 = 1$.

(d) $a_r + 6a_{r-1} + 9a_{r-2} = 3$ *with* $a_0 = 0$ *and* $a_1 = 1$.

(e) $a_r + a_{r-1} + a_{r-2} = 0$ *with* $a_0 = 0$ *and* $a_1 = 2$.

(f) $a_r - a_{r-1} - a_{r-2} = 0$ *with* $a_0 = 1$ *and* $a_1 = 1$.

(g) $a_r - 2a_{r-1} + 2a_{r-2} - a_{r-3} = 0$ *with* $a_0 = 2$, $a_1 = 1$ *and* $a_2 = 1$.

2 Given that $a_0 = 0$, $a_1 = 1$, $a_2 = 4$, and $a_3 = 12$, satisfy the recurrence equation

$$a_r + C_1 a_{r-1} + C_2 a_{r-2} = 0 \quad and$$

determine a_r.

In Exercises 3 through 8, identify whether the given recurrence equation is linear homogeneous or not. If the equation is a linear homogeneous, give its degree.

3. $a_n = 2.5 * a_{n-1}$
4. $b_n = -3b_{n-1} - 2b_{n-2}$
5. $c_n = 2^n * c_{n-1}$
6. $d_n = nd_{n-1}$
7. $e_n = 5e_{n-1} + 3$
8. $g_n = \sqrt{g_{n-1} + g_{n-1}}$

In Exercises 9 through 14, use the backtracking method to find an explicit formula for the sequence defined by the recurrence equation and initial condition(s).

9. $a_n = 2.5 * a_{n-1} \quad with \quad a_1 = 4$
10. $b_n = b_{n-1} - 2 \quad with \quad b_1 = 0$
11. $c_n = c_{n-1} + n \quad with \quad c_1 = 4$
12. $d_n = -1.1 * d_{n-1} \quad with \quad d_1 = 5$
13. $e_n = 5e_{n-1} + 3 \quad with \quad e_1 = 2$
14. $g_n = ng_{n-1} \quad with \quad g_1 = 6$

In Exercises 15 through 20, solve each of the recurrence equations.

15. $a_n = 4a_{n-1} + 5a_{n-2} \quad with \quad a_1 = 2 \quad and \quad a_2 = 6.$
16. $b_n = -3b_{n-1} - 2b_{n-2} \quad with \quad b_1 = -2 \quad and \quad b_2 = 4$
17. $c_n = -6c_{n-1} - 9c_{n-2} \quad with \quad c_1 = 2.5 \quad and \quad c_2 = 4.7$
18. $d_n = 4d_{n-1} - 4d_{n-2} \quad with \quad d_1 = 1 \quad and \quad d_2 = 7$
19. $e_n = 2e_{n-2} \quad with \quad e_1 = \sqrt{2} \quad and \quad e_2 = 6$
20. $g_n = 2g_{n-1} - 2g_{n-2} \quad with \quad g_1 = 1 \quad and \quad g_2 = 4$
21. Develop a general explicit formula for a non-homogeneous recurrence equation of the form

$$a_n = ra_{n-1} + s$$

where, r and s are constants.

22. Find the closed formula for the Fibonacci sequence

$$f_n = f_{n-1} + f_{n-2}$$

23. The solution of the recurrence relation $C_0 a_r + C_1 a_{r-1} + C_2 a_{r-2} = f(r)$ is

$$3^r + 4^r + 2$$

If f(r) = 6 for all r, determine C_0, C_1, and C_2.

24. Solve the difference equation

$$a_r - ra_{r-1} = r! \quad for\ r \geq 1 \quad and \quad given \quad that\ a_0 = 2$$

[Hint : Use change of variable method. Let $b_r = a_r/r!$]

25. Solve the difference equation

$$a_r^2 - 2a_{r-1}^2 = 1 \quad for\ r \geq 1 \quad and \quad given \quad that\ a_0 = 2$$

[Hint: Use change of variable method. Let $b_r = a_r^2$]

26. Solve the difference equation

$$ra_r + ra_{r-1} - a_{r-1} = 2^r \quad for\ r \geq 1 \quad and \quad given \quad that\ a_0 = 273$$

[Hint: Use change of variable method. Let $b_r = ra_r$]

27. Let d_r denote the number of ways of permuting r integers {1,2,3,4,...,r} so that the integer i will not be in the ith position for $1 \leq i \leq r$.

(a) Use combinatorial argument to show that

$$d_r = (r-1)(d_{r-1} + d_{r-2})$$

(b) Show that

$$d_r = r!\left[1 - \frac{1}{1!} + \frac{1}{2!} - \frac{1}{3!} + \cdots + (-1)^r \frac{1}{r!}\right]$$

satisfies the recurrence equation (a).

28. The solution of the recurrence equation

$$a_r = Aa_{r-1} + B3^r \quad for \quad r \geq 1$$

is

$$a_r = C2^r + D3^{r+1} \quad for \quad r \geq 0$$

Given that $a_0 = 19$ and $a_1 = 50$, determine the constants A, B, C, and D.

29. Consider the multiplication of bacteria in a controlled environment. Let a_r denote the number of bacteria there are on r^{th} day. If $a_r - 2a_{r-1}$ be the rate of growth of rth day and that the rate of growth doubles every day, determine a_r given that $a_0 = 1$.

30. How many r-bits long binary sequences are having no adjacent 0's?

31. Let a_r denote the total assets of a bank at the end of the rth month. This is equal to the sum of the total deposit in the rth month and 1.1 times the total assets at the end of the previous month. Given that the total deposit is a constant 100 (in thousand Rs.), determine a_r if $a_0 = 0$.

32. n circular discs are slipped onto a peg with the largest disc at the bottom. These rings are to be transferred one at a time onto another peg, and there is a third peg available on which discs can be left temporarily. During the course of transfer, no disc can be placed on top of the smaller one. Find how many moves are required to transfer all n discs from one peg to other.

33. A particle is moving in the horizontal direction. The distance it travels in each second is equal to twice of the distance it traveled in previous second. Let a_n denote the position of the particle in nth second, determine a_n given that $a_0 = 3$ and $a_3 = 10$.

34. Consider a sorting algorithm for an array containing $n \geq 2$ numbers.
(i) Use $2n - 3$ comparisons to determine the largest and the second largest of the n numbers.
(ii) Recursively, sort the remaining $n - 2$ numbers.
Let a_n denotes the number of comparisons used for sorting n numbers, determine a_n.

35. Let a_n denotes the number of partitions of a set of n elements. Show that

$$a_{n+1} = \sum_{i=0}^{n} {}^nC_i a_i$$

where $a_0 = 1$.

36. Let a_n denotes the total assets of a company in Rupees in the nth year. Clearly, $a_n - a_{n-1}$ is the increase in assets during nth year. If the increase in assets during each year is always **five** times the increase during the previous year, what are the total assets in the nth year? Given that $a_0 = 3$ and $a_1 = 7$.

37. Let a_n be the number of subsets of the set {1,2,3,….,n} that do not contain two consecutive numbers. Determine a_n.

38. Consider the operation of a factory whose average profit in every two successive months is equal to the average new order in that period. Let a_n denotes the new order received and b_n denotes the monthly profit in the nth month. Determine b_n when $a_n = 2^n$ for all $n \geq 0$ and $b_0 = 0$.
Determine b_n when $b_0 = 0$ and

$$a_n = \begin{cases} 2^n & 0 \leq n \leq 9 \\ 2^{10} & n \geq 10 \end{cases}$$

39. Determine the particular solution for the difference equations
$(a)\ \ a_n - 3a_{n-1} + 2a_{n-2} = 2^n$ and $(b)\ \ a_n - 4a_{n-1} + 4a_{n-2} = 2^n$

40. Determine the particular solution for the difference equations

$(a)\ \ a_n - 2a_{n-1} = 7n^2$ and $(b)\ \ a_n - 2a_{n-1} = 7n$

Solve the following recurrence equations 41 through 50 using generating function method.

41. $A_n = A_{n-1} + A_{n-2} + 2n$ given that $A_0 = 0$ and $A_1 = 1$

42. $a_n + a_{n-1} = 3n2^n$ given that $a_0 = 0$

43. $a_n = 5a_{n-1} - 3$ given that $a_0 = 1$

44. $h_n = 4h_{n-2}$ given that $h_0 = 0$ and $h_1 = 1$

45. $a_n = \begin{cases} 0 & \text{for } n = 0 \\ 2a_{n-1} + 3 & \text{for } n > 0 \end{cases}$

46. $a_n = \begin{cases} 16 & \text{for } n = 2 \\ 60 & \text{for } n = 3 \\ 8a_{n-1} - 16a_{n-2} & \text{for } n > 3 \end{cases}$

47. $a_n - 2a_{n-1} - 4a_{n-2} = 4^n$ given that $a_0 = 1$ and $a_1 = 7$

48. $a_n - 5a_{n-1} + 6a_{n-2} = 2^n + n$ given that $a_1 = 0$ and $a_2 = 10$

49. $3a_{n+2} - 8a_{n+1} - 3a_n = 3^n - 2n + 1$

50. $a_{n+2} - 6a_{n+1} + 8a_n = 3n^2 + 2 - 5 * 3^n$

51. Solve the following recurrence equations.

(i) $a_n = 2a_{\frac{n}{2}}$ with $a_1 = 1$

(ii) $a_n = 2a_{\frac{n}{2}} + n$ with $a_1 = 1$

(iii) $a_n = 2a_{\left\lfloor \frac{n}{2} \right\rfloor} + 3$ with $a_0 = 0$

CHAPTER FOUR
Combinatorics

Proofs are the heart of mathematics. If you are studying mathematics either as a major or as a subject to be used extensively in other major fields of study, then you must come to the terms with proofs. You must be able to read, understand and write them (proofs). **What is the secret**? What magic do you need to know? The straight answer is **there is no secret, no mystery and no magic**. All that is needed is some common sense and a basic understanding of a few trusted and easy to understand techniques.

The **permutation** and **combination,** as a method, has been used to prove many counting laws and to solve many counting problems. We shall discuss the basic concepts of permutation and combination and use of **generating functions** to solve problems related to it in this chapter. Whenever and wherever we discuss the topics of **combinatorics** the list of solution techniques seems to be incomplete without discussing the **pigeonhole principle**. We will learn about this principle and its application in this chapter.

4.1 Methods of proofs

The **basic structure** of a proof is easy. It is just a series of statements. Each such statement being either

- **An assumption** or
- **A conclusion, clearly following from an assumption or previously proved result**.

And that is all. Sometimes there will be some clarifying remarks, but that is for the reader and has no logical bearing on the structure of the proof. A well-written proof will flow. That is, the reader should feel as though they are being taken on a ride that takes them directly and inevitably to the desired conclusion without any distraction about irrelevant details. Each step should be clear or at least clearly justified. A good proof is easy to understand. When you are finished with the proof, apply the simple test listed below to every step (sentence): is it

- clearly an assumption, or
- a justified conclusion?

If the sentence fails the test, may be it is not needed in the proof. In this section, we will discuss the following techniques one by one.

- **Direct proofs**

- **Proof by contradiction**
- **Proof by contra-positive**
- **If, and Only If**
- **Mathematical Induction**
- **Counter Examples**
- **Proof by Exhaustion**
- **Constructive Verses Existential Proofs**

Direct proofs

If a and b are two natural numbers, we say that **a divides b** if ∃ a natural number **k** such that **b = a*k**. For example, 3 divides 21 because ∃ a natural number k (= 7) such that 21 = 3 k. Now let us proof the following theorem.

Theorem 1 If a divides b and b divides c then prove that a divides c.

Proof: By our assumption, and the definition of divisibility, ∃ natural numbers k_1 and k_2 such that **b** = **a** k_1 and **c** = **b** k_2. Consequently, **c** = **b** k_2 = **a** k_1 k_2. Let **k** = k_1 k_2. Since **k** is a natural number such that **c** = **a** k, so by the definition of divisibility a divides c.

Proved.

Most theorems that we want to prove are either explicitly or implicitly in the form **"If P, Then Q"**. In the theorem proved above, **P** is "If a divides b and b divides c" and **Q** is "a divides c". **This is the standard form of a theorem.** A **direct proof** should be thought of as a flow of implications beginning with "**P**" and ending with "**Q**" i.e., **P** → ... → **Q.**

Most of the proofs are direct proofs. Always try direct proofs first, while proving any theorems or statements, unless you have a good reason not to do so. If you find a simple proof, and you are convinced of its correctness, then do not feel shy about. Many times proofs are simple and short. See the theorem below.

Theorem 2 Every odd integer is the difference of two perfect squares.

Proof: Let 2n + 1 is an odd integer for any n ∈ I. Then

$$2n+1=(n+1)^2-n^2$$

Proved.

Wow! What is this? This is the proof and that too of one line.

Example 1 Prove that the number 1000...001 (with 3n –1 zeroes for n > 0) is a composite number.

Solution: We can rewrite the given number as

$$1000\cdots001 = 10^{3n}+1 \quad where\ n>0 \quad and\ n\in N$$
$$= (10^n)^3+1$$
$$= (10^n+1)(10^{2n}-10^n+1)$$

In the above expression, both the factors are integers and are greater than 1. Thus the given number is a composite number.

Example 2 If r_1 and r_2 are two distinct roots of the polynomial $p(x) = x^2 + bx + c$, then show that $r_1 r_2 = -b$ and $r_1 + r_2 = c$.

Solution: If r_1 and r_2 are two distinct roots of the polynomial p(x), then we can write $p(x) = (x - r_1)(x - r_2)$. Upon expansion of the right hand side, we get

$$p(x) = x^2 - (r_1 + r_2)x + r_1 r_2 \qquad \text{\_\_\_\_\_\_\_\_(1)}$$

We know that two polynomials are said to be equal if the corresponding coefficients are equal. Comparing the coefficients of the given polynomial with that of equation (1), we get

$$r_1 + r_2 = -b \quad \text{and } r_1 r_2 = c$$

Proved.

Proof by contradiction

In a proof by contradiction we assume, along with the hypothesis, the **logical negation** of the result to be proved. Then we reach some kind of contradiction. That is, if we want to prove "If **P** then **Q**", we assume **P** and Not **Q**. The contradiction we arrive at could be some conclusion contradicting one of our assumptions, or something obviously untrue like 6 = 2. Proof by contradiction is often used to prove the impossibility of something. You assume it is possible and arrive at some contradiction. Read the proof of the irrationality of the square root of 2 in the following theorem.

Theorem 3 Prove that the square root of 2 is an irrational number.

Proof:

$$Let \ \sqrt{2} = s$$
$$So \ s^2 = 2 \qquad \text{\_\_\_\_\_\_\_\_(1)}$$

If s were a rational number, we can write

$$s = \frac{p}{q} \qquad \text{\_\_\_\_\_\_\_\_\_\_\_\_\_\_\_\_(2)}$$

Where p is an integer and q is any positive integer. We can further assure that the **gcd**(p, q) =1 by driving out common factor between p and q, if any. Now using equation (2), in equation (1), we get

$$p^2 = 2q^2 \qquad \text{\_\_\_\_\_\_\_\_\_\_\_\_\_\_\_\_(3)}$$

Conclusion: From equation (3), **2** must be a prime factor of $\mathbf{p^2}$. Since **2** itself is a prime, 2 must be a factor of **p** itself. Therefore, 2*2 is a factor of $\mathbf{p^2}$. This implies that 2*2 is a factor of $2*\mathbf{q^2}$.

$\Rightarrow$ 2 is a factor of $\mathbf{q^2}$
$\Rightarrow$ 2 is a factor of **q**

Since, 2 is factor of both **p** and **q** **gcd**(p, q) = 2. This is contrary to our assumption that **gcd**(p, q) = 1. Therefore, √2 is an irrational number.

Proved.

Many such theorems are proved using the method of contradictions. One of the first proofs by contradiction is the following theorem attributed to Euclid.

Theorem 4 Prove that there are infinite many prime numbers.

Proof: Assume to the contrary that there are only finite number of prime numbers, and all of then can be listed as

$$p_1, p_2, p_3, \cdots\cdots, p_n$$

Now, consider the number

$$q = p_1 p_2 p_3 \cdots\cdots p_n + 1$$

This number is not divisible by any of the listed primes since if we divide q by p_r for each r = 1, 2, 3, …,n, we get remainder as 1. Well then, we must conclude that q has a prime factor different from the primes listed above. This is contrary to the assumption that all primes are in the list

$$p_1, p_2, p_3, \cdots\cdots, p_n$$

Hence, there are infinite many prime numbers.

Proved.

Example 3 Prove that there are no rational number solution to the equation

$$x^3 + x + 1 = 0$$

Solution: There is formula for solving the general cubic equation

$$ax^3 + bx^2 + cx + d = 0$$

But, in this example, we wish to prove there is no rational solution (roots) to the given cubic equation without using the general cubic formula. Let us assume that there is a rational number **p/q**, in reduced form, with **p** $\neq 0$ that satisfies the equation. Then, we have

$$\frac{p^3}{q^3} + \frac{p}{q} + 1 = 0 \qquad or, \quad p^3 + pq^2 + q^3 = 0$$

There are, now, three cases to consider. *Case 1:* If both p and q are odd, then the LHS of the above equation is odd. *Case 2:* If p is odd and q is even, then also the LHS is odd. *Case 3:* If p is even and q is odd, then again LHS is odd. The fourth case, p and q both being even does not arise because of our assumption that **p/q** is in reduced form. Since 0 is not an odd number, we arrive at a contradiction that **0** is an odd number. This contradiction is because of our assumption that the given equation has a rational root. Hence there are no rational solution to given cubic equation.

Proved.

Proof by contra-positive

Proof by contra-positive takes advantage of the logical equivalence between "**P implies Q**" and "**Not Q implies Not P**". For example, the assertion "If it is my car, then it is red" is equivalent to "If that car is not red, then it is not mine". So, to prove "If P, then Q" by the method of contra-positive means to prove "If Not Q, then Not P". At first it seems that both contra-positive method and contradiction method are similar, but there is subtle difference between them.

- Contradiction : Assume P and Not Q and prove some sort of contradiction
- Contra-positive: Assume not Q and prove Not P.

Read the proof of the following theorem to understand the contra-positive method.

Theorem 5 If n is a positive integer such that $n \equiv_4 2$ or $n \equiv_4 3$, then n is not a perfect square. Prove it.

Proof: We will prove the contra-positive version of the given statement. That is if n is a perfect square, then n mod (4) is neither 2 nor 3, i.e., if n is a perfect square, then n mod (4) is either 0 or 1. If n is a perfect square, then $\exists$ an integer k such that $n = k^2$. There are four possible cases to consider, corresponding to four possible remainders 0, 1, 2 and 3 when, k is divided by 4.

Case 1: If k mod (4) = 0, then k =4q, for some integer q. Then,

$$n = k^2 = 16q^2 = 4(4q^2) \quad \Rightarrow \quad n \bmod (4) = 0$$

Case 2: If k mod (4) = 1, then k =4q+1, for some integer q. Then,

$$n = k^2 = 16q^2 + 8q + 1 = 4(4q^2 + 2q) + 1 \quad \Rightarrow \quad n \bmod (4) = 1$$

Case 3: If k mod (4) = 2, then k =4q+2, for some integer q. Then,

$$n = k^2 = 16q^2 + 16q + 4 = 4(4q^2 + 4q + 1) \quad \Rightarrow \quad n \bmod (4) = 0$$

Case 4: If k mod (4) = 3, then k =4q+3, for some integer q. Then,

$$n = k^2 = 16q^2 + 24q + 9 = 4(4q^2 + 6q + 2) + 1 \quad \Rightarrow \quad n \bmod (4) = 1$$

Hence, it is proved that if n is perfect square, then n mod(4) cannot be equal to 2 or 3. That is if n mod(4) is 2 or 3, then n is not a perfect square.

Proved.

If, and Only If

Many theorems are stated in the form "P, if, and only if Q". Another way to say the same thing is : "Q is necessary, and sufficient for P". This means two things: one "If P, Then Q" and "If Q, Then P". So to prove an "If, and Only If" theorem, you must prove two implications. See the following example.

Example 4 Prove that if **a** is an integer, then **a** is not evenly divisible by 3 if, and only if, $\mathbf{a^2-1}$ is evenly divisible by 3.

Solution: Since this an "if, and only if" problem, we must prove two implications.
("If") We must prove that "a is not evenly divisible by 3 if $a^2 - 1$ is evenly divisible by 3". So we assume that 3 evenly divides

$$a^2 - 1 = (a - 1)(a + 1)$$

Since 3 is a prime number therefore, 3 must evenly divide either **a – 1** or **a + 1.** In either case, it should be apparent that 3 cannot evenly divide **a.**

("Only If") We must prove "**a** is not evenly divisible by 3 only if $a^2 - 1$ is evenly divisible by 3". This means "If **a** is not evenly divisible by 3, then $a^2 - 1$ is evenly divisible by 3". We can write **a = 3q + r**, where r = 0, 1 or 2. Our assumption that **a** is not divisible by 3 implies that r cannot be 0. If r = 1, then a – 1 = 3q and so 3 evenly divides $a^2 - 1 = (a-1)(a+1)$. Similarly if r = 2, then a+1 = 3q + 3 therefore 3 evenly divides $a^2 - 1$.

Proved.

Example 5 A positive integer n is evenly divisible by 3 if, and only if, the sum of the digits of n is divisible by 3.

Solution: Let **n** is a positive integer whose digit representation is $_0 a_1 a_2 \ldots a_k$. That means

$$n = a_0 + 10a_1 + 10^2 a_2 + \cdots\cdots + 10^k a_k \quad \textit{and digit sum is}$$

$$s = a_0 + a_1 + a_2 + \cdots\cdots + a_k$$

Now, we have

$$n - s = a_0 + 10a_1 + 10^2 a_2 + \cdots\cdots + 10^k a_k - (a_0 + a_1 + a_2 + \cdots\cdots + a_k)$$

$$= 9a_1 + 99a_2 + \cdots\cdots + (999\cdots 9)a_k$$

Where the last term has k 9's. So, clearly n – s is divisible by 3. It follows that n is divisible by 3 if, and only if, s is divisible by 3.

Proved.

Mathematical Induction

Suppose we have to prove a theorem or statement of the form "For all integers n greater than equal to **k, P(n)** is true". **P(n)** must be an assertion that we wish to be true for all **n = k, k+1, k+2, ...;** like a mathematical formula. We first verify the **initial step**. That is we must verify that **P(k)** is true. Next comes the **inductive step**. Here we must prove "if there is a **m**, greater than or equal to **k**, for which **P(k)** is true, then for this same **k, P(k+1)** is true".

Since we have verified **P(k)**, it follows from inductive step that **P(k+1)** is true, and hence **P(k+2)** is true, and hence **P(k+3)** is true, and so on. In this way the theorem is proved using mathematical induction. See the following examples to understand the working procedure of mathematical induction in proving any formula (theorem or statement).

Example 6: For any positive integer n,

$$1+2+3+\cdots\cdots+n=\frac{n(n+1)}{2}$$

Prove this using mathematical induction method.

Solution: Let the given formula be **P(n)**. Then we have to prove that **P(n)** is true for all positive integer n i.e., for n = 1, 2, 3, ...m,...; the prove consists of two steps: the **initial step** and **inductive step.**

Initial Step: For n =1, we have P(1) = 1(1+1)/2, which is clearly true. Thus the **P(1)** is verified.

Inductive Step: Here we must prove the assertion "if there is a **m** such that **P(m)** is true, then for this same **m, P(m+1)** is true". Thus, we assume that there is a **m** such that

$$1+2+3+\cdots\cdots+m=\frac{m(m+1)}{2}$$

We must prove, for this same **m,** the formula

$$1+2+3+\cdots\cdots+m+(m+1)=\frac{(m+1)(m+2)}{2}$$

We have

$$1+2+3+\cdots\cdots+m=\frac{m(m+1)}{2}$$ [*assumption that* **P(m)** *is true*]

Adding $(m+1)$ *to the both sides of the above equation, we get*

$$1+2+3+\cdots\cdots+m+(m+1)=\frac{m(m+1)}{2}+(m+1)=\frac{m^2+m+2m+2}{2}$$
$$=\frac{(m+1)(m+2)}{2}$$

i.e., $\mathbf{P(m+1)}=\dfrac{(m+1)((m+1)+1)}{2}$ *is true.*

Therefore, by induction **P(n)** is true for all positive integers.

Proved.

***Example** 7* Prove by mathematical induction that for all positive n ∈ N

$$3^{2^n} - 1 \textit{ is not divisible by } 2^{n+3}$$

Solution: Let the given statement be **P(n)**.
Initial Step: Let n =1, then, we have

$$3^{2^n} - 1 = 3^2 - 1 = 8 \quad \textit{and}$$

$$2^{n+3} = 2^{1+3} = 16$$

Since 8 is not divisible by 16, the statement is true for n =1 i.e., **P(1)** is true.

Inductive Step: Let the given statement is true for n = k i.e., **P(k)** is true. Then we have to prove that for the same **k, P(k+1)** is also true. Now,

$$3^{2^{k+1}} - 1 = 3^{2^k \cdot 2} - 1 = \left(3^{2^k}\right)^2 - 1$$

$$= \left(3^{2^k} - 1\right)\left(3^{2^k} + 1\right)$$

In the above expression, the RHS has two factors. The first factor is **P(k)**, which is assumed to be true. Further, the two factors are two consecutive even numbers. Therefore, when divided by **2,** one of the factors becomes an odd number, and thereafter this factor cannot be further divided by any power of **2.** Since **P(k)** is true, we have

$$3^{2^k} - 1 \textit{ is not divisible by } 2^{k+3}$$

$$\textit{and also any odd multiple of } 3^{2^k} - 1 \textit{ is not divisible by } 2^{k+3}.$$

Now, if we prove that the second factor is twice of an odd number then the task will be over.

$$3^{2^k} - 1 = (3^2)^{2^{k-1}} - 1$$

$$= (3^2 - 1)\left[(3^2)^{2^{k-1}-2} + (3^2)^{2^{k-1}-3} + \cdots\cdots + 1\right]$$

$$= 8\left[(3^2)^{2^{k-1}-2} + (3^2)^{2^{k-1}-3} + \cdots\cdots + 1\right]$$

Since the first factor has **8** as a factor, so it is not twice of an odd number. Hence, the second factor is twice of an odd number. Therefore,

$$3^{2^{k+1}} - 1 \quad \textit{is not divisible by } 2^{(k+1)+3}$$

This implies that **P(k+1)** is true. Therefore, by induction **P(n)** is true for all positive n $\in$ N.

Proved.

Example 8 Prove by mathematical induction that any integer composed of $\mathbf{3^n}$ identical digits is divisible by $\mathbf{3^n}$ for $n \geq 1$.

Solution: Let **P(n)** be the statement that that any integer composed of $\mathbf{3^n}$ identical digits is divisible by $\mathbf{3^n}$.

Initial Step: Let n =1. Then, the number consists of three identical digits. The sum of these digits is divisible by 3. Hence, the number is divisible by 3. This implies that **P(1)** is true.

Inductive Step: Let the given statement is true for n = k i.e., **P(k)** is true. Then we have to prove that for the same **k, P(k+1)** is also true. Before proceeding for proof, let us give a look on the following facts.

$$\left[\begin{array}{ll} 222 = 2(10^{2\cdot 3^0} + 10^{1\cdot 3^0} + 1) = 2\cdot(111) & \textit{for} \quad n = 1 \\ \| ry, \quad 222{,}222{,}222 = 222(10^{2\cdot 3^1} + 10^{1\cdot 3^1} + 1) = 222(1001001) & \textit{for} \quad n = 2 \\ \| ry,\ 222\cdots 27\ \textit{times} = 222{,}222{,}222(10^{2\cdot 3^2} + 10^{1\cdot 3^2} + 1) & \\ \qquad = 222(1000000001000000001) & \textit{for} \quad n = 3 \end{array}\right]$$

Let x is any integer consisting of 3^{n+1} identical digits. Then, x can be written as

$$\mathbf{x = y \times z}$$

Where y is an integer consisting of same digit 3^n times and z is of the form

$$z = (10^{2\cdot 3^k} + 10^{1\cdot 3^k} + 1)$$

We have assumed that y is divisible by 3^n. Since, z contains exactly 3 **ones** and remaining **zeroes,** z is divisible by 3. Therefore, the product **y×z** is divisible by 3^{n+1}. This implies that **P(k+1)** is also true. Therefore, by induction **P(n)** is true for all positive n ≥ 1.

Proved.

Example 9 Define a sequence **a** = $a_0, a_1, a_2, \ldots a_n, \ldots$ by the recurrence equation

$$a_{n+1} = 2a_n - a_n^{\,2}$$

Then a closed form formula for a_n is given by

$$a_n = 1 - (1 - a_0)^{2^n} \quad \textit{for } n = 0,1,2,3,\cdots$$

Prove this by mathematical induction.

Solution: Let the given statement be **P(n)**.

Initial Step: Let n = 0, then, we have from recurrence equation

$$a_1 = 2a_0 - a_0^{\,2} = 1 - (1 - a_0)^{2^1}$$

Therefore, **P(0)** is true.

Inductive Step: Let the given statement is true for n = k i.e., **P(k)** is true i.e., closed form formula for a_k is given as below.

$$a_k = 1 - (1 - a_0)^{2^k}$$

Then we have to prove that for the same **k, P(k+1)** is also true. Now from recurrence equation, we have

$$a_{k+1} = 2a_k - a_k^{\,2} \quad \textit{substituting the value of } a_k, \textit{ we get}$$

$$a_{k+1} = 2\left[1 - (1 - a_0)^{2^k}\right] - \left[1 - (1 - a_0)^{2^k}\right]^2$$

$$or, \quad a_{k+1} = \left[1 - (1 - a_0)^{2^k}\right]\left[2 - \left[1 - (1 - a_0)^{2^k}\right]\right]$$

$$or, \quad a_{k+1} = \left[1 - (1 - a_0)^{2^k}\right]\left[1 + (1 - a_0)^{2^k}\right] = \left[1 - \left\{(1 - a_0)^{2^k}\right\}^2\right]$$

$$or, \quad a_{k+1} = \left[1 - (1 - a_0)^{2^{k+1}}\right]$$

This implies that **P(k+1)** is also true. Therefore, by induction **P(n)** is true for all positive $n \geq 0$.

Proved.

Counter Examples

Counter Examples play an important role in mathematics. A complicated proof may be the only way to demonstrate the validity of a particular theorem, whereas a single counter example is enough to refute the validity of a proposed theorem. For example, numbers of the form $2^{2^n} + 1$, where n is positive integer, were once thought to be prime. Those numbers are prime for n =1, 2, 3 and 4. When n = 5, we get $2^{2^5} + 1 = 4294967297 = (641)(6700417)$, a composite number. Thus, we conclude that when faced with a number in the form $2^{2^n} + 1$, we are not to assume it either prime or composite, unless we know for sure for some other reason.

A natural place for counter examples to occur is when the converse of a known theorem comes into question. The **converse** of an assertion in the form **"If P, Then Q"** is **"If Q, then P"**.

Example 10 State the converse of "if a and b are even integers then a + b is an even integer". Show that converse is not true by giving a counter example.

Solution: The converse of the given statement is "if a + b is an integer then a & b are even". The counter example to disprove this hypothesis can be given by a = 3 and b = 5. Here, a + b = 8 is even but a and b both are odd.

Proved.

Example 11 In calculus we have learnt that if a function is differentiable at a point, then it is continuous at that point. What would the converse assert? It would state that if a function is continuous at a point, then it is differentiable at that point. But we know that this is not true. The counter example is $f(x) = |x|$. This function is continuous at x = 0, but it is not differentiable at x = 0. This counter example is all we need to disprove the converse of a theorem.

Proof by Exhaustion

Sometimes the most straightforward, if not the most elegant, way to construct a proof is by checking all possible cases and verifying the result in all such cases. This method of proof is called **proof by exhaustion** or **case by case proof.** Let us see the following example.

Example 12 Show that if n is a positive integer then $n^7 - n$ is divisible by 7.

Solution: The given expression can be written as

$$n^7 - n = n(n^6 - 1) = n(n^3 - 1)(n^3 + 1) = n(n-1)(n^2 + n + 1)(n+1)(n^2 - n + 1)$$

Now, when n is divided by 7, there are seven possible remainders depending on $n=7q + r$ where $r = 0, 1, 2, 3, 4, 5$ or 6. We will examine in all these cases to ensure that whichever positive integer n, we take, $n^7 - n$ is divisible by 7.

Case 1: When $r = 0$, i.e., $n = 7q$. Then $n^7 - n$ is divisible by 7 since n is a factor.

Case 2: When $r = 1$, i.e., $n = 7q+1$. Then $n - 1$ is divisible by 7. Since $(n-1)$ is a factor of $n^7 - n$, $n^7 - n$ is divisible by 7.

Case 3: When $r = 2$, i.e., $n = 7q+2$, then the factor $(n^2 + n + 1) = (7q+2)^2 + (7q+2) + 1 = 49q^2 + 35q + 7$ is clearly divisible by 7 and hence $n^7 - n$ is divisible by 7.

Case 4: When $r = 3$, i.e., $n = 7q+3$, then the factor $(n^2 - n + 1) = (7q+3)^2 - (7q+3) + 1 = 49q^2 + 35q + 7$ is clearly divisible by 7 and hence $n^7 - n$ is divisible by 7.

Case 5: When $r = 4$, i.e., $n = 7q+4$, then the factor $(n^2 + n + 1) = (7q+4)^2 + (7q+4) + 1 = 49q^2 + 63q + 21$ is clearly divisible by 7 and hence $n^7 - n$ is divisible by 7.

Case 6: When $r = 5$, i.e., $n = 7q+5$, then the factor $(n^2 - n + 1) = (7q+5)^2 - (7q+5) + 1 = 49q^2 + 63q + 21$ is clearly divisible by 7 and hence $n^7 - n$ is divisible by 7.

Case 7: When $r = 6$, i.e., $n + 1 = 7q + 7$ is clearly divisible by 7 and hence $n^7 - n$ is divisible by 7.

Therefore, we have proved that in all cases of n, $n^7 - n$ is divisible by 7.

Proved.

Constructive Verses Existential Proofs

How would you prove that $2^{99} + 1$ is a composite number? Certainly, we will try to exhibit a factorization: $2^{99} + 1 = (2^{33})^3 + 1 = (2^{33} + 1)(2^{66} + 2^{33} + 1)$. That is to say that we have constructed a factorization to show that $2^{99}+1$ is a composite number. We call such a method of proof as **proof by construction.**

Example 13 A **Pythagorean triple** is a triplet of positive integers (a, b, c) that satisfies the equation $a^2 + b^2 = c^2$. For example, (3, 4, 5) is a Pythagorean triple because $3^2 + 4^2 = 5^2$. Are there any more? Yes, there are infinitely many. The triplet (3m, 4m, 5m) for all positive integer m are Pythagorean triples. We call (3m, 4m, 5m) **one parameter** family of solutions. Here m is a parameter. *Similarly prove that there is a two parameters family of Pythagorean triples.*

Solution: Let us construct the solution. Let $a = m^2 - n^2$ and $b = 2\,m\,n$ where m and n are positive integers with m > n. Then $a^2 + b^2 = (m^2 - n^2)^2 + (2\,m\,n)^2 = (m^2 + n^2)^2$. Thus, $(m^2 - n^2,\ 2\,m\,n,\ m^2 + n^2)$, for m > n, is two parameters family of Pythagorean triples.

Proved.

Example 14 There is rational number that lies strictly between the square root of 10^{100} and the square root of $10^{100}+1$. Prove this using constructive proof method.

Solution: The square root of 10^{100} is 10^{50}. Let us assume that $x = 10^{50}+10^{-51}$. Clearly, x is a rational number greater than 10^{50}. Now we have to prove that x is less than the square root of $10^{100}+1$. Let us compute square of x, which is $(10^{50}+10^{-51})^2 = 10^{100}+(2).10^{-1}+10^{-102}$, which is clearly less than $10^{100}+1$.

Proved.

Sometimes it is possible to prove the existence of something mathematical without actually constructing it. The reason could be that you cannot think of a constructive proof or that a constructive proof is very long and tedious. In any case, existential proofs are another important method for proving a theorem. Let us see the following example.

Example 15 Prove that the polynomial $p(x) = x^3 + x - 1$ has exactly one real root.

Solution: We have to prove this in two parts: First there exist a real root and second there is not more than one real root to the given polynomial.

(**First Part:** Direct Existential proof.) We know from Intermediate value theorem that if p(x) changes sign in an interval [a, b] then there is a real number c in [a, b] such that p(c) = 0 i.e., p(x) has a real root in the interval [a, b]. Let a = 0 and b = 1, then p(a) = -1 < 0 and p(b) = 1 > 0, thus there exists a real number c in [0, 1] such that p(c) = 0. Hence, we have proved that there is a real root for the given polynomial p(x).

(Second Part: Only one real root) This we shall prove by contradiction method. Let us assume that there are more than one real root to the polynomial p(x). Let c_1 and c_2 be two real roots to the polynomial p(x), then $p(c_1) = p(c_2) = 0$. From Mean Value Theorem, there must exist a number c between c_1 and c_2 such that

$$p'(c) = \frac{p(c_2) - p(c_1)}{c_2 - c_1} = 0$$

But a direct calculation shows that $p'(x)=3x^2+1$, which can never be zero as $x^2 \geq 0$. Therefore, we have proved that there exists not more than one real roots.

Proved.

Theorem 6 If p is a prime number, then show that

$$(a+b)^p = a^p + b^p + multiple\ of\ p$$

Proof: Using binomial expansion, we can write

$$(a+b)^p = a^p + {}^pC_1 a^{p-1}b + {}^pC_2 a^{p-2}b^2 + {}^pC_3 a^{p-3}b^3 + \cdots\cdots + {}^pC_p b^p$$

$$or,\ (a+b)^p = a^p + {}^pC_1 a^{p-1}b + {}^pC_2 a^{p-2}b^2 + {}^pC_3 a^{p-3}b^3 + \cdots\cdots + {}^pC_{p-1} ab^{p-1} + b^p$$

$$or,\ (a+b)^p = a^p + b^p + \left\{ {}^pC_1 a^{p-1}b + {}^pC_2 a^{p-2}b^2 + {}^pC_3 a^{p-3}b^3 + \cdots\cdots + {}^pC_{p-1} ab^{p-1} \right\}$$

Here, every term inside {} has coefficient of the form ${}^pC_r$. It is also given that p is a prime number and $1 \leq r \leq p-1$. Now the general coefficient ${}^pC_r$ can be expanded as

$${}^pC_r = \frac{p(p-1)(p-2)\cdots(p-r+1)}{r!}$$

Since p is a prime number it is not divisible by any of the number from 2 to r. Further r is a number less than p, so p cannot divide r. This implies that p is factor of every term inside {} in the above expansion. Thus, we can write

$$(a+b)^p = a^p + b^p + multiple\ of\ p$$

Proved.

In general, if p is a prime number then, for $a_1, a_2, a_3, \ldots, a_n$, we can write

$$(a_1 + a_2 + \cdots + a_n)^p = a_1{}^p + a_2{}^p + \cdots + a_n{}^p + multiple\ of\ p$$

In particular, if $a_1 = a_2 = a_3 = \ldots = a_n = 1$ then, we have from above generalization $n^p = n +$ multiple of p i.e., $n^p - n$ is divisible by p.

Example 16 Show that $23^{47} - 23$ is divisible by 47.

Solution: Readers may apply the above theorem to prove this. This is left as an exercise to the reader.

4.2 Permutation

1 and 2 are two digits. The numbers 12 and 21 are formed with these two digits. Further, no other numbers can be formed with only these two digits. The only difference in 12 and 21 is the order of arrangement of digits 1 and 2. Next, let us suppose that we have three letters a, b, and c. Then, all possible arrangements of any two letters out of these three letters can be enumerated as: ab, ac, bc, ba, ca and cb. If we make an arrangement of all three letters out of these three, then we have abc, acb, bac, bca, cab and cba as possible arrangements. Each of the distinct order of arrangements of a given set of distinct objects, taking some or all of them at a time (with or without repetition), is called a **permutation** of the objects.

It is obvious from the above example that the total number of permutations of n distinct objects, taken r at a time ($r \leq n$), is equal to the total number of ways of placing n objects in r boxes. This is denoted as **P(n, r)** or $^{n}P_{r}$. This number is equal to n(n-1)(n-2)...(n-r+1). Since the first box can be filled by any of the n objects, so we have n choices. Once an object is selected, we are left with n-1 objects and r-1 empty boxes to fill. The second time we can select any one of the n–1 objects, so we have n–1 choices and so on. This is the direct result of the **multiplication principle of counting.**

Theorem 1 **(multiplication principle of counting)** Suppose that two tasks T_1 and T_2 are to be performed in sequence. If T_1 can be performed in n_1 ways, and for each of these n_1 ways, T_2 can be performed in n_2 ways, the sequence $T_1 T_2$ can be performed in n_1n_2 ways.

Proof: Each choice of a method of performing T_1 will result in a different way of performing the task sequence. There are n_1 ways to perform T_1, and for each of these we may choose n_2 ways of performing T_2. Therefore, there will be, in total, n_1n_2 ways to complete the tasks sequence $T_1 T_2$.

Proved.

This principle can be extended to the finite sequence of m > 2 tasks. Suppose that tasks $T_1, T_2, T_3, \ldots, T_m$ are to be performed in sequence. If T_1 can be performed in n_1 ways and for each of n_1 ways, T_2 can be performed in n_2 ways and for each of n_1n_2 ways of performing task sequence T_1T_2, T_3 can be performed in n_3 ways and so on, then the tasks sequence $T_1 T_2 T_3 \ldots T_m$ can be performed in $n_1n_2 n_3 \ldots n_m$ ways. Readers may recall **the**

principle of inclusion and exclusion, discussed in the chapter 1. This principle is also known as **addition principle of counting**. It is, sometimes, used in permutation of objects also. Let there are three flight services and four train services between two places **A** and **B**. A person can complete his (or her) journey between these two places in seven possible ways. Notice the use of **addition principle of counting**.

Example 1 A gentleman has 6 friends to invite. In how many ways can he send invitation cards to them, if he has three servants to carry the cards?

Solution: A card can be send to any one friend by any one of the three servants. Let us take the tasks of sending cards to six friends as T_1, T_2, T_3, T_4, T_5 and T_6. Each of the tasks can be completed in three distinct ways according to the number of servants to carry the cards. Thus, by the multiplication principle of counting the tasks $T_1T_2T_3T_4T_5T_6$ can be performed in $3\times3\times3\times3\times3\times3 = 729$ ways. **Ans.**

Example 2 A telegraph has 5 arms and each arm is capable of 4 distinct positions, including the position of rest. What is the total number of signals that can be made?

Solution: There are five arms say T_1, T_2, T_3, T_4 and T_5. Each arm can be in any one of the four positions. For each of the position of arm T_1, there are four possible positions for the second arm T_2, for each of the possible positions for T_1T_2, there are four possible positions for the third arm T_3 and so on. Thus, by the multiplication principle of counting the total possible positions for $T_1T_2T_3T_4T_5$ is $4\times4\times4\times4\times4= 1024$. Since each distinct position is a distinct signal, so total number of possible signals is 1024 including the signal (meaningless) corresponding to the situation when all arms are in rest position. Therefore, the total number of signals that can be generated is $1024 - 1 = 1023$. **Ans.**

Example 3 How many numbers of three digits can be formed with the digits 1, 2, 3, 4 and 5 if the digits in the same number are not repeated? How many such numbers are possible between 100 and 10000?

Solution: Here we have to find the number of permutations of 5 distinct objects (digits) taken 3 at a time. This is given by ${}^5\mathbf{P}_3 = 5\times4\times3 = 60$. **Ans.**

Second part: The numbers between 100 and 10000 are numbers of three digits and of four digits. The total number of three digits numbers, formed with the given digits, is calculated above and it is equal to 60. Similarly, the total number of four digits numbers, formed with 1, 2, 3, 4 and 5, is given by $^5P_4 = 5 \times 4 \times 3 \times 2 = 120$. Thus, the required number is 60 + 120 = 180.

Ans.

Notation ^nP_r

We know that ^nP_r is the number of permutation of n distinct objects taken r at a time, and this is equal to n(n-1)(n-2)...(n-r+1). Therefore,

$$^nP_r = \frac{n(n-1)(n-2)\cdots(n-r+1)*(n-r)!}{(n-r)!} = \frac{n!}{(n-r)!}$$

$$Therefore,\quad {}^nP_r = \frac{n!}{(n-r)!}$$

Meaning of 0! and 1/(-r)!

According to the definition of factorial, 0! is meaningless. But, permutation of 1 object taken 1 at a time is obviously equal to 1, i.e., $^1P_1 = 1$. Using the above formula, we have

$$^1P_1 = \frac{1!}{(1-1)!} \quad or, \quad 1 = \frac{1}{0!} \quad or, \quad 0! = 1$$

Thus, for consistency, we take 0! = 1. Similarly, in the definition of ^nP_r, if we put n = 0, we have the following result.

$$n(n-1)(n-2)\cdots(n-r+1) = \frac{n!}{(n-r)!}$$

$$\Rightarrow\ 0(0-1)(0-2)\cdots(0-r+1) = \frac{0!}{(0-r)!} \ \Rightarrow\ 0 = \frac{0!}{(-r)!}$$

In fact, factorial of a negative integer does not exist. However, we write 0 (zero) in place of 1/(-r)!, wherever 1/(-r)! occurs in a product.

Permutation of objects when all are not distinct

Suppose there are **n** objects in which **p** objects are of one type, **q** objects are of second type, **r** objects are of third type and rest all are distinct. Let us further assume that the required number of permutations in this case is **x.** If we replace **p** objects of the same type with **p** distinct objects, then we shall have total number of permutations equal to

x*p!. Similarly, if we replace all equal **q** objects and all equal **r** objects with the **q** numbers of distinct objects and **r** numbers of distinct objects respectively, then we shall have total number of permutations equal to **x*p!q!r!.** Further, all these replacement will make all the n objects distinct. We know that the number of permutations of n distinct objects, taken n at a time, is given by $^nP_n = n!$. Therefore, we have

$$x * p!q!r! = n! \Rightarrow x = \frac{n!}{p!q!r!}$$

Example 4 How many distinguishable permutations of the letters in the word BANANA are there?

Solution: The word 'BANANA' has 6 letters. All the letters are not distinct. Let us use subscript to distinguish them temporarily. Let the letters be B, A_1, N_1, A_2, N_2, A_3. Thus, the number of permutations are 6!=720. Some of the permutations are identical like A_1A_2 $A_3BN_1N_2$ and A_1A_2 $A_3BN_2N_1$ except the order in which the N's appear. This means that if we drop the subscripts, the total number of permutations will be 6!/2! = 360. Similarly, if we drop subscript with A's then total number of distinguishable permutations are 360/3! = 60. **Ans**.

In general, the above situation can be described as: "***The number of distinguishable permutations that can be formed from a collection of n objects, taken all n at a time, in which the first object appears k_1 times, the second object k_2 times, and so on, is given by***

$$\frac{n!}{k_1!k_2!k_3!\cdots k_r!}$$

Where $k_1 + k_2 + \ldots + k_r = n$".

Example 5 Find the number of positive integers greater than a million that can be formed with the digits 2, 3, 0, 3, 4, 2 and 3.

Solution: The numbers greater than a million must be of 7 digits. In the given set of digits, 2 appear twice, 3 appear thrice and all others are distinct. Thus, the total number of seven digit numbers that can be formed with given digits is

$$\frac{7!}{2!3!} = 420$$

The set of these 420 positive integers, include some numbers which begin with **0**. Clearly, these numbers are less than a million and they must not be counted in our answer. The number of such numbers is given by the permutations of 6 non-zero digits and is equal to

$$\frac{6!}{2!3!} = 60$$

Therefore, the number of positive integers greater than a million that can be formed with given digits is equal to 420 - 60 = 360. **Ans.**

Permutation when objects can be repeated

Suppose we are given five digits: 1, 2, 3, 4 and 5. A 3-digits number, that can be formed using these digits, may be 123 or 111 or 122 or 533 or 551 etc. Repetitions of digits are allowed here. To form a 3-digits number, we have to fill three places with digits. We have five digits. The first place can be filled in 5 ways. For each of these ways, the second place can be filled in 5 ways so we have 5×5 ways to fill first two places. Similarly, for each of these 25 ways, the third place can be filled in 5 ways, Thus we have $5 \times 5 \times 5 = 125$ ways to fill three places. This implies that we can form 125 only 3-digits numbers in this case. In general this result can be summarized as "*The number of permutations of n distinct objects, taken r at a time, when repetitions are allowed, is given by* n^r."

Example 6 There are 10 stalls for animals in an exhibition. Three animals: lion, pussycat and horse are to be exhibited. Animals of each kind are not less than 10 in number. What is the possible number of ways of arranging the exhibition.

Solution: There are three types of animals and 10 stalls. One stall can be filled by any of the three animals. Once the first stall is filled, the second stall can be filled again in three ways by placing any of the three animals in it. We have to fill 10 such stalls and number of each animal is greater than equal to 10, so we have $3^{10} = 59049$ ways to fill the stalls. Thus, we can arrange the exhibition in 59049 ways.

Ans.

Circular Permutation

A permutation is called a **linear permutation** if the objects are arranged in a row, i.e., there is a first object and a last object in the arrangement. For example, $\mathbf{G_1G_2G_3G_4}$, $\mathbf{G_2G_3G_4G_1}$, $\mathbf{G_3G_4G_1G_2}$ and $\mathbf{G_4G_1G_2G_3}$ are all different arrangements, if we take them as linear permutations of four objects. However, if the same four arrangements are viewed as a circular arrangement as shown in the following diagram, all the four are same. The reason is there is nothing like the first object or the last object in a **circular permutation**. Any object could be the first and then the last is determined. Thus *in a circular permutation, we consider one object as fixed and the remaining objects are arranged as in any linear permutation.* The number of ways of allowing four people to sit in a row is given by 4!, whereas the number of ways of allowing the same four people to sit in a circle is (4-1) ! =3!.

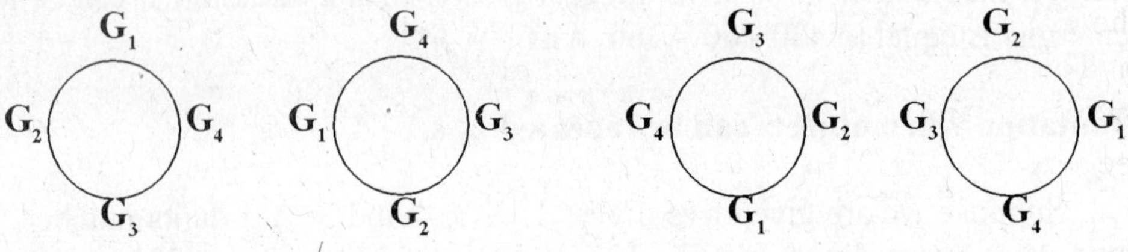

In circular permutation, **clockwise** and **anti-clockwise** arrangements of objects are possible and both are distinguishable. In the diagram shown below, the first is the permutation $\mathbf{G_1G_2G_3G_4}$ -anti-clockwise, and $\mathbf{G_1G_2G_3G_4}$ –clockwise. Therefore, when distinction is made between anti-clockwise and clockwise arrangements then the number of circular permutations of n objects taken all n at a time is (n-1)!. However, when there is no distinction between the direction of traversal along the circle, the number of circular permutations of n objects taken all n at a time is **half of the (n-1)!.**

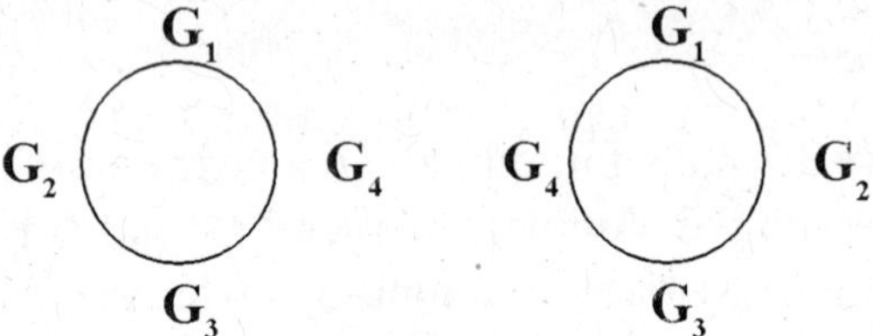

Example *7* In how many different ways can 5 men and 5 women sit around a table, if

(i) there is no restriction;

(ii) no two women sit together?

Solution: The problem is related to circular permutation of 10 objects (5 men and 5 women). If there is no restriction then the number of permutations is (10 –1)! = 9! = 362880. Notice here the difference in arrangement between clockwise and anti-clockwise. In the second case, there is a restriction that no two women are allowed to sit side by side. To meet this restriction each women should occupy a sit between two men. The number of ways five men can sit around a table is 4! = 24. Once these five men have sat on alternate chars, the five women can occupy the 5 empty chairs in 5! ways. Thus, total number of ways, in this case, will be 24∗5! =24∗120 = 2880. **Ans.**

Restricted Permutation

We have learned to find the number of permutations of given number of objects when choice of selection is arbitrary. In real life application, we encounter many situations in which counting is required under some constraints (restriction). For example, to find the number of positive even integers of 3 digits (distinct) that can be formed with digits 1,2,3,4 and 5. Here, to make a 3- digit number, we have to fill 3 places by the given digits. The unit place can be filled either by 2 or by 4 to make the number even. The remaining two places can be filled by any of the remaining four digits. Thus, count of 3-digits positive even integers, having all distinct digits, $= 2*{}^4P_2 = 2*12 = 24$. Let us see some examples to understand the concept of restricted permutation.

Example 8 In how many ways can 10 different examination papers be scheduled so that

(i) the best and the worst always come together;

(ii) the best and the worst never come together?

Solution: (i) Let us combine the best and the worst paper together and consider them as one object. We have, now, 9 objects (papers). These 9 objects can be arranged in 9! ways. And in each of these 9! arrangements, the best and the worst papers can be arranged in 2! ways. Therefore, the number of ways in which the 10 papers can be scheduled in this situation $= 2!*9! = 725760$

(ii) Without any restriction, the 10 papers can be scheduled in 10! ways. We have just calculated in part (i) that total number of ways in which the 10 papers can be scheduled so that the best and the worst always come together = 725760. Therefore, the number of ways of scheduling 10 papers so that the best and the worst never come together $= 10! - 725760 = 3628800 - 725760 = 2903040$

Ans.

In the example 8, we have used the notion that "*the number of ways of arranging objects under some restriction = the number of arrangements of the same number of object without restriction – the number of arrangements of the same number of arrangements with the opposite restriction*". Let us see a few more examples on restricted permutation.

Example 9 Find the sum of all the four-digit numbers that can be formed with the digits 3, 2, 3 and 4.

Solution: One thing is worth noticing here that a four-digit number so formed does not contain a repeated digit except the digit 3. This is implied from the question, because if it were not so, 3 should not have been repeated in the list of the digits.

To find the sum of the four-digit numbers formed with 3, 2, 3 and 4, we have to calculate the sum of digits at unit place in all such numbers. The sum of the digits at ten, hundred and thousand place will be the same, only theirs place value will change.

The number of four-digit numbers in which 2 appears at unit place is determined by the number of permutations of digits 3, 3 and 4 to fill ten, hundred and thousand place. And, this is $= 3! / 2! = 3$.

Similarly, the number of four-digit numbers in which 3 appears at unit place is $= 3! = 6$.

The number of four-digit numbers in which 4 appears at unit place is $= 3! / 2! = 3$.

Therefore, sum of the digits in the unit place of all the numbers

$$= 3\times2 + 6\times3 + 3\times4 = 36.$$

As stated above, the sum of the digits in all such numbers at ten, hundred and thousand places is 36 each. Thus the sum of all such numbers

$$= 36\times1000 + 36\times100 + 36\times10 + 36 = 39996$$

Ans.

Example 10 How many binary sequences of r-bits long have even number of 1's?

Solution: There will be 2^r possible binary sequences of r-bits long. This can be verified by the permutation of objects when repetitions are allowed. There are r places and two objects (0 and 1). The first place can be filled in 2 ways, for each of these, the second place can be filled in 2 ways, so we have 2×2 ways to fill first two places. Extending the sequence up to the rth steps, we have 2^r possible arrangements and hence 2^r possible binary sequences.

We can now make pairs of binary sequences in such a way that two sequences differ only at r^{th} place. There will 2^{r-1} such pairs, and in each pair one sequence will have even number of 1's. Thus, number of binary sequences of r bits long having even number of 1's $= 2^{r-1}$. **Ans.**

Example 11 We are asked to make slips for all numbers up to five-digit. Since the digits 0, 1, 6, 8 and 9 can be read as 0, 1, 9, 8 and 6 when they are read upside down, there are pairs of numbers that can share same slip if the slips are read upside down or right side up(e.g., 89166 and 99168). Find the number of slips required for all five-digit numbers.

Solution: We have 10 digits. We have to make all five-digit numbers. The total such numbers is equal to 10^5. Here we have to make slips for these many numbers. The numbers made of digits 0, 1, 6, 8 and 9 can be read upside down or right side up. And, there are 5^5 many such five-digit numbers (all those five-digit numbers made of digits 0, 1, 6, 8 and 9). Out of these 5^5 many numbers, however, there are some numbers that read the same either upside down or right side up. For example, 91816, and there are 3×5^2 such numbers (center place filled with 0, 1 or 8). Thus, there are $5^5 - 3\times5^2$ numbers that can be read upside down or right side up. And, for these numbers we need only

$$\frac{1}{2}\left(5^5 - 3*5^2\right)$$

number of slips. Therefore number of slips required to be made, is

$$10^5 - \frac{1}{2}\left(5^5 - 3*5^2\right) \qquad \textbf{Ans.}$$

Generating function for Permutation

Recall the exponential generating function that we have learnt in the chapter 2. The coefficient of $x^r / r!$ in a polynomial **p(x)** is $^n\mathbf{P}_r$. And $^n\mathbf{P}_r$ is the number of permutations of n objects taken r at a time. Generating function is a very handy technique for solving many counting problems. The general procedure for finding permutations is based on the exponential generating function.

Example 12 Find the generating function, also called enumerator, for permutations of n objects with the following specified conditions:

(a) Each object occurs at the most twice.
(b) Each object occurs at least twice.
(c) Each object occurs at least once and at the most k times.

Solution: (a) Each object occurs at the most twice implies that an object may occur 0, 1 or 2 times. The exponential generating function for an object under this condition is given as

$$1 + x + \frac{x^2}{2!}$$

There are n objects, so the generating function for the problem is written as

$$\left(1 + x + \frac{x^2}{2!}\right)^n$$

(b) In this case, an object occurs at least twice. This implies that an object may appear 2 or more times. The exponential generating function for an object under this condition is

$$\frac{x^2}{2!}+\frac{x^3}{3!}+\frac{x^4}{4!}+\cdots\cdots$$

Therefore, for the problem dealing with n objects, the generating function can be written as

$$\left(\frac{x^2}{2!}+\frac{x^3}{3!}+\frac{x^4}{4!}+\cdots\cdots\right)^n$$

(c) Here, each object occurs at least once and at the most k times. That is to say that an object may occur 1 or 2 or 3 or ... or k times. The exponential generating function for an object under this condition is

$$x+\frac{x^2}{2!}+\frac{x^3}{3!}+\cdots\cdots+\frac{x^k}{k!}$$

Since the problem is for n objects, the generating function for the problem is written as

$$\left(x+\frac{x^2}{2!}+\frac{x^3}{3!}+\cdots\cdots+\frac{x^k}{k!}\right)^n$$

Ans.

Example 13 A fair six-sided die is tossed four times and the numbers shown are recorded in a sequence. How many different sequences are there?

Solution: Let us assume, here, that each toss is an object and the number appearing on the face of a die is the number of times it occurs. Thus each object occurs at least once and at the most 6 times. Also the order of appearing of different 1's (conceptually) to make it 2 or 3 or 4 or 5 or 6 is fixed and is one and only one way. The generating function for first toss is given as

$$x + x^2 + x^3 + \ldots + x^6$$

The sum of all the coefficients, here, is 6 and number of possible sequences of numbers, on face of a die in one toss, is 6 only. Die is tossed four times, therefore generating function for the problem is

$$\{x + x^2 + x^3 + \ldots + x^6\}^4$$

The number of sequence is the sum of coefficients of all the terms in this generating function. This is equal to 6^4. Reader may verify this.

Ans.

Example 14 How many different words can be made from letters of the word '**committee**'.

Solution: The word committee contains letters c, o & i once and m, t & e twice. When a word is formed from these letters, a letter may appear at the most the number of times it appear in the word **committee** or not at all. So, generating function for c, o and i is given by (1+x) each, whereas, for c, m and e is given by $(1 + x + x^2/2!)$ each.

Thus generating function for the problem is then given by

$$(1+x)^3\left(1+x+\frac{x^2}{2!}\right)^3 = (1+3x+3x^2+x^3)\left(1+x+\frac{x^2}{2!}\right)^3$$

If words are to be formed taking all the letters at once, then the numbers of such words is given by the coefficient of $x^9/9!$, and this is equal to 9!/2!2!2!.

Ans.

Obviously, it is cumbersome to use generating functions to solve counting problems by hand calculation, because it's no fun multiplying polynomials. But it is very easy for computers to do the calculations, so generating functions provide a powerful tool for solving counting problems by machine. In addition, generating functions provide an excellent tool for studying counting problems in the abstract.

4.3 Combination

You may recall the difference in arrangement of terms in a sequence and that in a set. In the former, order of terms is important, whereas in the latter, order is **not important**. Generally, when order matters, we count the number of sequences or permutations and when order is not important, we count number of subsets or combinations.

Let A be a set such that |A| = n. Then count of all r-element ($r \leq n$) subsets of A is the combination of n elements taken r at a time. This is denoted as ${}^nC_r$. We know that ${}^nP_r$ is the number of permutations of n elements taken r at a time. This includes the different orders in which r elements can be arranged. It is also known that the r elements can be arranged in r! different ways. Thus, if we divide the ${}^{n}P_{r}$ by r!, then the order consideration is done away with. The result so obtained is the number of arrangements of n elements taken r at a time and that too without the consideration of order. This is, precisely, the definition of ${}^{n}C_{r}$. Therefore,

$$^{n}C_{r} = \frac{1}{r!}\,^{n}P_{r} = \frac{n!}{r!(n-r)!}$$

Example 1 A person has 8 children of them he takes 3 at a time to a circus. He does not take the same three children twice to the circus. How many times he will have to go to circus to ensure that every three children have seen the circus together? In this case find the number of times a particular child has visited the circus.

Solution: Here we have to find the number of combinations of 8 children taken 3 at a time. Note that order of selection of child is not important in this case. This selection can be made in ${}^{8}C_{3}$ ways. We can make ${}^{8}C_{3}$ =56 distinct groups of three children, and for each such group, the person will have to go to circus once. Therefore, the person will have to visit circus 56 times.

Second Part: A particular child goes to circus with every possible pair of two children out of remaining 7 children. Number of such possible pair is ${}^{7}C_{2}$= 21. Therefore, a particular child goes to circus 21 times.

Ans.

Example 2 From 8 men and 4 women a team of 5 is to be formed. In how many ways can this be done so as to include at least one woman?

Solution: This is a case of restricted combination. Total number of persons = 8 men + 4 women = 12 persons. A team of 5 has to be made, and this can be made in ${}^{12}C_{5}$ ways. This count includes the case of teams containing all five men (i.e. no women in the team) which is equal to ${}^{8}C_{5}$. Thus, number of ways in which the specified team can be selected $= {}^{12}C_{5} - {}^{8}C_{5} =$

$$\frac{12!}{5!7!} - \frac{8!}{5!3!} = \frac{12\times11\times10\times9\times8}{5\times4\times3\times2} - \frac{8\times7\times6}{3\times2} = 792 - 56 = 736 \quad \textbf{Ans}$$

Example 3 In a party of 30 people, each shakes hand with others. How many handshakes took place in the party?

Solution: In a normal case, a handshake involves two persons. This case is of counting 2-elements subsets of a set containing 30 elements. And this count is ${}^{30}C_2 = 30 \times 29/2 = 435$.

Ans.

Example 4 What is the number of diagonals that can be drawn in a polygon of n sides?

Solution: A polygon having n sides has n vertices. The total number of lines that can be drawn in a polygon of n vertices $= {}^{n}C_2$. There are already n sides (lines) which are not diagonal. The remaining lines will be diagonals. Therefore, the number of diagonals that can be drawn in a polygon of side n is equal to ${}^{n}C_2 - n =$

$$\frac{n!}{2!(n-2)!} - n = \frac{n(n-1)}{2} - n = \frac{n^2 - n - 2n}{2} = \frac{n(n-3)}{2} \quad \textbf{Ans}$$

Example 5 There are 10 points in a 2-D plane. Four of these are co-linear. Find the number of different straight lines that can be drawn by joining these points.

Solution: Any two points are always co-linear. So, a line can be drawn between any two points. If there are three **non-co-linear points** (a single line cannot be drawn joining all these three points), we can draw $3 = {}^{3}C_2$ distinct lines. A triangle is an example of this. Thus, we can draw ${}^{10}C_2$ distinct lines joining 10 points. Out of these 10 points, 4 are co-linear. So ${}^{4}C_2$ lines will be same and we consider them as one line. Therefore, actual number of lines that can be drawn $= {}^{10}C_2 - {}^{4}C_2 + 1 = 45 - 6 + 1 = 40$. **Ans.**

Combination when replacement is allowed

Consider the situation of placing 10 balls of different color in 3 pots, say A, B and C, with the flexibility that whenever required any number of balls of any color are available. A person places a red ball in pot A and is willing to place a red ball again in pot B and in pot C. Other balls may be placed according to the different possible combinations. After placing a red ball in pot A, only 9 balls are left and none is of red color. To place a red ball in pot B, the person needs a red ball, that is provided making the total number of balls 10 again. After placing this new ball in pot B, again a new red ball is required to place it in pot C. Once provided the number of balls is again 10. Once this new ball is placed in pot C, the person is left with 9 balls and 3 balls are in 3 pots. This implies that 3 balls have been selected from 9 + 3 balls. The number of combinations in this case is ${}^{9+3}C_3 = {}^{10+3-1}C_3$. This situation can be generalized in the following words.

" *Suppose that k selections are to be made from n objects without regard to order. Also assume that repeats are allowed and at least k copies of each of the n objects are available. Then, the number of ways selections of k objects can be made from n object is* ${}^{n+k-1}C_k$."

Example 6 A bookstore allows the recipient of a gift coupon to choose 6 books from the combined list of 10 best-selling fiction books and 10 best-selling non- fiction books. In how many different ways can the selection of 6 books be made?

Solution: The number of different types of books is 10+10 =20. A gift coupon recipient may select any 6 books, possibly 6 copies of a single book. This is a case of selection of 6 objects from 20 objects with repetitions allowed. The number of ways the selection can be made is ${}^{20+6-1}C_6 = {}^{25}C_6 = 177100$.

Ans.

Example 7 If three dice are rolled, and we make a set of numbers shown on the three dice, How many different sets are possible?

Solution: Rolling three dice is equivalent to selecting three numbers from the list of six numbers 1, 2, 3, 4, 5 and 6 with repetitions allowed. Because sequence 111, 121 etc are possible. Thus the different possible combinations is ${}^{6+3-1}C_3 = {}^{8}C_3 = 56$.

Ans.

Open Selection

Suppose a coin is tossed n times. Obviously there will be 2^n sequences of H (Heads) and T (Tails) each of the form $X_1X_2X_3X_4 \ldots X_n$. Each X_i is either H or T. If we are asked to find number of sequences having at least one head or sequences having one or more or all heads, then the count is $2^n - 1$. This is the result of the following sum

$${}^{n}C_1 + {}^{n}C_2 + {}^{n}C_3 + \ldots + {}^{n}C_n$$

Similarly, if we are asked to find the number of sequences having at least 5 heads and at the most 10 heads ($n \geq 10$), then the number is given by the sum ${}^nC_5 + {}^nC_6 + {}^nC_7 + \ldots + {}^nC_{10}$. In general, combinations of n items taking r at a time where r lies between a and b where $1 \leq a$ and $b \leq n$, is given by the sum

$$^nC_a + {}^nC_{a+1} + {}^nC_{a+2} + \ldots + {}^nC_b$$

Example 8 In an election, there are four candidates contesting for three vacancies; an elector can vote for any number of candidates not exceeding the number of vacancies. In how many ways can one cast his votes?

Solution: An elector may vote for any one or, any two or, any three candidates out of total 4. Therefore, an elector may vote in ${}^4C_1 + {}^4C_2 + {}^4C_3 = 4 + 6 + 4 = 14$ different possible ways. **Ans.**

Example 9 In an election the number of candidates is one more than the number of vacancies. If a voter can vote in 30 different ways, find the number of candidates.

Solution: Let the number of candidates be x. An elector may vote for any one or, any two or, any three up to maximum of any $x - 1$ candidates from total of x, because number of vacancies is $x - 1$ only. Therefore, number of ways in which an elector can cast his vote is ${}^xC_1 + {}^xC_2 + {}^xC_3 + \ldots + {}^xC_{x-1}$ and this value is given to be 30. Thus,

$$^xC_1 + {}^xC_2 + {}^xC_3 + \ldots + {}^xC_{x-1} = 30$$

$$\text{or, } {}^xC_0 + {}^xC_1 + {}^xC_2 + {}^xC_3 + \ldots + {}^xC_{x-1} + {}^xC_x - {}^xC_0 - {}^xC_x = 30$$

$$\text{or, } 2^x - 2 = 30 \Rightarrow x = 5$$

Therefore, the number of candidates is 5. **Ans.**

Example 10 In an examination a candidate has to pass in each of the 5 papers. How many different combinations of papers are there so that a student may fail?

Solution: For a student to pass the examination, he/she will have to pass in each of the five papers. To fail, a student may fail in any one or, in any two or so on including in all the five papers. Thus, a student may fail in as many as

$$^5C_1 + {}^5C_2 + {}^5C_3 + {}^5C_4 + {}^5C_5 = 2^5 - 1 = 31 \text{ ways.}$$ **Ans.**

Example 11 Find the total number of selections of at least one red ball from 4 red balls and 3 green balls, if

(a) the balls of the same color are different,
(b) the balls of the same color are identical.

Solution: (a) From 4 different red balls and 3 different green balls, we have to find number of selections taking at least one red ball and any number of (including 0) 3 green balls. The total number of ways of selecting at least one red ball from 4 different red balls $= {}^4C_1 + {}^4C_2 + {}^4C_3 + {}^4C_4 = 15$. Corresponding to each of these selections, the number of ways of selecting green balls $= {}^3C_0 + {}^3C_1 + {}^3C_2 + {}^3C_3 = 8$. Therefore, total number of different ways of selection $= 15 \times 8 = 120$.

Ans.

Distributing different objects into groups

Suppose 12 different books are to be equally distributed between two persons **A** and **B**. In how many different ways this task can be accomplished? When half of the books have been given to **A,** the rest 6 books have already been selected for **B**. Since the order of selection is not important, the task of selecting 6 books from 12 books can be done in ${}^{12}C_6$ different ways. If the task is to make two sets of 6 books each, then it can be performed in ${}^{12}C_6 \div 2!$ different ways. In the last case, interchange of two sets does not give any new combination, whereas, in the previous case it is important.

Now Consider the equal distribution of the same 12 books among 4 persons. We can select 3 books from 12 books in ${}^{12}C_3$ ways for first person. For each of these selections, 3 books can be selected from the remaining 9 in ${}^9C_3$ ways for second person. Similarly, we can select 3 books in ${}^6C_3$ ways for third person and in ${}^3C_3$ ways for last person. Therefore, total number of ways in which 12 books can be distributed equally among 4 persons is ${}^{12}C_3\,{}^9C_3\,{}^6C_3\,{}^3C_3 =$

$$\frac{12!}{3!9!} \times \frac{9!}{3!6!} \times \frac{6!}{3!3!} \times \frac{3!}{3!0!} = \frac{12!}{(3!)^4}$$

The above result is to be divided by 4!, if we have to find the number of ways to make 4 sets of 3 books each. Instead of divided equally, if A has to get 7 books and B has to get 5 books, then the number of ways $= {}^{12}C_7 = 12!/(7!5!)$. This discussion, in general, can be summarized as follows.

- *The number of ways of distributing p+ q different objects between two distinguishable groups in such a ways that one group gets p objects and other gets q objects is given by*

$$\frac{(p+q)!}{p!q!}$$

- *If the two groups are indistinguishable in the above case then, the number of ways of distribution is given by*

$$\frac{(p+q)!}{2!\,p!\,q!}$$

- *In general, if n different objects are to be distributed among m distinguishable groups containing* $p_1, p_2, p_3, \ldots, p_m$ objects, where $p_1 + p_2 + p_3 + \ldots + p_m = n$, The number of ways in which this tasks can be completed is given by

$$\frac{n!}{p_1!\,p_2!\,p_3!\cdots p_m!}$$

- *In the previous result if the m groups are indistinguishable, the number of ways of distribution is given by*

$$\frac{n!}{m!\,p_1!\,p_2!\,p_3!\cdots p_m!}$$

Example 12 In how many ways can a pack of 52 cards be equally divided into four groups? If the cards are to be distributed equally among four players, then find the number of ways of this distribution.

Solution: First Part: When 52 cards are distributed equally among four groups, each group contains 13 cards. Since groups are indistinguishable, the number of ways of distribution is given by

$$\frac{52!}{4!13!13!13!13!}$$

Second Part: Here four groups (players) are distinguishable, thus number of ways of distribution is given by

$$\frac{52!}{13!13!13!13!}$$

Ans.

Generating Function for Combinations

Recall the notion of a binomial generating function that we have studied in the chapter 2. The coefficient of x^r in $(1+x)^n$ is $^n\mathbf{C}_r$. And ^nC_r is the number of combinations of n objects taken r at a time. This implies that if we can find a binomial generating function for a given combination problem, the task of finding the number of combinations is reduced to getting the relevant coefficient. We shall use binomial generating function to solve a few problems here.

Example 13 A library has 5 indistinguishable black books, 4 indistinguishable red books and 3 indistinguishable yellow books. In how many distinguishable ways can a student take home (a) 6 books? (b) 6 books taking at least 1 of each color? (c) 6 books taking 2 of each color?

Solution: Here all books of a particular color are indistinguishable. A black (or red or yellow) book can be selected or, not selected. If selected then maximum number of black books that can be selected is 5, i.e., possible ways of selections for black books are 0, 1, 2, 3, 4 or 5. Similarly, the possible ways of selections for red books are 0, 1, 2, 3 or 4 and for yellow books are 0,1,2 or 3. If we take three variables x, y and z for the numbers of black, red and yellow books selected by a student, respectively then the number of solutions to the equation

$$\mathbf{x + y + z = 6} \text{ where } 0 \le x \le 5,\ 0 \le y \le 4 \text{ and } 0 \le z \le 3$$

is the required number of ways in which a student can take home 6 books. The generating function for x is $(1+x+x^2+x^3+x^4+x^5)$, for y is $(1+y+y^2+y^3+y^4)$ and for z is $(1+z+z^2+z^3)$. If we replace y and z by x, we get the generating function for the above problem as

$$f(x) = (1+x+x^2+x^3+x^4+x^5)(1+x+x^2+x^3+x^4)(1+x+x^2+x^3)$$

The coefficient of x^6 in f(x) is the required number, and this number is **18**.

In the second part, at least one book of each color has to be selected, so generating function f(x) is given as

$$f(x) = (x+x^2+x^3+x^4+x^5)(x+x^2+x^3+x^4)(x+x^2+x^3)$$

The coefficient of x^6 in f(x) is the required number, and this number is **9**.

In the third part, two books of each color are to be selected. So generating function f(x) is given as $f(x) = x^2 * x^2 * x^2$
The coefficient of x^6 in f(x) is the required number, and this number is **1**. **Ans.**

Example 14 A library has 5 black books, 4 red books and 3 yellow books, all with different titles. How many distinguishable ways can a student take home 6 books, 2 of each color?

Solution: Notice that this problem is different from that in the example 13. In this case, the books of a particular color are distinguishable by their title. Each book can be either selected or not, giving the possible number of selection for each book as 0 or 1. This can be written, in polynomial form, as 1+x. Thus, the generating function for 5 black books is $(1+x)^5$. Similarly, the generating function for 4 red books is $(1+y)^4$ and for 3 yellow books is $(1+z)^3$. To count the number of ways 6 books, 2 of each color, can be selected, we take coefficient of $\mathbf{x^2 y^2 z^2}$ in the generating function

$$f(x) = (1+x)^5(1+y)^4(1+z)^3$$

The coefficient of $\mathbf{x^2 y^2 z^2}$ is ${}^5C_2\,{}^4C_2\,{}^3C_2 = 10\times6\times3 = 180$.

Ans.

Example 15 In how many ways can one choose n pieces of fruit, assuming there are an infinitely large number of apples, bananas, oranges and pears, and he (she) wants an even number of apples, an odd number of bananas, not more than 4 oranges and at least two pears?

Solution: The generating function for the selection of apple can be written as $(1 + x^2 + x^4 + x^6 + \ldots\ldots)$. For the selection of bananas can be written as $(x + x^3 + x^5 + x^7 + \ldots\ldots)$. For the selection of oranges can be written as $(1+x + x^2 + x^3 + x^4)$ and for pears it can be written as $(x^2 + x^3 + x^4 + \ldots\ldots)$. Therefore, the generating function for this problem is

$$f(x) = (1+x^2+x^4+x^6+\ldots)(x+x^3+x^5+x^7+\ldots)(1+x+x^2+x^3+x^4)(x^2+x^3+x^4+\ldots)$$

The answer is simply the nth coefficient of this infinite series. *Here, although we cannot see the numerical answers, this function contains all the data: a computer could readily provide us with a table of the answers for values of n as large as we please, of course, within the limits of computer memory.*

Ans.

Example 16 In how many ways 2n + 1 items can be distributed among 3 persons so that the sum of the number of items received by any two persons should exceed the number of items received by the other?

Solution: Let us take three variables x, y and z for the number of items received by the three persons. Then number of ways 2n + 1 items can be distributed among three persons is equal to the number of positive integer solutions to the equation

$$x + y + z = 2n + 1 \text{ where } 1 \le x, y, z \le n$$

To ensure that the sum of the two variables must exceed the third one, it is important to define the range of values for each variable as above. Therefore, the generating function for the problem is written as

$$f(x) = (x + x^2 + x^3 + \ldots + x^n)(x + x^2 + x^3 + \ldots + x^n)(x + x^2 + x^3 + \ldots + x^n)$$

$$\text{or,} \quad f(x) = (x + x^2 + x^3 + \ldots + x^n)^3$$

$$= \left\{\frac{x(1-x^n)}{1-x}\right\}^3$$

$$= x^3(1 - 3x^n + 3x^{2n} - x^{3n})(1-x)^{-3}$$

In this function f(x), the coefficient of x^{2n+1} is the count we are looking for. And, this coefficient is equal to the coefficient of x^{2n-2} in $(1 - 3x^n + 3x^{2n} - x^{3n})(1-x)^{-3}$. This value is

$$^{3+2n-2-1}C_{2n-2} - 3 \times {}^{3+n-2-1}C_{n-2}$$

$$or, \quad ^{2n}C_{2n-2} - 3 \times {}^{n}C_{n-2} = \frac{2n(2n-1)}{2!} - 3\frac{n(n-1)}{2!}$$

$$= \frac{n}{2}(4n - 2 - 3n + 3) = \frac{n(n+1)}{2} \qquad \textbf{Ans.}$$

Mixed Problem

Some combinatorial problems require that the counting of permutations and combinations be combined or supplemented by the direct use of the addition or the multiplication principles of counting. See the following example.

Example 17 A valid password consists of seven symbols. Symbols are chosen from digits and Roman capital alphabets. The first symbol of the password must be a Roman capital alphabet. How many different passwords are possible?

Solution: Number of symbols = 26 +10 = 36. The first place of seven characters password can be chosen in 26 ways. The remaining 6 places can be filled in 36 ways each i.e., the second symbol for a password can be chosen in 36 ways, and for each of this, the third place can be filled in 36 ways and so on. Thus, the number of different possible passwords = $26 \times (36)^6 = 56596340736$.

Ans.

4.4 Pigeonhole Principle

"If **five** men (pigeonholes) are married to **six** women (pigeons) then at least one men has more than one wife."

Or

"If **five** men (pigeons) are married to **four** women (pigeonholes) then at least one women has more than one husband."

Or

"If we have **six** pigeons in **five** pigeonholes then at least one pigeonholes would have more than one pigeon."

Arguments of this kind are used in solution to many mathematical problems and quite a few beautiful theorems have been proven with their help. All these arguments share the name "**Pigeonhole Principle**". In this text we shall use this theorem to prove problems related to numerical, counting, integer division and geometry.

Theorem 1 If **n** objects (pigeons) are placed in **m** places (pigeonholes) for m < n, then one of the places (pigeonholes) must contains at least

$$\left\lfloor \frac{n-1}{m} \right\rfloor + 1$$

objects (pigeons).

Proof: The proof is very simple. Let us assume that none of the places (pigeonholes) contains more than **floor((n – 1)/m)** objects (pigeons). Then there are at the most **m** * **floor((n – 1)/m)** ≤ **m** * **(n – 1)/m** = **n-1** objects. This is contrary to the given fact that there are **n** objects. This contradiction is because of our assumption that "none of the places contains more than **floor((n – 1)/m)** objects". Thus our assumption is wrong. Therefore, one of the pigeonholes must contains at least

$$\left\lfloor \frac{n-1}{m} \right\rfloor + 1$$

objects.

Proved.

This so called '**pigeonhole principle'** is very simple. Its application, however, is very **intelligence intensive**. It needs skill, which can be developed by attempting on different types of examples. The core is to identify the **pigeons**, **pigeonholes** and theirs respective numbers. Once you have these figures, you have almost found the solution to the problem. Let us see some examples to understand the application of this principle.

Example 1 15 students wrote a dictation. Tarun made 13 errors. Each of the other students made less than that number of errors. Prove that at least two students made equal number of errors.

Solution: Since maximum number of errors committed by any student is 13, there are students who might have made no errors, some may have made one error and so on. But, none of the students has made more than 13 errors. Let there are 14 chairs (pigeonholes) marked 0, 1, 2, ... 13. We let the students be seated in the chair having mark equal to the number of errors committed by the student. After 14 students have been seated, there is one student left who will be sharing chair with any of the 14 students. This means that there are at least two students who have made same number of errors. Here, n = number of pigeons = number of students = 15. m = number of pigeonholes = 14 (count of number from 0 to 13, possible number of errors by different student). Thus using the pigeonhole principle we have at least one pigeonhole containing at least

$$\left\lfloor \frac{n-1}{m} \right\rfloor + 1 = \left\lfloor \frac{15-1}{14} \right\rfloor + 1 = 1 + 1 = 2$$

students.

Proved.

Example 2 There are 500 boxes with apples. Each contains no more than x apples. Find the maximum possible value of x, such that one can for sure find 3 boxes containing equal number of apples.

Solution: Let us place these 500 boxes (pigeons) in pigeonholes marked with number according to the **number of apples** in a box. Let there are pigeonholes marked with 0 for empty boxes, 1 for boxes containing one apple, 2 for boxes containing two apples... and so on up to **x**. Thus we have **x+1** pigeonholes. Applying the pigeonhole principle, we get

$$\left\lfloor \frac{500-1}{x+1} \right\rfloor + 1 = 3 \quad Or, \quad \left\lfloor \frac{500-1}{x+1} \right\rfloor = 2 \quad Or, \quad 499 = 2x + 2$$

$$Or, \quad x = \left\lfloor \frac{499-2}{2} \right\rfloor = 248.$$

Therefore, the maximum possible value for **x** is 248. **Ans.**

Example 3 There are 33 students in the class, and the sum of their ages is 430 years. Is it true that for some 20 students the sum of their ages is more than 260 years?

Solution: Consider the average age of students which is 430/33 = 13.03 years. Now some students may be of age less than this average, some may be of average age and some other may be of age greater than average. Let there be three pigeonholes **L**, **A** and **G** for less than, average and greater than ages respectively. If we put the 33 pigeons (students) in these three pigeonholes, then there are three possibilities.

- All three pigeonholes get equal number of pigeons. or
- The distribution is skewed toward **L**. or
- The distribution is skewed toward **G**.

In the first case, we can always pick up 20 students (all from **G** and 9 from **A**) so that their ages will total more than 260 years. In the third case too we can pick up 20 students (all from **G,** remaining from **A** and, if any more required then from **L**) so that the ages will total more than 260 years. In second case, we use the fact that **if any number is less than the average then there exists a number which will be greater than the average**. Thus if there are students in **L**, then there must be some students in **G** whose age(s) will maintain the average. By picking up 20 students as in the cases of first and third, theirs total age will exceed 260 years. Thus, we can always find 20 students such that the sum of their ages is more than 260 years.

Ans.

Example 4 There are pencils in the box: 10 red ones, 8 blue, 8 green, 4 yellow. Let us take, with eye closed, some number of pencils from the box. What is the least number of pencils we have to take in order to ensure that we get at least 4 pencils of the same color?

Solution: Let us consider 4 pigeonholes corresponding to four colors. Then, our problem is to find the number of pigeons (pencils) so that when placed in holes, one of the holes must contain at least four pigeons (pencils). In the theorem above, m = 4, then we have to find n such that

$$\left\lfloor \frac{n-1}{4} \right\rfloor + 1 = 4 \qquad or, \quad n-1=12 \qquad or, \quad n=13$$

Therefore, at least 13 pencils are to be picked up to ensure that we get at least 4 pencils of same color.

Ans.

The examples discussed above are related to counting problems. This principle can be used to solve many such counting problems. Let us now consider a few examples related to some other area of applications.

Example 5 There are 50 people in a room. Some of them are acquainted with each other, some not. Prove that there are two persons in the room who have equal numbers of acquaintances.

Solution: If there is a person in the room who has no acquaintances at all then each of the other persons in the room may have either 1, or 2, or 3, …., or 48 acquaintances, or do not have acquaintances at all. Therefore we have 49 pigeonholes numbered 0, 1, 2, 3, …..,48. If we put 50 people one by one in the pigeonholes according to the number of acquaintances the person has, we have one pigeonhole containing at least

$$\left\lfloor \frac{50-1}{49} \right\rfloor + 1 = 1 + 1 = 2$$

pigeons. Thus, there are at least two persons in the room who have equal number of acquaintances.

Proved.

Example 6 Prove that, given any 12 natural numbers, we can chose two of them such that their difference is divisible by 11.

Solution: "*If two numbers **a** and **b** have the same remainder upon division by a number **c** then the difference **a – b** is divisible by **c**.*" We shall use this property of integers to solve this type of problems. There are 11 possible remainders upon division by 11. These are 0, 1, 2, …., 10. Let these be the pigeonholes. We have 12 pigeons (natural numbers). If we put these pigeons in the 11 pigeonholes, then we have one pigeonhole containing at least

$$\left\lfloor \frac{12-1}{11} \right\rfloor + 1 = 2$$

integers yielding same remainder when divided by 11. Using the property of integers, as stated above, we conclude that we can choose two of the 12 natural numbers such that their difference is divisible by 11.

Proved.

Example 7 We are given five arbitrary natural numbers **a, b, c, d, e.** Prove that either one of them is divisible by 5, or the sum of several numbers in a row is divisible by 5.

Solution: Let us consider five sums **a, a + b, a + b + c, a + b + c + d, a + b + c + d + e.** When a number is divided by 5, the possible remainders are 0, 1, 2, 3, 4. Let the number of pigeonholes is 5. If we place the five sums one by one in these holes then either each hole will get one pigeon **or** some of holes will get more than one and some none. In the first case, the sum corresponding to holes **0** is divisible by 5. And, in the second case, the difference of the two sums falling in the same hole is divisible by 5. Also note that the difference of the sums is again one of the natural number or sum of one or more given natural numbers in row. Thus we have proved that either one of the numbers or the sum of several numbers in a row is divisible by 5.

Proved.

Example 8 Prove that there exists a multiple of 1997 whose decimal expansion contains only digits 1 and 0.

Solution: Let us write 1998 numbers like **1, 11, 111, 1111, ..., 111...111**. Each of these numbers yields 0 or 1 or 2 or ....1996 as remainder when divided by 1997. Let these 1997 possible remainders are the pigeonholes and the above 1998 written numbers are pigeons. If we place the pigeons one by one in the holes then we have al least one hole containing at least

$$\left\lfloor \frac{1998-1}{1997} \right\rfloor + 1 = 2$$

numbers from the 1998 numbers, listed as above. The difference of these two numbers is a number containing only digits 1 and 0.

Proved.

Example 9 51 points were placed in an arbitrary way into the square of side 1. Prove that some 3 of these points can be covered by a circle of radius 1/7.

Solution: Let us divide the square into 25 squares of side 1/5 each. Then at least one of these small squares will contain at least

$$\left\lfloor \frac{51-1}{25} \right\rfloor + 1 = 3$$

points (pigeons). Now the circle circumvented around the square with three points also contains these three point and has radius

$$r = \sqrt{\left(\frac{1}{10}\right)^2 + \left(\frac{1}{10}\right)^2} = \sqrt{\frac{2}{100}} = \sqrt{\frac{1}{50}} < \sqrt{\frac{1}{49}} = \frac{1}{7}$$

Hence, there are at least three points inside a circle of radius 1/7.

Proved.

Now, we shall use this theorem to solve a problem that is based on constructive prove method. See the example below and before looking at the solution, just try to solve on your own.

Example 10 A chess player wants to prepare for a championship match by playing some practice games in 77 days. She wants to play at least one game a day but no more than 132 games altogether. Show that there is a period of consecutive days within which she plays exactly 21 games, irrespective of the scheduling of her practice games.

Solution: Let a_i be the total number of games she plays up through the ith day. Obviously, the sequence $a_1, a_2, a_3, \ldots, a_{77}$ is a monotonically increasing sequence, where $a_1 \geq 1$ and $a_{77} \leq 132$. Now let us take another sequence $b_1, b_2, b_3, \ldots, b_{77}$ such that $b_i = a_i + 21$ and $b_{77} \leq 153$ (Notice, here, the construction of sequence **b**). This sequence **b** is also monotonically increasing. We have, now, 154 numbers (77 from sequence **a** and 77 from sequence **b**) and the values of these numbers range from 1 to 153. Therefore from pigeonhole principle, we have at least two numbers having same value. Since both **a** and **b** are monotonically increasing, two numbers of **a** and two numbers of **b** cannot be equal. So, there exists a number in the sequence **a,** which is equal to a number in the sequence **b**. Let a_i is equal to = b_j i.e., $a_i = a_j + 21$. This proves that there are consecutive days **i** through **j** during which she played exactly 21 games.

Proved.

Example 11 Prove that some integral power of 2 has the decimal expansion which starts with the digits 1999 i.e., there exists integer n such that

$$2^n = 1999....$$

Solution: Let there is a number the decimal expansion of which starts with the digits 1999.... If we prove that there exists a number starting with the digits 1999..., such that $\log_2 1999...$ is an integer, then the problem is solved. Readers may try to solve this. You may use the concept that $\log_2 10$ is an irrational number.

In computer science, we come across many situations where we are supposed to prove the correctness of the design and implementation of software and hardware. Methods of proof gives a first hand tool in performing these tasks. Suppose a design is proved mathematically right and if the final implemented software fails, it ensures that the implementation of the design has not been carried out correctly according to approved design. Thus the topics covered in this chapter equip readers in tackling the challenges of practical life when the task comes to actually implement a project.

Exercise

1. Population of New Delhi is above 600,000,00. Each has no more than 100,000 hairs on his or her head. Prove that some 599 residents of New Delhi have equal number of hairs.

2. Assume that there are n >1 people in the room. Prove that at least two of them have equal number of acquaintances.

3. Ten football teams play for a trophy. Every two of them have to meet in a game. Prove that at any given time there are two teams having played equal number of games.

4. Prove that, given three natural numbers, we can select two of them with even sum.

5. Prove that of any 100 natural numbers there is one number, or the sum of several numbers, which is divisible by 100.

6. Prove that there is a number of the form 19971997.... 199700...00 divisible by 1998.

7. Prove that for any natural number n there is a number written only by 5's and 0's and divisible by n.

8. Prove that of any 52 natural numbers one can find two numbers **m** and **n** such **m + n** or **m - n** is divisible by 100. Is the same statement true for 51 arbitrary natural numbers?

9. Given n integers, prove that there are several of them (may be one) with the sum divisible by n.

10. Prove that there is a natural number of the form 19971997... 1997 divisible by 1999.

11. We are given 7 straight lines on the plane. No two of them are parallel. Prove that there are two lines with the angle between them less than 26^0. Can we claim the same for 25^0?

12. Five points are positioned inside of the equilateral triangle of side 1 unit length. Prove that there are two of them at the distance less than 0.5 from each other.

13. A city has 10,000 different telephone lines numbered by 4-digit numbers. More than halve of these are in suburban. Show that there are two telephone numbers in the sub-urban whose sum is again the number of a sub-urban telephone line.

14. Ten people volunteer for a three-person committee. Every possible committee of three that can be formed from these 10 names is written on a slip of paper, one slip for each possible committee, and the slips are put in 10 hats. Show that at least one hat contains 12 or more slips of paper.

15. A store has introductory sales on 12 types of candy bars. A customer may choose one bar of any five different types and will be charged no more than Rs. 175. Show that, although different choices may cost different amounts, there must be at least two different ways to choose so that the cost will be the same for both choices.

16. Show that there must be at least 90 ways to choose six numbers from 1 to 15 so that all the choices have the same sum.

17. How many friends must you have to guarantee that at least five of them will have birthdays in the same month?

18. Prove that if any 14 numbers from 1 to 25 are chosen, then one of them is a multiple of another.

19. If a is an integer divisible by 4, then a is the difference of two perfect squares.

20. Prove that the sum of two rational numbers is a rational number.

21. If r_1, r_2 and r_3 are three distinct roots of polynomial $p(x) = x^3 + bx^2 + cx + d$, then show that $r_1 r_2 + r_1 r_3 + r_2 r_3 = c$.

22. Prove that the cube root of 2 is irrational.

23. Prove that there are no positive integer solutions to the diophantine equation $x^2 - y^2 = 10$.

24. Show that there is no rational number solution to the equation $x^5 + x^4 + x^3 + x^2 + 1 = 0$.

25. Show that the sum of a rational number **a** and an irrational number **b** is an irrational number.

26. Show that if product of two integers is even then at least one of them must be even. Show that if product of two integers is odd then both of them must be odd.

27. Show that if product of two real numbers is an irrational number then one of them must be an irrational number.

28. Prove that an integer **a** is not evenly divisible by 5 if, and only if, $a^4 - 1$ is evenly divisible by 5.

29. A **Diophantine equation** is an equation for which you seek integer solutions. For example, the triples (x, y, z) are positive integer solutions to the equation $x^2 + y^2 = z^2$. Show that there are no positive integer solution to the diophantine equation $x^2 - y^2 = 1$.

30. Two integers are said to have same **parity** if they both are odd or both are even. If x and y are two integers for which x + y is even, then show that x and y have the same parity.

31. Prove that a positive integer n is evenly divisible by 9 if and only if, the sum of the digits of n is divisible by 9.

32. Show that a positive integer n is evenly divisible by 11 if, and only if, the difference of the sum of the digits in the even and odd position in n is divisible by 11.

33. How many distinguishable permutations of the letters in the word (a) BANANA (b) ASSOCIATIVE (c) REQUIREMENTS are there?

34. In how many ways can seven people be seated in a circle?

35. In a psychological experiment, a person must arrange a square, a cube, a circle, a triangle and a pentagon in a row. How many different arrangements are possible?

36. A bookshelf is to be used to display six new books. Suppose that there are eight computer science books and five French books from which to choose. If we decide to show four books of computer science and two French books and we are required to keep the books in each subject together, how many different displays are possible?

37. Most of the programming languages allow variable names to be of eight letters or digits with restriction that first character must be a letter. How many eight character variable names are possible?

38. At a certain college, the housing office has decided to appoint, for each floor, one male and one female residential advisor. How many different pairs of advisors can be selected for a seven-story building from 12 male candidates and 15 female candidates?

39. In how many ways can a committee of 6 people be selected from a group of 10 people if one person is to be designated as chairperson of the committee?

40. How many ways can you choose three of seven fiction books and two of six nonfiction books to take with you on your vacation?

41. In an experiment, n fair six sided dice are tossed and the numbers showing on top are recorded. (a) How many sequences are possible? (b) How many sequences contain exactly one six? (c) How many sequences contain exactly four twos, assuming $n \geq 4$?

42. There are 10 points in a plane of which only 4 are collinear. How many different triangles can be formed with these points as vertices?

43. Find the number of ways of distributing 16 pens among 4 persons in such a way that every one must get at least three pens.

44. How many integers between 1 and 100000 have the sum of their digits equal to 16?

45. In how many ways can 52 cards be distributed amongst four persons sitting in a circle to play a game of cards?

46. A library has 5 indistinguishable black books, 4 indistinguishable red books and 3 indistinguishable yellow books. How many distinguishable ways can a student take home 6 books? Solve it using generating function.

47. In the exercise 46 above, how many ways can a student take home 6 books, 2 of each color?

48. In how many ways can two adjacent squares be selected from 8×8 chessboard?

49. In how many ways can two integers be selected from integers 1 to 100 so that their difference is (a) exactly seven? (b) less than equal to 7 ?

50. There are 10 pairs of shoes in a shoe rack. If eight shoes are chosen at random, what is the probability that no complete pair of shoes is chosen? That exactly one complete pair of shoes is chosen?

51. Two numbers are chosen one after the other from numbers between 1 and 100. What is the probability that the sum of the two numbers is divisible by 3 ?

52. Ten passengers got into an elevator on the ground floor of a 20-story building. What is the probability that they will all get off at different floors?

53. How many permutations of 10 digits 0, 1, 2, ..., 9 are there in which the first digit is greater than 1 and the last digit is less than 8?

54. How many permutations of 26 letters a, b, c,..., z are there in which the first letter is not a, b, or c and the last letter is not w, x, y, or z?

55. In how many ways can two integers be selected from 1, 2,..., n-1 so that their sum is larger than n?

56. Find the number of ways to place 2n+1 balls in three different boxes so that any two boxes together should contain more balls than the other one. Use generating function method.

Prove the following using mathematical induction.

57. $1 + 2 + 3 + \cdots + n = \dfrac{n(n+1)}{2}$

58. $\overline{\bigcap_{i=1}^{n} A_i} = \bigcup_{i=1}^{n} \overline{A_i}$

59. $1^2 + 2^2 + 3^2 + \cdots + n^2 = \dfrac{n(n+1)(2n+1)}{6}$

60. $1^3 + 2^3 + 3^3 + \cdots + n^3 = \left[\dfrac{n(n+1)}{2}\right]^2$

61. $a + ar + ar^2 + \cdots + ar^{n-1} = \dfrac{a(1-r^n)}{1-r}$ $\quad$ *for* $r \neq 1$

62. $1 + 2^n < 3^n$ $\quad$ *for* $n \geq 2$

63. $n < 2^n$ $\quad$ *for* $n \geq 0$

64. $\left(\bigcap_{i=1}^{n} A_i\right) \cup B = \bigcap_{i=1}^{n} (A_i \cup B)$

65. $\dfrac{1}{1*2} + \dfrac{1}{2*3} + \dfrac{1}{3*4} + \cdots\cdots + \dfrac{1}{n*(n+1)} = \dfrac{n}{n+1}$

66. Use Induction method to show that if **p** is a prime and **p | a**$^{\mathbf{n}}$ for $n \geq 1$, then **p | a**.

67. Find smallest positive integer **n** such that $\mathbf{2^n > n^2}$ and show that $\mathbf{2^m > m^2}$ for $m > n$.

68. Using Induction Method, prove that $10^{2n-1} + 1$ is divisible by 11 for all $n \in N$.

69. $2^{5n+5} - 31n - 32$ is divisible by 961 for all $n \in N$. Prove this by mathematical induction.

70. Prove that $7^{2n} + 2^{3n-3} * 3^{n-1}$ is divisible by 25 for all $n \in N$. Use mathematical induction method to prove this.

71. Using mathematical induction method, prove that $11^{n+2} + 12^{2n+1}$ is divisible by 133 for any $n \in N$.

72. Using induction or otherwise, prove that for any non-negative integers m, n, r and k

$$\sum_{m=0}^{k} (n-m)\frac{(r+m)!}{m!} = \frac{(r+k+1)!}{k!}\left[\frac{n}{r+1} + \frac{k}{r+2}\right]$$

Prove the following by induction method.

73. $\frac{1}{2}.\frac{3}{4}.\frac{5}{6}\cdots\frac{2n-1}{2n} < \frac{1}{\sqrt{2n+1}}$ *for all* $n \in N$

74. $10^{n-2} > 81n$ *for* $n \in N$ *and* $n \geq 5$

75. $\frac{n^5}{5}+\frac{n^3}{3}+\frac{7n}{15}$ *is a natural number for all* $n \in N$.

76. $p^{n+1}+(p+1)^{2n-1}$ *is divisible by* $p^2+p+1, n \in N$.

77. *Let* $s_1 = 2, s_2 = 0, s_3 = -14,$ *and* $s_{n+1} = 9s_n - 23s_{n-1} + 15s_{n-2},$ *for* $n \geq 3$, *then show that*

$$s_n = 3^{n-1} - 5^{n-1} + 2 \quad \text{for } n \geq 1.$$

78. *Let* $u_1 = u_2 = 5,$ *and* $u_{n+1} = u_n + 6u_{n-1},$ *for* $n \geq 2$, *then show that*

$$u_n = 3^n - (-2)^n \quad \text{for } n \geq 1.$$

Chapter Five
Group

Let G be a non-empty set. If f : $G \times G \rightarrow G$ be a function such that for any two elements a and b, the ordered pair (a, b) is in G, then we say that G is closed with respect to this binary function f and f is called a binary operation on G. If f is any binary operations like +, ∗,/ etc, then G is said to be closed with respect to a binary operation ∗ if and only if for any two elements a and b of G, a∗b is in G. For example, addition is a binary operation on the set of natural number N, because, the sum of any two natural numbers is a natural number. However, subtraction is not a binary operation on N as the $5 - 9 = -4 \notin N$. A binary operations on a set G is also called binary composition in a set G. In group theory, we are concerned with binary operation only. Therefore the phrase 'binary operation' and the word 'operation' are interchangeably used.

A non-empty set G is said to be equipped with a binary operation ∗ if ∗ is an operation on G, and this is denoted as (G, ∗). A non-empty set G equipped with one or more binary operations is called an algebraic structure. Thus, (G, ∗) is a representation of an algebraic structure with one operation. For example, (I, +), (N, +), (Q, –) etc, are algebraic structure with one operation, whereas, (R, +, ∗) is an algebraic structure with two operations.

5.1 Definitions and Properties

Let G be a non-empty set equipped with a binary operation denoted by ∗ i.e., a∗b or more precisely ab represents the element of G obtained by applying ∗ between a and b of G taken in that order. Then the algebraic structure (G, ∗), also called mathematical structure, is a group if the binary operation ∗ satisfies the following postulates:

1. **Closure property** i.e., $a * b \in G \ \forall\ a, b \in G$.
2. **Associative property** i.e., $(a * b) * c = a * (b * c) \ \forall\ a, b, c \in G$.
3. **Existence of Identity** i.e. there exists an element **e** $\in G$ s. t. $\mathbf{e} * a = a * \mathbf{e} = a \ \forall\ a \in G$. The element **e** is called the identity element of G if it exists.
4. **Existence of Inverse** i.e., each element of G possesses inverse. In other word, for $a \in G$ there exists an element b such that $a * b = b * a = \mathbf{e}$. The element b is then called the inverse of a and we write b as a^{-1}. Here a and b are inverse of each other.

Example 1 Show that (I, +) is a group, where I is the set of all integers and '+' is an integer addition operation.

Solution: To prove that (I, +) is a group, we shall have to show that it satisfies all the four postulates of a group.

Closure property: We know that the sum of two integers is also an integer i.e., for any a, $b \in I$, $a + b \in I$. So I is closed with respect to addition.

Associative property: We know that addition of integers is an associative operation. Thus, $a + (b + c) = (a + b) + c \ \forall \ a, b, c \in I$.

Existence of Identity: The number $0 \in I$. Also, we have $0 + a = a + 0 = a \ \forall a \in I$. Therefore, the integer 0 is the identity element for '+' and exists in I.

Existence of Inverse: If $a \in I$, then $-a \in I$. And, $(-a) + a = a + (-a) = 0$. This shows that every integer possesses additive inverse. Therefore, (I, +) is a group.

Proved.

Example 2 Prove that $(I_+, +)$, where I_+ is the set of positive integers is not a group with respect to addition.

Solution: Addition is obviously a binary composition in I_+ i.e., I_+ is closed with respect to addition. Also addition of positive integers is an associative operation. But, there exists no positive integer $e \in I_+$, such that $e + a = a + e = a \ \forall a \in I_+$. For the addition of positive integers, 0 is the identity and $0 \notin I_+$. Therefore, $(I_+, +)$ is not a group.

Proved.

It is superfluous to mention the **closure property** in the group definition as it is implied by the definition of binary operation. However, it is mentioned to the benefit of readers so that one should not forget to show this axiom while showing the group postulates in a problem.

A mathematical structure equipped with one binary operation is also called a **Groupoid** i.e., a mathematical structure (G, ∗) is said to be a **groupoid** if G is closed with respect to the binary operation ∗. Associativity, commutativity, etc. are not required. A groupoid can be empty. An associative groupoid is called a **Semi-group** i.e. if the binary operation ∗ is associative in G, then (G, ∗) is called a **Semi-group.** Further, if there exists an identity element '**e**' in an associative groupoid (G, ∗), then (G, ∗) is called a **Monoid.**

A group (G, ∗) is said to be an ***Abelian*** group if the binary operation ∗ is *commutative* in G i.e. for any two elements $a, b \in G$, $a * b = b * a$. For example, group (I, +) of example 1 is an abelian group because the addition of number is a commutative operation.

A group (G, $*$) is said to be a ***finite*** group if set G is a finite set, otherwise it is called an ***infinite*** group. *The number of elements in a finite group is called the* ***order*** *of the group.* An infinite group is called a group of infinite order.

Example 3 Let M be a set of all **m×n** matrices on real numbers and '+' be a binary operation of matrix addition. Show that (M, +) is an Abelian group.

Solution: To prove that (M, +) is an Abelian group, we shall have to show that it satisfies all the five postulates of an Abelian group.

Closure property: Let A and B be any two matrices of order **m×n** and belong to set M. Then A + B is again a matrix of order **m×n** and all its elements are real numbers, thus A + B is in M. Therefore, M is closed with respect to addition.

Associative property: We know that addition of matrices is an associative operation. Thus, $A + (B + C) = (A + B) + C \ \forall\ A, B, C \in M$.

Existence of Identity: 0 is a real number. A Null matrix **0** of order **m×n** belongs to M and for any matrix $A \in M$, we have

$$\mathbf{0} + A = A + \mathbf{0} = A \ \forall A \in M.$$

Therefore, the matrix **0** is the identity element for '+' and exists in M.

Existence of Inverse: Let A be any matrix of M. The matrix $-A \in M$. And,

$$-A + A = A + (-A) = \mathbf{0}.$$

This shows that every matrix possesses additive inverse. Therefore, (M, +) is a group.

Commutative property: We know that addition of matrices is a commutative operation. Thus, $A + B = B + A \ \forall\ A, B \in M$.

Hence (M, +) is an Abelian group.

Proved.

Example 4 Let P = {[0], [1], [2], ..., [m −1]} be the set of equivalence classes of I determined by the equivalence relation $\equiv_m$ on I. Here, m is any positive finite integer > 1. Show that (P, +) is an Abelian group where '+' is a binary operation of addition of equivalence classes.

Solution: The cardinality of P is m, which is a finite number. To show that (P, +) is an Abelian group, we have to prove that it satisfies all the five postulates of an Abelian group.

Closure property: Let [i] and [j] be any two equivalence classes from set P. Obviously, 0 ≤ i, j < m. When (i + j) is divided by m, it yields remainder say k. The k lies between 0 and m–1. Thus [i] + [j] = [k], and [k] ∈ P. It show that P is closed with respect to the addition of equivalence classes.

Associative property: We know that addition of number is an associative operation and so is the addition of equivalence classes [i]'s. Thus,

$$[i] + ([j] + [k]) = ([i] + [j]) + [k], \forall [i], [j], [k] \in P.$$

Existence of Identity: [0] ∈ M. Also we have

$$[0] + [k] = [k] + [0] = [k] \ \forall [k] \in P.$$

Therefore, the equivalence class [0] is the identity element for '+' and exists in P.

Existence of Inverse: If [k] be any equivalence class in P, then the equivalence class [m – k] also belongs to P. And,

$$[k] + [m - k] = [m - k] + [k] = [0].$$

This shows that every equivalence class in P possesses additive inverse. Therefore, (P, +) is a group.

Commutative property: We know that addition of number is a commutative operation and so is the addition of equivalence classes [i]'s. Thus,

$$[i] + [j] = [j] + [i], \forall [i], [j] \in P.$$

Hence, (P, +) is an Abelian group of finite order.

Proved

Example 5 Let Σ^* be a set of all strings on the set of alphabets Σ={a, b, c,..., z} and '*' be a binary operation of concatenation of two strings. Show that $(\Sigma^*, *)$ is not a group. Determine the type of algebraic structure of $(\Sigma^*, *)$.

Solution: The cardinality of Σ^* is infinite. To determine whether $(\Sigma^*, *)$ is a group, we have to test whether it satisfies each of the four postulates of a group.

Closure property: Let α and β be any two strings in Σ^*. Obviously, concatenated string α*β is also a string on the same alphabet Σ. It show that Σ^* is closed with respect to the concatenation of strings.

Associative property: Let α, β and γ be any three strings in Σ^*. Two strings (α*β)*γ and α*(β*γ) represent the same string on Σ obtained by concatenating β after α and then concatenating γ at the end. Thus,

$$\alpha * (\beta * \gamma) = (\alpha * \beta) * \gamma, \forall \alpha, \beta, \gamma \in \Sigma^*.$$

Existence of Identity: A null string is a string that contains no alphabets (symbols). It is denoted by ε or Λ. Obviously, ε is in Σ^*. Any string does not get altered when concatenated with ε in the left side or in the right side. Thus, we have

$$\varepsilon * \alpha = \alpha * \varepsilon = \alpha \ \forall \ \alpha \in \Sigma^*.$$

Therefore, null string ε is the identity element for the operation of concatenation '∗' and exists in Σ^*.

Existence of Inverse: Let α be any string in Σ^*. We cannot find any string β such that when concatenated with α, it nullifies α. This shows that there exists no inverse for strings in Σ^*.

Therefore, $(\Sigma^*, *)$ is a ***not*** a group. However, $(\Sigma^*, *)$ is a Monoid of infinite order because it is an associative groupoid of infinite order with identity element. This structure is also a semigroup. This semigroup is known as ***free semigroup.***

Proved.

Example 6 Show that the set of all positive rational number forms an Abelian group under the composition defined by

$$a * b = \frac{ab}{2}$$

Solution: Let Q_+ is the set of all positive rational numbers. We have to show that $(Q_+, *)$ satisfies the postulates for an Abelian group.

Closure property: Let a and b be any two elements of Q_+. Since a and b are positive rational number, (ab/2) is also a positive rational number i.e.,

$$\text{for any a, b} \in Q_+ \quad a * b = \frac{ab}{2} \in Q_+$$

So Q_+ is closed under ∗.

Associative property: Let a, b and c be any three elements of Q_+. Now, we have

$$(a * b) * c = \frac{(ab)c}{4} = \frac{a(bc)}{4} = a * (b * c)$$

This shows that ∗ is an associative binary operation on Q_+.

Existence of Identity: Let **e** be the identity element of ∗. Let us see whether it is in Q_+. If **e** is an identity element of ∗, then for any element $a \in Q_+$, we must have

$$a * \mathbf{e} = \mathbf{e} * a = a \Rightarrow \frac{a\mathbf{e}}{2} = a \Rightarrow \mathbf{e} = 2$$

Thus, 2 is the identity element of * and since 2 is a rational number, it is in Q_+. Therefore, identity element exists.

Existence of Inverse: Let a be any element of Q_+. Let b be an inverse of a. Then, we have

$$a * b = b * a = 2 \qquad [\because 2 \text{ is identity element}]$$

$$\Rightarrow \frac{ab}{2} = 2 \Rightarrow b = \frac{4}{a}$$

Since for any positive rational numbers a, 4/a is also a positive rational number, it is proved that every element of Q_+ has an inverse in Q_+.

Commutative property: Let a and b be any two elements of Q_+. Now, we have

$$a * b = \frac{ab}{2} = \frac{ba}{2} = b * a$$

This shows that * is a commutative binary operation on Q_+.

Hence, (Q_+, *) is an Abelian group. Its order is infinite.

Proved.

General Properties of a Group

Let (G, *) be a group. G possesses some properties regarding uniqueness of its identity element and uniqueness of inverse. In our school day, we have learnt that if 2x = 4y, then we can write x = 2y. This is simply because 2 has been cancelled from both sides. Is this cancellation allowed everywhere? The answer is no. It is possible only if we are dealing with an algebraic structure like group. Since (I, *) is not a group, we can not write, $0.x = 4.y \Rightarrow$ something meaningful. Properties like this and more will be discussed in this section.

Theorem 1 Let **e** be an identity element in a group (G, *), then **e** is unique. (This property is known as ***uniqueness of identity)***

Proof: Suppose **e** and **e′** are two identity elements in G. Then, we have

ee′ = **e** if **e′** is identity

ee′ = **e′** if **e** is identity

Since **ee′** is a unique element in G, **e** and **e′** are same and hence the identity element is unique in a group

Proved.

Theorem 2 Inverse of each element of a group (G, *) is unique. (This property is known as ***uniqueness of inverse)***

Proof: Let a be any element of G and **e** be the identity element of G. Suppose b and c be two different inverses of a in G, then we have

$$ba = \mathbf{e} = ab \quad \text{if b is inverse of a}$$
$$ca = \mathbf{e} = ac \quad \text{if c is inverse of a}$$

Also, we have $b(ac) = b\mathbf{e} = b$ and $(ba)c = \mathbf{e}c = c$. By associative property, we also have $b(ac) = (ba)c$. Thus, $b = c$. Therefore, a has a unique inverse. *It is important to note that an identity element is inverse of itself.*

Proved.

Theorem 3 If a^{-1} is the inverse of an element a of group $(G, *)$, then $(a^{-1})^{-1} = a$.

Proof: Let **e** be the identity element of group $(G, *)$. Then, we have

$$a^{-1}a = \mathbf{e} \Rightarrow (a^{-1})^{-1}(a^{-1}a) = (a^{-1})^{-1}\mathbf{e}$$
$$\Rightarrow ((a^{-1})^{-1}a^{-1})a = (a^{-1})^{-1}$$
$$\Rightarrow \mathbf{e}a = (a^{-1})^{-1} \Rightarrow a = (a^{-1})^{-1}$$

Proved.

*It is important to note that an additive inverse of an element **a** is denoted by **–a**.*

Theorem 4 If $(G, *)$ be a group then for any two elements a and b of $(G, *)$ prove that $(a*b)^{-1} = b^{-1}*a^{-1}$. (*This theorem is known as **rule of reversal***)

Proof: It is given that a^{-1} and b^{-1} are inverses of a and b respectively. If **e** is the identity element of $(G, *)$ then we have

$$a^{-1}*a = \mathbf{e} = a*a^{-1}$$
$$b^{-1}*b = \mathbf{e} = b*b^{-1}$$

Now, $$(a*b)*(b^{-1}*a^{-1}) = [(a*b)*b^{-1}]*a^{-1}$$
$$= [a*(b*b^{-1})]*a^{-1}$$
$$= (a*\mathbf{e})*a^{-1}$$
$$= a*a^{-1} = \mathbf{e}$$

Similarly, $(b^{-1}*a^{-1})*(a*b) = \mathbf{e}$

This shows that $b^{-1}*a^{-1}$ is inverse of $a*b$. Thus, $(a*b)^{-1} = b^{-1}*a^{-1}$

Proved.

Theorem 5 If a, b and c be any three elements of a group $(G, *)$, Then

$$ab = ac \Rightarrow b = c$$

and $$ba = ca \Rightarrow b = c$$

This theorem is known as ***Cancellation laws***.

Proof: $a \in G \Rightarrow \exists\, a^{-1} \in G$ such that $a^{-1}a = a\,a^{-1} = e$ where, e is the identity element of G.
Now

$$ab = ac \Rightarrow a^{-1}(ab) = a^{-1}(ac)$$ [Multiplying both sides on the left by a^{-1}]

$$\Rightarrow (a^{-1}a)b = (a^{-1}a)c$$ [By Associativity]

$$\Rightarrow eb = ec$$ [Since $a^{-1}a = e$]

$$\Rightarrow b = c$$ [Since e is identity element]

Also,

$$ba = ca \Rightarrow (ba)a^{-1} = (ca)a^{-1}$$ [Multiplying both sides on the right by a^{-1}]

$$\Rightarrow b(aa^{-1}) = c(aa^{-1})$$

$$\Rightarrow be = ce$$

$$\Rightarrow b = c$$

Proved.

Example 7 If a and b are any two elements of a group (G, ∗) then show that equations ax =b and ya =b have unique solutions in G.

Solution: $a \in G \Rightarrow \exists\, a^{-1} \in G$ such that $a^{-1}a = a\,a^{-1} = e$ where, e is the identity element of G. Therefore, $a \in G, b \in G \Rightarrow a^{-1} \in G, b \in G \Rightarrow a^{-1}b \in G$. Now substituting $x = a^{-1}b$ in the left hand side of the equation ax = b, we have

$$a\,(a^{-1}b) = (aa^{-1})b = eb = b$$

Thus $(a^{-1}b)$ is a solution in G of the equation ax = b. To show that the solution is unique, let us suppose that $x = x_1$ and $x = x_2$ are two solutions of the equation ax = b. Then $ax_1 = b$ and $ax_2 = b$ i.e. $ax_1 = ax_2$. Therefore, by left cancellation law, we have $x_1 = x_2$. This shows that the solution is unique. Similarly we can prove for the other equation.

Proved.

Example 8 Show that the set $G = \{a + b\sqrt{2} \mid a, b \in Q\}$ is a group with respect to addition.

Solution: We have to show that (G, +) satisfies the postulates of a group.

Closure property: Let $x = a + b\sqrt{2}$, $y = c + d\sqrt{2}$ be any two elements of G where a, b, c and d ∈ Q. Then, $x + y = (a + c) + (b + d)\sqrt{2} \in G$ because (a + c) and (b + d) are rational numbers. So G is closed with respect to addition.

Associative property: We know that addition of real numbers is an associative operation. Thus, $x + (y + z) = (x + y) + z\ \forall\ x, y, z \in I$.

Existence of Identity: Number $0 + 0\sqrt{2} \in G$ since $0 \in Q$. If $x = a + b\sqrt{2}$ be any element of G, then

$$(0 + 0\sqrt{2}) + (a + b\sqrt{2}) = (a + b\sqrt{2}) = (a + b\sqrt{2}) + (0 + 0\sqrt{2}).$$

Therefore, real number $(0 + 0\sqrt{2})$ is the identity element for '+' and exists in G.

Existence of Inverse: If $x = a + b\sqrt{2}$ be any element of G, then $-x = -a - b\sqrt{2}$ is also an element of G. And, $(-x) + x = x + (-x) = 0 + 0\sqrt{2}$. This shows that every element of G possesses additive inverse in G. Therefore, (G, +) is a group.

Proved.

Example 9 Show that the set of matrices

$$A_\alpha = \begin{bmatrix} \cos\alpha & -\sin\alpha \\ \sin\alpha & \cos\alpha \end{bmatrix}$$

where α is a real number from set R, forms a group under matrix multiplication.

Solution: Let G be set of A_α as defined above. We have to show that (G, $*$) satisfies the postulates of a group.

Closure property: Let A_α and A_β be two elements of G where, $\alpha, \beta \in R$. Then, we have

$$A_\alpha * A_\beta = \begin{bmatrix} \cos\alpha & -\sin\alpha \\ \sin\alpha & \cos\alpha \end{bmatrix} \begin{bmatrix} \cos\beta & -\sin\beta \\ \sin\beta & \cos\beta \end{bmatrix}$$

$$= \begin{bmatrix} \cos(\alpha+\beta) & -\sin(\alpha+\beta) \\ \sin(\alpha+\beta) & \cos(\alpha+\beta) \end{bmatrix} = A_{\alpha+\beta}$$

Since $\alpha+\beta \in R$, $A_\alpha * A_\beta \in G$. So G is closed with respect to matrix multiplication.

Associative property: We know that matrix multiplication is an associative operation. Thus, $A_\alpha * (A_\beta * A_\gamma) = (A_\alpha * A_\beta) * A_\gamma \ \forall\ A_\alpha, A_\beta, A_\gamma \in G$

Existence of Identity: Since 0 is a real number, the matrix

$$I = \begin{bmatrix} \cos 0 & -\sin 0 \\ \sin 0 & \cos 0 \end{bmatrix} = \begin{bmatrix} 1 & 0 \\ 0 & 1 \end{bmatrix} \in G$$

And for any matrix $A_\alpha \in G$, we have $A_\alpha * I = I * A_\alpha = A_\alpha$.
Therefore, matrix I is the identity element for matrix multiplication and exists in G.

Existence of Inverse: We know that whenever $\alpha \in R$, $-\alpha \in R$. As shown in closure property, we have $A_\alpha * A_{-\alpha} = A_{\alpha-\alpha} = A_0 = I = A_{-\alpha+\alpha} = A_{-\alpha} * A_\alpha$. This shows that every matrix in G possesses its matrix multiplicative inverse in G. Therefore, (G, *) is a group.

Proved.

Example 10 Let S be any non-empty set and let A(S) be the set of all one-to-one functions of the set S onto itself. Show that A(S) is a group with respect to composition of functions as the binary operation on A(S). Is it an abelian group?

Solution: Let the binary operation of composition of function be denoted as 'o'. Now we have to show that (A(S), o) satisfies all the postulates of a group.

Closure property: Let f and g be any two elements of A(S). Then f and g are one-to-one functions from S onto itself and hence fog is also one-to-one function from S onto itself. This shows that fog $\in$ A(S) $\forall$ f, g $\in$ A(S). Therefore, A(S) is closed with respect to o.

Associative property: We know that composition of function is an associative operation. Thus, f o (g o h) = (f o g) o h $\forall$ f, g, h $\in$ A(S).

Existence of Identity: Let I be identity function from S to itself. Clearly, I is one-to-one onto function so I $\in$ A(S). Also Iof = foI = f $\forall$ f $\in$ A(S). Therefore, the function I is the identity element for o and exists in A(S).

Existence of Inverse: Let f be any element of A(S). Then $f^{-1} \in$ A(S). And,

$$fof^{-1} = f^{-1}of = \mathbf{I}.$$

This shows that every element has an inverse in A(S). Therefore, (A(S), o) is a group.

We know that function composition is not a commutative operation, (A(S), o) is not an Abelian group.

Proved.

Composition Table

Let S = $\{a_1, a_2, a_3, a_4, \ldots a_n\}$ is a finite set having n elements. Let there be a binary operation defined on S multiplicatively. All possible binary composition of elements of S can be arranged in a tabular form as:

- Write elements of S in a horizontal row, say it ***column header***
- Write elements of S in a vertical column, say it ***row header***
- The element $a_i * a_j$ associated with the ordered pair (a_i, a_j) is placed at the intersection of the row headed by a_i and column a_j.

The table so obtained is called ***composition table*** for a finite set under a given binary operation. The composition table for a finite group contains each element exactly once in each of its rows and columns.

Example 11 Show that the four fourth roots of unity namely: 1, -1, i, -i; form a group with respect to multiplication.

***Solution*:** Let G = {**1, -1, i, -i**}. Let $*$ be the binary operation of multiplication in G. To prove that (G, $*$) is a group, we form the composition table as below and then we shall show that it satisfies all the postulates of a group.

*	1	-1	i	-i
1	1	-1	i	-i
-1	-1	1	-i	i
i	i	-i	-1	1
-i	-i	i	1	-1

Table 5.1.1

Closure property: Since all the entries in the composition table are elements from the set G, G is closed with respect to the binary operation of multiplication.

Associative property: The elements of G are all complex numbers and the multiplication of complex numbers is associative.

Existence of Identity: From the composition table, it is obvious that the row and column headed by 1 in the table just coincide with row and column headers respectively. Thus, we have

$$1*a = a*1 = a \ \forall \ a \in G$$

Thus 1 is the identity element of G.

Existence of Inverse: Since every row and every column contains the identity element, and that too only once, each element of G has an inverse in G.

Hence (G, *) is a group. Since the table shown is a symmetric, * is a commutative operation on G, thus (G, *) is an Abelian group also.

Proved.

Example 12 Show that the set $G = \{1, \omega, \omega^2\}$, where ω is an imaginary cube root of unity is a group with respect to multiplication.

Solution: Let us form composition table for (G, *) as shown below.

*	1	ω	ω^2
1	1	ω	ω^2
ω	ω	ω^2	1
ω^2	ω^2	1	ω

Table 5.1.2

Closure property: Since all the entries in the composition table are elements from the set G, G is closed with respect to the binary operation of multiplication.

Associative property: The elements of G are all complex numbers and we know that multiplication of complex numbers is associative.

Existence of Identity: From the composition table, it is obvious that the row and column headed by 1 in the table just coincide with row and column headers respectively. Thus, we have

$$1*a = a*1 = a \ \forall \ a \in G$$

Thus 1 is the identity element of G.

Existence of Inverse: Since every row and every column contains the identity element, and that too only once, each element of G has an inverse in G.

Hence (G, *) is a group. Since the table shown is a symmetric, * is a commutative operation on G, thus (G, *) is an Abelian group also of order 3.

Proved.

Example 13 Show that the set of six functions f_1, f_2, f_3, f_4, f_5 and f_6 defined by

$$f_1(z) = z, \quad f_2(z) = \frac{1}{z}, \quad f_3(z) = 1 - z, \quad f_4(z) = \frac{z1}{1-z},$$

$$f_5(z) = \frac{1}{1-z}, \quad f_6(z) = \frac{1-z}{z}$$

on the set of complex numbers forms a finite non-abelian group of order six with respect to the composition of functions.

Solution: Let G = $\{f_1, f_2, f_3, f_4, f_5, f_6\}$. Let the binary operation of function composition is represented by o. We have to prove that (G, o) is a non-abelian group. The composition table for (G, o) can be given as shown in the table 5.1.3.

o	f_1	f_2	f_3	f_4	f_5	f_6
f_1	f_1	f_2	f_3	f_4	f_5	f_6
f_2	f_2	f_1	f_5	f_6	f_3	f_4
f_3	f_3	f_6	f_1	f_5	f_4	f_2
f_4	f_4	f_5	f_6	f_1	f_2	f_3
f_5	f_5	f_4	f_2	f_3	f_6	f_1
f_6	f_6	f_3	f_4	f_2	f_1	f_5

Table 5.1.3

Closure property: Since all the entries in the composition table are elements from the set G, G is closed with respect to the binary operation of multiplication.

Associative property: We know that composition of functions is an associative binary operation hence o is associative in G

Existence of Identity: From the composition table, it is obvious that the row and column headed by f_1 in the table just coincide with row and column headers respectively. Thus, we have

$$f_1 * a = a * f_1 = a \ \forall \ a \in G$$

Thus f_1 is the identity element of G.

Existence of Inverse: Since every row and every column contains the identity element f_1, and that too only once, each element of G has an inverse in G.

Hence (G, o) is a group. Since the table shown is not symmetric, o is not a commutative operation on G, thus (G, *) is a non-abelian group of order six.

Proved.

Example 14 Let A = {[0], [1], [2]} be a partition of Z –set of integers, by the equivalence relation '*congruence modulo 3*'. Show that (A, +) is an Abelian group, where + is a binary operation of '*addition of residue classes*'.

Solution: The composition table for the structure (A, +) can be given as in the table 5.1.4.

+	[0]	[1]	[2]
[0]	[0]	[1]	[2]
[1]	[1]	[2]	[0]
[2]	[2]	[0]	[1]

Table 5.1.4

Closure property: Since all the entries in the composition table are elements from the set A, A is closed with respect to the binary operation of *addition of residue classes*.

Associative property: Since addition of integers is an associative operation, *addition of residue classes* is also an associative operation.

Existence of Identity: From the composition table, it is obvious that the row and column headed by [0] in the table just coincide with row and column headers respectively. Thus, we have

$$[0]*a = a*[0] = a \ \forall \ a \in A$$

Thus [0] is the identity element of A.

Existence of Inverse: Since every row and every column contains the identity element, and that too only once, each element of A has an inverse in A.

Hence (A, +) is a group. Since the table shown is a symmetric, + is a commutative operation on A, thus (A, +) is an Abelian group also of order 3.

Proved.

Example 15 Let T = {±1, ±i, ±j, ±k}. Define a multiplicative binary operation '*' on T by setting

$$i * i = j * j = k * k = 1$$
$$i * j = -j * i = k$$
$$j * k = -k * j = i$$
$$k * i = -i * k = j$$

Show that (T, *) is a non-abelian group. This group is called '***Quaternion group***'.

Solution: We can form a composition table to show this. Readers are suggested to do this. After a careful analysis of elements of T, we can say that T is a set of following matrices:

$$\pm\begin{bmatrix} 1 & 0 \\ 0 & 1 \end{bmatrix}, \pm\begin{bmatrix} i & 0 \\ 0 & -i \end{bmatrix}, \pm\begin{bmatrix} 0 & i \\ -i & 0 \end{bmatrix} \pm\begin{bmatrix} 0 & 1 \\ -1 & 0 \end{bmatrix}$$

The structure (T, +), where + is a binary operation of matrix addition on T, is a non-abelian group. This can be proved as in the example 3.

Ans.

5.2 Subgroup

Let (G, *) be a group. *Any non-empty subset of G is called a* ***complex*** *of the group G.* Let H be any non-empty subset of G, then H is a complex of G. Further, H is said to be a **Stable** complex of G if

$$x \in H \text{ and } y \in H \Rightarrow x * y \in H$$

Otherwise, H is called an **unstable** complex of G. The binary operation * is called an ***induced*** operation on H by G. If H is a group with respect to this induced binary operation then H is called a subgroup of G. A subgroup can be defined as

*"A non-empty subset H of a group (G, *) is said to be a subgroup of G if (H, *) is also a group, where * is an induced binary operation in H by G."*

It is important to note that every subgroup H of a group G is a complex of G but the reverse is not always true. Since every set is a subset of itself, G is itself a subgroup of G. Further, if e is the identity element of G, then the singleton {e} is also a subgroup of G.

These two subgroups of any group G are called **trivial** or **improper** subgroup of G. Otherwise a subgroups is called a **proper** subgroup.

In general, A non-empty subset H of a mathematical structure (G, *) is said to be a sub-structure of G if (H, *) satisfies all postulates satisfied by G. For example, if (G, *) is a semi-group then (H, *) is said to be a sub-semi-group of G if (H, *) is a semi-group. The same argument is true for Monoid and Abelian group.

Example 1 The multiplicative group {1, -1} is a subgroup of the multiplicative group {1, -1, -i, -i}. ♦

Example 2 The multiplicative group of positive rational number is a subgroup of the multiplicative group of all non-zero rational numbers. ♦

Example 3 Let G be the additive group of integers. Then show that the set of all multiples of integers by a fixed integer **m** is a subgroup of G.

Solution: Here G = {…, –3, –2, –1, 0, 1, 2, 3, …} is the additive group of integers. Let H be the set of all multiples of integers by a fixed number say m, then, we have

$$H = \{\ldots, -3m, -2m, -m, 0, m, 2m, 3m, \ldots\}$$

Obviously, $H \subseteq G$. Now we have to show that (H, +) is a group. This can be proved as is done in the example 1 of 5.1. ♦

Theorem 1 Let (G, *) be a group and H be any subgroup of G. Show that

(i) The identity of H is the same as that of the G.

(ii) The inverse of any element **a** of H is the same as the inverse of the same element regarded as an element of G.

Proof: (i) Let e and e' be identities of G and H respectively. Let a be any element of H. Then, we have

$$a*e' = e'*a = a \quad \text{[e' is the identity in H]}$$

Since $a \in H \Rightarrow a \in G$, we have

$$a*e = e*a = a \quad \text{[e is the identity in G]}$$

Therefore, in G we have $a*e = a*e'$

$$\Rightarrow e = e' \quad \text{[By left cancellation law in G]}$$

Thus, identity of H is the same as that of the G.

Proved.

(ii) Let b is the inverse of **a** in H and c be the inverse of a in G. Since H and G have the same identity e, we have

$$a*b = b*a = e \text{ and}$$
$$a*c = c*a = e$$

Therefore, $a*b = a*c \Rightarrow b = c$ [By left cancellation law in G]

Proved.

Order of an element in a Group

Suppose (G, *) be a group. Let **a** be any element of G and **e** be the identity element in G. *Order of **a** is the least positive integer n such that*

$$a*a*a*\ldots*a \text{ (n times)} = \mathbf{e}$$

This is also written as $\mathbf{a}^n = \mathbf{e}$, when binary operation is represented multiplicatively and **na** = **e**. If there exists a positive integer n, then we say that **a** is of finite order otherwise it is of infinite order. Order of any element a is denoted by O(a). In any group identity element is of order one and it is the only element of order one.

Theorem 2 Let (G, *) be a group and H be any subgroup of G. Show that the order of any element of H is the same as the order of the element regarded as a member of G.

Proof: Let $a \in H$. Let O(a) = n in H. Thus, we have $a^n = e$, where e is the identity element in H. Since e is the identity element in G also, so we have $a^n = e$ in G. Hence order of any element of H is the same as the order of the element regarded as a member of G.

Proved.

Product of Complexes

Let H and K be two complexes of a group (G, *). The product HK is defined as

$$\{x \mid x \in G \text{ and } x = h*k \text{ for } h \in H, k \in K\}$$

Obviously $HK \subseteq G$. Thus HK is a complex of G consisting of the elements of G obtained on multiplying each member of H with each member of K. It is easy to verify that multiplication of complexes is an associative operation. Let h, k, l be any arbitrary elements of H, K, L respectively so that $(hk)l \in (HK)L$.

Since (hk)l = h(kl), we have $(hk)l \in H(KL)$. Thus $(HK)L \subseteq H(KL)$. Similarly, we can show that $H(KL) \subseteq (HK)L$. Therefore H(KL) = (HK)L i.e. multiplication of complexes is an associative operation. It is important to note that HK = KH does not necessarily mean that hk = kh for all h in H and for all k in K. What it means actually that each element of HK is present in KH and vice versa.

Inverse of a Complex

Let H be any complex of G. Then we define $H^{-1} = \{h^{-1} \mid h \in H\}$ i.e. H^{-1} is a complex of G consisting of the inverses of the elements of H.

Example 4 Let H and K be any two complex of a group G, then show that

$$(HK)^{-1} = K^{-1}H^{-1}.$$

Solution: Let x be any element of $(HK)^{-1}$. Then

$$x = (hk)^{-1} \text{ where } h \in H \text{ and } k \in K$$
$$= k^{-1}h^{-1} \in K^{-1}H^{-1}.$$
$$\Rightarrow (HK)^{-1} \subseteq K^{-1}H^{-1} \quad \text{———— (1)}$$

Similarly, let y be any element of $K^{-1}H^{-1}$. Then

$$y = k^{-1}h^{-1} \text{ where } k \in K \text{ and } h \in H$$
$$= (hk)^{-1} \in (HK)^{-1}$$
$$\Rightarrow K^{-1}H^{-1} \subseteq (HK)^{-1} \quad \text{———— (2)}$$

Therefore, from equation (1) and (2), we have

$$(HK)^{-1} = K^{-1}H^{-1}$$

Ans.

Example 5 Let H be any subgroup of G then H^{-1}=H. Also show that the converse is not true.

Solution: Let h^{-1} be any element of H^{-1}, then $h \in H$. Since H is a subgroup of G, $h \in H \Rightarrow h^{-1} \in H$. Thus, $H^{-1} \subseteq H$. Now let $h \in H$. Then,

$$h \in H \Rightarrow h^{-1} \in H$$
$$\Rightarrow (h^{-1})^{-1} \in H^{-1}$$
$$\Rightarrow h \in H^{-1}$$

Therefore, we have $H \subseteq H^{-1}$. This shows that $H = H^{-1}$.

Next, let us suppose that H is a complex of G such that $H = H^{-1}$. Then we have to show that H is not necessarily a subgroup of G. *It can be proved by method of counter example.* Let H = {–1} be a complex of a multiplicative group G = {1, –1}. Since inverse of –1 is –1, H^{-1} = {–1}. This shows that though $H=H^{-1}$, H is not a subgroup of G.

Ans.

Example 6 Show that HH = H, where H is any subgroup of a group G.

Solution: Let h_1h_2 be any element of HH, where $h_1 \in H$ and $h_2 \in H$. Since H is a subgroup of G, thus $h_1 \in H$ and $h_2 \in H \Rightarrow h_1h_2 \in H$. Therefore $HH \subseteq H$.

Now let h be any element of H, then we can write h = he, where e is the identity element of G. Also $he \in HH$ since $h \in H$ and $e \in H$. This implies that $H \subseteq HH$.

Hence, we have HH = H.

Proved.

Theorem 3 A nonempty subset H of a group G is a subgroup of G if and only if

$a \in H\ b \in H \Rightarrow ab^{-1} \in H$ where b^{-1} is inverse of the element b in G.

Proof: The proof is divided into parts:

1. If H is a subgroup of G then $a \in H\ b \in H \Rightarrow ab^{-1} \in H$ and
2. If $a \in H\ b \in H \Rightarrow ab^{-1} \in H$

1st Part: Since H is a subgroup, $b \in H \Rightarrow b^{-1} \in H$ because inverse of every element exists in H. Also, a and $b \in H \Rightarrow a, b^{-1} \in H \Rightarrow ab^{-1} \in H$ because H is closed with respect to the binary operation of G.

2nd Part: Let us suppose that for any two elements a and b of G, we have $ab^{-1} \in H$. Then we have to show that H is a subgroup of G.

Existence of Identity We have from the given condition $a \in H$, $a \in H \Rightarrow aa^{-1} \in H$. And $aa^{-1} = e$, where e is the identity element of group G. Thus e is in H.

Existence of Inverse Let a be any element of H. Then by the given condition, we have

$$e \in H \text{ and } a \in H \Rightarrow ea^{-1} \in H \Rightarrow a^{-1} \in H.$$

Thus each element of H possesses inverse.

Closure Property Let a and b be any two elements of H. As shown above

$$b \in H \Rightarrow b^{-1} \in H$$

Now, $a \in H$ and $b^{-1} \in H \Rightarrow a(b^{-1})^{-1} \in H \Rightarrow ab \in H$. Hence H is closed with respect to the binary operation of group G.

Associative Property The elements of H are also elements of G. The composition in G is associative, thus it is also associative in H.

Hence H is itself is a group under the composition in G. Therefore H is a subgroup of G.

Proved.

Example 7 Let G be the set of all ordered pairs (a, b) of real numbers for which $a \neq 0$. Let a binary operation $*$ on G be defined by the formula

$$(a, b) * (c, d) = (ac, bc + d)$$

Show that (G, $*$) is a non-abelian group. Does the subset H of all those elements of G which are of the form (1, b) form a subgroup of G?

Solution: It can be easily shown that (G, $*$) is a non-abelian group. The ordered pair (1, 0) is the identity element and for any ordered pair (a, b), its inverse is given by

$$(a,b)^{-1} = \left(\frac{1}{a}, -\frac{b}{a}\right)$$

To find whether H is a subgroup of G, let us assume that (1, a) and (1, b) be any two elements in H. The inverse of (1, b) under the given composition is (1, –b). Now

$$(1, a) * (1, b)^{-1} = (1, a) * (1, -b) = (1, a - b)$$

Since (1, a – b) definitely belongs to H. Thus from the *theorem 3,* H is a subgroup of G.

Ans.

Example 8 Let H be a subgroup of a group G and define $T = \{x \mid x \in G \text{ and } xH = Hx\}$. Prove that T is a subgroup of G.

Solution: Let x and y be any two elements of T. Then, we have

$$xH = Hx \text{ and } yH = Hy.$$

Let y^{-1} be the inverse of y. we shall show that $y^{-1} \in T$.

We have $yH = Hy \Rightarrow y^{-1}(yH)y^{-1} = y^{-1}(Hy)y^{-1}$

$$\Rightarrow (y^{-1}y)Hy^{-1} = y^{-1}H(yy^{-1}) \text{ [By associative property]}$$

$$\Rightarrow Hy^{-1} = y^{-1}H$$

This shows that y^{-1} is in T. Now

$(xy^{-1})H = x(y^{-1}H) = x(Hy^{-1}) = (xH)y^{-1} = (Hx)y^{-1} = H(xy^{-1})$ i.e

$$(xy^{-1})H = H(xy^{-1}) \Rightarrow xy^{-1} \in T$$

Therefore T is a subgroup of G.

Ans.

Example 9 Show that the union of two subgroups is a subgroup if and only if one is contained in the other.

Solution: Let H and K be any two subgroups of a group G. Let either $H \subseteq K$ or $K \subseteq H$. Then $H \cup K = H$ or $H \cup K = K$. Since H and K both are subgroup of G, $H \cup K$ is also a subgroup of G.

Now, let us suppose that $H \cup K$ is a subgroup of G, then we have to show that either $H \subseteq K$ or $K \subseteq H$. Let us assume that neither $H \subseteq K$ nor $K \subseteq H$. Here

H is not a subset of K $\Rightarrow \exists\, a \in H$ such that $a \notin K$ ----------(1)

Similarly, K is not a subset of H $\Rightarrow \exists\, b \in K$ such that $b \notin H$ ---------(2)

From hypothesis (1) and (2), we have $a, b \in H \cup K$. Since $H \cup K$ is a group, ab = c(say) $\in H \cup K$. Now,

$$ab = c(\text{say}) \in H \cup K \Rightarrow c \in H \text{ or } c \in K$$

Suppose $c \in H$. Since $a \in H$ and H is a group, we have $a^{-1} \in H$. Also

$$a^{-1} \text{ and } c \in H \Rightarrow a^{-1}c \in H$$

This implies that $b = a^{-1}c \in H$. This is contrary to the hypothesis (2) that $b \notin H$. Similarly, we can prove that $a \in K$ contrary to the hypothesis (1). Hence either H is contained in K or K is contained in H.

Ans.

Example 10 Let H and K be any two subgroups of a group G. Prove that $H \cap K$ is also a subgroup of G.

Solution: Readers may try.

Note: It can also be shown that arbitrary intersection of subgroups of a group G is again a subgroup.

Coset

Suppose G is a group and H is any subgroup of G. Let a be any element of G. The set $Ha = \{ha \mid h \in H\}$ is called ***right coset*** of H in G generated by the element a. Similarly the set $aH = \{ah \mid h \in H\}$ is called a ***left coset*** of H in G generated by a.

It is obvious that aH and Ha both are subsets of G. If e is the identity element of G then aH = Ha. Therefore H itself is a right as well as a left coset. Since H is a subgroup and e is the identity element of G then $e \in H$ and $ea = a \in Ha$. This shows that a right coset Ha is non-empty. Similarly a left coset of a subgroup H is always non-empty.

If group G is an abelian group, then we have $ah = ha \ \forall \ h \in H$. Thus right coset Ha will be equal to the corresponding left coset aH. However, in the case of a non-abelian group, we may not have the same result always.

It is obvious that cosets of a subgroup are special type of complexes. These are also known as ***residue classes modulo the subgroup***.

Example 11 Let G be the additive group of integers i.e. $G = \{\ldots, -3, -2, -1, 0, 1, 2, 3, \ldots\}$. Let $H = \{\ldots, -9, -6, -3, 0, 3, 6, 9, \ldots\}$ be a subgroup of G. Find the right cosets of H in G.

Solution: It is easy to see that G is an abelian group. It is therefore, any right coset will be equal to the corresponding left coset. Let us form different right cosets of H in G.

We have 0 in G, so $H+0 = H = \{\ldots, -9, -6, -3, 0, 3, 6, 9, \ldots\}$
Also 1 is in G, so $H+1 = \{\ldots, -8, -5, -2, 1, 4, 7, 10, \ldots\}$
And 2 is in G, so $H+2 = \{\ldots, -7, -4, -1, 2, 5, 8, 11, \ldots\}$

It can be easily verified that the coset $H+3 = H+0$, $H+4 = H+1$, $H+5 = H+2$ and so on. This implies that there are three distinct right cosets: $H+0$, $H+1$, $H+2$. All these three are mutually disjoint. Obviously,

$$G = (H+0) \cup (H+1) \cup (H+2)$$

Ans.

Example 12 For any subgroup H of a group G, show that if $h \in H$ then

$$hH = H = Hh.$$

Solution: Let $h \in H$ and k be any arbitrary element of H. Then hk is an element of hH. Also we have $h \in H$ and $k \in H \Rightarrow hk \in H$. Thus $hH \subseteq H$.

Next, we have $k = (hh^{-1})k = h(h^{-1}k)$. This is because $h \in H \Rightarrow h^{-1} \in H$. Also $h^{-1} \in H$ and $k \in H \Rightarrow h^{-1}k \in H$. Therefore, $k = h(h^{-1}k) \in hH$. This implies that every element of H are also in hH, thus $H \subseteq hH$.

Hence finally $hH \subseteq H$ & $H \subseteq hH \Rightarrow H = hH$. Similarly, we can prove that $H = Hh$.

Ans.

Theorem 4 Any two cosets (right or left) of a subgroup are either disjoint or identical.

Proof: Let H is a subgroup of any group G and let Ha and Hb be any two right cosets of H in G, where a, b be any element of G. If Ha and Hb are disjoint then we have nothing to prove. Let Ha and Hb are not disjoint i.e. $\exists$ and element c such that c is in both Ha and Hb.

Let $c = h_1a$ and $c = h_2b$, where $h_1, h_2 \in H$. Now, we have $h_1a = h_2b \Rightarrow a = (h_1^{-1}h_2)b$. Since $(h_1^{-1}h_2) \in H$, we have $H(h_1^{-1}h_2) = H$. Therefore, $Ha = H(h_1^{-1}h_2)b = (Hh_1^{-1}h_2)b = Hb$.

Proved.

Theorem 5 If H is a subgroup of G, then G is equal to the union of all right cosets of H in G i.e.,

$$G = H \cup Ha \cup Hb \cup Hc \cup \ldots$$

Where a, b, c ... are elements of G.

Proof: G is a group. Each element of any right coset of H in G is an element of G. Hence union of all right cosets of H is a subset of G.

Now let y be any element of G then $y \in Hy$. Thus every element of G belongs to union of right cosets of H in G. Thus G is a subset of union of all right cosets of H in G. Therefore G is equal to the union of all right cosets of H in G.

Proved.

Example 13 Let H be any subgroup of a group G. Prove that there is a one-to-one correspondence between any two right cosets of H in G.

Solution: The reader is advised to do this problem using the concept of functions given in the chapter 2.♦

Normal Subgroup

"A subgroup H of G is said to be a ***normal subgroup*** *of a group G if and only if for every x in G we have xH = Hx i.e. for every h of H, we have* $xhx^{-1} \in H$."

If a group G is an abelian group, then by the definition, every subgroup of G is a *normal subgroup.*

Every group G possesses at least two normal subgroups: the group G itself and the subgroup consisting of identity element e alone. *These two subgroups are called* ***improper normal subgroups***. In practice we encounter many group for which these two are the only normal subgroups. Such a group is called ***Simple Group*** i.e. *"A group having no proper normal subgroup is called a simple group".*

Example 14 Show that every subgroup of an abelian group is normal.

Solution: Let x be any element of G and h be any element of H. Thus, we have

$$xhx^{-1} = xx^{-1}h = eh. \quad \text{[Since G is an abelian group]}$$

Therefore $\forall x \in G$ and $\forall h \in H$, we have $xhx^{-1} \in H$ i.e. xH = Hx. This shows that H is a normal subgroup of G.

Ans.

Example 15 If N and M are normal subgroups of a group G, prove that NM is also a normal subgroup of G.

Solution: Let us first prove that NM is a subgroup of G. To prove that NM is a subgroup of G, it is sufficient to prove that NM = MN.

Let $nm \in NM$ where $n \in N$ and $m \in M$. We write $nm = mm^{-1}nm = m(m^{-1}nm)$. Since N is a normal subgroup, we have $m^{-1}nm \in N$. Thus $m(m^{-1}nm) \in MN$. Hence

$$NM \subseteq MN$$

Similarly, we can show that

$$MN \subseteq NM$$

Combining these two, we have

$$NM = MN$$

Therefore NM is a subgroup of G.

Now let us show that NM is a normal subgroup of G. Let x be any element of G and nm be any element of NM, where $n \in N$ and $m \in M$. Then, we have

$$x(nm)\,x^{-1} = (xn\,x^{-1})(xm\,x^{-1}) \in NM.$$

Therefore NM is a normal subgroup of G.

Ans.

Theorem 6 A subgroup H of a group G is normal if and only if

$$xHx^{-1} = H \ \forall x \in G$$

Proof: Let $xHx^{-1} = H \ \forall x \in G$. Then $xHx^{-1} \subseteq H \ \forall x \in G$. Thus H is a normal subgroup of G.

Conversely, let H be a normal subgroup of . Then $xHx^{-1} \subseteq H \ \forall x \in G$. ----(1)
Also $x \in G \Rightarrow x^{-1} \in G$. Therefore, we have

$$x^{-1}H(x^{-1})^{-1} \subseteq H \ \forall x \in G \Rightarrow x^{-1}Hx \subseteq H \ \forall x \in G$$

$$\Rightarrow x(x^{-1}Hx)x^{-1} \subseteq xHx^{-1} \ \forall x \in G$$

$$\Rightarrow H \subseteq xHx^{-1} \ \forall x \in G \text{ -------(2)}$$

From results (1) and (2), we have $xHx^{-1} = H \ \forall x \in G$.

Proved.

Theorem 7 A subgroup H of a group G is a normal subgroup of G if and only if the product of two right cosets of H in G is again a right coset of H in G.

Proof: Let H be a normal subgroup of a group G. Let x, y be any two elements of G. The Hx and Hy are two right cosets of H in G. Now, we have

$$\begin{aligned}(Hx)(Hy) &= H(xH)y \\ &= H(Hx)y \quad [\text{H is normal subgroup so } Hx = xH] \\ &= HH(xy) \quad [\text{His normal so } HH = H] \\ &= Hxy\end{aligned}$$

Since $x \in G$ and $y \in G \Rightarrow xy \in G$. Thus Hxy is also a right coset of H in G.

Conversely, let us assume that H is a subgroup of G such that product of two right cosets of H in G is again a right cosets of H in G. Let x be any element of G, then $x^{-1} \in G$ and Hx and Hx^{-1} are two right cosets of H in G. Consequently, by hypothesis, $HxHx^{-1}$ is again a right coset of H in G. Since $e \in H$, we have $exex^{-1} = e$ in $HxHx^{-1}$

We know that two right cosets are either disjoint or identical, thus

$$e \in H \text{ and } e \in HxHx^{-1} \Rightarrow H = HxHx^{-1} \ \forall x \in G$$

Now,

$$HxHx^{-1} = H \ \forall x \in G \Rightarrow hxkx^{-1} \in H \ \forall x \in G \text{ and } \forall h, k \in H$$

$$\Rightarrow h^{-1}hxkx^{-1} \in h^{-1}H \ \forall x \in G \text{ and } \forall h, k \in H$$

$$\Rightarrow xkx^{-1} \in H \ \forall x \in G \text{ and } \forall k \in H$$

Therefore, H is a normal subgroup.

Proved.

Example 16 Let H is the only subgroup of finite order m in the group G. Prove that H is a normal subgroup of G.

Solution: Let H is a subgroup of G and O(H) = m. If x is any element of G then it can be easily proved that xHx^{-1} is also a subgroup of G. Now let

$$H = \{h_1, h_2, h_3, \ldots, h_n\}$$

Then,

$$xHx^{-1} = \{xh_1x^{-1}, xh_2x^{-1}, xh_3x^{-1}, \ldots, xh_nx^{-1}\}$$

Obviously, the number of distinct element in xHx^{-1} is m. Thus $O(xHx^{-1}) = m$. But is given that H is the only subgroup of finite order m, thus $H = xHx^{-1}$ $\forall\, x \in G$. Thus H is a normal subgroup of G.

Ans.

5.3 Homomorphism of groups

Let $(G_1, *_1)$ and $(G_2, *_2)$ be any two groups. Let $f: G_1 \rightarrow G_2$ be a function. The function f is said to be a ***homomorphism*** of G_1 ***into*** G_2 if

$$f(a *_1 b) = f(a) *_2 f(b) \ \forall\, a, b \in G_1$$

i.e. f preserves the composition in G_1 and G_2. A function $f: G_1 \rightarrow G_2$ is said to be ***homomorphism*** of G_1 ***onto*** G_2 if

1. f is an onto function and
2. $f(a *_1 b) = f(a) *_2 f(b) \ \forall\, a, b \in G_1$

A group G_2 is called ***homomorphic image*** of G_1 under f if f is a homomorphism onto from G_1 to G_2. *A homomorphism of a group* G_1 *onto itself is called* ***endomorphism.*** Let us see now the following example.

Example 1 Let $(G_1, *_1)$ be a group of all ordered pairs (a, b) of real numbers with the binary operation $*_1$ defined by

$$(a, b) *_1 (c, d) = (a + c, b + d)$$

Next, let $(G_2, *_2)$ be the group of real numbers, where $*_2$ is a binary operation of addition of real number. Let f be a function $f: G_1 \rightarrow G_2$ defined by

$$f(a, b) = a \ \forall\, (a\ b) \in G_1$$

Show that f is a homomorphism of G_1 onto G_2.

Solution: We have shown in the beginning of this chapter that (R, +) is a group, thus (G_2, $*_2$) is a group. Also it can be easily verified that (G_1, $*_1$) is a group. Only thing required to prove is that f is homomorphism of G_1 onto G_2. The given function f is onto function because for any real number $x \in G_2$, we have an ordered pair $(x, y) \in G_1$. To show that f preserves the composition, let (a, b) and (c, d) be any two ordered pairs in G_1. From the definition of f, we have

$$f(a, b) = a \text{ and } f(c, d) = c.$$

And from the definition of $*_1$ and $*_2$, we have

$$(a, b) *_1 (c, d) = (a + c, b + d)$$

$$a *_2 c = a + c$$

Thus, $f((a, b) *_1 (c, d)) = f((a + c, b + d)) = a + c = a *_2 c = f(a, b) *_2 f(c, d)$. This proves that f preserves the composition. Hence f is a homomorphism of a G_1 onto G_2.

Ans.

Example 2 Let G be a group and let e be the identity element of G. Show that the function f: G→G defined by $f(a) = e \ \forall \ a \in G$ is an endomorphism of G.

Solution: Let a and b be any two elements of G. Thus, we have

$$f(a) = e \text{ and } f(b) = e$$

Also we have

$$f(ab) = e = ee = f(a)f(b)$$

This shows that f preserves the composition and it is a function from G to G. Therefore, f is an endomorphism of G.

Ans.

Theorem 1 If f is a homomorphism of a group G_1 into a group G_2, then

(i) $f(e_1) = e_2$, where e_1 is identity element of G_1 and e_2 is the identity element of G_2.

(ii) $f(a^{-1}) = [f(a)]^{-1} \ \forall \ a \in G_1$

(iii) If the order of $a \in G_1$ is finite, then the order of f(a) is a divisor of the order of a.

Proof: (i) Let a be any element of G_1, then $f(a) \in G_2$. Now, we have

$$f(a)\, e_2 = f(a) = f(a\, e_1) = f(a)f(e_1)$$

Now, by left cancellation law in G_2, we have

$$f(a)\, e_2 = f(a)f(e_1) \Rightarrow e_2 = f(e_1)$$

(ii) Let a be any element of G_1, then $a^{-1} \in G_1$ and $f(a), f(a^{-1}) \in G_2$. We have just proved above that

$$f(e_1) = e_2 \Rightarrow f(a\, a^{-1}) = e_2$$
$$\Rightarrow f(a)\, f(a^{-1}) = e_2$$

This shows that $f(a^{-1})$ is the inverse of f(a) in G_2. Therefore,

$$f(a^{-1}) = [f(a)]^{-1}\ \forall\ a \in G_1$$

(iii) Let a be any element of G_1 such that o(a) = m. Thus $a^m = e_1$. Now

$$f(a^m) = f(e_1) \Rightarrow f(a\ a\ a\ a \ldots \text{m times}) = e_2$$
$$\Rightarrow f(a)^m = e_2$$
$$\Rightarrow o(f(a)) \text{ is a divisor of m.}$$

Proved.

Isomorphism of groups

Let $(G_1, *_1)$ and $(G_2, *_2)$ be any two groups. The group G_1 is said to be ***isomorphic*** to the group G_2 if there exists a function f: $G_1 \rightarrow G_2$ such that

1. f is one to one,
2. f is onto and
3. f preserves the composition i.e. $f(a *_1 b) = f(a) *_2 f(b)\ \forall\ a, b \in G_1$

The function f itself is called ***isomorphic function*** or ***isomorphic mapping***. If the group G_1 is isomorphic to the group G_2 then they are called ***abstractly identical*** and symbolically it is written as $\mathbf{G_1 \cong G_2}$ **or** $\mathbf{G_1 \approx G_2}$. From point of view of abstract algebra, the two isomorphic groups are treated as same groups and not two different groups.

It is easy to interpret that if f: $G_1 \rightarrow G_2$ is a homomorphism of G_1 into G_2 and f is one to one, then f is called ***isomorphism of*** $\boldsymbol{G_1}$ ***into*** $\boldsymbol{G_2}$. In the same way, if f: $G_1 \rightarrow G_2$ is a homomorphism of G_1 onto G_2 and f is one to one, then f is called ***isomorphism of*** $\boldsymbol{G_1}$ ***onto*** $\boldsymbol{G_2}$.

If G_1 and G_2 are two isomorphic groups then there may be many isomorphic mappings from G_1 to G_2. Further, if one group is finite then other must be finite and the number of elements in both the groups must be equal for them to be isomorphic. Now let us see some examples.

Example 3 Let **R** be the additive group of real numbers and $\mathbf{R}_+$ be the multiplicative group of positive real numbers. Prove that the function f: $\mathbf{R} \to \mathbf{R}_+$ defined by $f(x) = e^x \ \forall\, x \in \mathbf{R}$ is an isomorphism from **R** onto $\mathbf{R}_+$.

Solution: For any real x (positive, negative or zero) e^x is always positive. Thus, f is everywhere defined. Let x and y be any two real number $\in \mathbf{R}$. Then e^x and $e^y \in \mathbf{R}_+$ and both are distinct. Now

$$f(x) = f(y) \Rightarrow e^x = e^y \Rightarrow \log_e e^x = \log_e e^y \Rightarrow x = y$$

Thus f is one to one. Also for any $y \in \mathbf{R}_+$, we have $\log_e y \in \mathbf{R}$ such that

$$f(\log_e y) = y$$

That means f is onto also.

Next, let us suppose that x and y be any two elements in **R**, then $(x + y) \in \mathbf{R}$. And

$$f(x + y) = e^{x+y} = e^x e^y = f(x)f(y)$$

That is f preserves the composition in **R** and $\mathbf{R}_+$. Therefore f is an isomorphism from **R** onto $\mathbf{R}_+$.

Ans.

Example 4 Show that the additive group of integers Z = {...,–2, –1, 0, 1, 2, ...} is isomorphic to the additive group of Z_m = {...,–2m, –1m, 0, 1m, 2m, ...}, where m is any fixed integer not equal to zero.

Solution: Let x be any element of Z, then $mx \in Z_m$. Let f : $Z \to Z_m$ be a function defined by $f(x) = mx \ \forall\, x \in Z$. Now, we have to prove that f is one to one and onto function and it preserves the composition.

f is one to one: Let x, y ∈ Z. Then,

$$f(x) = f(y) \Rightarrow mx = my \Rightarrow x = y$$

This shows that f is one to one.

f is onto: Let y be any element in Z_m. Then obviously y is a multiple of m and hence y/m is an integer. Thus y/m ∈ Z and f(y/m) = y. This shows that f is onto also.

f preserves the composition: Again, let us suppose that x and y be any two elements of Z. Then,

$$f(x + y) = m(x + y) = mx + my = f(x) + f(y)$$

Thus, f preserves the composition. Therefore f is an isomorphism of Z onto Z_m.

Ans.

Properties of Isomorphic groups

Let G_1 and G_2 be two isomorphic groups under the isomorphic mapping f. Since f is an isomorphism of G_1 onto G_2 f is a homomorphism of G_1 onto G_2 also. Therefore, the theorem 1 of this section holds good for two isomorphic groups also. In the case of isomorphism, ***transference of group properties*** is possible as shown in the following theorem.

Theorem 2 If $(G_1, *_1)$ is a group and G_2 be a set equipped with a binary composition $*_2$ and if there exists a one to one function from G_1 onto G_2 such that

$$f(a *_1 b) = f(a) *_2 f(b) \ \forall\ a, b \in G_1$$

then $(G_2, *_2)$ is also a group isomorphic to $(G_1, *_1)$.

Proof: In order to prove that the mathematical structure $(G_2, *_2)$ is a group, it has to be shown that $(G_2, *_2)$ satisfies all the group postulates.

Closure Property: It is followed from the hypothesis that G_2 is a set equipped with the binary composition $*_2$.

Associativity: Let a_2, b_2 and c_2 be any three elements in G_2. Since function f is an onto function from G_1 to G_2, there exists three elements a_1, b_1 and c_1 such that

$$f(a_1) = a_2,\ f(b_1) = b_2 \text{ and } f(c_1) = c_2$$

Now,

$$\begin{aligned}
(a_2 *_2 b_2) *_2 c_2 &= (f(a_1) *_2 f(b_1)) *_2 f(c_1) \\
&= (f(a_1 *_1 b_1) *_2 f(c_1) && \text{[By the definition of f]} \\
&= f((a_1 *_1 b_1) *_1 c_1) && \text{[By the definition of f]} \\
&= f(a_1 *_1 (b_1 *_1 c_1)) && \text{[By Associativity of } *_1\text{]} \\
&= f(a_1) *_2 f(b_1 *_1 c_1) && \text{[By the definition of f]} \\
&= f(a_1) *_2 (f(b_1) *_2 f(c_1)) \\
&= a_2 *_2 (b_2 *_2 c_2)
\end{aligned}$$

Thus, $*_2$ is an associative operation in G_2.

Existence of Identity: Let e is the identity element in G_1, then f(e) is in G_2. Further, let y be any element in G_2 then there exists an element x in G_1 such that f(x) = y. Now,

$$f(e) *_2 y = f(e) *_2 f(x) = f(e *_1 x) = f(x) = y$$

And

$$y *_2 f(e) = f(x) *_2 f(e) = f(x *_1 e) = f(x) = y$$

This shows that f(e) is the identity element in G_2.

Existence of Inverse: Let y be any element of G_2 then there exists an element x in G_1 such that f(x) = y. Also $x \in G_1 \Rightarrow x^{-1} \in G_1$ and $f(x^{-1}) \in G_2$. Now

$$y *_2 f(x^{-1}) = f(x) *_2 f(x^{-1}) = f(x *_1 x^{-1}) = f(e)$$

And

$$f(x^{-1}) *_2 y = f(x^{-1}) *_2 f(x) = f(x^{-1} *_1 x) = f(e)$$

This implies that $f(x^{-1})$ is inverse of y and it exists in G_2. Hence $(G_2, *_2)$ is a group.

Proved.

Now let us consider **Ğ** as the set of all groups. A relation R defined on **Ğ** is said to be a ***relation of isomorphism*** if for any two groups X and Y of **Ğ**, (X, Y) ∈ R implies that group X is isomorphic to the group Y. Then the relation R is an equivalence relation as shown in the following theorem.

Theorem 3 The relation of isomorphism on the set of all groups is an equivalence relation.

Proof: Let R be the relation defined on the set **Ğ** of all groups. Then in order to prove that R is an equivalence relation, it is to be shown that R is reflexive, symmetric and transitive on **Ğ**.

Reflexivity: Let X be any group in **Ğ**. Since every group is isomorphic to itself, X is R related to X i.e. (X, X) ∈ **R** ∀ X ∈ **Ğ**. Hence R is reflexive on **Ğ**.

Symmetry: Let X and Y be any two elements in **Ğ** such that (X, Y) ∈ R. This shows that X is isomorphic to Y. We can say now that Y is isomorphic to X also i.e. (Y, X) ∈ R. Hence R is a symmetric relation on **Ğ**.

Transitivity: Let X, Y and Z be any three elements in **Ğ** such that ordered pairs (X, Y) and (Y, Z) are in R. That is group X is isomorphic to group Y and group Y is isomorphic to group Z. This implies that group X is isomorphic to group Z and hence (X, Z) ∈ R. Thus, R is a transitive relation on **Ğ**.

Therefore R is an equivalence relation on **Ğ**.

Proved.

5.4 Cyclic Group

*A group (G, *) is said to be a* ***cyclic group*** *if there exists an element $a \in G$ such that every element $x \in G$ can be written in the form a^n, where n is some integer. The element a is then called a generator of G.*

Possibly, there may be more generators for a group G. If $a \in G$ is a generator of G, then G is written as G = {a} or G = (a). The element of G will be of the form $a^0 = e$, a^1, a^2, a^3, a^4, a^5,.... All of them may not be distinct. The power may be integral including the negative one.

Example 1 The multiplicative group G = {1, –1, i, –i}is cyclic. We can write G = {i^1, i^2, i^3, i^4}. Thus G is a cyclic group and 'i' is a generator. The element –i is also a generator of G.♦

Example 2 The multiplicative group G = {1, ω, ω^2} is cyclic. We can write G = {ω^1, ω^2, ω^3}. Thus G is a cyclic group and 'ω' is a generator. The element ω^2 is also a generator of G.♦

Example 3 The additive group G of integers is cyclic. A generator of G is the element 1. We have (1 added zero times)$1^0 = 0$, (1 added one time)$1^1 = 1$, (1 added two times)$1^2 = 2$, and so on. Similarly, (1 added –1 times) $1^{-1} = -1$, (1 added –2 times)$1^{-2} = -2$ and so on. Thus G is a cyclic group and '1' is a generator. Is there any other generator of G here?♦

Properties of Cyclic Groups

Theorem 1 Every cyclic group is an abelian group.

Proof: Let G be a cyclic group and $*$ be the binary operation on G. Let G = {a} and x, y be any two elements of G. Then there exist integers r and s such that

$$x = a^r \text{ and } y = a^s$$

Now $x*y = a^r * a^s = a^{r+s} = a^{s+r} = a^s * a^r = y*x$.

This shows that $x*y = y*x \ \forall\ x, y \in G$. Hence G is an abelian group.

Proved.

Theorem 2 If an element a is a generator of a group G then its inverse a^{-1} is also a generator of G.

Proof: Let G be a cyclic group and $*$ be the binary operation on G. Let G = {a}. Let a^r be any element of G, where r is some integer. We can write $a^r = (a^{-1})^{-r}$. Since $-r$ is also an integer, each element of G is generated by a^{-1} also. Therefore, a^{-1} is also a generator of G.

Proved.

Theorem 3 If a finite group of order n contains an element of order n, the group must be cyclic.

Proof: Let G be a finite group of order n and a be an element of G of order n. Let H be a cyclic subgroup of G generated by a, then order of H is n. This shows that H is a cyclic subgroup of G such that order of H is equal to the order of G. Hence G is itself a cyclic group and a is generator of G.

Proved.

Theorem 4 Every group of prime order is cyclic.

Proof: Let G be a finite group of prime order p. Then G must contain at least two elements. Thus, there exists an element '**a**' such that **a** is not the identity element. Since **a** is not an identity element, its order is definitely ≥ 2. Let O(a) = m. Let H = {a}, then H is a cyclic subgroup of G of order m. By Lagrange's theorem, m must be a divisor of p. But it is given that p is a prime, so m = p. Thus G = H. Hence G is cyclic group and a is generator of G.

Proved.

Theorem 5 Every finite group of composite order possesses proper subgroup.

Proof: Let G be a finite group of composite order say mn, where neither m nor n is equal to 1. Let G is a cyclic group and a is its generator. Thus, O(a) = mn.

$$O(a) = mn \Rightarrow (a^n)^m = e \Rightarrow O(a^n) \text{ is finite and is } \leq m.$$

Let $O(a^n) = r < m$, then $a^{nr} = e$. Since $r < m$, we have $nr < nm$. Since order of a is mn, $r < m$ is not possible. Thus, r = m. That is $O(a^n) = m$. Let $H = \{a^n\}$. Thus H is a cyclic group of order m and since $m < mn$, H is a proper subgroup of G.

Now let G is not a cyclic group. Then there exists at least one element a of order p such that $2 \leq p < mn$. Thus H = {a} is a proper subgroup of G.

Proved.

Theorem 6 Every subgroup of a cyclic group is cyclic.

Proof: Let H be any subgroup of a cyclic group G = {a}, where a is generator of G. If H = {e}, then the result is obvious i.e. H is cyclic. Thus let us assume that H is a proper subgroup. Obviously, elements of H are some integral power of a. Let m is the least integer such that $a^m \in H$.

Let a^n be any element of H. By division algorithm, there exists two integers q and r such that n = mq + r. We have $a^m \in H \Rightarrow a^{-m} \in H \Rightarrow (a^{-m})^q \in H \Rightarrow a^{-mq} \in H$. Further, $a^n \in H$ and $a^{-mq} \in H \Rightarrow a^{n-mq} \in H \Rightarrow a^r \in H$.

But this is contrary to the assumption that m is the least element such that $a^m \in H$. Thus r = 0. This shows that n = mq and $a^n \in H$. Thus, H = $\{a^m\}$. Therefore, H is a cyclic subgroup of G.

Proved.

Theorem 7 Every proper subgroup of an infinite cyclic group is infinite.

Proof: Let G = {a} be an infinite cyclic group and H be any proper subgroup of G. Obviously, H is also cyclic and let H = {b}, where b is generator of H. Thus there exists an integer m such that $b = a^m$. Let order of H is p. Thus, $a^{mp} = e$. This shows that order of a is finite. This is contrary to the fact that G is an infinite cyclic group. Hence, there does not exists such finite p. Therefore H is of infinite order.

Proved.

Example 4 Show that ({1, 2, 3, 4, 5, 6}, x_7} is a cyclic group, where x_7 is multiplication modulo 7 i.e. a x_7 b is equal to the remainder when a x b is divided by 7. How many generators are there?

Solution: Let G be the given group. If there exists an element **a** in G such that O(**a**) = 6, then G is cyclic and **a** is a generator of G. Here, we have

$$3^1 = 3,$$
$$3^2 = 3 \; x_7 \; 3 = 2,$$
$$3^3 = 3^2 \; x_7 \; 3 = 2 \; x_7 \; 3 = 6,$$
$$3^4 = 3^3 \; x_7 \; 3 = 6 \; x_7 \; 3 = 4,$$
$$3^5 = 3^4 \; x_7 \; 3 = 4 \; x_7 \; 3 = 5,$$
$$3^6 = 3^5 \; x_7 \; 3 = 5 \; x_7 \; 3 = 1 \text{ the identity element of G.}$$

Thus O(3) = 6 = O(G). Hence G is cyclic and G = {a}.

Ans.

Example 5 Let G be a group, G ≠ {e}. Then G has no proper subgroup iff G is a finite cyclic group of prime order.

Solution: **If Part** – Let G be a finite cyclic group of finite order, say p. If H be any proper subgroup of G, then O(H) must be a divisor of O(G). But p is a prime number, it is not divisible by any integer x, where $1 < x < p$. Therefore G cannot have a proper subgroup.

Only if Part – Let G has no proper subgroup. Then, we have to prove that G is **cyclic, finite** and of **prime** order. Let $x \neq e$ be any element of G, then {x} is a cyclic subgroup of G generated by x. Since G has no proper subgroup, G = {x}. Hence G is a cyclic group. If it is infinite, then $\{x^2\}$ is a proper subgroup of G. Since it is given that G has no proper subgroup, O(x) is finite and hence G is a finite group. At the end, let O(G) = n. If n is not prime then we can write n = rs, where r and s are positive integers greater than 1 and less than n. In this case, we have $O(x^r) = s < n$, i.e. G has a proper subgroup generated by x^r. Since G has no proper subgroup, n cannot be a composite number. Hence n is a prime. Therefore G is a cyclic group of finite prime order.

Ans.

Example 6 A cyclic group G with generator of finite order n, is isomorphic to the multiplicative group of nth roots of unity.

Solution: Let G = {a} such that $a^n = e$. Since a is a generator of finite order n, the group G has n distinct elements. Let U be the multiplicative group of nth roots of unity. The elements of U can be written as $\exp^{2\pi ir/n}$, where $0 \leq r \leq n - 1$. Now let us define a function f: G→U as

$$f(a^r) = \exp^{2\pi ir/n} \text{ for } 0 \leq r \leq n - 1$$

We shall now show that f is one to one onto and it preserves the composition. Let r and s be any two integers such that $0 \leq r, s \leq n - 1$. Then we have

$$f(a^r) = f(a^s) \Rightarrow \exp^{2\pi ir/n} = \exp^{2\pi is/n}$$
$$\Rightarrow r = s$$
$$\Rightarrow a^r = a^s$$

This implies that f is one to one. Since both G and U are finite and have same cardinality, f is onto also. Next,

$$f(a^r a^s) = f(a^{r+s}) = f(a^{nq+k}) \qquad [r + s = nq + k \text{ by division algorithm }]$$
$$= f(a^{nq} a^k) = f(a^k) = \exp^{2\pi ik/n}$$
$$= \exp^{2\pi ik/n} = \exp^{2\pi inq/n} \exp^{2\pi ik/n} = \exp^{2\pi i(nq + k)/n}$$
$$= \exp^{2\pi i(r + s)/n} = \exp^{2\pi ir/n} \exp^{2\pi is/n}$$
$$= f(a^r) f(a^s)$$

Therefore f preserves the composition in G and U and hence G is isomorphic to U.

Ans.

***Example** 7* A cyclic group G with generator of finite order n, is isomorphic to the dditive group of residue classes modulo n.

olution: The proof is similar to that of the previous example 6. Thus it is left as an xercise to the reader.

Ans.

xample 8 If a cyclic group G is generated by an element a of finite order n, then a^m is a enerator of G iff the greatest common divisor of m and n is 1 i.e. m and n are relative rimes.

olution: Let m be any positive integer relatively prime to n. Let H = $\{a^m\}$ be a subgroup f G generated by a^m. Since each integral power of a^m is also an integral power of a, H $\subseteq$. Since m is relatively prime to n, there exists two integers u and v such that

$$mu + nv = 1$$

$$\text{Therefore, } a^{mu + nv} = a \Rightarrow a^{mu}\, a^{nv} = a$$

$$\Rightarrow a^{mu} = a$$

$$\Rightarrow (a^m)^u = a$$

his implies that each integral power of a^m is also an integral power a. Thus G $\subseteq$ H. Thus = H = $\{a^m\}$. Hence a^m is also a generator of G.

onversely, let us suppose that a^m is also a generator of G and m is not relatively prime to that is gcd(m, n) = d $\neq$ 1. Then m/d and n/d must be an integer. Thus,

$^m)^{n/d} = (a^n)^{m/d} = (e)^{m/d} =$ e. Obviously, n/d is an integer less than n, thus a^m cannot a generator of G. This is a contradiction. Hence gcd(n, m) = 1.

Ans.

xample 9 If G is an infinite cyclic group, then G has exactly two generators and G is omorphic to the additive group of integers.

lution: Let G = $\{a\}$ is an infinite cyclic group generated by a. The elements of G can expressed as some integral power of a. Further no two distinct integral powers of a will equal. Because for r $\neq$ s and r > s,

$$a^r = a^s \Rightarrow a^r a^{-s} = a^s a^{-s} = a^0$$

$$\Rightarrow a^{r-s} = a^0 = e$$

This shows that order of a is a finite number r – s, which is not possible. Next if a^r is any element of G then it can also be written as $(a^{-1})^r$. Thus a^{-1} is also a generator of G. Since a and a^{-1} are two distinct element of G, we have proved that G has two generators. Now to prove that G has exactly two generators, we shall prove that no other integral power of a can generate G.

Let a^m be any element of G, where m is not equal to either 1 or –1. This cannot be a generator of G because if it is so, there must exist an element k such that

$$(a^m)^k = a$$

This shows that two distinct power of a is equal contrary to the fact just proved above.

Now let I be the additive group of integers. Let f: G → I be a function defined as

$$f(a^m) = m \ \forall \ m \in I$$

This can easily be proved that f is an isomorphism from G to I. Hence G is isomorphic to I.

Ans

5.5 Permutation Group

A ***permutation*** *is defined as a* ***bijection*** *from a set A to itself.* Let A = {1, 2, 3} be a set. We can define the following bijections (one to one onto functions) on A.

$$f_1 = \begin{pmatrix} 1 & 2 & 3 \\ 1 & 2 & 3 \end{pmatrix}, \quad f_2 = \begin{pmatrix} 1 & 2 & 3 \\ 2 & 3 & 1 \end{pmatrix}, \quad f_3 = \begin{pmatrix} 1 & 2 & 3 \\ 3 & 1 & 2 \end{pmatrix}, \quad f_4 = \begin{pmatrix} 1 & 2 & 3 \\ 1 & 3 & 2 \end{pmatrix}$$

$$f_5 = \begin{pmatrix} 1 & 2 & 3 \\ 2 & 1 & 3 \end{pmatrix}, \quad f_6 = \begin{pmatrix} 1 & 2 & 3 \\ 3 & 2 & 1 \end{pmatrix}$$

In each of the above functions, elements in the second row are images of the corresponding elements in the first row under the respective functions. That is to say, $f_1(2) = 2$, $f_2(2) = 3$, $f_3(2) = 1$ and so on. Further, the elements in second row are only some permutations of the elements {1, 2, 3}. If |A| = n then we have n! permutations of these elements and hence there are **n!** bijections on A. Here n –number of elements in finite set A, is called ***degree*** of permutation and collection of all the distinct bijections on A is called set of permutations and is denoted by P_n. Obviously, $|P_n| = n!$. The set P_n is also called **symmetric set of permutations** and sometimes denoted as S_n. The name symmetric is drawn from the symmetry of polyhedron of n vertices.

Any two permutations f and g of degree n on a set A are said to be equal if

$$f(a) = g(a) \ \forall \ a \in A$$

For example let A = {1, 2, 3, 4} and f and g are defined as below.

$$f = \begin{pmatrix} 1 & 2 & 3 & 4 \\ 2 & 3 & 4 & 1 \end{pmatrix} \qquad g = \begin{pmatrix} 2 & 3 & 1 & 4 \\ 3 & 4 & 2 & 1 \end{pmatrix}$$

Since $\forall$ a $\in$ A f(a) = g(a), we have f = g. The degree of f and g is 4 and both are member of P_4. It is very easy to find that permutation g has been obtained from f just be interchanging some columns. This conveys that the interchange of columns will not change the permutation.

Identity Permutation

Let A = {1, 2, 3, ...,n} be a finite set with |A| = n. A permutation *I* is said to be an identity permutation if *I* maps each element of A to itself. For example,

$$I = \begin{pmatrix} 1 & 2 & 3 \ldots n \\ 1 & 2 & 3 \ldots n \end{pmatrix}$$

is an identity permutation of degree n. Similarly if B = {a_1, a_2, a_3, a_4,... a_n}be a finite set then the permutation

$$I = \begin{pmatrix} a_1 & a_2 & a_3 & \ldots a_n \\ a_1 & a_2 & a_3 & \ldots a_n \end{pmatrix}$$

is an identity permutation of degree n.

Composition of Permutation

Since permutation is a function, composition of two permutations is similar to the composition of functions as discussed in the chapter two. Let f and g be any two permutations of degree n given as below:

$$f = \begin{pmatrix} a_1 & a_2 & a_3 & \ldots a_n \\ b_1 & b_2 & b_3 & \ldots b_n \end{pmatrix} \qquad g = \begin{pmatrix} b_1 & b_2 & b_3 & \ldots b_n \\ c_1 & c_2 & c_3 & \ldots c_n \end{pmatrix}$$

Then the product fg is given as

$$fg = \begin{pmatrix} a_1 & a_2 & a_3 & \ldots a_n \\ c_1 & c_2 & c_3 & \ldots c_n \end{pmatrix}$$

This is obtained by first carrying out the operation defined by f and then by g. Similarly, we can find gf.

Example 1 Let f and g be two permutations of degree 3 as given below. Find fg and gf.

$$f = \begin{pmatrix} 1 & 2 & 3 \\ 1 & 3 & 2 \end{pmatrix} \qquad g = \begin{pmatrix} 1 & 2 & 3 \\ 2 & 3 & 1 \end{pmatrix}$$

Solution: Using the procedure as discussed above, we have

$$fg = \begin{pmatrix} 1 & 2 & 3 \\ 2 & 1 & 3 \end{pmatrix} \quad and \quad gf = \begin{pmatrix} 1 & 2 & 3 \\ 3 & 2 & 1 \end{pmatrix}$$

It is obvious here that fg ≠ gf i.e. product of permutation is not a **commutative** operatior

An

Example 2 Let f and g be two permutations of degree 5 as given below. Find fg and gf.

$$f = \begin{pmatrix} 1 & 2 & 3 & 4 & 5 \\ 2 & 3 & 4 & 5 & 1 \end{pmatrix} \qquad g = \begin{pmatrix} 1 & 2 & 3 & 4 & 5 \\ 1 & 2 & 4 & 5 & 3 \end{pmatrix}$$

Solution: Here,

$$fg = \begin{pmatrix} 1 & 2 & 3 & 4 & 5 \\ 2 & 3 & 4 & 5 & 1 \end{pmatrix}\begin{pmatrix} 1 & 2 & 3 & 4 & 5 \\ 1 & 2 & 4 & 5 & 3 \end{pmatrix} = \begin{pmatrix} 1 & 2 & 3 & 4 & 5 \\ 2 & 3 & 4 & 5 & 1 \end{pmatrix}\begin{pmatrix} 2 & 3 & 4 & 5 & 1 \\ 2 & 4 & 5 & 3 & 1 \end{pmatrix}$$

$$= \begin{pmatrix} 1 & 2 & 3 & 4 & 5 \\ 2 & 4 & 5 & 3 & 1 \end{pmatrix}$$

And

$$gf = \begin{pmatrix} 1 & 2 & 3 & 4 & 5 \\ 1 & 2 & 4 & 5 & 3 \end{pmatrix}\begin{pmatrix} 1 & 2 & 3 & 4 & 5 \\ 2 & 3 & 4 & 5 & 1 \end{pmatrix} = \begin{pmatrix} 1 & 2 & 3 & 4 & 5 \\ 1 & 2 & 4 & 5 & 3 \end{pmatrix}\begin{pmatrix} 1 & 2 & 4 & 5 & 3 \\ 2 & 3 & 5 & 1 & 4 \end{pmatrix}$$

$$= \begin{pmatrix} 1 & 2 & 3 & 4 & 5 \\ 2 & 3 & 5 & 1 & 4 \end{pmatrix}$$

The simple way to find fg is to arrange the first row of g to match the second row o Then copy the first row of f and second row of rearranged g. Whatever we have is permutation fg.

A

Group of Permutation

Let A be a finite set and |A| = n. Let P_n be the set of all permutations of degre on A. Let $*$ be the binary operation of composition (product) of permutations. Then mathematical structure $(P_n,*)$ is a group. This group is called ***permutation*** group. This is proved in the following theorem.

Theorem 1 The set P_n of all permutations on n symbols is a finite group of order n! with respect to the binary operation of composition of permutations. Further for $n \leq 2$, the group is abelian and for $n > 2$ the group is always non-abelian.

Proof: Let $S = \{a_1, a_2, a_3, a_4, \ldots a_n\}$ be a finite set. Clearly $|S| = n$. The elements of S can be arranged in n! distinct ways and hence there are n! distinct permutations of degree n. Thus $|P_n| = n!$. Let us now show that $(P_n,*)$ is a group.

Closure Property: Let f and g be any two permutations in P_n. Let f and g are given as

$$f = \begin{pmatrix} a_1 & a_2 & a_3 & \ldots a_n \\ b_1 & b_2 & b_3 & \ldots b_n \end{pmatrix} \qquad g = \begin{pmatrix} b_1 & b_2 & b_3 & \ldots b_n \\ c_1 & c_2 & c_3 & \ldots c_n \end{pmatrix}$$

Where $(b_1, b_2, b_3, b_4, \ldots, b_n)$ and $(c_1, c_2, c_3, c_4, \ldots, c_n)$ are different arrangements of elements $(a_1, a_2, a_3, a_4, \ldots, a_n)$. Then fg is a permutation of degree n and is given by

$$f * g = \begin{pmatrix} a_1 & a_2 & a_3 & \ldots a_n \\ c_1 & c_2 & c_3 & \ldots c_n \end{pmatrix}$$

Obviously, f*g is in P_n. Thus P_n is closed with respect to *.

Associativity: A composition of permutations (functions) is an associative operation.

Existence of Identity: Let I be the identity permutation of degree n then $I \in P_n$. Then I can be written as

$$I = \begin{pmatrix} a_1 & a_2 & a_3 & \ldots a_n \\ a_1 & a_2 & a_3 & \ldots a_n \end{pmatrix} \quad or \quad \begin{pmatrix} b_1 & b_2 & b_3 & \ldots b_n \\ b_1 & b_2 & b_3 & \ldots b_n \end{pmatrix}$$

and f be any permutation given as above then, we have

$$f * I = I * f = \begin{pmatrix} a_1 & a_2 & a_3 & \ldots a_n \\ b_1 & b_2 & b_3 & \ldots b_n \end{pmatrix} = f$$

Therefore, I is an identity element.

Existence of Inverse: Let f be any permutation of degree n then f is a bijection on S and thus its inverse f^{-1} exists and it is also a bijection. Therefore f^{-1} is a permutation of degree n. Hence $f^{-1} \in P_n$.

Therefore $(P_n,*)$ is a group of order n! with respect to composition of permutations. Also we know that a group of order one or of order two is abelian. Therefore, $(P_n,*)$ is an abelian group for $n \le 2$. For $n > 2$, $(P_n,*)$ is not an abelian group as composition of permutation is not a commutative operation.

Proved.

The group $(P_n,*)$ is also called symmetric group. It is also important to note that inverse of a permutation is obtained by just interchanging the two rows. While writing a permutation of degree n, it is immaterial that what symbols we use to denote the elements of the set S. We can use the numbers 1, 2, 3, ..., n or we can use the letters a_1, a_2, a_3, a_4, ..., a_n.

Let f is a permutation of degree n on a set S having n distinct elements. Let it be possible to arrange k elements of the set S in such a way that the f-image of each element in the row is the element which follows it, and the f-image of the last element is the first element. Also the remaining n – k elements are left unchanged by f. *Such a permutation is called a **cyclic permutation** of length k or simply a cycle of length k or k-cycle.* The length of cycle means simply the number of elements permuted by the cycle.

Example 3 Let f is a permutation of degree 6 given by

$$f = \begin{pmatrix} 1 & 2 & 5 & 3 & 6 & 4 \\ 2 & 4 & 5 & 1 & 6 & 3 \end{pmatrix}$$

Then it can be also written as

$$f = \begin{pmatrix} 1 & 2 & 4 & 3 & 5 & 6 \\ 2 & 4 & 3 & 1 & 5 & 6 \end{pmatrix}$$

Here elements 1, 2, 4, 3 are arranged in such a way that 2 is f-image of 1, 4 is f-image of 2, 3 is f-image of 4 and 1 is f-image of 3. The remaining elements 5 and 6 are left unchanged by f. Thus f is a cyclic permutation of length 4 and it is written as (1 2 4 3). ♦

Example 4 Let g is a permutation of degree 6 given by

$$g = \begin{pmatrix} 1 & 2 & 4 & 3 & 6 & 5 \\ 2 & 1 & 3 & 4 & 6 & 5 \end{pmatrix}$$

Since it cannot be arranged in the way that $k \le n$ elements form a cycle and remaining n – k elements are left unchanged, g is not a cyclic permutation. ♦

A cycle does not change by changing the places of its elements provided their cyclic order is not changed. Thus (1 2 4 3), (2 4 3 1), (3 1 2 4) etc all represent the same permutation. Also (5 6) and (6 5) are the same permutation. The last two are cyclic permutations of length two. *A cycle of length two is called* ***transposition***. Thus (5 6) is a transposition. It represents a permutation in which the image of 5 is 6, the image of 6 is 5 and the remaining missing elements are left unchanged.

A cycle of length one implies that the image of the element involved is the element itself and the remaining elements are left unchanged. Thus all the elements are left unchanged. This conveys that every cycle of length one represents the identity permutation.

Two cycles are said to be ***disjoint*** if they have no element in common. For example, (1 3 5) and (2 4 6 9) are disjoint cycles of length 3 and 4 respectively. Both represent permutation of degree 9. However, (1 4 6) and (2 5 6 8) are not disjoint cycles.

Multiplication of Cycles

Multiplication of cycles means composition of the corresponding permutations. For example if the cycles (1 2 3) and (5 6 4 1) represent permutations of degree 6 on symbols 1, 2, 3, 4, 5 and 6, then

$$(1\ \ 2\ \ 3)(5\ \ 6\ \ 4\ \ 1) = \begin{pmatrix} 1 & 2 & 3 & 4 & 5 & 6 \\ 2 & 3 & 1 & 4 & 5 & 6 \end{pmatrix}\begin{pmatrix} 1 & 2 & 3 & 4 & 5 & 6 \\ 5 & 2 & 3 & 1 & 6 & 4 \end{pmatrix}$$

$$= \begin{pmatrix} 1 & 2 & 3 & 4 & 5 & 6 \\ 2 & 3 & 5 & 1 & 6 & 4 \end{pmatrix} = \begin{pmatrix} 1 & 2 & 3 & 5 & 6 & 4 \\ 2 & 3 & 5 & 6 & 4 & 1 \end{pmatrix}$$

$$= (1\ \ 2\ \ 3\ \ 5\ \ 6\ \ 4)$$

Theorem 2 If f and g are two disjoint cycles then fg = gf i.e. the product of disjoint cycles is commutative.

Proof: Let f and g are two disjoint cycles of length r and s respectively. Let both represent permutations of degree n such that $r + s \leq n$. Let a_1, a_2, a_3, a_4, ... and a_n be n symbols. Without loss of any generality, we can assume that f permutes first r elements and g permutes next s element and remaining n – (r + s) elements are left unchanged by both. Thus f and g can be written as

$$f = \begin{pmatrix} a_1 & a_2 & a_3 & \ldots a_r & a_{r+1} & a_{r+2} & \ldots a_n \\ a_2 & a_3 & a_4 & \ldots a_1 & a_{r+1} & a_{r+2} & \ldots a_n \end{pmatrix} \quad \text{and}$$

$$g = \begin{pmatrix} a_1 & a_2 & a_3 & \ldots a_r & a_{r+1} & a_{r+2} & \ldots a_{r+s} & a_{r+s+1} & a_{r+s+2} & \ldots a_n \\ a_1 & a_2 & a_3 & \ldots a_r & a_{r+2} & a_{r+3} & \ldots a_{r+1} & a_{r+s+1} & a_{r+s+2} & \ldots a_n \end{pmatrix}$$

It is obvious that elements changed by f are left unchanged by g and vice versa. Thus fg and gf can be given as

$$fg = gf = \begin{pmatrix} a_1 & a_2 & a_3 & \dots a_r & a_{r+1} & a_{r+2} & \dots a_{r+s} & a_{r+s+1} & a_{r+s+2} & \dots a_n \\ a_2 & a_3 & a_4 & \dots a_1 & a_{r+2} & a_{r+3} & \dots a_{r+1} & a_{r+s+1} & a_{r+s+2} & \dots a_n \end{pmatrix}$$

Thus product of disjoint cycles is a commutative operation.

Proved.

A cycle is a permutation. As we know a permutation is a bijection on a finite set. Since every bijection is invertible, ***a cycle has always an inverse***. Let $(a_1, a_2, a_3, \dots, a_r)$ be a r-cycle representing a permutation of degree n on the symbols $a_1, a_2, a_3, a_4, \dots$ and a_n. Then f can be written as

$$f = \begin{pmatrix} a_1 & a_2 & a_3 & \dots a_r & a_{r+1} & a_{r+2} & \dots a_n \\ a_2 & a_3 & a_4 & \dots a_1 & a_{r+1} & a_{r+2} & \dots a_n \end{pmatrix}$$

Then its inverse can be given by just interchanging the rows. This is written then as

$$f^{-1} = \begin{pmatrix} a_2 & a_3 & a_4 & \dots a_{r-1} & a_r & a_1 & a_{r+1} & a_{r+2} & \dots a_n \\ a_1 & a_2 & a_3 & \dots a_{r-2} & a_{r-1} & a_r & a_{r+1} & a_{r+2} & \dots a_n \end{pmatrix}$$

$$= \begin{pmatrix} a_r & a_{r-1} & a_{r-2} & \dots a_2 & a_1 & a_{r+1} & a_{r+2} & \dots a_n \\ a_{r-1} & a_{r-2} & a_{r-3} & \dots a_1 & a_r & a_{r+1} & a_{r+2} & \dots a_n \end{pmatrix}$$

$$= (a_r \quad a_{r-1} \quad a_{r-2} \quad \dots a_2 \quad a_1)$$

From this it can be easily verified that $f f^{-1} = f^{-1} f = I$ where, I is the identity permutation of degree n. *Therefore inverse of a cycle can be obtained by arranging the elements of the cycle in reverse order. In particular, every* ***transposition*** *is its own inverse.* For example inverse of (1 4) is (4 1) and both represent the same permutation. The inverse of product of cycles is obtained in the same way as the inverse of a composition of functions i.e. $(fg)^{-1} = g^{-1} f^{-1}$ where, f and g are any two cycles. If f and g are disjoint cycles then we have

$$(fg)^{-1} = f^{-1} g^{-1} = (gf)^{-1}$$

Theorem 3 Every permutation can be expressed as a product of disjoint cycles.

Proof: Let f be any permutation of degree n on symbols $a_1, a_2, a_3, a_4, \dots$ and a_n and is given by

$$f = \begin{pmatrix} a_1 & a_2 & a_3 & \dots a_n \\ b_1 & b_2 & b_3 & \dots b_n \end{pmatrix}$$

Where, (b_1, b_2, b_3, b_4, ..., b_n) is an arrangement of elements (a_1, a_2, a_3, a_4, ..., a_n). We can express this permutation as a product of disjoint cycles as follows:

1. First we put down cycles of length one with the help of elements which remain unchanged under f.
2. Now start with a symbol that is not left unchanged. Say it is a_1, and its f-image is a_2, f–image of a_2 is, say a_3 and so on and f-image of a_r is a_1. Arrange this in a cycle as (a_1 a_2 a_3 ... a_r).
3. Now start a new bracket. Here write an element that has not yet been included and proceed as in step 2.
4. Repeat step 3 until every symbol has been taken care of.

Proceeding in this way we get disjoint cycles product of which is the permutation f. This is illustrated in the following examples.

Proved.

Example 5 Write the following permutations as a product of disjoint cycles.

(*i*) $f = \begin{pmatrix} 1 & 2 & 3 & 4 & 5 & 6 \\ 6 & 5 & 4 & 3 & 1 & 2 \end{pmatrix}$

(*ii*) $g = \begin{pmatrix} 1 & 2 & 3 & 4 & 5 & 6 & 7 & 8 & 9 \\ 2 & 3 & 4 & 5 & 1 & 6 & 7 & 9 & 8 \end{pmatrix}$

Solution: *(i)* From the step 1 of the proof of the theorem 3 above, we have no element in f which has not been changed by f. Thus there is no cycle of length one. From step 2, we have (1 6 2 5) and then from step 3 we have (3 4). Now no element is left. Thus we have two disjoint cycles (1 6 2 5) and (3 4). Therefore f can be written as f = (1 6 2 5) (3 4).

(ii) In this permutation elements 6 and 7 are left unchanged by g. Thus we have two cycles (6) and (7) of length one. Other cycles are (1 2 3 4 5) and (8 9). All these cycles are disjoint and theirs product is the permutation g i.e.

g =(1 2 3 4 5)(8 9)(6)(7)

Ans.

Example 6 Express the following permutations as the product of disjoint cycles:

(a) f = (1 2 3)(4 5)(1 6 7 8 9)(1 5)

(b) g = (1 2)(1 2 3)(1 2)

(c) h = (1 3 2 5)(1 4 3)(2 5 1)

Solution: Whenever a permutation is given as a product of cycles and it is asked to express it as a product of disjoint cycles, first find the product and then proceed as in the previous example. This is illustrated below.

(a) Let f = $C_1C_2C_3C_4$ where C_1 = (1 2 3), C_2 = (4 5), C_3 = (1 6 7 8 9) and C_4 = (1 5). Then we have

$$C_1C_2 = \begin{pmatrix} 1 & 2 & 3 & 4 & 5 & 6 & 7 & 8 & 9 \\ 2 & 3 & 1 & 4 & 5 & 6 & 7 & 8 & 9 \end{pmatrix}\begin{pmatrix} 1 & 2 & 3 & 4 & 5 & 6 & 7 & 8 & 9 \\ 1 & 2 & 3 & 5 & 4 & 6 & 7 & 8 & 9 \end{pmatrix}$$

$$= \begin{pmatrix} 1 & 2 & 3 & 4 & 5 & 6 & 7 & 8 & 9 \\ 2 & 3 & 1 & 5 & 5 & 6 & 7 & 8 & 9 \end{pmatrix}$$

And

$$C_1C_2C_3 = \begin{pmatrix} 1 & 2 & 3 & 4 & 5 & 6 & 7 & 8 & 9 \\ 2 & 3 & 1 & 5 & 4 & 6 & 7 & 8 & 9 \end{pmatrix}\begin{pmatrix} 1 & 6 & 7 & 8 & 9 & 2 & 3 & 4 & 5 \\ 6 & 7 & 8 & 9 & 1 & 2 & 3 & 4 & 5 \end{pmatrix}$$

$$= \begin{pmatrix} 1 & 2 & 3 & 4 & 5 & 6 & 7 & 8 & 9 \\ 2 & 3 & 6 & 5 & 4 & 7 & 8 & 9 & 1 \end{pmatrix}$$

Finally, we have

$$C_1C_2C_3C_4 = \begin{pmatrix} 1 & 2 & 3 & 4 & 5 & 6 & 7 & 8 & 9 \\ 2 & 3 & 6 & 5 & 4 & 7 & 8 & 9 & 1 \end{pmatrix}\begin{pmatrix} 1 & 2 & 3 & 4 & 5 & 6 & 7 & 8 & 9 \\ 5 & 2 & 3 & 4 & 1 & 6 & 7 & 8 & 9 \end{pmatrix}$$

$$= \begin{pmatrix} 1 & 2 & 3 & 4 & 5 & 6 & 7 & 8 & 9 \\ 2 & 3 & 6 & 1 & 4 & 7 & 8 & 9 & 5 \end{pmatrix}$$

$$= (1 \; 2 \; 3 \; 6 \; 7 \; 8 \; 9 \; 5 \; 4)$$

Therefore, f = (1 2 3 6 7 8 9 5 4).♦

(b) Let g = $C_1C_2C_3$ where C_1 = (1 2), C_2 = (1 2 3) and C_3 = (1 2). Then we have

$$g = C_1C_2C_3 = \begin{pmatrix} 1 & 2 & 3 \\ 2 & 1 & 3 \end{pmatrix}\begin{pmatrix} 1 & 2 & 3 \\ 2 & 3 & 1 \end{pmatrix}\begin{pmatrix} 1 & 2 & 3 \\ 2 & 1 & 3 \end{pmatrix}$$

$$= \begin{pmatrix} 1 & 2 & 3 \\ 3 & 2 & 1 \end{pmatrix}\begin{pmatrix} 1 & 2 & 3 \\ 2 & 1 & 3 \end{pmatrix} = \begin{pmatrix} 1 & 2 & 3 \\ 3 & 1 & 2 \end{pmatrix} = (1 \; 3 \; 2)$$

Therefore g = (1 3 2). ♦

(c) Similarly we can find expression for h. Reader may verify that the answer is

h = (1 2)(3 5 4).

Ans.

Transposition is a cycle of length two. Let us consider a cycle (1 2 3) of length three. We can write (1 2 3) = (1 2)(1 3). Similarly cycle (1 2 3 4) can be written as (1 2)(1 3)(1 4) or (1 2)(2 3)(3 2)(1 3)(1 4) and so on. This implies that ***every cycle can be expressed as a product of transposition in many possible ways***. An r-cycle may represent a permutation of degree n, where n is any integer greater than and equal to r. Thus a r-cycle may be represented as a product of transpositions in infinite many ways. And, in fact, there are infinite many ways to express a cycle as a product of transpositions.

Since every permutation can be expressed as a product of disjoint cycles, every permutation can be expressed as product of transposition in infinite many ways. *A permutation is said to be an* ***even permutation*** *if it is expressed as a product of* ***even*** *number of transpositions otherwise it is called* ***odd permutation.***

As discussed above, a 3-cycle can be expressed as a product of two transpositions, a 4-cycle can be expressed as a product of 3 transpositions and so on. In general, an r-cycle may be expressed as a product of (r – 1) transpositions. If r is odd then r-cycle represents an even permutation and if r is even then the corresponding permutation is odd. Thus we can say, "*Every transposition represents an odd permutation.*"

Every identity permutation can be expressed as a product of even number of transpositions. For example, Identity permutation I of degree 2 can be written as (1 2) (2 1), I of degree 3 can be written as (1 2)(2 1) or (1 2)(2 1)(1 3)(3 1) and so on. Similarly, product of any two even permutations is an even permutation and of any two odd permutations is also an even permutation. To prove this, let us assume that f and g be any two permutations expressed as product of r and s transpositions respectively. Then the product fg can be expressed as a product of (r + s) transpositions. If r and s both are odd or even, then (r + s) is always even. Therefore, fg is an even permutation. If one is odd and other is even permutation, then fg will be an odd permutation. *The set containing all even permutations of degree n is called* ***Alternating set of permutations*** *and it is denoted by* A_n. In the following theorem it is shown that the set A_n forms a finite group. Before that it is important to know the cardinality of the set A_n.

Theorem 4 Of the n! permutations on n symbols half of it are even permutations and other half are odd permutations.

Proof: Let P_n be the set of all permutations on n symbols. Let the number of odd and even permutations in P_n be m and k respectively. Thus m + k = n!. Let $o_1, o_2, o_3, \ldots, o_m$ be m number of distinct odd permutations and $e_1, e_2, e_3, \ldots, e_k$ be k number of distinct even permutations, then

$$P_n = \{o_1, o_2, o_3, \ldots, o_m, e_1, e_2, e_3, \ldots, e_k\}$$

Let t be any transposition in P_n. Since P_n forms a group with respect to the composition of permutations, permutations: $to_1, to_2, to_3, \ldots, to_m, te_1, te_2, te_3, \ldots,$ and te_k, all belongs to P_n and all are distinct. Otherwise, $to_1 = to_2 \Rightarrow o_1 = o_2$, which is contrary to the fact that o_1, o_2 are distinct.

Now $te_1, te_2, te_3, \ldots, te_k$ are odd permutations and $to_1, to_2, to_3, \ldots, to_m$ are even permutations. Since there are m number of odd permutations, thus k cannot be > m, thus k ≤ m. Similarly from the count of even permutations, we have m ≤ k. Therefore, we have m = k. Thus

$$m = k = \frac{n!}{2}$$

Proved.

Theorem 5 The set A_n of all even permutations of degree n forms a finite group of order n!/2 with respect to the composition of permutation.

Proof: Let f and g be any two even permutations on n symbols. Then fg is again an even permutation on n symbols. This implies that A_n is closed with respect to the composition of permutation.

It is known that composition of permutation is an associative operation.

It is also known that an identity permutation is an even permutation. Let I be an identity permutation on n symbols, then $I \in A_n$. Also for any $f \in A_n$, we have

$$If = fI = f$$

Therefore I is the identity element.

Let f be any even permutation in A_n, then $f \in P_n$. Thus there exists an inverse f^{-1} in P_n such that $f f^{-1} = f^{-1} f = I$. Since f is an even permutation so f^{-1} is also an even permutation and so $f^{-1} \in A_n$. This show that every element of A_n has an inverse in A_n.

Therefore A_n forms a group under the composition of permutation. This group is called an ***Alternating group***. Obviously the order of this group is **n!/2**.

Proved.

It is important to mention that the set O_n of all odd permutations does not form a group with respect to the composition of permutation. As the product of two odd permutations is an even permutation, the set O_n is not closed. Now let us see some examples based on composition table.

Example 8 Show that the $(P_3,*)$ of all permutations on three symbols 1, 2, 3 is a finite non-abelian group of order 6 with respect to the composition of permutations.

Solution: The number of permutations in P_3 is 3! = 6. These 6 permutations can be expressed as cycles. Thus, we can write P_n={I, (1 2), (1 3), (2 3), (1 2 3), (1 3 2)}. Let us represent permutation I as f_1, (1 2) as f_2, (1 3) as f_3, (2 3) as f_4, (1 2 3) as f_5 and (1 3 2) as f_6. The composition table for P_3 is shown in the table 5.5.1. Now let us show that $(P_3,*)$ satisfies all the four postulates of group.

1. **Closure property:** Since all the entries in the table are elements of P_3, P_3 is closed with respect the composition of permutation.
2. **Associativity:** It is known that composition of permutation is an associative operation.
3. **Existence of Identity:** Since the entries in the table corresponding to the column f_1 and row f_1 matches with row and column header respectively, the element f_1 is the identity element.
4. **Existence of Inverse:** Since every row and every column contains the identity element f_1 as an entry, every element of P_3has an inverse.

*	f_1	f_2	f_3	f_4	f_5	f_6
f_1	f_1	f_2	f_3	f_4	f_5	f_6
f_2	f_2	f_1	f_6	f_5	f_4	f_3
f_3	f_3	f_5	f_1	f_6	f_2	f_4
f_4	f_4	f_6	f_5	f_1	f_3	f_2
f_5	f_5	f_3	f_4	f_2	f_6	f_1
f_6	f_6	f_4	f_2	f_3	f_1	f_5

Table 5.5.1

Therefore $(P_3,*)$ is a group however, it is not a commutative group. Reader may verify this.

Ans.

Example 9 Show that the function f of the symmetric group P_n onto the multiplicative group G = {−1, 1} defined by

$$f(\alpha)=\begin{cases}1 & \text{if } \alpha \text{ is an even permutation}\\ -1 & \text{if } \alpha \text{ is an odd permutation}\end{cases}$$

is a homomorphism of P_n onto G.

Solution: Let α and β be any two permutations from P_n. Then we have the following four possible situations:

- Both α and β are even ⇒ αβ is even. Therefore, f(αβ) = 1, f(α) = 1, f(β) = 1. Thus f(αβ) =f(α) f(β).
- Both α and β are odd ⇒ αβ is even. Therefore, f(αβ) = 1, f(α) = −1 and f(β) = −1. Thus f(αβ) =f(α) f(β).
- α is even and β odd ⇒ αβ is odd. Therefore, f(αβ) = −1, f(α) = 1, f(β) = −1. Thus f(αβ) =f(α) f(β).
- α is odd and β even ⇒ αβ is odd. Therefore, f(αβ) = −1, f(α) = −1, f(β) = 1. Thus f(αβ) =f(α) f(β).

Thus, f(αβ) =f(α) f(β) ∀ α, β ∈ P_n. Also, obviously, f is onto function. Therefore, f is a homomorphism of P_n onto G.

Ans.

Example 10 Show that the multiplicative group G = $\{1, \omega, \omega^2\}$ is isomorphic to the Alternating group A_3 = {I, (a b c), (a c b)} on three symbols a, b, c.

Solution: Let f be a function from G to A_3 defined as:

$$f(1) = I,\ f(\omega) = (a\ b\ c) \text{ and } f(\omega^2) = (a\ c\ b)$$

The composition table for the groups G and A_3 can be shown as in table 5.5.2 and table 5.5.3. For simplicity we have used x, y and z to represent the element I, (a b c) and (a c b) of A_3 respectively.

*	1	ω	ω^2
1	1	ω	ω^2
ω	ω	ω^2	1
ω^2	ω^2	1	ω

Table 5.5.2

*	x	y	z
x	x	y	z
y	y	z	x
z	z	x	y

Table 5.5.3

The two composition tables are identical i.e. if we replace the elements in the table 5.5.2 by the corresponding images, we get the table 5.5.3. The function is one to one and onto. Thus the two groups are isomorphic to each other.

Ans.

A group -(G, ∗), may be considered as an object. The set G contains all the data and * is the operation applied on the data in G. The four postulates imply that certain functionalities are applicable to the data in G and all these are contained in (G, *). If we add some additional operations, it yields another object. **Ring, Field, Integral Domain** etc are examples of mathematical structures derived from group. *Interested reader may find details about these mathematical structures in the chapter nine of this book.* Like basic logic theory has evolved from set theory which led to circuit designing, the group theory has led to the integrated circuit designing and Object-oriented approach in computer industry.

Exercise

1. Prove the following:

   (a) The set **C** of all complex numbers with respect to the operation of addition of complex numbers is group.

   (b) The set of all rational numbers with respect to addition is a group.

   (c) The set of all real numbers with respect to addition is a group

   (d) The set R_0 of all non-zero real numbers with respect to multiplication is a group..

   (e) The set C_0 of all non-zero complex numbers with respect to multiplication is a group.

2. Define the order of a group. Show that the set of all even integers with zero is an abelian group wit respect to addition.

3. Show that the set M of complex numbers z with the condition $|z| = 1$ forms a group with respect to th operation of multiplication of complex numbers.

4. Let Q_+ be the set of all positive rational numbers and $*$ a binary operation on Q_+ defined by $a * b$ ab/3. Determine the identity element in Q_+ and inverse of the element 'a'.

5. Let R be the set of all real numbers and $*$ a binary operation on R defined by $a*b = a + b + a$ Determine the identity element in R and determine the inverse of a.

6. Show that the set of all rational numbers of the form $2^a 3^b$ where a and b are integers, is a group wi respect to the multiplication of rational numbers.

7. Consider the binary operation $*$ defined on the set A = {a, b, c, d} by the following table.

*	**a**	**b**	**c**	**d**
a	a	c	b	d
b	d	a	b	c
c	c	d	a	a
d	d	b	a	c

Compute

(a) c*d and d*c

(b) b*d and d*b

(c) a*(b*c) and (a*b) *c

(d) Is * commutative; associative?

In exercises 8 and 9, complete the given table so that the binary operation * is associative.

8.

*	**a**	**b**	**c**	**d**
a	a	b	c	d
b	b	a	d	c
c	c	d	a	b
d				

9.

*	**a**	**b**	**c**	**d**
a	a	b	c	d
b	b	a	c	d
c				
d	d	c	c	d

0. Show that the set of all m×n matrices with the operation of matrix addition is a group.

1. Determine whether (P(S), ∗) is a group or not, P(S) is the power set of a non-empty set S and ∗ on P(S) is defined as $A*B = A \oplus B$ for A, B ∈ P(S).

2. Let S ={x | x is a real number and x ≠ 0, x ≠ -1}. Consider the following functions $f_i : S \to S$, i = 1,2,3,…,6:

$f_1(x) = x$, $f_2(x) = 1 - x$, $f_3(x) = 1/x$, $f_4(x) = 1/(1 - x)$, $f_5(x) = 1 - 1/x$, $f_6(x) = x/(x - 1)$.
how that G = $\{f_1, f_2, f_3, f_4, f_5, f_6\}$ is a group under the operation of function composition. Find the mposition table for (G, ∗).

3. Prove that the set G of all translations of a fixed point (x, y) in the plane given by x' = x + a and y' = y + b, where a and b are real numbers, is a group under the operation of 'composition of translation'.

. (a) Prove that the set {1, -1} is an abelian group with respect to multiplication.
(b) Show that {x | x is cube roots of unity} form an abelian group under multiplication. Write down the multiplication table.

. (a) How many elements of the cyclic group of order 7 can be used as generators of the group?
(b) How many elements of the cyclic group of order 8 can be used as generators of the group?
(c) If G ={a} be an infinite cyclic group, prove that a and a^{-1} are the only generators of G.

. Is the multiplicative group of residue classes {1}, {2}, {3}, {4}, {5},{6} (mod 7) cyclic ?

. Let G is the multiplicative group of all non-singular square matrices of order n with real elements and S is the set of non-singular matrices of order n with real elements whose determinant is 1. Show whether S is a subgroup of G or not?

Show that the set of all elements x of a group G such that ax = xa for every element a of G is a subgroup of G.

Find all the subgroups of the cyclic group G = (a, a^2, a^3, a^4, a^5, a^6 = e}.

20. Let G be the group of complex numbers under multiplication. Determine whether S = {1, –1, i, –i} a subgroup of G or not?

21. Prove that the additive group R of real numbers is mapped homomorphically onto the multiplicati group C of complex numbers z such that |z| = 1, under function f: R→C defined as f(x) = cos(x) + sin(x).

22. Prove that the multiplicative group of I*/(5) is cyclic and then show that it is isomorphic to t additive group of I/(4).

23. Let {1, w, w^2} be the multiplicative group of the cube roots of unity. Let G' = {0, a, b} be the group rotations of an equilateral triangle about its centroid. Show that G and G' are isomorphic groups.

24. Prove that every abelian group of order 15 is cyclic.

25. Prove that any abelian group of order p*q where p and q are distinct primes, is necessarily cyclic a hence isomorphic to additive group I/(p* q).

26. Let G be the group of all translations of a fixed point (x, y) in the plane and H be the subgroup of translations parallel to the x - axis. Find the cosets of H in G.

27. Prove that the intersection of two sub monoids of a monoid (S, *) is a sub monoid of (S, *).

28. Let R^+ be the set of all positive real numbers. Show that the function f: R^+→R defined by f(x) = ln is an isomorphism of the semi-group (R^+, ×) to the semigroup(R,+), where × and + are ordin multiplication and addition of real numbers, respectively.

29. Let (S_1, *), (2_1, *') and (S_3, *'') be semi groups, and let f: S_1 → S_2 and g: S_2 →S_3 be isomorphis Show that gof:S_1→S_3 is an isomorphism.

30. An element x in a monoid is called an ***idempotent*** if x^2 = x * x = x. Show that the set of idempotents in a commutative monoid S is a sub-monoid of S.

31. Which of the following tables defines a semi-group?

(a)

*	a	b	c
a	c	b	a
b	b	c	b
c	a	b	c

(b)

*	a	b	c
a	c	b	a
b	b	c	b
c	a	b	c

In exercise 32 to 36, determine whether the relation R on the semigroup S is a congruence relation.

32. S = I, the set of all integers, under the operation of ordinary addition; aRb iff 2 does not divide a –

33. S = any semi-group; aRb iff a = b.

4. S = the set of all rational numbers under the operation of addition; a/b R c/d iff ad = bc.

5. S = I, the set of all integers, under the operation of integer addition of integers; aRb iff $a \equiv_3 b$.

6. S = set of all positive integers, under the operations of ordinary multiplication; aRb iff $|a - b| \leq 2$.

7. Let A = {0,1} and consider the free semi-group A* generated by A under the operation of catenation. Let N be the semi-group of all nonnegative integers under the operation of ordinary addition.
   (a) Verify that the function $f: A^* \to N$, defined by $f(\alpha)$ = the number of bits in α, is a homomorphism.
   (b) Let R be the relation on A* defined as $\alpha R \beta$ iff $f(\alpha) = f(\beta)$, where f is defined as in (a). Show that R is a congruence relation on A*.
   (c) Show that A*/R and N are isomorphic.

8. Let G be a group. Show by mathematical induction that if ab = ba, then $(ab)^n = a^n b^n$ for $n \in Z^+$.

9. Let G be a finite group with identity e, and let a be an arbitrary element of G. Prove that there exists a nonnegative integer n such that $a^n = e$.

0. Let H and K be subgroups of a group G. Prove that (a) $H \cap K$ is a subgroup of G and (b) $H \cup K$ need not be a subgroup of G.

1. Let G be an abelian group with identity element e and let $H = \{x \mid x^2 = e\}$. Show that H is a subgroup of G.

2. Show that the group in exercise 12 is isomorphic to S_3.

3. Let G be a group. Show that the function $f: G \to G$ defined by $f(a) = a^{-1}$ is an isomorphism iff G is abelian.

4. Write the multiplication table of the group $Z_2 \times Z_3$.

5. Let G_1 and G_2 be groups. Prove that $G_1 \times G_2$ and $G_2 \times G_1$ are isomorphic.

6. Let G = S_3. For each of the following subgroups H of G, determine all the left cosets of H in G.
   (a) H = $\{f_1, g_1\}$ (b) H = $\{f_1, g_3\}$
   (c) H = $\{f_1, f_2, f_3\}$(d) H = $\{f_1\}$ (e) H = $\{f_1, f_2, f_3, g_1, g_2, g_3\}$

. Let G = Z_8. For each of the following subgroups H of G, determine all the left cosets of H in G.
   (a) H = {[0], [4]} (b) H = {[0], [2],[4], [6]}

. Let N be a subgroup of a group G, and let $a \in G$. Let us define $a^{-1}Na = \{a^{-1}na \mid n \in N\}$. Prove that N is a normal subgroup of G iff $a^{-1}Na = N$ for all $a \in G$.

. Find all the normal subgroups of S_3.

. Let G be an abelian group and N a subgroup of G. Prove that G/N is and abelian group.

. Let (G,∗) be a mathematical structure such that for all a, b, c, d in G
   (i) a ∗ a = a and (ii) (a ∗ b) ∗ (c ∗ d) = (a ∗ c) ∗ (b ∗ d).
   Show that a ∗ (b ∗ c) = (a ∗ b) ∗(a ∗ c).

. Let (G,∗) be a semigroup. Show that for a, b, c in G if (i) a ∗ c =c ∗ a and (ii) b ∗ c = c ∗ b then
   (a ∗ b) ∗ c = c ∗ (a ∗ b).

53. Let $(G,*)$ be a mathematical structure such that for all a, b in G (i) $(a * b) * a = a$, (ii) $(a * b) * b = (b * a) * a$ and (iii) $a * b = (a * b) * b$. Show that $*$ is idempotent and commutative.

54. A *central groupoid* is a mathematical structure $(G,*)$ where $*$ is a binary operation such that $(a * b) * (b * c) = b$ for all a, b, c in G. Show that

$$a * ((a * b) * c) = a * b \quad \text{and}$$
$$(a * (b * c)) * c = b * c$$

in a central groupoid.

CHAPTER SIX
Ordered Set

Recall the relation on set discussed in the chapter 1. We discussed different types of relation in that chapter. In the chapter five, we have learnt about mathematical (algebraic) structure with one binary operation. A relation between two sets is also a binary operation. Any relation R defined on a non-empty set A is said to be a **Partial Order Relation,** if R is

- Reflexive on A i.e., xRx $\forall x \in A$
- Anti-symmetric on A i.e., xRy and yRx $\Rightarrow x = y$ and
- Transitive on A i.e., xRy and yRz $\Rightarrow$ xRz for x, y, z $\in$ A.

A partial order relation is denoted by the symbol $\leq$. For example, the relation 'less than equal to' is a partial order relation on the set N of natural numbers. Similarly, the relation 'greater than equal to' is also a partial order relation on the set N of natural number. It is for the reader to note that only equivalence relation that is also a partial order, is the *Identity relation* or *Diagonal relation*. The relation 'less than equal to' is referred as **usual partial order**.

6.1 Poset

Let R be a relation defined on a non-empty set A. The mathematical structure (A, R) is set to be a *Partial order set* or **poset** if the relation R is a partial order relation on A. A general notation for a poset is (A, $\leq$), where A is any non-empty set and '$\leq$' is any partial order relation defined on the set A. A poset is said to be **finite poset** if A is a **finite set** and **infinite poset** if A is an **infinite set**.

Example 1 Let S be any non-empty set and P(S) be the power set of S. If '$\subseteq$'(a subset of) is a relation defined on P(S), then show that (P(S), $\subseteq$) is a poset.

Solution: P(S) is obviously a non-empty set. The relation $\subseteq$ on P(S) is reflexive, anti-symmetric and transitive on P(S). So, $\subseteq$ is a partial order relation on P(S), and hence (P(S), $\subseteq$) is a poset.

Proved.

Example 2 Let A be a non-empty set and R be the set of all equivalence relation defined on A. If '$\subseteq$' (a subset of) is a relation defined on R, then show that the structure (R, $\subseteq$) is a poset.

Solution: Here, R is a non-empty set, because there always exists some relations (identity, universal etc.) on a set that are equivalence relations. The relation $\subseteq$ on R is reflexive, anti-symmetric and transitive, hence $\subseteq$ is a partial order relation on R. Therefore, (R, $\subseteq$) is poset.

Proved.

Theorem 1 Let R be a partial order relation on a set A and let R^{-1} be the inverse of relation R, then R^{-1} is also a partial order.

Proof: Since R is a partial order relation on A, R is reflexive, anti-symmetric and transitive.

Reflexivity of $R \Rightarrow \forall x \in A, (x, x) \in R \Rightarrow \forall x \in A, (x, x) \in R^{-1} \Rightarrow R^{-1}$ is reflexive.

Let (x, y) and (y, x) $\in R^{-1}$, then (y, x) and (x, y) $\in$ R. Since R is anti-symmetric, it implies that x = y. Thus R^{-1} is anti-symmetric.

Let (x, y) and (y, z) $\in R^{-1}$, then (z, y) and (y, x) $\in$ R. Since R is transitive, (z, x) $\in$ R. This implies that (x, z) $\in R^{-1}$. Hence, R^{-1} is transitive. Therefore R^{-1} is a partial order relation.

Proved.

In the theorem 1, it is proved that if R is a partial order relation on a non-empty set A, then R^{-1} is also a partial order relation on the same set A. Therefore, if (A, R) is a poset then (A, R^{-1}) is also a poset. The poset (A, R^{-1}) is called **Dual** of the poset (A, R) and vice versa. In the example 1, '$\subseteq$' is a partial order relation on P(S). The inverse of this relation is '$\supseteq$' (a superset of). Thus, (P(S), $\supseteq$) is a dual of the poset (P(S), $\subseteq$).

Example 3 Prove that (I, $\leq$) is a poset, where I is the set of all integers and $\leq$ is a relation on I defined as $a \leq b$ iff $a = b^k$ for some $k \in I^+$. Find the dual of this poset.

Solution: Since $1 \in I^+$, we have $a = a^1$, $\forall a \in I$. i.e., pairs (a, a) $\in \leq \forall a \in I$. Therefore, $\leq$ is reflexive. Let (a, b) and (b, a) are in $\leq$, then $a = b^k$ and $b = a^j$ for some k and $j \in I^+$. This is possible iff k = j =1, i.e., a = b. Thus, $\leq$ is anti-symmetric. Finally, let (a, b) and (b, c) are in $\leq$. Then, by definition of $\leq$, $a = b^k$ and $b = c^j$ for some k and $j \in I^+$. This implies that $a = c^{kj}$, and hence (a, c) is in $\leq$. Thus, $\leq$ is transitive relation. Therefore the structure (I, $\leq$) is a poset.

In the second part of the given problem, we have to find dual of the poset (I, $\leq$). The inverse of the relation $\leq$, say $\geq$ can be defined on I as: for any a, b of I, $a \geq b$ iff $a^k = b$ for some $k \in I^+$. This can be easily proved that $\geq$ is a partial order relation on I, and hence, (I, $\geq$) is dual of the poset (I, $\leq$) and vice versa.

Proved.

Example 4 Let A = {1, 2, 3, 4} and R ={(1, 1), (2, 2), (3, 3), (4, 4), (1, 2), (2, 4), (1, 3), (3, 4), (1, 4)} is relation defined on A. Show that (A, R) is a poset and find its dual.

Solution: The relation R is reflexive and transitive. It is anti-symmetric also as the matrix for R is upper triangular. ***If the matrix for a given relation is either upper triangular or lower triangular or diagonal then the relation is anti-symmetric.*** The reverse is not always true. Therefore, R is a partial order relation on A and hence (A, R) is a poset. Now, R^{-1} = {(1, 1), (2, 2), (3, 3), (4, 4), (2, 1), (4, 2), (3, 1), (4, 3), (4, 1)}. This inverse relation of R is also a partial order relation on A, thus (A, R^{-1}) is dual of (A, R).

Proved.

Example 5 Determine whether the relation R defined as aRb iff a = 2b, on the set I of all integers, is a partial order relation or not.

Solution: The relation R is not reflexive because no integer, other than zero, is equal to twice of itself. Since R is not reflexive, it is not a partial order relation. Note that there is no need for testing anti-symmetry and transitivity of R. **Ans.**

Let (A, ≤) be a poset. Any two elements a and b of A are said to be **comparable,** if either pair (a, b) or pair (b, a) is in ≤. That is either a ≤ b or b ≤ a. It is important to mention that in a partial order relation, every pair of element need not be comparable. For example, let N be the set of all positive integers and ≤ on N is defined as a ≤ b iff a divides b. Obviously, ≤ is a partial order relation on N and hence, (N, ≤) is a poset. But, there exists elements like 3 and 8 which are not **comparable** because neither 3 ≤ 8 nor 8 ≤ 3. *This implies that in a partially ordered set, every pair of elements may not be comparable and that's why it is called **partially ordered.*** If every pair of elements in a poset (A, ≤) is comparable, then (A, ≤) is called **linearly ordered**, or **totally ordered** or **simple ordered** or **complete ordered** set. A linearly ordered set is also called a **chain**. A set is said to be **anti-chain** if no two distinct elements of the set are related.

Example 6 The set inclusion relation ⊆ on a power set P(S) of a set S, is a partial order relation on P(S) (as proved in the example 1), but not a linear order relation. Let S = {1, 2, 3}. The sets {1} and {2} are in P(S) and neither {1}⊆{2} nor {2}⊆{1}. Therefore, (P(S), ⊆) is poset but not a **chain**.

Example 7 The relation ≤ (less than equal to) on the set R of all real numbers is a partial order relation on R and hence (R, ≤) is poset. Also for any two real numbers x and y, either x ≤ y or y ≤ x, i.e., every pair of real numbers are comparable with respect to the relation ≤. Therefore, (R, ≤) is also a **chain**.

Diagrammatic Representation of Poset

We have learnt in the chapter 1 about the diagrammatic representation of a relation on a finite set. The same concept we can use to represent any finite partial order relation $\leq$, on a set A, in diagrammatic way. Let A = {a, b, c} and R = {(a, a), (b, b), (c, c), (a, b), (b, c), (a, c)}. Obviously, R is a partial order relation on A and hence (A, R) is poset. The diagram for R is shown in the figure 6.1.1.

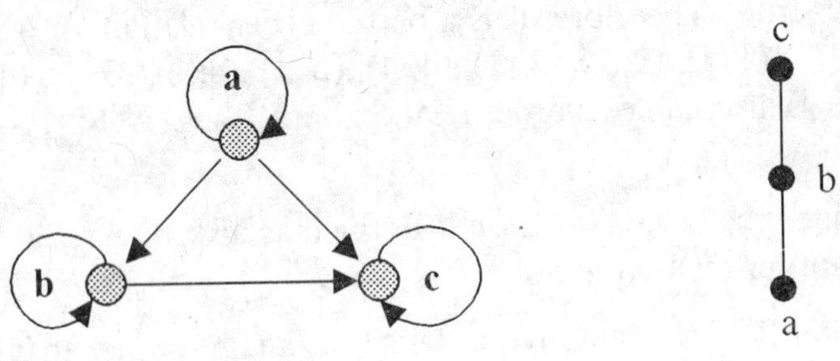

Figure 6.1.1 **Figure 6.1.2**

Let us simplify this representation for a finite poset. In a poset the reflexivity and transitivity of the relation is implied. The anti-symmetry of the relation we are not considering right now. Since we know that a partial order relation is reflexive, loop on each node in the diagram can be dropped. The arc showing the implied transitivity can also be dropped in a poset. Next, all the elements of poset (nodes in diagram) can be arranged in such a way that all edges points upward. The resulting diagram is a simplified diagram for a given a poset. This simplified diagram is called **Hasse Diagram** of a poset. We shall use this form of representation in this chapter for poset, lattice and for finite Boolean algebra. The above procedure to draw a Hasse diagram can be summarized as below.

- Draw a digraph for the partial order relation.
- Drop loops on all nodes.
- Drop all edges implied by the transitivity of the relation.
- Arrange all edges pointing in upward direction and then replace each directed arc with undirected arc.
- Represent each node with **thick dots (•)**

If we apply this procedure on the diagram shown in the figure 6.1.1, we get the Hasse diagram for the poset (A, R) as shown in the figure 6.1.2. In the figure 6.1.2 arc (a, c) is implied transitively from (a, b) and (b, c). Thus it is removed from the diagram of R. Other things are obvious.

Example 8 Let us define a set D_n, where n is any positive integer > 1, as the collection of all the positive integers which can evenly divide n. Thus, D_{12} = {1, 2, 3, 4, 6, 12}, D_6 = {1, 2, 3, 6} etc. Let us define a relation ≤ on D_{12} as, for any two integers a and b ∈ D_{12}, a ≤ b iff a divides b. Show that (D_{12}, ≤) is a poset. Draw a Hasse diagram for this poset.

Solution: Since every integer divides itself, the relation ≤ is reflexive on D_{12}. For any two integers a and b, if a divides b and b divides a, then both a and b are equal. Thus ≤ is anti-symmetric. Similarly, for any three integers a, b and c if a divides b and b divides c then a divides c. Thus, ≤ is transitive. Therefore (D_{12}, ≤) is a poset. Since | D_{12}| = 6, the poset is finite. The ordered pairs of the relation ≤ are {(1, 1), (1, 2), (1, 3), (1, 4), (1, 6), (1, 12), (2, 2), (2, 4), (2, 6), (2, 12), (3, 3), (3, 6), (3, 12), (4, 4), (4, 12), (6, 6), (6, 12), (12, 12)}. The diagram for this relation is shown in the figure 6.1.3.

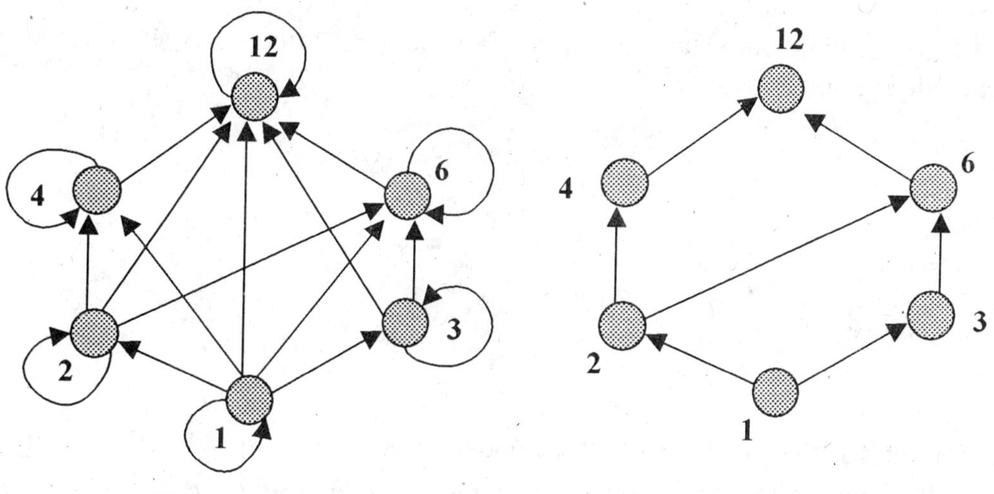

Figure 6.1.3 **Figure 6.1.4**

To draw a Hasse diagram for the poset (D_{12}, ≤), first drop all the loops in the figure 6.1.3 and then drop all the arcs that are transitively implied. If we look at the figure 6.1.3, arcs (1, 4), (1, 12), (1, 6), (2, 12) and (3,12) are transitively implied. Dropping them together with all loops from figure 6.1.3, we get the diagram as shown in the figure 6.1.4. All the arcs are arranged in upward direction, so only thing left is to represent the elements by thick dots and replacement of all the directed arcs with undirected arcs. Applying this on the figure 6.1.4, we get the Hasse diagram as shown in the figure 6.1.5.

Ans.

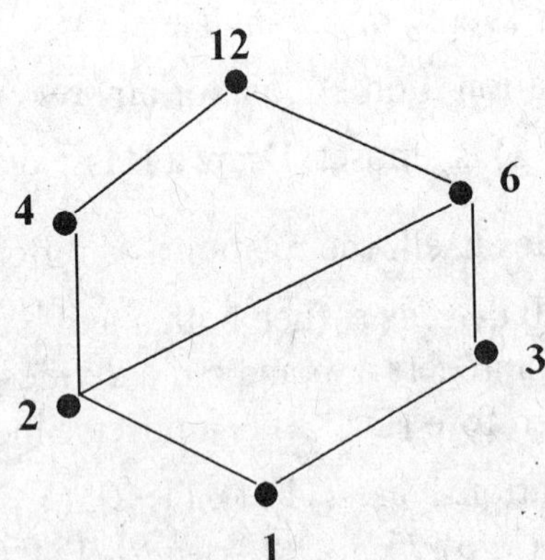

Figure 6.1.5

Example 9 From the following Hasse diagram of a poset (A, ≤) find set A and the partial order relation defined on A.

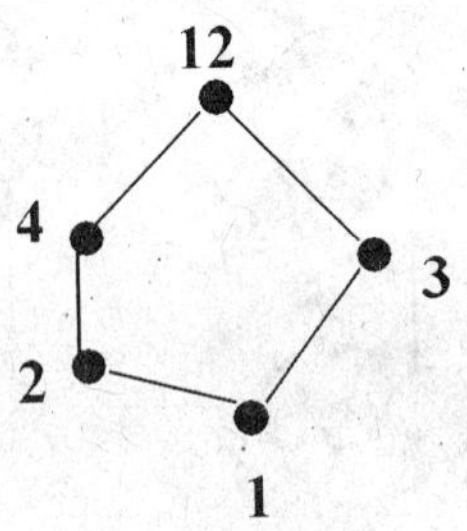

Figure 6.1.6

Solution: The set A is the collection of all the nodes i.e. A = {1, 2, 3, 4, 12}. The relation on A is a partial order. Thus All the reflexive pairs: (1, 1), (2, 2), (3, 3), (4, 4), (12, 12) must be in ≤. Next, all the edges when converted to upward directed arcs give the pairs of the relation. These are (1, 2), (1, 3), (2, 4), (4, 12) and (3, 12). The transitively implied arcs are (1, 4), (1, 12) and (2, 12). Therefore, the partial order relation on A as shown in the given Hasse diagram of figure 6.1.6, is {(1, 1), (2, 2), (3, 3), (4, 4), (12, 12), (1, 2), (1, 3), (2, 4), (4, 12) (3, 12), (1, 4), (1, 12), (2, 12)}.

Ans.

Theorem 2 The digraph of a partial order has no cycle of length greater than 1.

Proof: Suppose that there exists a cycle of length $n \geq 2$ in the digraph of a partial order ≤ on a set A. This implies that there are n distinct elements $a_1, a_2, a_3, \ldots, a_n$ such that $a_1 \leq a_2$, $a_2 \leq a_3, \ldots, a_{n-1} \leq a_n$ and $a_n \leq a_1$. Applying the transitivity n–1 times on $a_1 \leq a_2$, $a_2 \leq a_3, \ldots,$

$a_{n-1} \le a_n$, we get $a_1 \le a_n$. Since relation $\le$ is anti-symmetric $a_1 \le a_n$ and $a_n \le a_1$ together implies that $a_1 = a_n$. This is contrary to the fact that all $a_1, a_2, a_3, \ldots, a_n$ are distinct. Thus, our assumption that there is a cycle of length $n \ge 2$ in the digraph of a partial order relation is wrong. Hence **proved**.

Product Partial Order

Let us see the following theorem before delving into the concept of product partial order.

Theorem 3 If $(A, \le_1)$ and $(B, \le_2)$ are two posets, then $(A\times B, \le)$ is a poset, where $\le$ is defined on $A\times B$ as $(a, b) \le (a_1, b_1)$ if, and only if $a \le_1 a_1$ in A and $b \le_2 b_1$ in B, where (a, b) & $(a_1, b_1) \in A\times B$.

Proof: Let (a, b) be a pair of $A\times B$.

Then, $(a, b) \Rightarrow a \in A$ and $b \in B$.

$\Rightarrow a \le_1 a$ in A and $b \le_2 b$ in B since $(A, \le_1)$ & $(B, \le_2)$ are posets

$\Rightarrow (a, b) \le (a, b)\ \forall\ (a, b) \in A\times B$ by the definition of $\le$

$\Rightarrow \le$ is reflexive on $A\times B$.

Next, if $(a_1, b_1) \le (a_2, b_2)$ and $(a_2, b_2) \le (a_1, b_1)$ for any two pairs (a_1, b_1) & (a_2, b_2) of $A\times B$, Then,

$(a_1, b_1) \le (a_2, b_2) \Rightarrow a_1 \le_1 a_2$ in A and $b_1 \le_2 b_2$ in B and

$(a_2, b_2) \le (a_1, b_1) \Rightarrow a_2 \le_1 a_1$ in A and $b_2 \le_2 b_1$ in B

Now, $(a_1 \le_1 a_2)$ & $(a_2 \le_1 a_1) \Rightarrow a_1 = a_2$ because $\le_1$ is a partial order on A. Similarly, $(b_1 \le_2 b_2)$ & $(b_2 \le_2 b_1) \Rightarrow b_1 = b_2$. Therefore $(a_1, b_1) = (a_2, b_2)$. Thus, $\le$ is anti-symmetric on $A\times B$.

Finally, if $(a_1, b_1) \le (a_2, b_2)$ and $(a_2, b_2) \le (a_3, b_3)$ for any three pairs (a_1, b_1), (a_2, b_2) and (a_3, b_3) of $A\times B$, Then,

$(a_1, b_1) \le (a_2, b_2) \Rightarrow a_1 \le_1 a_2$ in A and $b_1 \le_2 b_2$ in B and

$(a_2, b_2) \le (a_3, b_3) \Rightarrow a_2 \le_1 a_3$ in A and $b_2 \le_2 b_3$ in B

Now, $(a_1 \leq_1 a_2)$ & $(a_2 \leq_1 a_3) \Rightarrow a_1 \leq_1 a_3$ because $\leq_1$ is a partial order on A. Similarly, $(b_1 \leq_2 b_2)$ & $(b_2 \leq_2 b_3) \Rightarrow b_1 \leq_2 b_3$. Therefore $(a_1, b_1) \leq (a_3, b_3)$. Thus, $\leq$ is a transitive relation on A×B. Therefore, $\leq$ is a partial order relation on A×B and hence (A×B, $\leq$) is a poset.

Proved.

A partial order as defined in the theorem 3 is called a **product partial order**, and poset (A×B, $\leq$) is called **product poset** of posets (A, $\leq_1$) and (B, $\leq_2$). The same concept can be extended for any finite number of posets. If $(A_1, \leq_1), (A_2, \leq_2), (A_3, \leq_3), \ldots, (A_n, \leq_n)$ are n posets, then $(A_1 \times A_2 \times A_3 \times \ldots \times A_n, \leq)$ is a poset, where $\leq$ is defined on $A_1 \times A_2 \times A_3 \times \ldots \times A_n$ as

$$(a_1, a_2, a_3, \ldots, a_3) \leq (b_1, b_2, b_3, \ldots, b_3) \text{ if, and only if}$$

$$a_1 \leq_1 b_1 \text{ in } A_1;\ a_2 \leq_2 b_2 \text{ in } A_2; \text{ and so on up to n.}$$

Let x and y are any two real numbers. **If $x \leq y$ and $x \neq y$, we say that $x < y$.** Let A be a set of natural numbers and let < be a relation "strictly less than" on A. Obviously, < is irreflexive and transitive on A. Any relation that is irreflexive and transitive is called a **Strong order relation** or **Quasi-order relation**. Let us modify the above definition of product partial order such that

$(a, b) < (a_1, b_1)$ if, and only if

either $a < a_1$ in A **or**

$a = a_1$ in A and $b \leq_2 b_1$ in B.

An ordering of this type is called **Dictionary Order** or **Lexicographic Order.** If we reason out with some sample words from a dictionary, we can easily make it out that all words in a dictionary are arranged in lexicographic order. All string sorting algorithms are based on this concept only.

Example 10 Let $\Sigma = \{a, b, c, \ldots, z\}$ be the set of ordinary alphabets arranged in usual order like $a < b < c \ldots < z$. Let $\Sigma^n = \Sigma \times \Sigma \times \Sigma \times \ldots$ n times. Obviously, Σ^n is set of all words of n characters. Let $w_1 = x_1x_2x_3 \ldots x_n$ and $w_2 = y_1y_2y_3 \ldots y_n$ be two words from Σ^n, where x_i's and y_i's are alphabets from Σ. We define $w_1 < w_2$ if, and only if $x_1 < y_1$ **or** $x_1 = y_1$ and $x_2 < y_2$ **or** $x_1 = y_1$; $x_2 = y_2$ and $x_3 < y_3$;... **or** $x_1 = y_1$; $x_2 = y_2$ and $x_3 = y_3$; ...; $x_n \leq y_n$. This relation determines whether word w_1 or word w_2 comes first in a dictionary. For example, 'past' and 'part' $\in \Sigma^4$ and 'part' < 'past' so in a dictionary we get 'part' before we get 'past', if we search for the word from first page without skipping. This is an example of **Lexicographic Order.**

Example 11 Let $(I_o, \leq)$ and $(I_e, \leq)$ be two posets, where I_o and I_e are sets of all odd and even integers respectively and $\leq$ is usual partial order. Show that $(I_o \times I_e, \leq^*)$ is a product poset of posets $(I_o, \leq)$ and $(I_e, \leq)$.

Solution: To prove that $(I_o \times I_e, \leq^*)$ is the product poset, it is sufficient to prove that $\leq^*$ is a product partial order i.e., the relation $\leq^*$ defined as

$$(a, b) \leq^* (x, y) \text{ if, and only if } (a \leq x) \text{ in } I_o \text{ and } (b \leq y) \text{ in } I_e$$

is a partial order relation on $I_o \times I_e$. I leave it to the reader to prove this as proved in the theorem 3.

Ans.

Bounding Elements of a Poset

Some elements in a poset are of special importance for understanding certain characteristics of a poset. Whether a subset of a poset is bounded within the poset? Can we designate any element of the poset as **first** one? Can we find a **last** element of the poset? And to get answer to many such queries, it is important to know about these special types of elements of a poset.

Let $(A, \leq)$ be a poset. *An element $x \in A$ is said to be **maximal** element of A if $\exists$ no element $y \in A$ such that $x < y$ i.e., there exists no element in A that strictly dominates x.* Similarly, *An element $x \in A$ is said to be **minimal** element of A if $\exists$ no element $y \in A$ such that $y < x$ i.e., there is no element in A that strictly precedes x.*

Example 12 Consider the poset of example 9 as shown in the figure 6.1.6. Find the minimal and maximal element of the poset.

Solution: In this poset 1 is the **minimal** element as there is no element in the poset which precedes 1 i.e., $\exists$ no element y in the poset such that $y < 1$. Similarly, 12 is the **maximal** element as there is no element in the poset which succeeds 12 i.e., $\exists$ no element y in the poset such that $12 < y$. **Ans.**

Example 13 Let $(C, \subseteq)$ be a poset, where C is the collection all non-empty sub-sets of a set $X=\{a, b, c\}$ and $\subseteq$ is 'set inclusion' relation defined on C. Find the maximal and minimal elements of $(C, \subseteq)$.

Solution: Here $C = \{\{a\}, \{b\}, \{c\}, \{a, b\}, \{a, c\}, \{b, c\}, \{a, b, c\}\}$, collection of all non-empty subsets of X. No elements of C are strictly (properly) contained in either $\{a\}$ or $\{b\}$ or in $\{c\}$, so all the singletons are **minimal** elements of poset $(C, \subseteq)$. Similarly, no element of C strictly (properly) contains $\{a, b, c\}$, so $\{a, b, c\}$ is **maximal** element of $(C, \subseteq)$.

Ans.

Example 14 Let (I, ≤) be a poset, where I is the set of all integers and ≤ is the usual partial order. In this poset there is no minimal element because for any integer x we can always find another integer (x–1) such that x–1 < x. Similarly, there is no maximal element in this poset. **Ans.**

Theorem 4 If (A, ≤) is a finite non-empty poset, then A has at least one maximal and at least one minimal element.

Proof: Let |A| = n, where n is finite number. Let x_1 be any element of A. If a is not maximal element, then we can find an element x_2 such that $x_1 < x_2$. Again, if x_2 is not maximal, we can find another element x_3 in A such that $x_1 < x_2 < x_3$. If we proceed in this way, the sequence may look like $x_1 < x_2 < x_3 ... < x_n$, which cannot be extended beyond n. Therefore, the element that is not strictly succeeded by any element, i.e. which appears in the end of the sequence, is the maximal element of the poset.

Similarly, we can prove that there exists at least one minimal element in a finite poset.

Proved.

Let (A, ≤) be a poset. *An element x ∈ A is said to be **greatest or largest** element of A if ∀ y∈ A y ≤ x i.e., every element in A precedes x.* Similarly, *An element x ∈ A is said to be **least or smallest** element of A if ∀ y∈ A x ≤ y i.e., every element in A succeeds x.*

Example 15 Let (A, ≤) be a poset, where A is the set of all non-negative real numbers and ≤ is usual partial order on A. Since 0 ∈ A and 0 ≤ x ∀ x ∈ A, 0 is the least element of A. There is no greatest element in A, because for any real number x we can always find another real number y such that x < y.

Ans.

Example 16 Consider the poset in example 9. Since the element 1 is succeeded by all the elements of the poset, I is the least element. Similarly, element 12 succeeds every element of the poset so 12 is the greatest element. Now look at the following posets obtained by removing 1, 12 and both 1 and 12 from the poset of example 9 and shown in figures 6.1.7, 6.1.8 and 6.1.9 respectively.

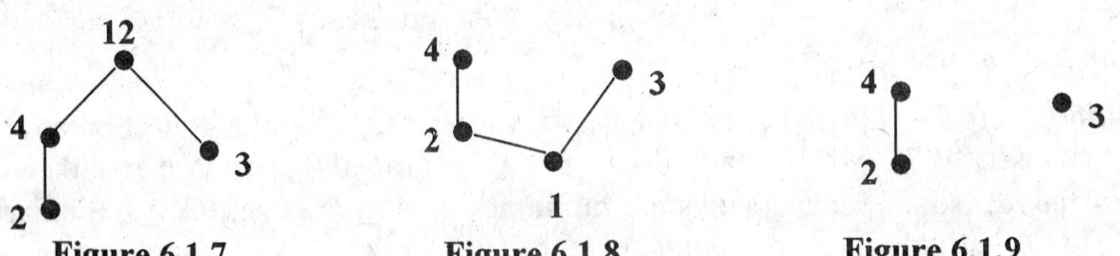

Figure 6.1.7 Figure 6.1.8 Figure 6.1.9

In the poset of figure 6.1.7, 12 is the greatest element but there is no least element. Neither 2 succeeds 3 nor 3 succeeds 2 in this poset. However, both 2 and 3 are minimal elements. In the poset of figure 6.1.8, 1 is the least element but there is no greatest element. However, there are two maximal elements 3 and 4 in this poset. Finally in figure 6.1.9, we have a poset in which there is no least and no greatest element. However, there are two minimal elements 2 and 3; two maximal elements 3 and 4. This is an example of a poset in which an element can be both minimal and maximal.

Ans.

Example 17 Let (P(A), ⊆) be a poset where A is any non-empty finite set. Find the least and the greatest element of (P(A), ⊆).

Solution: Since ϕ ∈ P(A) and every element of P(A) contains ϕ, ϕ is the least element. Similarly A ∈ P(A) and every element of P(A) are contained in A, A itself is the greatest element of P(A). ♦

Theorem 5 A poset has at the most one greatest and at the most one least element.

Proof: Let (A, ≤) be a poset. Suppose the poset A has two least elements x and y. Since x is the least element x ≤ y. Using the same argument, we can say that y ≤ x, since y is supposed to be another least element of the same poset. ≤ is an anti-symmetric relation, so x ≤ y and y ≤ x ⇒ x = y. Thus, there can be at the most one least element. Similarly, there can be at the most one greatest element.

Proved.

It is mentioned that if (A, ≤) is a poset then (A, ≥) is its **dual** poset. It is easy to infer from this ordering that maximal (minimal) element of a poset is minimal (maximal) element of its dual. Similarly, least (greatest) element of a poset is greatest (least) element of its dual. Further, it is important to note that there may not be any greatest element in a poset, however, if it exists, it is called **Unit** element of the poset and is denoted by **1.** The least element of a poset is called **Zero** element and is represented by **0**.

Let us consider the following poset (A, ≤), where A = {a, b, c, d, e, f} and ≤ is shown by the Hasse diagram in the figure 6.1.10. Let B = {a, b} and C = {c, d, e} be two subsets of A. In the diagram, there is no element in A, which precedes either a or b of B. However, there are **four** elements: c, d, e and f, which succeed both a and b of subset B. This illustration is to convey that there are elements in a poset that forms lower bounds (preceding elements) and there are some that form upper bound (succeeding elements) of a subset. Let us define these terms now.

Let (A, ≤) be a poset. Let B be any non-empty subset of A. *An element x ∈ A is said to be* ***Lower bound of B*** *if ∀ y∈ B x ≤ y i.e., every element of B is preceded by x.* Similarly, *An element x ∈ A is said to be* ***Upper bound of B*** *if ∀ y∈ B y ≤ x i.e., every element of B is succeeded by x.*

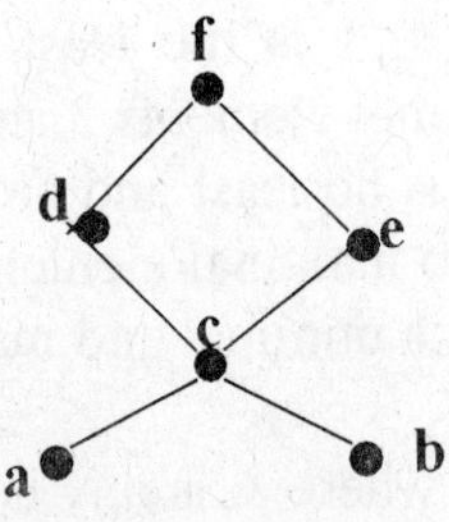

Figure 6.1.10

In this example, subset B has no lower bound i.e., **Lower Bound** (B) = ϕ. **Upper Bound** (B) = {c, d, e, f}. Let us take subset C now. C has three elements: c, d and e. There are elements a and b in A that precede all these three elements of C. Besides a and b, there is one element c that precedes c, d and e. Thus, **Lower Bound** (C) = {a, b, c}. So far as Upper bound is concerned, there exists only one element f $\in$ A, which succeeds every element of C. Thus **Upper Bound** (C) = {f}.

In the set of lower bounds of a subset, there may be an element that is the greatest among all of the elements. This element is called ***greatest lower bound*** **(glb)** or ***infimum*** and is defined as:

"*An element a* $\in$ (A, $\leq$) is **glb** of a non-empty subset B of A if, and only if

- *a is a lower bound of B and*
- *If there exists any b which is lower bound of B, then b$\leq$a.*"

In the continuing example based on the figure 6.1.10, the **glb**(B) = NIL, since there is no lower bound so there is no question of any greatest among them. However, **glb**(C) = c, because c is the greatest element in {a, b, c}. Likewise, in the set of upper bounds of a subset, there may be an element that is least among all of the elements. This element is called ***least upper bound*** **(lub)** or ***supremum*** and is defined as:

"*An element a* $\in$ (A, $\leq$) is **lub** of a non-empty subset B of A if, and only if

- *a is an upper bound of B and*
- *If there exists any b which is upper bound of B, then a$\leq$b.*"

In the poset of the figure 6.1.10, **lub**(B) = c and **lub**(C) = f.

Example 18 Let (R, $\leq$) be a poset, where R is the set of real numbers and $\leq$ is the usual partial order. Let B = {x | x is a real number such that $1 < x < 2$}. Find the **lub** and **glb** of B in R.

Solution: Here Lower bound (B) = {x | x is real such that x ≤ 1} and Upper bound (B) = {x | x is real such that x ≥ 2}. The greatest element in Lower Bound (B) is 1 and least element in Upper Bound (B) is 2. Therefore, **glb**(B) =1 and **lub**(B) = 2.

Ans.

Theorem 6 If (A, ≤) is a poset, then any subset B of A has at the most one **lub** and at the most one **glb**.

Proof: Let us first take the case of **lub**. A subset B may not have any upper bounds and hence no **lub.** If it has only one upper bound, then that will be the **lub**. In case of many upper bounds, we have to show that if there is any greatest, then it is at the most one only. The proof is similar to that in theorem 5.

Proved.

Isomorphic Posets

Let $(X, \leq_x)$ and $(Y, \leq_y)$ be two posets. A function f: X → Y is said to be **isomorphism** from X to Y, if

- f is one to one onto function and
- For any a, b ∈ A, $a \leq_x b \Leftrightarrow f(a) \leq_y f(b)$, i.e. images preserve the order of pre-images.

If f is an isomorphism, we say that $(X, \leq_x)$ and $(Y, \leq_y)$ are **isomorphic** posets.

Example 19 Let (X, ≤) and (Y, ≤) be two posets, where X is the set of all integers and Y be the set of all even integers. The partial order ≤ is the usual partial order. Let f: X → Y be a function defined by f(x) = 2x. Show that (X, ≤) and (Y, ≤) are isomorphic posets.

Solution: To prove that the two posets (X, ≤) and (Y, ≤) are isomorphic, it is sufficient to prove that f: X → Y is an isomorphism. Let x and y be any two elements of X, then x ≠ y ⇔ 2x ≠ 2y ⇔ f(x) ≠ f(y). Thus, f is **one to one**. Let z be any element of Y, then there exists an integer x ∈ X such that z = 2x, since Y is the set of all even integers, i.e. for every even integer z in Y, ∃ an integer x in X such that f(x) = 2x = z. Hence f is **onto** also.

Next, x ≤ y ⇔ 2x ≤ 2y ⇔ f(x) ≤ f(y). This implies that images preserve the order. Therefore, f: X → Y is an isomorphism and hence (X, ≤) and (Y, ≤) are isomorphic posets.

Proved.

Theorem 7 Let $(X, \leq_x)$ and $(Y, \leq_y)$ are isomorphic posets under the isomorphism f: X→Y, then show that

(a) If x is a minimal (maximal) element of the poset $(X, \leq_x)$, then f(x) is a minimal (maximal) element of $(Y, \leq_y)$.

(b) If x is the greatest (least) element of $(X, \leq_x)$, then f(x) is the greatest (least) element of $(Y, \leq_y)$.

(c) If x is an upper bound (lower bound, lub, glb) of a subset A of $(X, \leq_x)$, then f(x) is an upper bound (lower bound, lub, glb) of subset f(A) of $(Y, \leq_y)$.

(d) If every subset of $(X, \leq_x)$ has a lub (glb), then every subset of $(Y, \leq_y)$ has a lub(glb).

Proof: Since $(X, \leq_x)$ and $(Y, \leq_y)$ are isomorphic posets under the isomorphism $f: X \rightarrow Y$, all elements of X and theirs images in Y preserves the order under the respective partial order relation. If $x \in X$ is a minimal element of $(X, \leq_x)$, then according to the definition of 'minimal element', $\exists$ no y in X such that $y < x$. Now, $x \in X \Rightarrow f(x) \in Y$. Let z be any element of Y, then $\exists$ an element y in X such that $f(y) = z$ because f is onto. And $x \leq_x y \Leftrightarrow f(x) \leq_y f(y)$. Since there exists no y in X such that $y < x \Leftrightarrow \exists$ no f(y) in Y such that $f(y) < f(x)$. This implies that f(x) is a minimal element of $(Y, \leq_y)$.

Similarly other parts of the theorem can be proved. It is left as an exercise to the reader.

Proved.

Topological Sorting

Let $(A, \leq)$ be a poset. Sometimes it is required to extend the given partial order to a linear order. The *process of constructing linear order from a given partial order is called **Topological Sorting.*** Let us consider a poset as shown in the figure 6.1.11 (a) This poset is not a linear ordered set (since, 2 and 3 are not related, find the other pairs that are not related). The extension of partial order to a linear order simply means, an element must be designated as **first** one, a **second** element must follow it, then **third** then **fourth**, and so on. The algorithm for topological sorting is as below.

- **Step 0** Initialize A with the poset and SORT = ϕ.
- **Step 1** Select a minimal element from A, say it is a. If there is only one minimal element then that will be selected, otherwise, select any one of the minimal elements.
- **Step 2** Make $A \leftarrow A - \{a\}$ and SORT $\leftarrow$ SORT $\cup \{a\}$
- **Step 3** Repeat step 1 and step 2 until $A = \phi$.

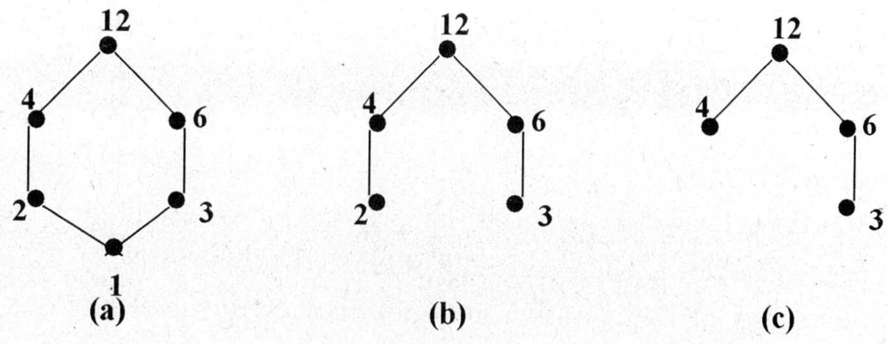

Figure 6.1.11

Let us consider poset of figure 6.1.11(a) and see how this algorithm works. Step 0 initializes SORT to {} and A to {1,2,3,4,6,12}. Step 1 picks 1 as the minimal element from A, since it is the only minimal element. Step 2 now sets A to {2,3,4,6,12} and SORT = {1}. The resulting Hasse diagram for A is shown in the figure 6.1.11(b). Now there are two minimal elements: 2 and 3. We can pick up any one, say 2. Then Sort = {1,2} and A = {3,4,6,12}. The resulting Hasse diagram for A, now, is shown in the figure 6.1.11(c). Had we selected 3, the resulting Hasse diagram would have been as shown in the figure 6.1.12(a). Again, there are two minimal elements: 3 and 4. We select 4. It makes Sort = {1,2,4} and A={3,6,12}. See diagram 6.1.13(a). Now there is only one minimal element 3. Selecting 3 makes SORT = {1,2,4,3} and A ={6,12}. Proceeding in the same way, we get SORT = {1,2,4,3,6,12} after two iterations from now.

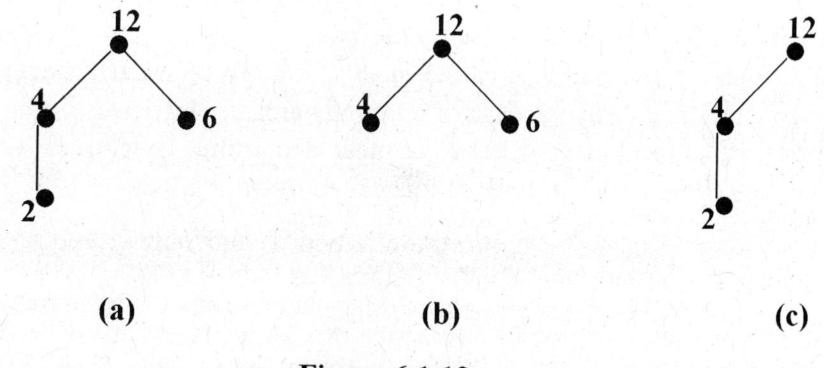

Figure 6.1.12

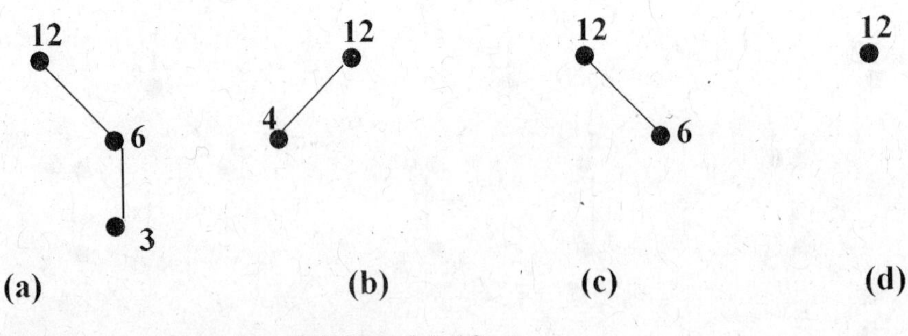

Figure 6.1.13

Diagrams for all possible cases of selection of minimal elements in different steps are given. We have obtained one linear order 1 < 2 < 4 < 3 < 6 < 12 for the given partial order. There is many more such linear order for the same poset. Reader may try to find those.

A poset (A, ≤) is said to be **well ordered set** if every non-empty subset of A contains a least element. For example, (N, ≤) is a well ordered set, where N is the set of all natural numbers and ≤ is usual partial order. The poset of figure 6.1.11(a) is **not well ordered set**, because ∃ a subset {2,3} such that it has no least element. However, posets of figures 6.1.12(c), 6.1.13(a), 6.1.13(b) and 6.1.13(c) are all **well ordered**. In the next section, we will learn another type of ordered set called **Lattice** in which each pair of element will have a **glb** and a **lub**.

6.2 Lattice

Let (L, ≤) be a poset. If every subset {x, y} containing any two elements of L, has a **glb** (Infimum) and a **lub** (Supremum), then the poset (L, ≤) is called a **lattice**. A **glb** ({x, y}) is represented by $x \wedge y$ and it is called ***meet*** of x and y. Similarly, **lub** ({x, y}) is represented by $x \vee y$ and it is called ***join*** of x and y. Therefore, **lattice** is a mathematical structure equipped with two binary operations **meet** and **join.** If set L is a finite set, lattice (L, ≤) is called ***finite*** lattice otherwise it is called ***infinite*** lattice.

A lattice (L, ≤) is said to be a **complete** lattice if, and only if every non-empty subset S of L has a **glb** and a **lub**.

A non-empty subset S is called a **sublattice** of a lattice (L, ≤) if $x \wedge y \in S$ and $x \vee y \in S$ whenever x, y ∈ S i.e., S is closed with respect to the binary operations of meet and join.

We know that join (meet) of a subset {x, y} in a poset is a meet (join) of the same subset {x, y} in the corresponding dual poset. If a poset (L, ≤) is a lattice, then its dual

(L, $\geq$) is also a lattice called **dual lattice** of (L, $\leq$). The direct implication of this properties is that whenever a statement is valid for a lattice (L,$\leq$), the same statement can be made valid for its dual lattice by replacing **join operations** with **meet operations** and **meet operations** with **join operations**. This is referred as the **principle of duality for lattice.**

Example 1 Consider the set R of all real numbers. This set is partially ordered by the usual partial order relation $\leq$. Thus (R, $\leq$) is a poset. If x and y be any elements of R, then $x \vee y = \max(x, y)$ and $x \wedge y = \min(x, y)$. For any subset {x, y} containing two elements x and y of R, the join and meet are real numbers. Therefore, (R, $\leq$) is closed with respect to join and meet and hence it is a lattice. However, (R, $\leq$) is not a **complete** lattice since A = $\{x \in R \mid x \geq 1\}$ a subset of R has no LUB. In fact it has no upper bound. Here, (A, $\leq$) is sub-lattice of (R, $\leq$). ♦

Example 2 Consider the set I_+ of all positive integers. This set is partially ordered with relation $\leq$ defined by setting $x \leq y$ if, and only if x divides y. Thus, (I_+, $\leq$) is a poset. If x and y are any two elements of I_+, then $x \wedge y$ is the highest common factor (HCF) of x & y and $x \vee y$ is the least common multiple (LCM) of x & y. Both HCF and LCM of two positive integers are positive integers. Therefore, (I_+, $\leq$) is closed with respect to join and meet and hence it is a lattice. It is however not a complete lattice. ♦

Example 3 Let S be a set and L = P(S). The mathematical structure (L, $\subseteq$) is a poset as it is shown in the previous section. Let A and B be any two subsets of S i.e. A and B be any two elements of P(S). We can define $A \vee B = A \cup B$ and $A \wedge B = A \cap B$. The poset (L, $\subseteq$) is closed with respect to join and meet operations i.e. for any subset {A, B} of P(S) there is a join and meet in P(S). Thus (L, $\subseteq$) is a lattice. A Hasse diagram is shown in the figure 6.2.1 for the lattice (L, $\subseteq$) where L= P(S) and S = {a, b, c}.

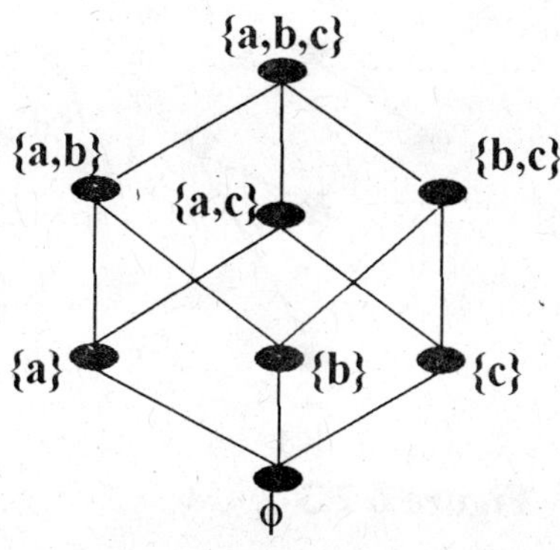

Figure 6.2.1

From this diagram, it is obvious that every pair of elements of P(S), there is a lub (join) and a glb (meet) in the P(S). A Hasse diagram is, sometimes, very handy to test whether a poset is lattice or not. ♦

Example 4 Let n be a positive integer and D_n be the set of all positive divisors of n. Let $\leq$ be a relation of divisibility defined on D_n. As proved in the previous section 6.1, $(D_n, \leq)$ is a poset. For any x and y $\in D_n$, the join and meet are defined as $x \vee y = LCM(x, y)$ and $x \wedge y = HCF(x, y)$. The poset $(D_n, \leq)$ is closed with respect to join and meet operations and hence $(D_n, \leq)$ is a lattice. The figures 6.2.2(a) and 6.2.2(b) are Hasse diagrams for the lattices $(D_{20}, \leq)$ and $(D_{105}, \leq)$ respectively. Here $D_{20} = \{1, 2, 4, 5, 10, 20\}$ and $D_{105} = \{1, 3, 5, 7, 15, 21, 35, 105\}$. A lattice $(D_n, \leq)$ is a **sublattice** of lattice of example 2. **Ans.**

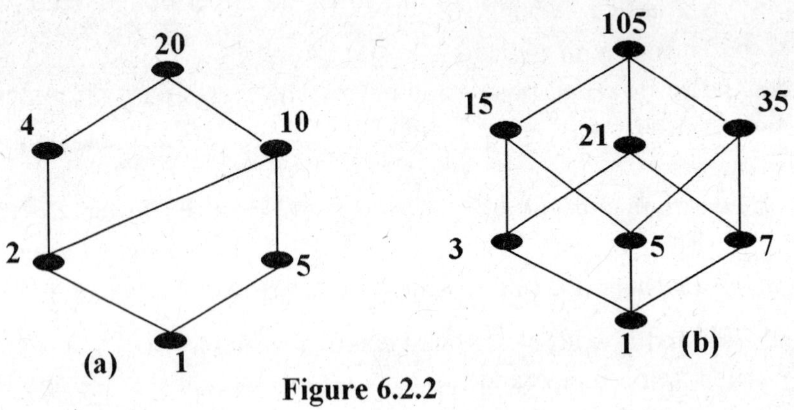

Figure 6.2.2

Example 5 Which of the Hasse diagrams in figure 6.2.3 represent lattices?

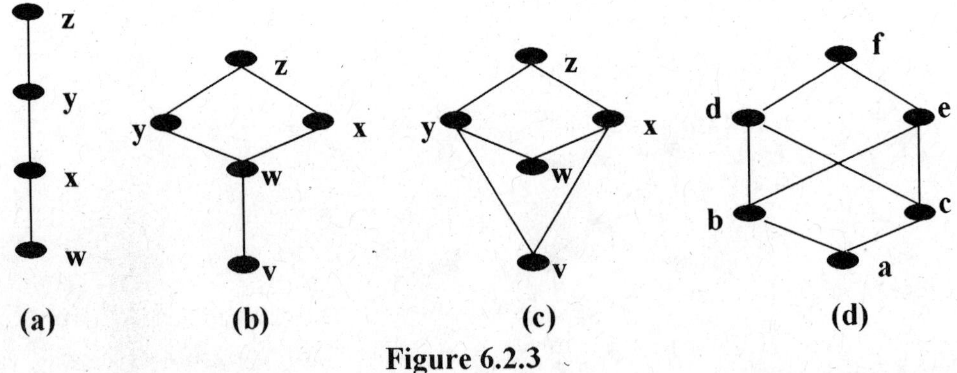

Figure 6.2.3

Solution: In this example, (a) and (b) are lattices, whereas (c) and (d) are not lattice. In (c), {x, y} has lower bounds {v, w}. Since v and w are not related, we cannot find greatest of the two, and hence **glb (meet)** does not exist for the element x and y. In (d), {d, e} has lower bounds {a, b, c} and no **glb(meet)** can be determined.

Ans.

Product Lattice

Let $(L_1, \leq_1)$ and $(L_2, \leq_2)$ be two lattices then $(L_1 \times L_2, \leq)$, where $\leq$ is the product partial order of $(L_1, \leq_1)$ and $(L_2, \leq_2)$, is called **product lattice** of two lattices $(L_1, \leq_1)$ and $(L_2, \leq_2)$. See the following theorem for the proof that $(L_1 \times L_2, \leq)$ is a lattice.

Theorem 1 Let $(L_1, \leq_1)$ and $(L_2, \leq_2)$ be two lattices then $(L, \leq)$ is a lattice, where $L = L_1 \times L_2$ and $\leq$ is the product partial order of $(L_1, \leq_1)$ and $(L_2, \leq_2)$.

Proof: Since $(L_1, \leq_1)$ is a lattice, for any two elements a_1 and a_2 of L_1, $a_1 \vee a_2$ is join and $a_1 \wedge a_2$ is meet of a_1 and a_2 and both exist in L_1. Similarly for any two elements b_1 and b_2 of L_2, $b_1 \vee b_2$ is join and $b_1 \wedge b_2$ is meet of b_1 and b_2 and both exist in L_2. Thus, $(a_1 \vee a_2, b_1 \vee b_2)$ and $(a_1 \wedge a_2, b_1 \wedge b_2) \in L$. Also (a_1, b_1) and $(a_2, b_2) \in L$ for any $a_1, a_2 \in L_1$ and any $b_1, b_2 \in L_2$.

Now if we prove that

$$(a_1, b_1) \vee (a_2, b_2) = (a_1 \vee a_2, b_1 \vee b_2) \text{ and}$$

$$(a_1, b_1) \wedge (a_2, b_2) = (a_1 \wedge a_2, b_1 \wedge b_2)$$

then we can conclude that for any two ordered pairs (a_1, b_1) and $(a_2, b_2) \in L$, its join and meet exist and thence $(L, \leq)$ is a lattice.

We know that $(a_1, b_1) \vee (a_2, b_2) =$ **lub** $(\{(a_1, b_1), (a_2, b_2)\})$. Let it be (c_1, c_2). Then,

$$(a_1, b_1) \leq (c_1, c_2) \text{ and } (a_2, b_2) \leq (c_1, c_2)$$

$$\Rightarrow a_1 \leq_1 c_1,\ b_1 \leq_2 c_2;\ a_2 \leq_1 c_1,\ b_2 \leq_2 c_2$$

$$\Rightarrow a_1 \leq_1 c_1,\ a_2 \leq_1 c_1;\ b_1 \leq_2 c_2,\ b_2 \leq_2 c_2$$

$$\Rightarrow a_1 \vee a_2 \leq_1 c_1;\ b_1 \vee b_2 \leq_2 c_2$$

$$\Rightarrow (a_1 \vee a_2, b_1 \vee b_2) \leq (c_1, c_2)$$

This shows that $(a_1 \vee a_2, b_1 \vee b_2)$ is an upper bound of $\{(a_1, b_1), (a_2, b_2)\}$ and if there is any other upper bound (c_1, c_2) then $(a_1 \vee a_2, b_1 \vee b_2) \leq (c_1, c_2)$. Therefore, **lub** $(\{(a_1, b_1), (a_2, b_2)\}) = (a_1 \vee a_2, b_1 \vee b_2)$ i.e.

$$(a_1, b_1) \vee (a_2, b_2) = (a_1 \vee a_2, b_1 \vee b_2)$$

Similarly, we can show for the meet.

Proved.

Example 6 Let $L_1 = \{a, b\}$ and $L_2 = \{x, y, z, w\}$ and lattices $(L_1, \leq_1)$ and $(L_2, \leq_2)$ are given by the Hasse diagrams of the figure 6.2.4(a) and 6.2.4(b) respectively. Find the product lattice $((L_1 \times L_2, \leq)$ where $\leq$ is the product partial order.

Solution: Here $L_1 \times L_2 = \{((a, x), (a, y), (a, z), (a, w), (b, x), (b, y), (b, z), (b, w)\}$. The Partial order relation $\leq$ on $L_1 \times L_2$ is defined as product partial order. The resulting product poset has 8 elements (ordered pairs) and the order relationships among them is shown as in the Hasse diagram of the figure 6.2.4 (c). We have $(a, x) \leq (b, x)$ [since $a \leq_1 b$ and $x \leq_2 x$]; $(a, x) \leq (a, y)$ [since $a \leq_1 a$ and $x \leq_2 y$]; $(a, x) \leq (b, z)$ [since $a \leq_1 b$ and $x \leq_2 z$]; and so on. An arc for $(a, x) \leq (b, z)$ is not shown in the diagram as it is transitively implied by the presence of relationships $(a, x) \leq (a, z)$ and $(a, z) \leq (b, z)$. **Ans.**

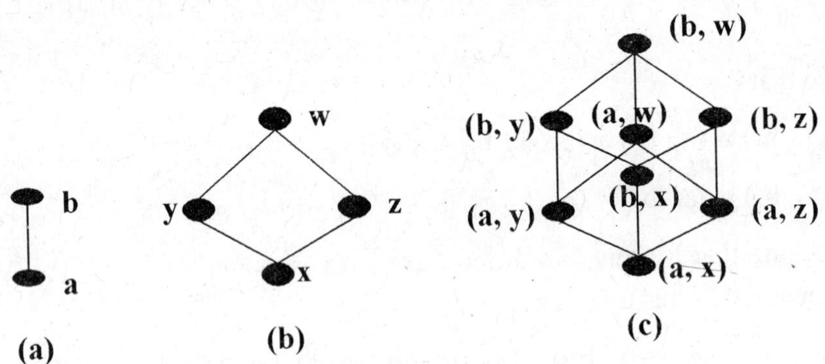

Figure 6.2.4

Properties of Lattice

Let $(L, \leq)$ be a lattice and a, b, c and d be any elements of L. The following properties hold good in a lattice. The proof of some of the properties are given at the end. Rest are left as an exercise to the reader.

1. $a \leq a \vee b$ and $b \leq a \vee b$ i.e. $a \vee b$ is an upper bound of a and b. Also for any element c, if $a \leq c$ and $b \leq c$ then $a \vee b \leq c$. This implies that $a \vee b$ is least upper bound of a and b. Recall the definition of join of two elements a and b.

2. $a \wedge b \leq a$ and $a \wedge b \leq b$ i.e. $a \wedge b$ is a lower bound of a and b. Also for any element c, if $c \leq a$ and $c \leq b$ then $c \leq a \wedge b$. This implies that $a \wedge b$ is greatest lower bound of a and b. Recall the definition of meet of two elements a and b.

3. (i) $a \vee b = b$ if, and only if $a \leq b$
(ii) $a \wedge b = a$ if, and only if $a \leq b$
$\Rightarrow$ **$a \wedge b = a$ if, and only if $a \vee b = b$**

4. (i) $a \vee a = a$
(ii) $a \wedge a = a$
Idempotent law

5. (i) $a \vee b = b \vee a$
(ii) $a \wedge b = b \wedge a$
Commutative law

6. (i) $a \vee (b \vee c) = (a \vee b) \vee c$
(ii) $a \wedge (b \wedge c) = (a \wedge b) \wedge c$
Associative law

7. If $a \leq b$ and c be any element of $(L, \leq)$, then $a \vee c \leq b \vee c$ and $a \wedge c \leq b \wedge c$.

8. $a \leq c$ and $b \leq c \Leftrightarrow a \vee b \leq c$

9. $c \leq a$ and $c \leq b \Leftrightarrow c \leq a \wedge b$

10. (i) $a \vee (a \wedge b) = a$
(ii) $a \wedge (a \vee b) = a$
Absorption law

11. (i) $(a \vee b)' = a' \wedge b'$
(ii) $(a \wedge b)' = a' \vee b'$
DeMorgan's law

12. If $a \leq b$, $c \leq d$, and $c \leq a \wedge b$, then $a \vee c \leq b \vee d$ and $a \wedge c \leq b \wedge d$.

Proof: Here, properties 3(i), 3(ii), 6(i) and 11(i) have been proved. Other properties can be proved easily using almost similar arguments, so is left as exercise to the reader.

3(i) Let $a \vee b = b$. Since $a \vee b$ is an upper bound of a, $a \leq a \vee b$. Therefore, $a \leq b$.

Next, let $a \leq b$. Since $\leq$ is a partial order relation, $b \leq b$. Thus, $a \leq b$ and $b \leq b$ together implies that b is an upper bound of a and b. We know that $a \vee b$ is least upper bound of a and b, so $a \vee b \leq b$. Also $b \leq a \vee b$ because $a \vee b$ is an upper bound of b. Therefore, $a \vee b \leq b$ and $b \leq a \vee b \Leftrightarrow a \vee b = b$ by the anti-symmetry property of partial order relation $\leq$. Hence, it is proved that $a \vee b = b$ if and only if $a \leq b$.

3(ii) Let $a \wedge b = a$. Since $a \wedge b$ is a lower bound of b, $a \wedge b \leq b$. Therefore, $a \leq b$.

Next, let $a \leq b$. Since $\leq$ is a partial order relation, $a \leq a$. Thus, $a \leq b$ and $a \leq a$ together implies that a is a lower bound of a and b. We know that $a \wedge b$ is greatest lower bound of a and b, so $a \leq a \wedge b$. Also $a \wedge b \leq a$ because $a \wedge b$ is a lower bound of a. Therefore, $a \wedge b \leq a$ and $a \leq a \wedge b \Leftrightarrow a \wedge b = a$ by the anti-symmetry property of partial order relation $\leq$. Hence, it is proved that $a \wedge b = a$ if and only if $a \leq b$.

6(i) Here we have to prove that $a \vee (b \vee c) = (a \vee b) \vee c$.
Let $a \vee (b \vee c) = x$ and $(a \vee b) \vee c = y$. Since $a \vee (b \vee c)$ is join (least upper bound) of a and $(b \vee c)$, we have $a \leq x$ and $b \vee c \leq x$. Also $b \vee c$ is join of b and c, so $b \leq b \vee c \leq x$ and $c \leq b \vee c \leq x$. Therefore, $a \leq x$, $b \leq x$ and $c \leq x$. Since $a \vee b$ is join of a and b, we have $a \vee b \leq x$. Similarly, $(a \vee b) \vee c$ is join of $(a \vee b)$ and c, so $(a \vee b) \vee c \leq x$. We have assumed that $(a \vee b) \vee c = y$, thus $y \leq x$.

Using the same argument, we can show that $x \leq y$. Since $\leq$ is an anti-symmetric relation, $y \leq x$ and $x \leq y$ together implies that $x = y$. Hence proved.

11(i) We have to prove that $(a \vee b)' = a' \wedge b'$, where a' and b' are complements of a and b respectively. To prove that $(a \vee b)' = a' \wedge b'$, we simply prove that $(a' \wedge b')$ is complement of $(a \vee b)$ i.e.,

$$(a \vee b) \vee (a' \wedge b') = I \quad \text{and}$$
$$(a \vee b) \wedge (a' \wedge b') = 0$$

We have
$$\begin{aligned}(a \vee b) \vee (a' \wedge b') &= [(a \vee b) \vee a'] \wedge [(a \vee b) \vee b'] \\ &= [(a \vee a') \vee b] \wedge [a \vee (b \vee b')] \\ &= [I \vee b] \wedge [a \vee I] \\ &= I\end{aligned}$$

and
$$\begin{aligned}(a \vee b) \wedge (a' \wedge b') &= [(a \wedge (a' \wedge b')] \vee [b \wedge (a' \wedge b')] \\ &= [(a \wedge a') \wedge b'] \vee [a' \wedge (b \wedge b')] \\ &= [0 \wedge b'] \vee [a' \wedge 0] \\ &= 0\end{aligned}$$

This shows that $(a' \wedge b')$ is complement of $(a \vee b)$. Hence **proved**.

Types of Lattice

There are some more structured classes of mathematical structures that are very useful in computer science. This gradual structuring of a mathematical structure, called lattice, leads us to the concept of ***Finite Boolean Algebra***, which we shall learn about in the next section of this chapter.

A lattice (L, ≤) is said to be a ***bounded*** if it has a greatest element ***I*** and a least element ***0***.

Example** 7* Let (I_+, |) be a lattice, where I_+ is a set of all positive integers and '|' is the relation of divisibility i.e. a | b if and only if a divides b. This lattice is not bounded since it has no greatest element ***I. Reader may note that this lattice has a least element *(zero element)*, the number 1.

Ans.

***Example** 8* Let (P(S), ⊆) be a lattice where S is any non-empty set. This lattice is bounded since its greatest element is S and its least element is ϕ and both are in P(S).

In a bounded lattice (L, ≤) for any element x in L, we have

(i) $0 \le x \le 1$

(ii) $x \vee 0 = x$ and $x \wedge 0 = 0$

(iii) $x \vee I = I$ and $x \wedge I = I$

A finite lattice is always bounded. Its greatest element is the **join** of all of its elements and its least element is given by the **meet** of all its elements. A Hasse diagram can be drawn for a finite poset. If the Hasse diagram corresponds to a lattice, the lattice is bounded. In other words, if there is a Hasse diagram for a lattice, then the lattice is bounded.

A lattice (L, ≤) is said be ***distributive*** if the **meet** operation distributes over **join** operation and **join** operation distributes over **meet** operation. That is, for any x, y, and z of L

1. $x \wedge (y \vee z) = (x \wedge y) \vee (x \wedge z)$
2. $x \vee (y \wedge z) = (x \vee y) \wedge (x \vee z)$

If (L, ≤) is not distributive then L is called a **non-distributive** lattice.

***Example** 9* The lattice (I, ≤) is a distributive lattice, where ≤ is the usual partial order on the set of all integers I. This lattice is a distributive lattice. It is important to note however, that it is not bounded because neither the least element nor the greatest element exists in this lattice. ♦

Example 10 Let (P(S), ⊆) be a lattice where S is any non-empty set. Then this lattice is a distributive lattice since union is distributive over intersection and intersection is distributive over union. ♦

Example 11 The lattice (D_{12}, |) is distributive. The lattices (A, |) and (B, |), where A = {1, 2, 3, 4, 12}, B = {1, 2, 3, 5, 30} and | is the partial order of divisibility are not distributive lattice as shown below. The Hasse diagrams are given in the figures 6.2.5.

Solution: The Hasse diagram of figures 6.2.5(a) is for the lattice (D_{12}, |).

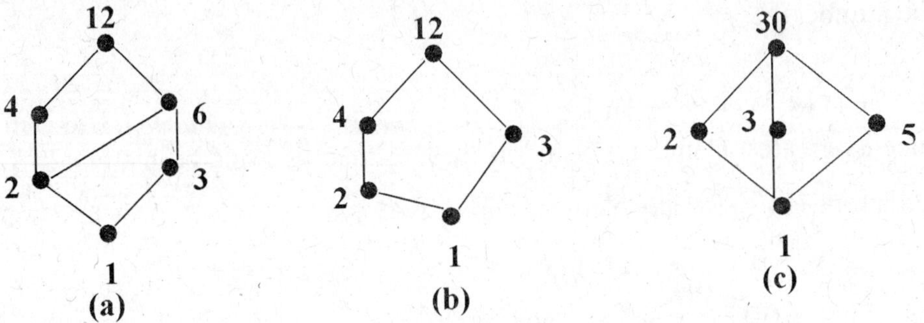

Figure 6.2.5

It can be easily verified that for every three elements of (D_{12}, |), the distributive properties are satisfied. However, in the case of (A, |), as shown in the diagram 6.2.5(b), we have

$$4 \wedge (2 \vee 3) = 4 \wedge 12 = 4 \text{ and}$$
$$(4 \wedge 2) \vee (4 \wedge 3) = 2 \vee 1 = 2$$

Since 4 ≠ 2, the meet is not distributive over join and hence, the lattice (A, |) is a non-distributive lattice. This lattice is bounded with zero element **1** and Unit element **12**. In the case of (B, |), as shown in the diagram 6.2.5(c), we have

$$2 \wedge (3 \vee 5) = 2 \wedge 30 = 2 \text{ and}$$
$$(2 \wedge 3) \vee (2 \wedge 5) = 1 \vee 1 = 1$$

Since 1 ≠ 2, the meet is not distributive over join and hence, the lattice (B, |) is a non-distributive lattice. This lattice, like (A, |), is bounded with zero element **1** and Unit element **30**.

Ans.

Let (L, ≤) be a bounded lattice having unit element **I** and zero element **0**. Sometimes a zero element of a lattice is also called *universal lower bound* and unit element is called *universal upper bound* of the lattice. Let $a \in L$ be any element. An element a' of L is said to be complement of a in L, if

$$a \vee a' = a' \vee a = \mathbf{I} \text{ and}$$
$$a \wedge a' = a' \wedge a = \mathbf{0}$$

It is important to note here that $\mathbf{0}' = \mathbf{I}$ and $\mathbf{I}' = \mathbf{0}$.

Example 12 Refer to the lattice of example 10 where S = {a, b, c}. The Hasse diagram for this lattice is shown in figure 6.2.1. In this lattice, every element $A \subseteq S$ has a complement S – A in this lattice.

Example 13 The lattice (D_{105}, |) is another example of a bounded distributive lattice in which every element has a complement. The unit element is 105 and zero element is 1 in this case. The Hasse diagram is given in the figure 6.2.2(b).

Example 14 Refer to the lattice of figure 6.2.5(a). It has an element 6 such that it has no complement. It can be tested that $6 \wedge 2 = 2$, $6 \vee 2 = 6$; $6 \wedge 4 = 2$, $6 \vee 4 = 12$; $6 \wedge 3 = 3$, $6 \vee 3 = 6$. But for any element b to be complement of 6 we must have $6 \wedge b = 1$, $6 \vee b = 12$. There exists no such b in this lattice. Similarly, element 2 does not have a complement in this lattice. In the case of the lattice of figure 6.2.5(b), every element has a complement. In fact the element 3 has two complements: 2 and 4. Finally, see the lattice of figure 6.2.5(c). Here, every element except unit element (12) and zero element (1), has two complements. 3 has two complements: 5 and 7; 5 has two complements: 3 and 7 and 7 has two complements: 3 and 5.

It is interesting to note from the above examples that in a bounded distributive lattice if a complement exists it is unique whereas in a non-distributive lattice an element may have more than one complements. This is true not only in a particular case but is essentially valid for all bounded and distributive lattice as stated in the following theorem.

Theorem 2 Let (L, ≤) be a bounded distributive lattice. Prove that if a complement for an element $a \in L$ exists, then it is unique.

Proof: Let I and 0 are the unit and zero elements of L respectively. Let b and c be two complements of an element $a \in L$. Then from the definition, we have

$$a \wedge b = 0 = a \wedge c \text{ and}$$
$$a \vee b = I = a \vee c$$

We can write $b = b \vee 0 = b \vee (a \wedge c)$

$$= (b \vee a) \wedge (b \vee c) \quad \text{[since lattice is distributive]}$$
$$= I \wedge (b \vee c)$$

$= (b \vee c)$

Similarly, $c = c \vee 0 = c \vee (a \wedge b)$

$= (c \vee a) \wedge (c \vee b)$ [since lattice is distributive]

$= I \wedge (b \vee c)$ [since $\vee$ is a commutative operation]

$= (b \vee c)$

The above two results show that b = c.

Proved.

A lattice (L, ≤) is said to be a ***complemented*** lattice if it is bounded and if every element of L has a complement in L. Lattices of examples 12 and 13 are examples of complemented lattice as they are bounded and every element has a complement in the respective lattice. Both these lattices are distributive. A non-distributive lattice may also be complemented. Lattices of figures 6.2.5(b) and 6.2.5(c) are the examples of bounded non-distributive and complemented lattice. A bounded, distributive and complemented lattice is of special concern in computer science and is known as ***finite Boolean algebra.*** Before taking up this topic, let us know about isomorphic lattice.

Let $(L_1, \leq_1)$ and $(L_2, \leq_2)$ be two lattices. They are said to be ***isomorphic*** to each other if they are isomorphic as posets i.e. if there exists a function f: $L_1 \rightarrow L_2$ such that

- f is one to one onto and
- it preserves the partial order i.e., for $a \leq_1 b \Leftrightarrow f(a) \leq_2 f(b)$

If $(L_1, \leq_1)$ is isomorphic to a lattice $(L_2, \leq_2)$, then $(L_1, \leq_1)$ possesses all the properties of $(L_2, \leq_2)$. For example, L_2 is bounded iff L_1 is bounded, L_2 is distributive iff L_1 is distributive, L_2 is complemented iff L_1 is complemented and so on. In general, any formula that is valid in L_1 can be made valid in L_2 by replacing elements of L_1 in the formula by its corresponding images in L_2. Isomorphic lattices have identical Hasse diagrams.

6.3 Finite Boolean Algebra

Let B = {0, 1}. This is the set of Boolean symbols: 1 for **true** and 0 for **false**. Then, B×B = {(0, 0), (0, 1), (1, 0), (1, 1)}. If we write (0, 0) as 00, (0, 1) as 01 and so on and B×B as B_2 then, we have B_1= {0, 1}; B_2 = {00, 01, 10, 11}; ... and $B_n = \{x \mid x$ is a binary string of length n$\}$. Also, $|B_n| = 2^n$. Let ≤ be a partial order relation on B_n defined as, if $x = a_1 a_2 a_3 \ldots a_n$ and $y = b_1 b_2 b_3 \ldots b_n$, where a_i's and b_i's are either 0 or 1,

1. $x \leq y$ if and only if $a_k \leq b_k$ for $k = 1,2,3, \ldots ,n$. For example, $0101 \leq 0111$ because for $k = 1$: $0 \leq 1$; for $k = 2$: $1 \leq 1$; for $k = 3$: $0 \leq 1$ and for $k = 4$: $1 \leq 1$. The two strings 0101 and 1000 are not related since neither $0101 \leq 1000$ nor $1000 \leq 0101$. Also, every n-bit long string is related to itself i.e. for all x, $x \leq x$.
2. $x \wedge y = c_1\ c_2\ c_3 \ldots c_n$ where $c_k = \min(a_k, b_k)$, e.g., $0101 \wedge 0110 = 0100$ and $1001 \wedge 1101 = 1001$. Notice that here meet ($\wedge$) is a Boolean AND operation.
3. $x \vee y = c_1\ c_2\ c_3 \ldots c_n$ where $c_k = \max(a_k, b_k)$, e.g., $0101 \vee 0110 = 0111$ and $1001 \vee 1101 = 1101$. Notice here too that join ($\vee$) is a Boolean OR operation.
4. x has a complement $x' = c_1\ c_2\ c_3 \ldots c_n$ where $c_k = 0$ if $a_k = 1$ and $c_k = 1$ if $a_k = 0$, e.g., $(0101)' = 1010$; $(1001)' = 0110$ and so on. In this case, the complement is just one's complement of the binary string.

Obviously, this relation $\leq$ is a partial order relation on B_n. Thus $(B_n, \leq)$ is a poset. Since every pair of elements has a join and a meet as defined above, $(B_n, \leq)$ is a lattice. This lattice is bounded because it is a finite lattice. Since Boolean AND is distributive over Boolean OR and Boolean OR is distributive over Boolean AND, the join distributes over meet and the meet distributes over join in this lattice. (Refer to the definition of join and meet above). Therefore, $(B_n, \leq)$ is a distributive lattice. The one's complement of n – bit long string is again a n-bit long string, every element in $(B_n, \leq)$ has a complement. Hence, $(B_n, \leq)$ is a ***finite Boolean algebra*** since it is a bounded, distributive and complemented lattice. The figure 6.3.1 shows the Hasse diagrams of $(B_n, \leq)$ for n = 1, 2 and 3.

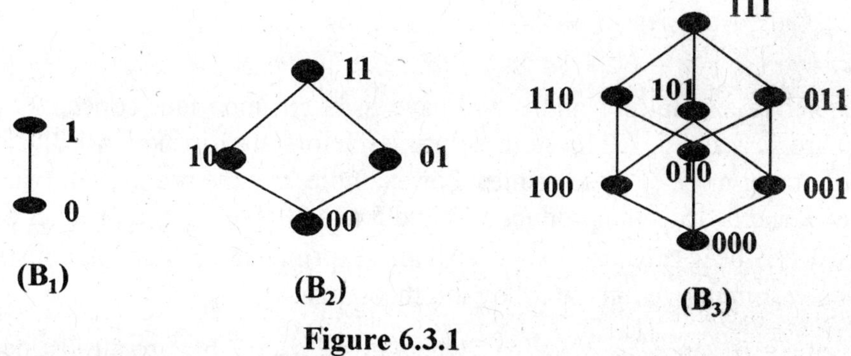

Figure 6.3.1

Let S be any set with $|S| = n$ then $|P(S)| = 2^n$. Thus we can always find a one to one onto function from P(S) to B_n. We know that (P(S), $\subseteq$) is a bounded, distributive and complemented lattice and hence (P(S), $\subseteq$) is a ***finite Boolean algebra.*** See the Hasse diagram in the figure 6.2.1. If we replace each subset by its binary representation, we get the Hasse diagram of B_3 as shown in the figure 6.3.1. Therefore, for any set S such that $|S| = 3$, (P(S), $\subseteq$) is isomorphic to B_3. In general, lattice (P(S), $\subseteq$) is isomorphic to $B_{|S|}$. Since isomorphic lattices possess same properties, if one is a finite Boolean algebra then other is also a finite Boolean algebra. A finite Boolean algebra may be defined in terms of B_n as "*any finite lattice is said to be a **finite Boolean algebra** if and only if it is isomorphic to some B_n.*"

Example 1 Show that D_{30} is a finite Boolean algebra with the partial order of divisibility.

Solution: Here $D_{30} = \{1, 2, 3, 5, 6, 10, 15, 30\}$. $|D_{30}| = 8 = |B_3|$. Let us define a function f: $D_{30} \to B_3$ as f(1) = 000, f(2) = 100, f(3) = 010, f(5) = 001, f(6) = 110, f(10) = 101, f(15) = 011 and f(30) = 111. Now a Hasse diagram for (D_{30}, |) can be obtained by just replacing the corresponding pre-images in the Hasse diagram of B_3 in figure 6.3.1. This shows that (D_{30}, |) is isomorphic to B_3. Hence (D_{30}, |) is a finite Boolean algebra.

Proved.

Example 2 Show that D_{20} is **not** a finite Boolean algebra with the partial order of divisibility.

Solution: Here $D_{20} = \{1, 2, 4, 5, 10, 20\}$. $|D_{20}| = 6$. There exists no n such that $2^n = 6$ and therefore (D_{20}, |) is not isomorphic to any B_n. Hence (D_{20}, |) is not a finite Boolean algebra. In fact this lattice is not even distributive lattice as it is isomorphic to a non-distributive lattice shown in the figure 6.2.5(b).

Proved.

In the above examples 1 and 2, we have used an important concept. The prime factors of 30 are 2, 3 and 5. All these three prime factors are distinct. All divisors of 30 are product of one or more of these primes. For example, 6 is the product of 2 and 3, 10 is the product of 2 and 5, 15 is the product of 3 and 5 and 30 itself is the product of 2,3 and 5. Whereas 20 can be written as 2×2×5. All these prime factors are not distinct. The concept has been summarized in the following theorems.

Theorem 1 Let $n = p_1 \times p_2 \times p_3 \times \ldots \times p_m$; where $p_1, p_2, p_3, \ldots, p_m$ are all distinct primes. Prove that D_n is a Boolean algebra.

Proof: Let $S = \{p_1, p_2, p_3, \ldots, p_m\}$. P(S) is the collection of all subsets of S including Null set. The product of all primes in a subset of S is a divisor of n. This is true for all subsets if we take 1 corresponding to the $\phi \subseteq S$. There is no divisor for n other than those obtained from the subsets of S. Since there are 2^m subsets of S, there will be 2^m distinct divisors of n. Therefore,

$$D_n = \{x \mid x \text{ is a divisor of } n\}$$

$$= \{x \mid x \text{ is the product of elements of subsets of S and } x = 1 \text{ for } \phi\}$$

Here, $|D_n| = 2^m$. Thus $(D_n, |)$ is isomorphic to B_m where m is a positive integer. Therefore, $(D_n, |)$ is finite Boolean algebra.

Proved.

Theorem 2 Let n be a positive integer greater than 1. If in the prime factorization of n, all the factors are not distinct i.e. p^2 divides n where p is a prime number, then prove that D_n is not a finite Boolean algebra.

Proof: Let us suppose that D_n is a finite Boolean algebra. Then it is bounded and complemented lattice according to the definition of a finite Boolean algebra. Since p^2 divides n, we can write $n = p^2 q$ for some integer q. Also, $p \in D_n$ because p is a divisor of n. This implies that p has a complement p' in D_n such that LCM(p, p') = n and HCF(p, p') =1. Both p and p' are divisors of n and p is a prime number, we have

$$\text{LCM}(p, p') = n \Rightarrow pp' = n$$

$$\Rightarrow pp' = p^2 q \Rightarrow p' = pq$$

$$\Rightarrow \text{LCM}(p, p') = pq \neq n \text{ and HCF}(p, p') = p \neq 1$$

This is obviously a contradiction that we arrive at because of our assumption that D_n is a finite Boolean algebra. Therefore D_n is **not** a finite Boolean algebra.

Proved.

Example 3 Determine whether D_n is a finite Boolean algebra where (a) n = 12 (b) n = 40 (c) n = 75 (d) n = 21 (e) n = 70?

Solution: From the theorems 1 and 2, it is obvious that if n can be written as product of distinct primes then corresponding D_n is a finite Boolean algebra otherwise it is not. (a) Since all the prime factors of 12 (2×2×5) are not distinct, D_{12} is not a finite Boolean algebra. Similarly, D_{40} (40 = 2×2×2×5), D_{75} (75 = 3×5×5) are not finite Boolean algebra. However, D_{21} (21 = 3×7) and D_{70} (70 = 2×5×7) are finite Boolean algebra as both 21 and 70 have distinct prime factors.

Ans.

Since a Boolean algebra is essentially a lattice, a finite Boolean algebra possesses all the properties possessed by a lattice. Thus, all the properties listed in 6.2 under heading "**properties of Lattice**" are valid for any finite Boolean algebra.

Let us consider a function f: $B_n \to B$ where B_n and B are sets as explained before. Obviously this is a **Boolean function** in n Boolean variables. The schematic diagram is shown is the figure 6.3.2. If we construct a table showing every n-tuple (n-bit long binary string) of B_n and its image in B under f then the table will contain 2^n rows. Each row contains an n-tuple and its image in B under f. A table so obtained is called **truth table** for the Boolean function f.

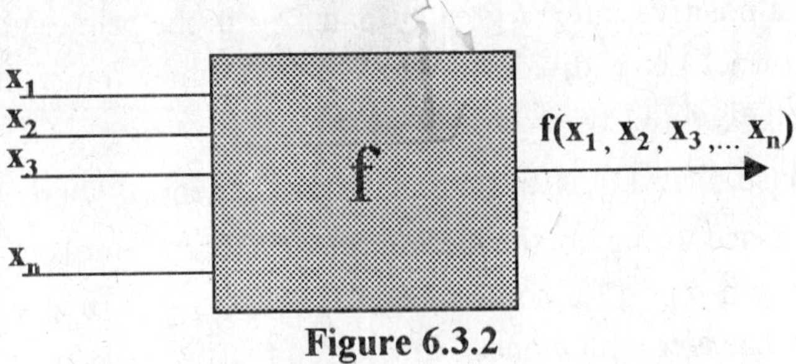

Figure 6.3.2

A typical function f: $B_3 \to B$ in three Boolean variables x, y and z is given in the table 6.3.1.

x	y	z	f(x, y, z)
0	0	0	0
0	0	1	0
0	1	0	1
0	1	1	0
1	0	0	0
1	0	1	0
1	1	0	0
1	1	1	0

Table 6.3.1

Here, if we write x' as complement of x i.e. whenever x is 1 x' is 0 and whenever x is 0 x' is 1, then $f(x, y, z) = x' \wedge y \wedge z'$. Consider a two variable function f: $B_2 \to B$ as given in table 6.3.2. We can write $f(x, y) = (x' \wedge y) \vee (x' \wedge y') = x'$. Similarly, we can write $f(x, y) = (x \wedge y) \vee (x' \wedge y) = y$ for the function $f: B_2 \to B$ as given in table 6.3.3. The function f: $B_3 \to B$ as given in table 6.3.4 below, can be written as $f(x, y, z) = (x' \wedge y \wedge z) \vee (x \wedge y \wedge z)$.

x	y	f(x, y)
0	0	1
0	1	1
1	0	0
1	1	0

Table 6.3.2

x	y	f(x, y)
0	0	0
0	1	1
1	0	0
1	1	1

Table 6.3.3

x	y	z	f(x, y, z)
0	0	0	0
0	0	1	0
0	1	0	0
0	1	1	1
1	0	0	0
1	0	1	0
1	1	0	0
1	1	1	1

Table 6.3.4

This shows that a Boolean function can be expressed in terms of Boolean variables. Any expression that involves only Boolean variables, Boolean constants and Boolean operators is called a **Boolean expression** or **Boolean Polynomial**. The Boolean operators are $\wedge$ **(AND)**, $\vee$ **(OR)** and $\neg$ (NOT). $\neg$ is an unary operator whereas $\wedge$ & $\vee$ are binary operators. The operator $\neg$ is also called complement or negation.

Boolean Polynomial and Functions

A Boolean variable is defined over the set B = {0, 1}. If a Boolean function is of n variables, we can say that f is a function from B_n to B. Any Boolean function can be written as a Boolean polynomial as shown in the above examples. This we will prove as a theorem here. Before that let us define what is a Boolean polynomial (Boolean expression).

Let $x_1, x_2, x_3, \ldots, x_n$ be a set of n Boolean variables. A **Boolean polynomial** $p(x_1, x_2, x_3, \ldots, x_n)$ in n Boolean variables x_k's is defined recursively as follows:

1. $x_1, x_2, x_3, \ldots, x_n$ are all Boolean polynomial.
2. The symbols 0 and 1 (Boolean constants) are Boolean polynomial.

3. If **p** and **q** are two Boolean polynomials then **p∧q** and **p∨q** are Boolean polynomial.
4. If **p** is a Boolean polynomials then **negation of p (¬p or p')** is a Boolean polynomial.
5. A Boolean polynomial can also be obtained by repeated use of any or more or all of the above four rules 1 to 4. There exists no other way to define a Boolean polynomial.

Again look at the table 6.3.2. There are four entries in the table under columns **x** and **y**. These are {(0 and 0) or (0∧0)}, {(0 and 1) or (0∧1)}, {(1 and 0) or (1∧0)} and {(1 and 1) or (1∧1)}. The same entries can be written involving variable names as (x'∧y'), (x'∧y), (x∧y') and (x∧y). Each of these Boolean expressions corresponding to entries in a truth table is called a **minterm**. There are eight minterms in tables 6.3.1 and eight in table 6.3.4. In general, there are 2^n minterms in a truth table for a function f from B_n to B. If b is any n-tuple of B_n, then $b = (x_1, x_2, x_3, \ldots, x_n)$, where each x_i is either 0 or 1. Generally, when $x_i = 0$, we say that the variable is complemented in b and is denoted by $\mathbf{x_i'}$ (complement of x_i) and when $x_i=1$, we say that the variable is not complemented in b and is denoted by $\mathbf{x_i}$. The minterm corresponding to n-tuple b is then given by the Boolean expression

$$z_1 \wedge z_2 \wedge z_3 \wedge \ldots \wedge z_n$$

Where $z_i = x_i$ if $x_i=1$ and $z_i = x_i'$ if $x_i = 0$. Notice that a minterm is the meet of all the variables in present form (complemented or non-complemented) in an n-tuple. Similarly, a **maxterm** is defined as the join of all the variables in present form in an n-tuple. The maxterm corresponding to n-tuple b is then given by the Boolean expression

$$z_1 \vee z_2 \vee z_3 \vee \ldots \vee z_n$$

Where $z_i = x_i$ if $x_i = 0$ and $z_i = x_i'$ if $x_i = 1$.

The meet operator ∧ (referred as AND) is also called **conjunction**. Therefore a minterm is also called **conjunctive**. Similarly, the join operator ∨ (referred as OR) is called **disjunction**. Thus, a maxterm is also called **disjunctive**. A Boolean function may be expressed in **Conjunctive Normal Form (CNF)** of Boolean polynomial or in **Disjunctive Normal Form (DNF)** of a Boolean polynomial. A Boolean polynomial is said to be in CNF, if it is expressed as the meet of those maxterms that produce output 1 for the function. Notice that a minterm that produces 0 for a function corresponding maxterm produces 1 for the function. **(minterm)' = (maxterm).** For example, function f(x, y) corresponding to the truth table 6.3.2 can be written in CNF as

$$f(x, y) = (x' \vee y) \wedge (x' \vee y')$$

And f(x, y, z) corresponding to the truth table 6.3.4 can be written in CNF as

$$f(x, y, z) = (x \vee y \vee z) \wedge (x \vee y \vee z') \wedge (x \vee y' \vee z) \wedge (x' \vee y \vee z) \wedge (x' \vee y \vee z'') \wedge (x' \vee y' \vee z)$$

Likewise, a Boolean polynomial is said to be in DNF, if it is expressed as the join of those minterms that produce output 1 for the function. For example, function f(x, y) corresponding to the truth table 6.3.2 can be written in DNF as

$$f(x, y) = (x' \wedge y') \vee (x' \wedge y)$$

And f(x, y, z) corresponding to the truth table 6.3.4 can be written in DNF as

$$f(x, y, z) = (x' \wedge y \wedge z) \vee (x \wedge y \wedge z)$$

Example 4 Let f: $B_3 \rightarrow B$ is given as $f(x, y, z) = x' \wedge (y \vee z) \wedge (x \vee y \vee z)$. Find truth table for this function and then reduce this into CNF and DNF representation.

Solution: The truth table for the given function is shown in the table 6.3.5. This table contains three minterms producing 1 for the function and five minterms that produce 0 for the function. To represent the given function in DNF, we have to take only those minterms that produce 1 for the function. The join of these minterms is the DNF representation for f. Thus, f can be written in DNF as

$$f(x, y, z) = (x' \wedge y' \wedge z) \vee (x' \wedge y \wedge z') \vee (x' \wedge y \wedge z)$$

For the CNF representation, we take only those minterms that produces 0 for f. Convert the minterm to maxterm and then meet of these maxterms is the CNF representation for f. Thus, f can be written in CNF as

$$f(x, y, z) = (x \vee y \vee z) \wedge (x' \vee y \vee z) \wedge (x' \vee y \vee z') \wedge (x' \vee y' \vee z) \wedge (x' \vee y' \vee z')$$

Ans.

x	y	z	$x' \wedge (y \vee z) \wedge (x' \vee y \vee z)$
0	0	0	0
0	0	1	1
0	1	0	1
0	1	1	1
1	0	0	0
1	0	1	0
1	1	0	0
1	1	1	0

Table 6.3.5

Let us define a set $M_t(f) = \{b \in B_n \mid f(b) = 1\}$ of all minterms that produce a 1 for a function $f: B_n \to B$. If $|M_t(f)| = 1$ then f equal to the minterm $b \in B_n$. This implies that f can be produced by a Boolean expression. In general, this is true for any Boolean function.

Theorem 3 Let f, g and h are three functions from B_n to B and $M_t(f)$, $M_t(g)$ and $M_t(h)$ are set of minterms for f, g and h respectively. Prove that

(a) If $M_t(f) = M_t(g) \cup M_t(h)$ then $f(b) = g(b) \vee h(b)\ \forall\ b \in B_n$

(b) If $M_t(f) = M_t(g) \cap M_t(h)$ then $f(b) = g(b) \wedge h(b)\ \forall\ b \in B_n$

Proof: (a) Let b be any n-tuple of B_n. If $b \in M_t(f)$ then $f(b) = 1$.

$$\begin{aligned} b \in M_t(f) &\Rightarrow b \in M_t(g) \cup M_t(h) \\ &\Rightarrow b \in M_t(g) \text{ or } b \in M_t(h) \\ &\Rightarrow g(b) = 1 \text{ or } h(b) = 1 \\ &\Rightarrow g(b) \vee h(b) = 1 \end{aligned}$$

Now, if $b \notin M_t(f)$ then $f(b) = 0$.

$$\begin{aligned} b \notin M_t(f) &\Rightarrow b \notin M_t(g) \cup M_t(h) \\ &\Rightarrow b \notin M_t(g) \text{ and } b \notin M_t(h) \\ &\Rightarrow g(b) = 0 \text{ and } h(b) = 0 \\ &\Rightarrow g(b) \vee h(b) = 0 \end{aligned}$$

This shows that $f(b) = g(b) \vee h(b)$

(b) Let b be any n-tuple of B_n. If $b \in M_t(f)$ then $f(b) = 1$.

$$\begin{aligned} b \in M_t(f) &\Rightarrow b \in M_t(g) \cap M_t(h) \\ &\Rightarrow b \in M_t(g) \text{ and } b \in M_t(h) \\ &\Rightarrow g(b) = 1 \text{ and } h(b) = 1 \\ &\Rightarrow g(b) \wedge h(b) = 1 \end{aligned}$$

Now, if $b \notin M_t(f)$ then $f(b) = 0$.

$$b \notin M_t(f) \Rightarrow b \notin M_t(g) \cap M_t(h)$$

$$\Rightarrow b \notin M_t(g) \text{ or } b \notin M_t(h)$$

$$\Rightarrow g(b) = 0 \text{ or } h(b) = 0$$

$$\Rightarrow g(b) \wedge h(b) = 0$$

This shows that $f(b) = g(b) \wedge h(b)$

Proved.

Theorem 4 Let $f: B_n \rightarrow B$ be a Boolean function then prove that it is produced by a Boolean expression.

Proof: Let $M_t(f) = \{b_1, b_2, b_3, \ldots, b_m\}$. We can define a function f_i for each of b_i, where $1 \leq i \leq m$ such that $f_i\ (b_i) = 1$ and $f_i\ (b_i) = 0$ if $b_i \neq b$. This definition implies that $M_t(f_i) = \{b_i\}$ and so

$$M_t(f) = M_t(f_1) \cup M_t(f_2) \cup M_t(f_3) \cup \ldots \cup M_t(f_m)$$

Therefore from the theorem 3 above, we have

$$f = f_1 \vee f_2 \vee f_3 \vee \ldots \vee f_m$$

If M_{b_i}'s are the corresponding minterms for each f_i, then f can be written as join of these minterms as below.

$$f = M_{b_1} \vee M_{b_2} \vee M_{b_3} \vee \ldots \vee M_{b_m}$$

Proved.

In the table 6.3.4 above, $f(x, y, z) = (x' \wedge y \wedge z) \vee (x \wedge y \wedge z)$. This function can also be written, in simple form as $f(x, y, z) = y \wedge z$. The process of expressing a Boolean expression as join of minterms and then simplifying the resulting expression can be formalized. There are different procedures to simplify a Boolean expression so that minimum number of **gates** (fundamental circuits for AND, OR and NOT) can be used to achieve the same output as desired by the function. In this text, we shall discuss Boolean Expression Simplification by Karnaugh Map and Quine Mcluskey Method.

6.4 Simplification of Boolean Expression

The information contained in a truth table may be expressed in compact form by listing decimal equivalent of those minterms that produce a **1** for the function. For example, for the truth table 6.3.5, we can write

$$f(x, y, z) = \Sigma(1,2,3)$$

Because in the truth table 6.3.5, the minterms $x'\wedge y'\wedge z$, $x'\wedge y\wedge z'$ and $x'\wedge y\wedge z$ produce a **1** for the function, the decimal equivalent for these minterms are 1, 2 and 3 respectively.

Sometimes a minterm is irrelevant for a function. Consider a situation of a circuit having 10 input lines numbered 1 to 10. It is designed to work in such a way that it takes only first 8 input lines and ignores the last two input lines. Thus, out of 2^{10} total possible minterms, only 2^8 are relevant and others are of no use. There are many such situations in which one or more minterms are not taken into consideration while determining the nature of a function. *A minterm that is irrelevant for a function output belongs to a set of **don't care conditions** of the function.* If there is any minterm belonging to the set of ***don't care conditions,*** it is mentioned together with the function definition. For example, if the output of a Boolean function in four variables: x, y, z and w, is 1 for minterms 1, 3, 5, 6, 12 & 15 and 2, 7, 9 & 11 belong to ***don't care conditions***, then function is represented as

$$f(x, y, z, w) = \Sigma(1, 3, 5, 6, 12, 15)$$
$$d(x, y, z, w) = \Sigma(2, 7, 9, 11)$$

Simplification of a Boolean Expression means finding of essential ***prime implicants***. What is this prime implicant and how to find it? Let $A = \{a_1, a_2, a_3, \ldots, a_n\}$ be a set of Boolean expressions. We say that the Boolean expression **b** covers or subsumes A, precisely, when

$$\mathbf{b} + x = \mathbf{b} \ \forall\ x \in A$$

i.e., x is a product of **b** with some other terms such that x = **bc**,

$$\therefore\ \mathbf{b} + x = \mathbf{b} + \mathbf{bc} = \mathbf{b}(1 + c) = \mathbf{b}$$

In general, covering term contains fewer variables than the terms that it covers. For example, both **ab** and **ab'** are covered by **a,** so **a** is the covering term for the set of Boolean expressions {**ab, ab'**}. A ***prime implicant*** is a minimal covering term i.e., one that covers a maximal set of terms. There are different ways to find essential prime implicants. In this text, we shall discuss **Karnaugh** Map and **Quine Mcluskey** method.

Karnaugh Map (K-Map)

The K-Map is a visual tool to minimize a Boolean expression involving a few Boolean variables. It consists of a rectangle divided into squares. Each square corresponds to a minterm. These squares are arranged in a way that the contiguous squares differ in exactly one variable being complemented or uncomplemented. The K-MAP must be thought of as covering the surface of a sphere, so that, squares on opposite edges of the map are contiguous and all four corner squares of the map have a point in common. K-Map for 2, 3 and 4 variables are shown in the figure 6.4.0(a), 6.4.0(b) and 6.4.0(c) respectively. Minterm and its decimal equivalent that correspond to a square are also shown in the respective figure.

This K- Map is for two variables p and q. A square in right hand side figure contains decimal equivalent of a minterm.

	q'	q
p'	00	01
p	10	11

	q'	q
p'	0	1
p	2	3

Figure 6.4.0(a)

The following K- Map is for three variables p, q and r. A square in right hand side figure contains decimal equivalent of a minterm of corresponding square of left hand side.

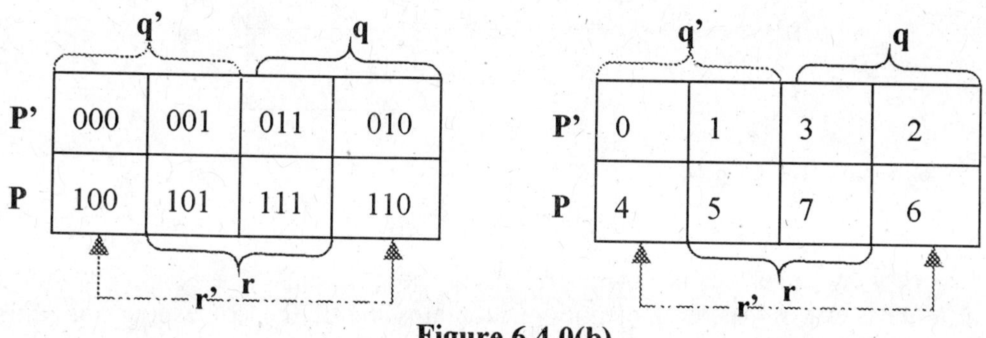

Figure 6.4.0(b)

The K- Map shown in figure 6.4.0(c_1) and 6.4.0(c_2) are for four variables p, q, r and s. A square in figure 6.4.0(c_2) contains decimal equivalent of a minterm of corresponding square of figure 6.4.0(c_1).

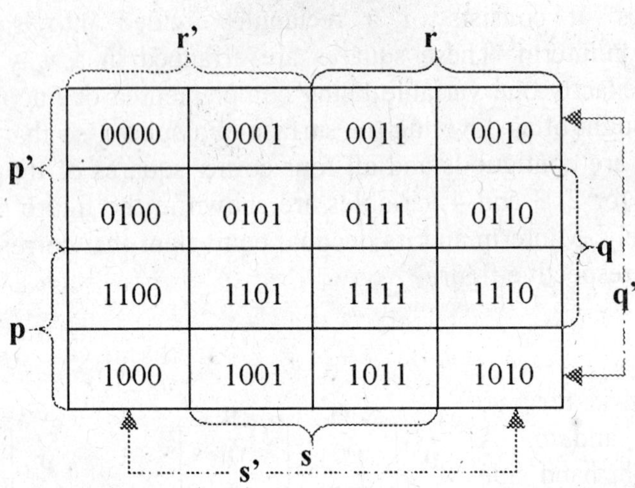

Figure 6.4.0(c_1)

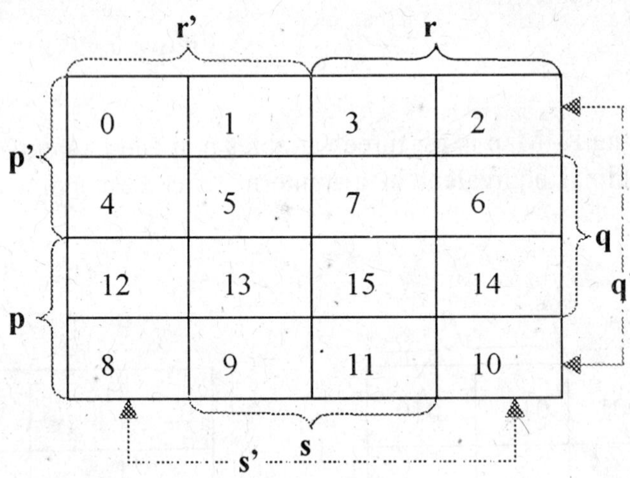

Figure 6.4.0(c_2)

A square in a K-Map for a function contains a **1** if the corresponding minterm produces **1** for the function, if a minterm produces **0** for the function then the corresponding square contains **0**. If the minterm belongs to ***don't care conditions***, the corresponding square contains an 'X'. This is the way to fill a K-Map for a function.

The next step involves finding of one or more **duet** or **quad** or **octet** and so on. A **duet** is contiguous two squares either horizontally or vertically, but not diagonally in any case, having **1** as entry. A **quad** is contiguous four squares either horizontally or vertically or in square or rolling having **1** as entry. An **octet** is contiguous eight squares either horizontally or vertically or in rectangle or rolling having **1** as entry. Each duet, quad and octet can be expressed in terms of product of variables. The sum of these products is the simplified expression for the given function. The procedure is explained through the following examples.

Example 1 Minimize the Boolean expression for a function given as

$$f(p, q) = \Sigma(1, 3)$$

Solution: The Given function is of two variables. The K-Map for this function is given as in figure 6.4.1. Minterms of two variables having decimal equivalent 1 and 3 produce **1** as output for the function, so squares corresponding to minterm 01 and 11 contain an entry 1. The remaining squares contain 0 as an entry. There is no don't care condition so no square contains an entry 'X'.

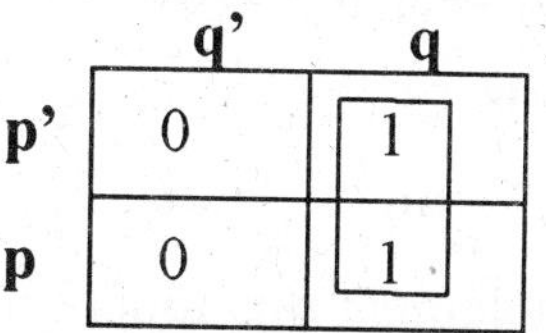

Figure 6.4.1

The duet corresponds to column q, so the given function f(p, q) = p'q + pq can be expressed in **simplified** form as **f(p, q) = q**. No gate is required to implement this function. Only a simple wire connected to power supply is needed. However, in the initial expression we need four gates (one inverter, two AND gates and one OR gate). **Ans.**

Example 2 Minimize the Boolean expression for a function given as

$$f(p, q, r) = \Sigma(2, 6, 7)$$

Solution: In this example, the given function is of three Boolean variables. The K-Map for this function is given as in figure 6.4.2. Minterms of three variables having decimal equivalent 2, 6 and 7 produce **1** as output for the function, so squares corresponding to minterm 010, 110 and 111 contain an entry 1.

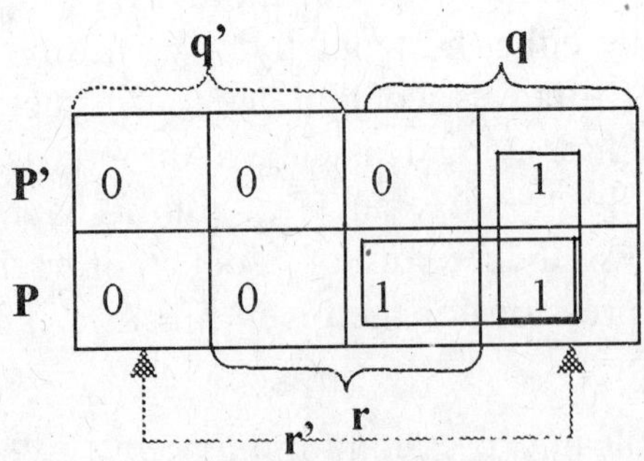

Figure 6.4.2

The remaining squares contain 0 as an entry. There is no don't care condition so no square contains an 'X'.

There are two overlapping duets marked with rectangles. The vertical rectangle corresponds to the intersection of q and r'. The horizontal rectangle corresponds to the intersection of p and q. Therefore the given function

$$f(p, q, r) = p'qr' + pqr' + pqr$$

can be expressed in **simplified** form as

$$\mathbf{f(p, q, r) = qr' + pq}.$$

Now number of gates required to implement this function is four (one inverter, two AND gates and one OR gate). Compare this with the original requirement of ten gates (two inverters, six AND gates and two OR gates). **Ans.**

Example 3 Minimize the Boolean expression given by the function

$$f(p, q, r, s) = \Sigma(5, 8, 9, 10, 11, 12, 13, 14, 15)$$

Solution: The function is of four Boolean variables. The K-Map for this function is given as in figure 6.4.3. Minterms of four variables having decimal equivalent 5, 8, 9, 10, 11, 12, 13, 14 and 15 produce **1** as output for the function, so squares corresponding to minterm 0101, 1000, 1001, 1010, 1011, 1100, 1101, 1110 and 1111 contain an entry 1. The remaining squares contain 0 as an entry. There is no don't care condition so no square contains an 'X'.

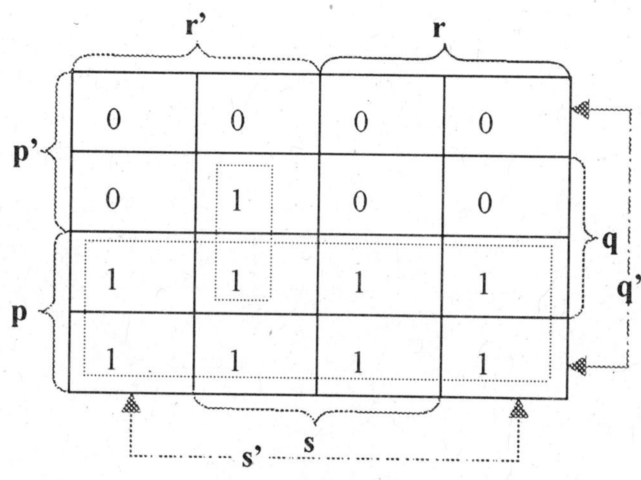

Figure 6.4.3

There is one duet and one octet marked with rectangles. Duet is overlapping with octet. If overlapping duet (or quad or octet) reduces the number of variables in an expression, it is worth considering. The octet corresponds to p and duet corresponds to qr's. Therefore the given function can be expressed in **simplified** form as

$$\mathbf{f(p, q, r, s) = p + qr's}.$$

Compare the requirement of gates in this simplified expression with that in the original expression.

Ans.

Example 4 Minimize the Boolean Expression given by the function

$$f(p, q, r, s) = \Sigma(1, 2, 3, 6, 8, 9, 10, 12, 13, 14)$$

Solution: The function is of four Boolean variables. The K-Map for this function is given as in figure 6.4.4. Minterms of four variables having decimal equivalent 1, 2, 3, 6, 8, 9, 10, 12, 13 and 14 produce **1** as output for the function, so squares corresponding to minterm 0001, 0010, 0011, 0110, 1000, 1001, 1010, 1100, 1101 and 1110 contain an entry 1. The remaining squares contain 0 as an entry. There is no don't care condition so no square contains an 'X'.

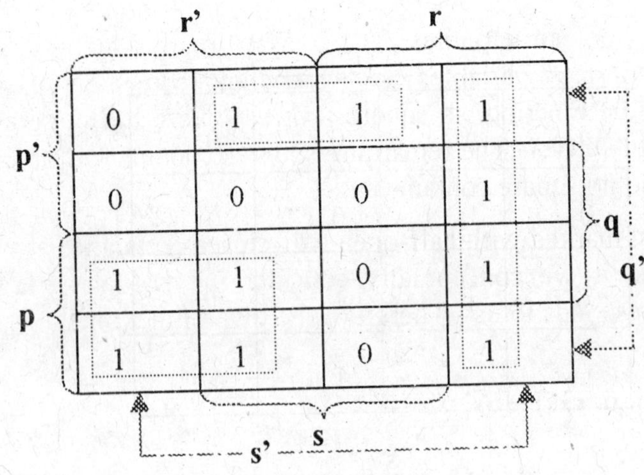

Figure 6.4.4

There is one duet and two quads marked with rectangles. The duet corresponds to p'q's. The vertical quad corresponds to rs'. The other quad corresponds to pr'. Therefore the given function can be expressed in **simplified** form as

$$\mathbf{f(p, q, r, s) = p'q's + rs' + pr'}.$$

Ans.

Example 5 Minimize the Boolean Expression given by the function

$$f(p, q, r, s) = \Sigma(4, 6, 12, 14)$$

Solution:

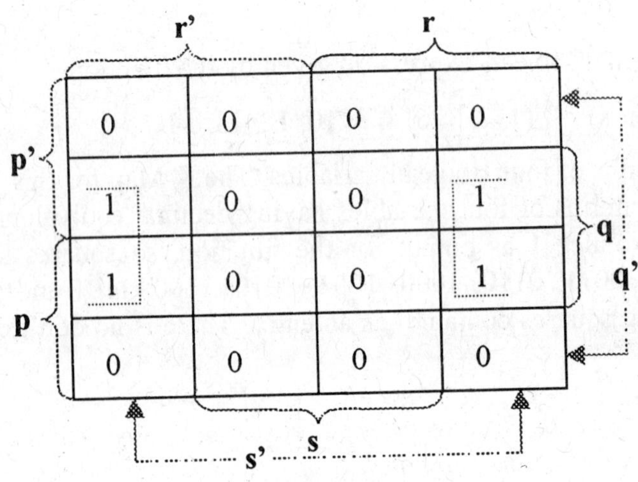

Figure 6.4.5

The function is of four Boolean variables. The K-Map for this function is given as in figure 6.4.5. Minterms of four variables having decimal equivalent 4, 6, 12 and 14 produce **1** as output for the function, so squares corresponding to minterm 0100, 0110, 1100 and 1110 contain an entry 1. The remaining squares contain 0 as an entry. There is no don't care condition so no square contains an 'X'.

There is one **rolling** quad marked with half-open half-closed rectangles at two ends. This rectangle gets completed if we conceptually consider the K-Map as a surface of a cylinder. This quad corresponds to qs'. Therefore the given function can be expressed in **simplified** form as

f(p, q, r, s) = qs'.

Ans.

Example 6 Minimize the Boolean Expression given by the function

$$f(p, q, r, s) = \Sigma(5, 11, 13, 15)$$

Solution: The function is of four Boolean variables. The K-Map for this function is given as in figure 6.4.6. Minterms of four variables having decimal equivalent 5, 11, 13 and 15 produce **1** as output for the function, so squares corresponding to minterm 0101, 1011, 1101 and 1111 contain an entry 1. The remaining squares contain 0 as an entry. There is no don't care condition so no square contains an 'X'.

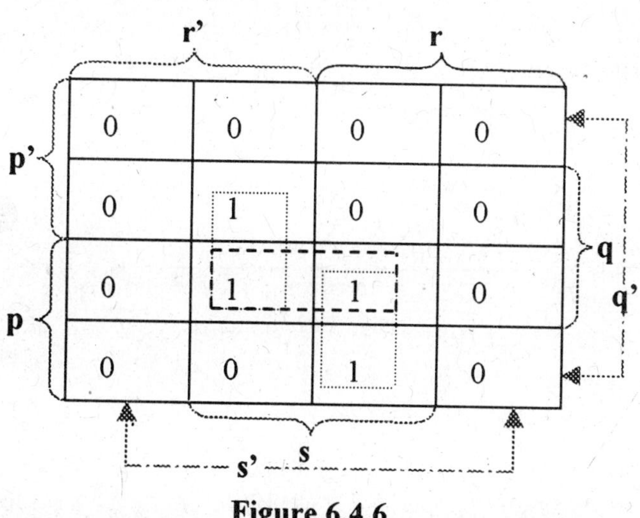

Figure 6.4.6

There are three **duets** marked with rectangles. Here one duet marked with dashed rectangle is **redundant**. We do not take into consideration the product term corresponding to a redundant duet (or quad or octet) in the simplified expression of the function. The non-redundant duets correspond to qr's and prs. Therefore the given function can be expressed in **simplified** form as

$$\mathbf{f(p, q, r, s) = qr's + prs}$$

Ans.

Example *7* Minimize the Boolean Expression given by the function

$$f(p, q, r, s) = \Sigma(0, 3, 4, 5, 7)$$

$$d(p, q, r, s) = \Sigma(8, 9, 10, 11, 12, 13, 14, 15)$$

Solution: The function is of four Boolean variables. The K-Map for this function is given as in figure 6.4.7. Minterms of four variables having decimal equivalent 0, 3, 4, 5 and 7 produce **1** as output for the function, so squares corresponding to minterm 0101, 1011, 1101 and 1111 contain an entry 1. There are minterms having decimal equivalent of 8, 9, 10, 11, 12, 13, 14 and 15 that belong to don't care condition. Thus, squares corresponding to minterm 1000, 1001, 1010, 1011, 1100, 1101, 1110 and 1111 contain an entry X. The remaining squares contain 0 as an entry.

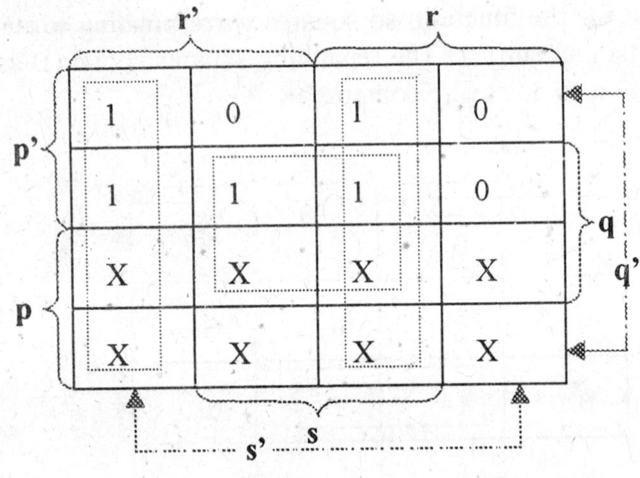

Figure 6.4.7

Whenever, there is a square with entry X in a K-Map, we can consider that square as having entry either 1 or 0 depending upon our requirement to form duet, quad or octet Thus we have three quads and corresponding product terms are r's', rs and qs. Therefore the given function can be expressed in **simplified** form as

$$\mathbf{f(p, q, r, s) = r's' + rs + qs}$$

Ans

K-Map method may be used for the simplification of a Boolean expression involving five or more variables, but the visualization of the result becomes increasingly difficult. To deal with such problems, we use Quine-Mcluskey method.

Quine-Mcluskey Method

This programmable deterministic algorithm based method is used to automate the technique of circuit minimization. This method is very robust and best suitable in the situation when number of Boolean variables involved are four or more. This method has two phases:

Phase 1: Distributive law i.e., ax + ax' = a(x + x') = a. This is used repetitively to get prime implicants.

Phase 2: Selection Phase i.e., only those prime implicants are selected which are necessary to represent Boolean function.

The sum of these prime implicants is the result. The result so obtained, at the end, is the simplest sum of the products of Boolean variables for the given function. The step by step working procedure is explained below with an example.

Example 1 Find the minimal expression for the Boolean expression

wx + xy + yz + wz + w'x'yz' + w'x'y'z

Solution: There are four variables: w, x, y and z, used in the given Boolean expression. The initial task is to find all the prime implicants. This is what is done in the phase 1 and this involves a number of steps as outlined below.

Step 1: Represent each term in the given expression by a string of 1, 0 and –, where '1' stand for a variable, '0' for complemented variable and '–' for absent variable in the term. Thus, the given terms can be written as follows:

wx = 11– – ; **xy = –11–;** **yz = – – 11;**

wz = 1– – 1; **w'x'yz' = 0010** and **w'x'y'z = 0001**

Step 2: An absent variable in a Boolean term implies that it is ignored while finding the output. Expand all terms containing '–' in the **step 1**, with all possible combinations of '1' and '0'. Expanding the above terms, we get

wx = 11– – :	**1111**	**1110**	**1101**	**1100**
xy = –1 1– :	**1111**	**1110**	**0111**	**0110**
yz = – – 11:	**1111**	**1011**	**0111**	**0011**
wz = 1– –1:	**1111**	**1101**	**1011**	**1001**

w'x'yz' = **0010**

w'x'y'z = **0001**

Step 3: Select the distinct terms from the exhaustive list of terms as obtained in the Step 2. Arrange them in the decreasing order of the number of 1's in the term. If there are many terms having same number of 1's then arrange them in decreasing order of magnitude within the same group. Applying this on the terms obtained in the step 2, we get

$$1111\} \ \textit{Group 1 containing 4 1's}$$

$$\left.\begin{matrix}1110\\1101\\1011\\0111\end{matrix}\right\} \ \textit{Group 2 containing 3 1's}$$

$$\left.\begin{matrix}1100\\1001\\0110\\0011\end{matrix}\right\} \ \textit{Group 3 containing 2 1's}$$

$$\left.\begin{matrix}0010\\0001\end{matrix}\right\} \ \textit{Group 4 containing 1 1's}$$

Step 4: Combine those terms that differ by a 0 or 1 in exactly one position e.g. 1111 and 1110 are combined to yield 111– i.e., wxyz + wxyz' = wxy. It is important to note that a term can combine with a term in the group either immediately above or immediately below its group. The result of this step applied on the arranged terms in the step 3 is given below.

$$\left.\begin{matrix}111- & (1111,1110)\\11-1 & (1111,1101)\\1-11 & (1111,1011)\\-111 & (1111,0111)\end{matrix}\right\} A_1 \ \textit{Group 1 combined with Group 2}$$

$$\left.\begin{array}{ll} 11-0 & (1110, 1100) \\ -110 & (1110, 0110) \\ 110- & (1101, 1100) \\ 1-01 & (1101, 1001) \\ 10-1 & (1011, 1001) \\ -011 & (1011, 0011) \\ 011- & (0111, 0110) \\ 0-11 & (0111, 0011) \end{array}\right\} A_2 \quad \textit{Group 2 combined with Group 3}$$

$$\left.\begin{array}{ll} -001 & (1001, 0001) \\ 0-10 & (0110, 0010) \\ 001- & (0011, 0010) \\ 00-1 & (0011, 0001) \end{array}\right\} A_3 \quad \textit{Group 3 combined with Group 4}$$

Step 5: Repeat step 4 on the terms currently obtained until no more terms can be combined. It is important to mention here that number of groups of terms containing different number of 1's decreases by one, after every such repetition. During the process, we may get a term that cannot combine with any other terms. Those terms that do not combine further are *prime implicants*. Keep on collecting all such terms, if any, from every repetition. Repeating the step 4, we get

$$\left.\begin{array}{ll} 11-- & (111-, 110-) \\ -11- & (111-, 011-) \\ 11-- & (11-1, 11-0) \\ 1--1 & (11-1, 10-1) \\ 1--1 & (1-11, 1-01) \\ --11 & (1-11, 0-11) \\ -11- & (-111, -110) \\ --11 & (-111, -011) \end{array}\right\} B_1 \quad \textit{By combining } A_1 \textit{ and } A_2$$

$$\left.\begin{array}{ll} -0-1 & (10-1, 00-1) \\ 0-1- & (011-, 001-) \\ -0-1 & (-011, -001) \\ 0-1- & (0-11, 0-01) \end{array}\right\} B_2 \quad \textit{By combining } A_2 \textit{ and } A_3$$

Step 6: Steps 4 and 5 give all the prime implicants for the given Boolean expression. In this step, collect all the unique prime implicants. These are, in this case, 11– –, –11–, 1– –1, – –11, –0 –1 and 0 –1–. In terms of the variable, these prime implicants can be written as wx, xy, wz, yz, x'z and w'y respectively. All these prime implicants may not be essential for representing the given Boolean expression in minimized form. Therefore, the selection phase follows.

Step 7: This is **selection phase**. Create a table of m+1 rows and n+1 columns, where m is the number of expanded terms uniquely identified in the step 3 and n is the number of prime implicants uniquely identified in the step 6. One extra row and column is taken for headings. Each cell in the table is uniquely determined by ordered pairs (Expanded term, Prime implicant). We put an 'X' in a cell (Expanded term, Prime implicant), when the prime implicant covers the expanded term. For example, 11– – covers 1111, so there is an 'X' in the cell (1111,11– –), i.e., first row and first column. Since 1111 is not covered by – 0–1, so there is no 'X' in the cell (1111, – 0 – 1), i.e., fifth column of first row. We have the following table.

Expanded Terms	11––	–11–	1––1	––11	–0–1	0–1–
1. 1111	X	X	X	X		
2. 1110	X	X				
3. 1101	X		X			
4. 1011			X	X	X	
5. 0111		X		X		X
6. 1100	**X**					
7. 1001			X		X	
8. 0110		X				X
9. 0011				X	X	X
10. 0010						**X**
11. 0001					**X**	

Table 6.4.1

From the table 6.4.1, prime implicants are selected as follows:

Step 7.1 Look at the rows from top to bottom and select the row, if any, having single entry 'X'. The prime implicant corresponding to the column, in which 'X' is present, is essential prime implicant.

Step 7.2 Mark all the rows in which there is a 'X' in the column corresponding to the essential prime implicants determined in the step 7.1. This prime implicant covers all the expanded terms for which there is a 'X' in this column.

Step 7.3 Repeat steps 7.1 and 7.2 on the table created in step 7 until we get a covering prime implicant for all the expanded terms (rows).

In the table 6.4.1, we have selected row 6. Thus the prime implicant 11–– is essential. Also there are 'X' in the column corresponding to 11–– for rows 1, 2 and 3 besides for row 6. Thus, 11–– covers the expanded terms 1111, 1110, 1101 and 1100 corresponding to the rows 1, 2,3 and 6 respectively. Next, rows 10 and 11 have single entry 'X'. The corresponding prime implicants are 0–1– and –0–1 respectively. It can be verified that 0–1– covers expanded terms 0111, 0110, 0011 and 0010 corresponding to the rows 5, 8, 9 and 10 respectively. And, 0–1– covers expanded terms 1011, 1001, 0011 and 0001 corresponding to the rows 4, 7, 9 and 11 respectively. The expanded term 0011 is covered by both 0–1– and –0–1. This does not create any problem as long as both are essential. Thus out of the six prime implicants, only three are essentially required to represent the given Boolean expression. These are 11––, 0–1– and –0–1 i.e., wx, w'y and y'z. Therefore, the minimized Boolean expression is

$$wx + w'y + y'z$$

Ans.

It is important to mention that, in most of the cases, the answer does not come so easily as above. There may not be any row having single entry 'X'. How to proceed then? A tie has to be broken. Select a row, if any, having two 'X's otherwise look for three 'X's and so on. After selecting such row, one prime implicant has to be selected. We select the prime implicant that covers maximum number of expanded terms. In case of tie, select the prime implicant having minimum number of variables. If there is a tie, again, select having minimum number of complemented variables. See the following example.

Example 2 Find the minimal expression for the Boolean expression

$$xyz + x'z + y'z'$$

Solution: There are three variables: x, y and z, used in the given Boolean expression.

Step 1: Representing each term in the given expression by a string of 1, 0 and –, we get

xyz = 111; **x'z = 0–1;** **y'z' = –00;**

Step 2: Expanding all terms containing '–' in the **step 1**, with all possible combinations of '1' and '0', we get

xyz = 111
x'z = 0–1: 011 001
y'z' = – 00: 100 000

Step 3: Arranging the distinct terms in the decreasing order of the number of 1's in the term, we get

$$\left.111\right\}\ \textit{Group}\ 1\ \textit{containing}\ 3\ \text{1's}$$
$$\left.011\right\}\ \textit{Group}\ 2\ \textit{containing}\ 2\ \text{1's}$$
$$\left.\begin{matrix}100\\001\end{matrix}\right\}\ \textit{Group}\ 3\ \textit{containing}\ 1\ \text{1's}$$
$$\left.000\right\}\ \textit{Group}\ 4\ \textit{containing}\ \textit{no}\ \text{1's}$$

Step 4: Combining the terms that differ by a 0 or 1 in exactly one position, we get

$$-11\ \ (111, 011)\}\,A_1\quad \textit{Group 1 combined with Group 2}$$
$$0-1\ \ (011, 001)\}\,A_2\quad \textit{Group 2 combined with Group 3}$$
$$\left.\begin{matrix}-00\ \ (100, 000)\\00-\ \ (001, 000)\end{matrix}\right\}A_3\quad \textit{Group 3 combined with Group 4}$$

Step 5: None of the terms of the step 4 can combine with the other terms, so the repetition of the step 4 is not required in this case.

Step 6: The unique prime implicants are –11, 0–1, –00 and 00–. In terms of variables, these prime implicants can be written as yz, x'z, y'z' and x'y' respectively.

Step 7: In the **selection phase**, we now create a table of 5+1 rows and 4+1 columns for 5 expanded terms uniquely identified in the step 3 and prime implicants uniquely identified in the step 6.

Expanded Terms	–11	0–1	–00	00–
1. 111	<u>X</u>			
2. 011	X	X		
3. 100			<u>X</u>	
4. 001		X		X
5. 000			X	X

Table 6.4.2

In the table 6.4.2, we select row 1. Thus the prime implicant –11 is essential. Also there is 'X' in the column corresponding to –11 for row 2 besides for row 1. Thus –11 covers the expanded terms **111** and **011.** Next, row 3 has single entry 'X' in it. The corresponding prime implicant is –00. It covers expanded terms **100** and **000**. Thus expanded terms corresponding to rows 1, 2, 3 and 5 have been covered by the prime implicants –11 and –00. The expanded term **001** is covered either by 0–1 or 00–. We have to select one of them. Both the terms have two variables. 0–1 has one complemented variable whereas 00– has two. According to the selection criteria mentioned above, the term having minimum number of complemented variables is selected. So, we select here 0–1. Thus the essential prime implicants are –11, –00 and 0–1. When expressed in terms of variables, these prime implicants are yz, y'z' and x'z respectively. Therefore, the minimized Boolean expression is

yz + y'z' + x'z

Ans.

6.5 Propositional Calculus

Consider statements like:

1. Today is Monday.
2. It will rain tomorrow.
3. For any positive integer x, x + 3 > 3.
4. You are at home.

These statements are either true or false at a given point of time. None of them can be both true and false at the same time. Also, the answer cannot be like 'may be' or 'probably' etc in any of the above case, i.e., there is no ambiguity in the answer. Statements like above are known as declarative sentences.

A ***proposition*** is any declarative sentence that is either true or false but never both at the same time. The two valued (T, F) Boolean algebra is very closely related to the system of basic logic called ***propositional calculus***. This logical system is concerned with propositions. It is important to note that this system does not allow for such ideas as " possibly true" or "indeterminate". A proposition that corresponds to a single declarative sentence is called ***atomic proposition*** and is generally represented by lower case alphabets **a, b, ... q, r, ..., x, y, z.** The truth values are denoted by **T**(True) and **F**(False). Value **T** & **F** directly correspond to Boolean constants **1 and 0** respectively. The concept of propositional calculus can be extended to **predicate calculus** that allows us to use quantifiers (∀, ∃) and propositional functions in a Boolean expression. In this text, our discussion is restricted to propositional calculus.

An atomic proposition in itself is not sufficient to deal with many situations of a logical system. Generally atomic propositions are combined to give a logical meaning to a compound sentence. For example, let p be the proposition "you are at home" and let q be the proposition " you are in office". The compound sentence "you are either at home or in office" can be represented as **(p or q)**. *A proposition that is obtained from the combination of other propositions is called* ***compound proposition***. In the propositional calculus we use connectives to combine propositions. Sometimes we need to use negative of a proposition also in compound proposition. The connectives are shown in the table 6.5.1 with its corresponding meaning in Boolean Algebra

Sl. No.	Connective	Called as	Boolean equivalent	Set equivalent
1.	$\vee$	**Disjunction**	**Inclusive OR**	**Union**
2.	$\wedge$	**Conjunction**	**AND**	**Intersection**
3.	$\rightarrow$	**Implies that**	**If .. Then..**	$\Rightarrow$
4.	$\equiv$	**Equivalence**	**Equality**	**Equality**
5.	**~(or ¬)**	**Negation**	**1's complement**	**Complement**

Table 6.5.1

Let us correlate atomic propositions with Boolean variables and compound proposition with Boolean function. Like, every Boolean function has a truth table, every compound proposition has a truth table showing the truth values of the compound proposition in terms of its component part (atomic proposition). Truth table for the negation of a proposition **p** is shown in the table 6.5.2. The possible values (truth) for the compound proposition **p** $\vee$ **q** and **p** $\wedge$ **q** are listed in the table 6.5.3. Similarly, table 6.5.4 shows possible values (truth) for propositions **p→q, q→p** and **p** $\equiv$ **q** under the respective columns of the table.

p	~p
F	**T**
T	**F**

Table 6.5.2

p	q	p∨q	p∧q
F	**F**	**F**	**F**
F	**T**	**T**	**F**
T	**F**	**T**	**F**
T	**T**	**T**	**T**

Table 6.5.3

p	q	p→q	q→p	(p→q)∧(q→p) or (p≡q)
F	F	T	T	T
F	T	T	F	F
T	F	F	T	F
T	T	T	T	T

Table 6.5.4

If a compound proposition has two atomic propositions as components, then the truth table for the compound proposition contains four entries. These four entries may be all T, may be all F, may be one T and three F and so on. There are in total 16 (2^4) possibilities as shown in the table 6.5.5 in the 16 different columns. Column 1 contains all T. This implies that if the column corresponding to a compound proposition in a truth table contains all T, then the proposition is always true. This is called **tautology.** Now look at column 16. It contains all F. This implies that if the column corresponding to a compound proposition in a truth table contains all F, then the proposition is **never true**. This situation is referred as **contradiction.** If the column corresponding to the compound proposition in a truth table contains both T and F, then the proposition is true for some combination of its component and false for some other combination. This is called **contingency.**

Out of the 16 columns, columns from 9 to 16 are just negation of the columns from 1 to 8 in reverse order i.e. column I ($1 \le I \le 8$) is just negation of the column 17 –I. Two propositions p and q are said to be equal if and only if we have a **tautology** from the truth table of p = q (or p ≡ q).

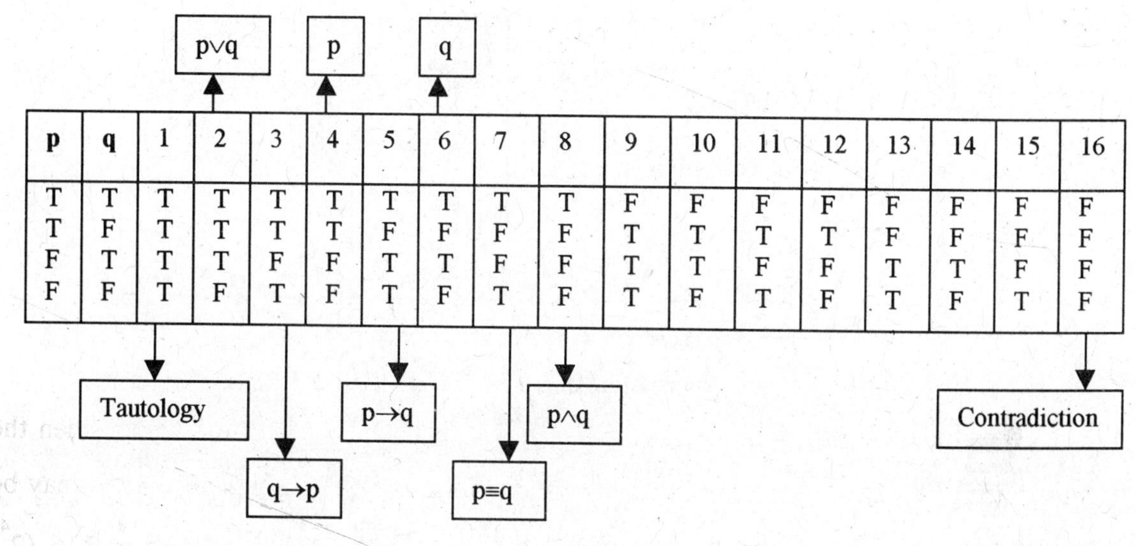

p	q	1	2	3	4	5	6	7	8	9	10	11	12	13	14	15	16
T	T	T	T	T	T	T	T	T	T	F	F	F	F	F	F	F	F
T	F	T	T	T	T	F	F	F	F	T	T	T	T	F	F	F	F
F	T	T	T	F	F	T	T	F	F	T	T	F	F	T	T	F	F
F	F	T	F	T	F	T	F	T	F	T	F	T	F	T	F	T	F

Table 6.5.5

Example 1 Find the truth table for the proposition p→(q∨~p)

Solution: The given compound proposition has two atomic proposition p and q as its components. The truth table will have four entries as shown in the table 6.5. 6.

p	q	p→(q∨~p)
F	F	T
F	T	T
T	F	F
T	T	T

Table 6.5.6

Ans.

Example 2 Determine the compound proposition (logical expression) f(p, q, r) having truth values as shown in the following table 6.5.7

p	q	r	f(p, q, r)
T	T	T	T
T	T	F	F
T	F	T	T
T	F	F	T
F	T	T	F
F	T	F	F
F	F	T	T
F	F	F	F

Table 6.5.6

Solution: The compound statement f(p, q, r) has three atomic propositions as components. To find an expression we use the concept of minterms discussed in the section 6.4 of this chapter. Look for those combinations of propositions p, q and r that produce a truth value T for f. The minterms are $p\wedge q\wedge r$, $p\wedge\sim q\wedge r$, $p\wedge\sim q\wedge\sim r$ and $\sim p\wedge\sim q\wedge r$. These are called conjunctive. A conjunctive is defined as conjunction of atomic proposition or negation of an atomic proposition. The disjunction of these conjunctives is the required expression. Thus,

$$\mathbf{f(p, q, r) = (p\wedge q\wedge r) \vee (p\wedge\sim q\wedge r) \vee (p\wedge\sim q\wedge\sim r) \vee (\sim p\wedge\sim q\wedge r)}$$

Ans.

Properties of Propositional Calculus

Let p, q and r be any three propositions then following are true.

1. Conjunction and disjunction are commutative i.e.

$$\mathbf{p\wedge q = q\wedge p}$$
$$\mathbf{p\vee q = q\vee p}$$

2. Conjunction and disjunction are associative i.e.

$$\mathbf{p\wedge(q\wedge r) = (p\wedge q)\wedge r}$$
$$\mathbf{p\vee(q\vee r) = (p\vee q)\vee r}$$

3. Conjunction distributes over disjunction and vice versa i.e.

$$\mathbf{p\wedge(q\vee r) = (p\wedge q) \vee(p\wedge r)}$$
$$\mathbf{p\vee(q\wedge r) = (p\vee q) \wedge(p\vee r)}$$

4. Idempotent law is obeyed by proposition with respect to conjunction and disjunction i.e.

$$\mathbf{p\wedge p = p}$$

$$p \vee p = p$$

5. De Morgan's law i.e. complement (negation) of conjunction is disjunction of complements (negations) and complement of disjunction is conjunction of complements.

$$\sim(p \wedge q) = \sim p \vee \sim q$$
$$\sim(p \vee q) = \sim p \wedge \sim q$$

6. Involution i.e. negation of a negation of a proposition p is p itself

$$\sim(\sim p) = p$$

7.
$$p \rightarrow q = (\sim p) \vee q$$
$$p \rightarrow q = ((\sim q) \rightarrow \sim p)$$
$$p \leftrightarrow q = (p \rightarrow q) \wedge (q \rightarrow p)$$ [$\leftrightarrow$ is also called equivalence i.e. $\equiv$]
$$\sim(p \rightarrow q) = p \wedge \sim q$$
$$\sim(p \leftrightarrow q) = (p \wedge \sim q) \vee (q \wedge \sim p)$$

All these properties can be proved using truth table. To prove any of these properties and other compound propositions like this, we simply find truth table for that proposition. If the truth table is a tautology then the given proposition is true and hence treated as proved. The proof of these properties is left as exercise to the reader.

We have used atomic propositions and different connectives to form compound propositions. There is a rule to form any compound proposition from atomic propositions. Any arbitrary combination may not be a valid proposition. Thus, any proposition to be a valid proposition, it must be a **well-formed formula (wff)**. Any proposition is said to be a well-formed formula if it is formed using the following rule

1. **<Atomic proposition>::= a|b|c|...|y|z**
2. **<Connectives>::= →|∧|∨|≡**
3. **<wff>::=<Atomic proposition>|~(<wff>)|(<wff>) <Connectives>(<wff>).**

The above rule is interpreted as

- An atomic proposition is a well-formed formula.
- Negation of a well-formed formula is a well-formed formula
- If p and q are any well-formed formula then $p \wedge q$, $p \vee q$, $p \rightarrow q$, and $p \equiv q$ are well-formed formula.

To determine that whether a given proposition is well-formed or not, we use Polish notations. First we convert the given proposition in infix form into **Polish prefix notation (operator operand (operand))** or into **Polish postfix notation (operand**

(operand) operator). Then we apply a partial and total sum rule to determine the validity of proposition. This is explained in the following examples. The following notations, as shown in the table 6.5.7, are used to denote different connectives in a polish notation of a proposition. The connectives have the precedence as shown in precedence column. Lower number means higher precedence. Parentheses are taken into account on top priority. Deeper is the parentheses, higher is its precedence.

Connective	Precedence	Polish Symbol
~	1	N
∧	2	K
∨	3	A
→	4	C
≡	5	E

Table 6.5.7

Example 3 Convert the proposition

(p→(q→r)) ≡(~(p∨r)∧~q)

into polish prefix notation.

Solution: Recall the concept you have studied in your fundamental courses of data structure for conversion of expression from infix to prefix, infix to postfix, prefix to postfix and so on. See the steps followed here.

Step 1 Here we first take the **(q→r)** and **(p∨r)** as they have higher precedence. Converting them to Polish notation, we have

(p→ Cqr) ≡ (~Apr ∧~q)

Step 2 Now we take the ~**Apr** and ~**q** as they have now higher precedence. Converting them to Polish notation, we have

(p→ Cqr) ≡ (NApr ∧Nq)

Step 3 Next we take **p→ Cqr** and **NApr ∧Nq** as they have now higher precedence. Converting them to Polish notation, we have

CpCqr ≡ KNAprNq

Step 4 Now, only connective left is ≡. Converting them, we have

ECpCqrKNAprNq

Ans.

Example 4 Convert the proposition

$$(\sim(p \equiv \sim(q \to r))) \equiv ((p \vee r) \vee \sim(\sim p \to \sim q)) \text{ and}$$
$$((p \vee \sim q) \to (p \wedge \sim r)) \equiv (\sim p \to (q \vee r))$$

into polish prefix and postfix notation.

Solution: It is left as an exercise to the reader.

Now, once we have a Polish prefix or Polish postfix notation we can apply the partial and total sum rule to determine the validity of the given expression. This is a three steps method.

Step 1 Write the expression in Polish prefix (postfix) notation in a line and in the next line assign a value **+1** to all **atomic proposition or constant**, **0** to all **unary connective** (there is only one unary connective ~) and **-1** to all **binary connectives.**

Step 2 Find the partial sum for all symbols (atomic proposition, constants and connectives) from **RHS** if the notation is in Polish prefix and from **LHS** if the notation is in Polish postfix.

Step 3 The sum corresponding to left most symbol is called a **total** sum and all other are called **partial sums**. If all the partial sums are **positive** and total sum is equal to **1**, then the given expression is valid otherwise the expression is not valid i.e. not a well-formed formula.

Example 5 Determine whether the proposition given below in Polish prefix notation is valid.

CECpEpNqCprNq

Solution: It is mentioned here that the notation is in prefix form. Even if it is not mentioned it is not difficult to know that whether the given expression is in prefix or postfix notation. (*Look at first symbol, if it is an operator the expression is in prefix, if the very last symbol is an operator the expression is in postfix otherwise the expression is in infix.*)

C	E	C	p	E	p	N	q	C	p	r	N	q
-1	-1	-1	+1	-1	+1	0	+1	-1	+1	+1	0	+1
1	2	3	4	3	4	3	3	2	3	2	1	1

Total Sum (arrow to the leftmost sum, 1) — Partial sums (arrow to the remaining sums)

Figure 6.5.1

In the figure 6.5.1, we have arranged the symbols of the given expression in first line. In the second line each symbol has been assigned a value according to the rule described in the step 2. Since the notation is in Polish prefix, we have calculated the partial sums from RHS. Thus, the partial sum for the first symbol from RHS is 1, for second symbol it is 1 (previous partial + its own assigned value = 1 + 0 =1), for third it is 2 and so on. The last symbol (left most side) has partial sum 1. **We call this partial sum as total sum**. From the table it is obvious that all the partial sums are positive and total sum is equal to 1. Therefore, the given expression is a valid well-formed formula.

Ans.

Example 6 Determine whether the proposition given below is valid.

KpACpqNqp

Solution: Look at first symbol, if it is an operator so the expression is in Polish prefix notation.

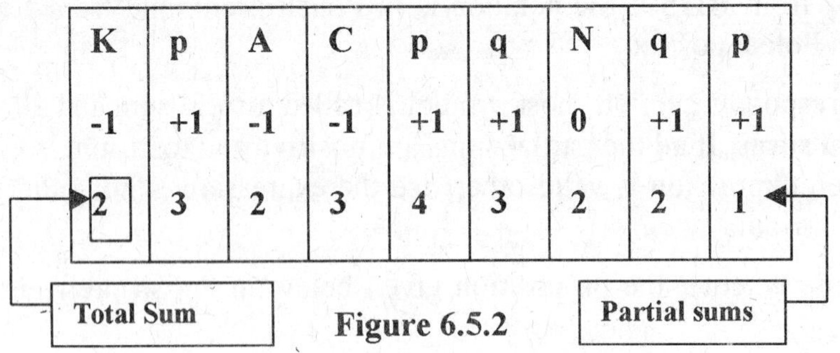

K p A C p q N q p
-1 +1 -1 -1 +1 +1 0 +1 +1
2 3 2 3 4 3 2 2 1
Total Sum
Partial sums

Figure 6.5.2

In the figure 6.5.2, we have arranged the symbols of the given expression in first line. In the second line each symbols have been assigned a value according to the rule described in the step 2. Since the notation is in Polish prefix, We have calculated the partial sums from RHS. Thus, the partial sum for the first symbol from RHS is 1, for second symbol it is 2 (previous partial + its own assigned value = 1 + 1 =1), for third it is 2 and so on. The total sum is 2. From the table it is obvious that all the partial sums are positive, but the total sum is equal to 2. Therefore, the given expression is **not** a valid well-formed formula.

Ans.

Example 7 Determine whether the proposition given below is valid.

pqNCNAprNA

Solution: Look at first symbol, if it is not an operator so the expression is not in Polish prefix notation. The last symbol is an operator so it is in Polish postfix notation.

In the figure 6.5.3, we have arranged the symbols of the given expression in first line. In the second line each symbols have been assigned a value according to the rule described in the step 2. Since the notation is in Polish postfix, We have calculated the partial sums from LHS. Thus, the partial sum for the first symbol from LHS is 1, for second symbol it is 2 (previous partial + its own assigned value = 1 + 1 =1), for third it is 2 and so on. The total sum is 1. From the table it is obvious that all the partial sums are **not** positive. One partial sum is **zero**. Though the total sum is equal to 1, the given expression is **not** a valid well-formed formula.

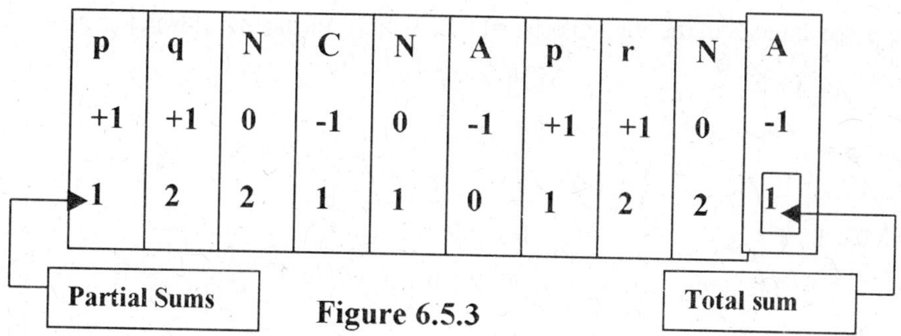

p q N C N A p r N A
+1 +1 0 -1 0 -1 +1 +1 0 -1
1 2 2 1 1 0 1 2 2 1
Partial Sums
Total sum

Figure 6.5.3

Ans.

It is worth mentioning, as a conclusion, that sets, lattices, finite Boolean algebra and propositions have many properties in common. These similarities have been outlined in this chapter to make the reader aware that how the concepts of hardware circuit design has evolved over the period of time. This shows how a basic mathematical theory can bring a revolutionary change in industry when supported by an appropriate technology e.g. integration techniques.

Exercise

1. Determine whether the relation R is a linear order on the set A.

(a) A = R –the set of real numbers, and aRb iff a is less than or equal to b ($a \leq b$).

(b) A = I –the set of all integers, and aRb iff a divides b^2.

(c) A = R –the set of real numbers, and aRb iff $a \geq b$.

(d) A = R × R, where R is the set of real numbers, and (a, b) R (a', b') iff $a \leq b$ and $b \leq b'$, where $\leq$ is usual partial order.

2. What can you say about the relation R on a set A if R is a partial order and an equivalence relation?

3. If (A, ≤) is a poset and A' is a subset of A, show that (A', ≤') is also a poset, where ≤' is the restriction of ≤ to A'.

4. Show that if R is a linear order on A then its inverse R^{-1} is also a linear order on A.

5. Let A = {x | x is a real number and -5 ≤ x ≤ 20}. Show that the usual relation < is a quasi-order.

6. Let A = {1, 2, 3, 5, 6, 10, 15, 30} and consider the partial order ≤ of divisibility on A. That is a ≤ b iff a divides b (a | b). Let A' = P(S), where S = {x, y, z}, bet the poset with partial order ⊆. Show that (A, ≤) and (A', ⊆) are isomorphic.

7. Show that if R is a quasi-order on A then its inverse R^{-1} is also a quasi-order on A.

8. Determine the Hasse diagram of the relation on A ={1, 2, 3, 4, 5}, the matrix of which is given below.

$$(a)\ \begin{bmatrix} 1 & 1 & 1 & 1 & 1 \\ 0 & 1 & 1 & 1 & 1 \\ 0 & 0 & 1 & 1 & 1 \\ 0 & 0 & 0 & 1 & 1 \\ 0 & 0 & 0 & 0 & 1 \end{bmatrix} \qquad (b)\ \begin{bmatrix} 1 & 0 & 1 & 1 & 1 \\ 0 & 1 & 1 & 1 & 1 \\ 0 & 0 & 1 & 1 & 1 \\ 0 & 0 & 0 & 1 & 0 \\ 0 & 0 & 0 & 0 & 1 \end{bmatrix}$$

9. Determine the matrix of the partial order relation of which Hasse diagram is given by the following figures.

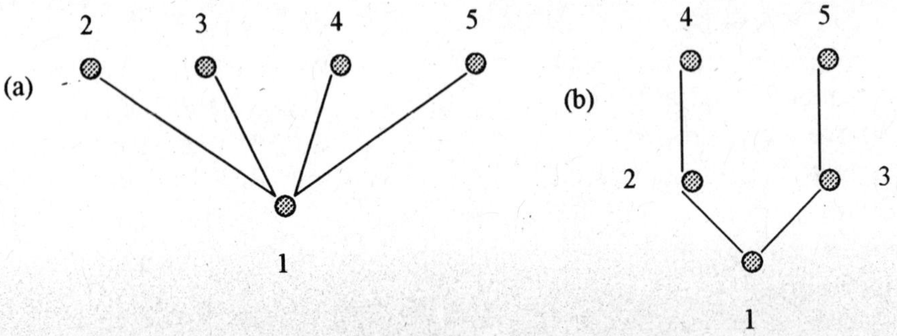

10. Let A = $I^+ \times I^+$ have lexicographic order. Mark each of the following as true or false.
(a) (2, 12) < (5, 3) (b) (3, 6) <(3, 24)
(c) (4, 8) < (4, 6) (d) (15, 92) < (12, 3)

11. Let R = {(a, a), (b, b), (c, c), (d, d), (e, e), (c, a), (c, b), (d, a), (d, b), (d, e), (b, a), (e, a)} be a relation defined on set A = {a, b, c, d, e}. Draw the Hasse diagram for R.

12. Let A = {1, 2, 3, 5, 6, 10, 15, 30} be a set and R be a relation of divisibility on A. Show that R is a partial order on A and draw a Hasse diagram of R. Is R a linear order relation? What about if A = {2, 4, 8, 16, 32}?

13. Find all the maximal and minimal elements of the poset shown by the following Hasse diagrams.

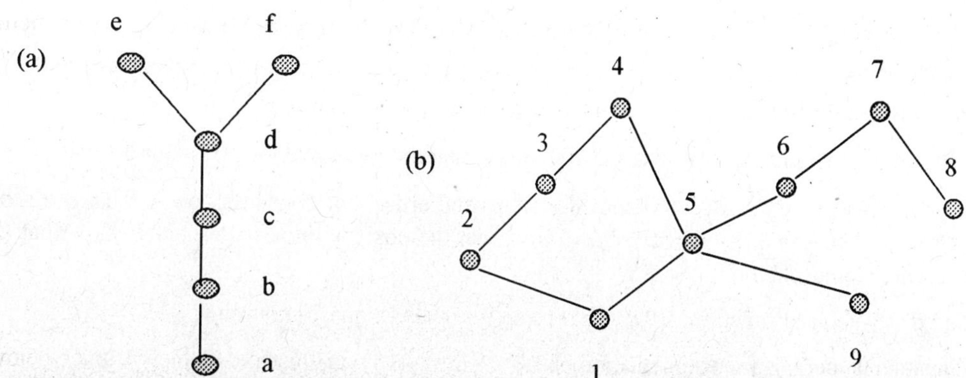

14. (a) Find all the maximal and minimal elements of the poset (A, ≤), where ≤ is the usual partial order and A is the set of all real numbers in the intervals
   (i) [0,1] (ii) [0, 1) (iii) (0, 1] (iv) (0, 1)
   (b) Find all the maximal and minimal elements of the poset (A, ≤), A = {2, 3, 4, 6, 8, 24, 48} and ≤ is defined as the partial order of divisibility.

15. Find the greatest and least element, if they exist, of the poset the Hasse diagram of which is given as:

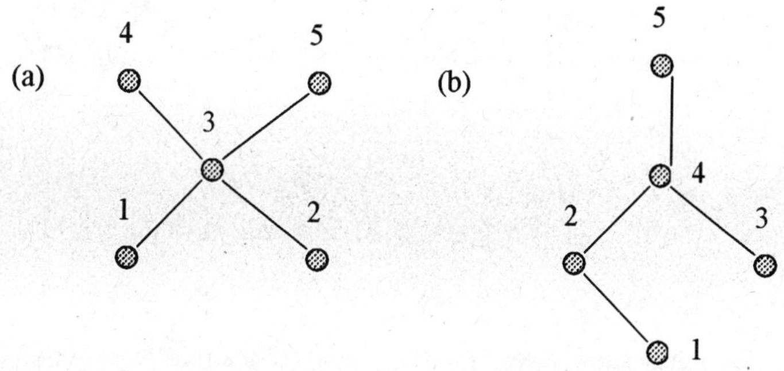

16. Find the greatest and least elements in poset given in exercise 14.

17. Find all the upper bounds and lower bounds of B = {x | x is a real number in [1, 2] } in the poset (A, ≤), where A = {x | x is any real number} and ≤ is usual partial order.

18. Construct the Hasse diagram of a topological sorting of the poset whose Hasse diagram is shown in 15 (a).

19. Let A = {2, 3, 4, 6, 8, 12, 24, 48} and ≤ denotes the relation of divisibility i.e., a ≤ b iff a divides b. Show that (A, ≤) is a poset and find all the lower bounds, upper bound, glb and lub, if any, of subset B = {4, 6, 12}.

20. Let A = R –the set of all real numbers, and ≤ is the usual partial order on A. Find all the lower bounds, upper bounds, glb and lub, if any, of subset B = {x | x is a real number in [1, 2)}.

21. Let A = R –the set of all real numbers, and $\leq$ is the usual partial order on A. Find all the lower bounds, upper bounds, glb and lub, if any, of subset $B = \{x \mid x \text{ is a real number in } (1, 2)\}$.

22. Let L and M are two lattices shown by the following Hasse diagram. Draw the Hasse diagram of $L \times M$ with product partial order.

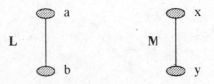

23. Let L = P(S) be the lattice of all subsets of a set S under the relation of set containment. If T is a subset of S then show that P(T) is a sub-lattice of L

24. Let S = {a, b, c} and L = P(S). Prove that $(L, \subseteq)$ is isomorphic to D_{42}.

25. Let S is a set. Let $\mathfrak{R}$ be the set of all equivalence relations on S and $\prod$ be the set of all equivalence partitions of S. We know that every equivalence relation induces a partition, and there is an equivalence relation corresponding to every equivalence partition, we can define a function f from $\mathfrak{R}$ to $\prod$ which is one to one and onto. We know that $(\mathfrak{R},\subseteq)$ is a poset. Answer the following:

(a) Show that $(\mathfrak{R},\subseteq)$ is a lattice.

(b) Let us define a relation $\leq$ on $\prod$ as for any P_1 and P_2 of $\prod$, $P_1 \leq P_2$ iff $R_1 \subseteq R_2$, where R_1 and R_2 are equivalence relations corresponding to partition P_1 and P_2 respectively. Show that $(\prod, \leq)$ is a lattice.

(c) Let $P_1 = \{A_1, A_2, \ldots\}$ and $P_2 = \{B_1, B_2 \ldots\}$ be two partitions of S. Show that $P_1 \leq P_2$ iff each A_m is contained in some B_1.

27. Let L be a bounded lattice with at least two elements. Show that no element of L is its own complement.

28. A lattice is said to be **modular** if, for all a, b and c, $a \leq c \Rightarrow a \vee (b \wedge c) = (a \vee b) \wedge c$. Show that a distributive lattice is modular. Further show that the lattice shown in the following Hasse diagram is a non-distributive lattice and is modular.

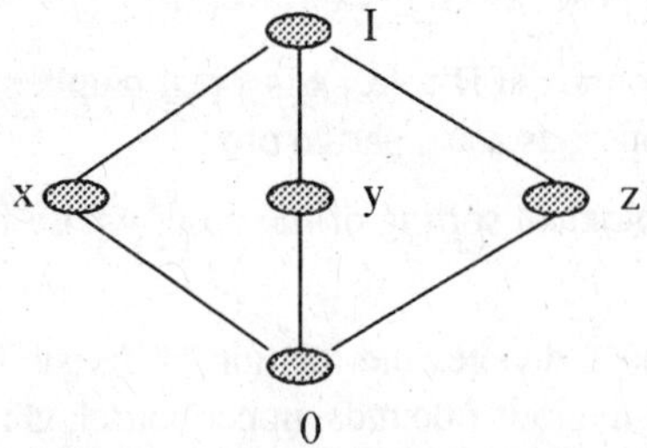

29. Prove that if a and b are elements of a bounded distributive lattice and if a has a complement a', then

(a) $a \vee (a' \wedge b) = a \vee b$ (b) $a \wedge (a' \vee b) = a \wedge b$

30. Show that if L_1 and L_2 are two distributive lattices, then $L = L_1 \times L_2$ is also distributive, where partial order in L is the product of partial orders in L_1 and L_2.

31. Find the complement of each element in D_{42}.

32. Find the complement of each element in D_{105}.

33. Show that set of all positive integers together with the usual partial order 'Less than or equal to" forms a distributive lattice.

34. Show that the lattice D_n is distributive for any n.

35. Let L be a lattice and let a and b be elements of L such that $a \le b$. The interval [a, b] is defined as the set of all $x \in L$ such that $a \le x \le b$. Prove that [a, b] is a sub-lattice of L.

36. Find all sub-lattices of D_{24} that contains at least five elements.

37. Show that a subset of a linearly ordered poset is a sub-lattice.

38. Let L is a distributive lattice. Show that if there exists an **a** with $a \wedge x = a \wedge y$ and $a \vee x = a \vee y$ then $x = y$.

39. Let a, b and c be elements of a lattice $(A, \le)$. Show that

(a) $a \wedge b = b$ if and only if $a \vee b = b$.

(b) If $a \le b$, then $a \vee (b \wedge c) \le b \wedge (a \vee c)$.

(c) $a \vee (b \wedge c) \le (a \vee b) \wedge (a \vee c)$.

40. Prove that a relation corresponding to a triangular matrix is anti-symmetric relation. Show by the method of counter example that the reverse of the above statement is not true.

41. Simplify the following Boolean expressions.

(a) $(a \wedge b) \vee (a \wedge b \wedge c) \vee (b \wedge c)$.

(b) $(a \wedge b) \vee (a' \wedge b \wedge c') \vee (b \wedge c)$.

(c) $((a \wedge b') \vee c) \wedge (a \vee b') \wedge c$.

42. Show that in a finite Boolean algebra the following statements are equivalent for any a and b.

(a) $a \vee b = b$ (b) $a \wedge b = a$ (c) $a' \vee b = I$ (d) $a \wedge b' = 0$ and (c) $a \le b$

43. Show that in a finite Boolean algebra, for any a, b and c $((\mathbf{a} \vee \mathbf{c}) \wedge (\mathbf{b}' \vee \mathbf{c}))' = (\mathbf{a}' \vee \mathbf{b}) \wedge \mathbf{c}'$.

44. Show that in a finite Boolean algebra, for any a, b and c, if $a \le b$, then $\mathbf{a} \vee (\mathbf{b} \wedge \mathbf{c}) = \mathbf{b} \wedge (\mathbf{a} \vee \mathbf{c})$.

45. Consider the Boolean polynomial $p(x, y, z) = (\mathbf{x} \vee \mathbf{y}) \wedge (\mathbf{z} \vee \mathbf{x}')$. If $B = \{0, 1\}$, compute the truth table of the function $f: B_3 \to B$ defined by p.

46. Consider the Boolean polynomial $p(x, y, z) = (\mathbf{x} \wedge \mathbf{y}) \vee (\mathbf{x}' \wedge (\mathbf{y} \wedge \mathbf{z}'))$. If $B = \{0, 1\}$, compute the truth table of the function $f: B_3 \to B$ defined by p.

47. Show that the following Boolean polynomials are equivalent

(a) $(x \vee y) \wedge (x' \vee y)$ and y (b) $x \wedge (y \vee (y' \wedge (y \vee y')))$ and x

(c) $(z' \vee x) \wedge ((x \wedge y) \vee z) \wedge (z' \vee y)$ and $x \wedge y$.

48. Reduce the following Boolean expressions, defined over the two-valued Boolean algebra, to Conjunctive Normal Form and Disjunctive Normal Form.

(a) $f(x, y, z) = (x \wedge y) \vee (x \wedge z) \vee (y' \wedge z)$

(b) $f(x, y, z) = \{(x \vee y)' \vee (x' \wedge z)\}'$

(c) $f(w, x, y, z) = (w \wedge x \wedge y') \vee (w \wedge x' \wedge z) \vee (x \wedge y' \wedge z')$

49. Using Karnaugh Map minimize the following Boolean expressions:

(a) $xyz + x'z + y'z'$

(b) $x (y'z + y(xz + y')) + x'(z + y (z + x'))$

(c) $w (x + y (z + x') + y') + w'x'y'z'$

(d) $wx'y + wx'z + w (y' + z) + w'x (y' + z')$

(e) $v'wxy + vw'xy + vwx'y + vwxy' + vxz + x'y'z + w'xy$.

50. Using Quine Mcluskey method minimize the following Boolean expressions:

(a) $x (y'z + y(xz + y')) + x'(z + y(z + x'))$

(b) $wx'y + wx'z + w(y' + z) + w'x(y' + z')$

(c) $v'wxy + vw'xy + vwx'y + vwxy' + vxz + x'y'z + w'xy$

(d) $u'xy' + v'wz' + u'xz' + v'wy + x'y'z + uvw$

(e) $u(v + w (x + y (z + x) + v) + w') + u'vx'z + u'wx'y' + v'wz'$.

51. Convert the following formula into Polish prefix and Polish postfix notation.

(a) $((p \vee \neg q) \rightarrow (p \wedge \neg r)) \equiv (\neg p \rightarrow (q \vee r))$.

(b) $\neg(\neg p \equiv \neg(q \rightarrow \neg r)) \wedge (p \rightarrow (q \vee r) \equiv (\neg r \vee \neg p))$.

(c) $((p \rightarrow \neg r) \wedge ((q \vee s) \equiv (p \wedge \neg s))) \rightarrow (p \vee \neg q \vee r \wedge s)$.

(d) $((p \vee \neg q \vee s) \wedge (q \vee r \vee \neg s)) \rightarrow (((\neg p \vee q) \rightarrow (\neg r \vee s)) \equiv (p \equiv s))$.

52. Test the formula to determine whether it is well–formed or not.

(a) ACpEpNqAKprCNrNEpq.

(b) pqrAqNpNsKCNpNCrpNqACE.

(c) EACpqrNpsNAKpqNsKACpNCqs.

(d) spEpNqAKNprCApNprsAKspCqKApKCE.

(e) pqKrNpAqsNrNKApNqAK.

(f) **CAsEpACqNpAKrNpqs.**

53. Convert the following formula to infix notation:

(a) **CAKEpqNrAKpNrNqp.**

(b) **KAEpEqrENpApqKANprNq.**

(c) **pqrKANpNqrEEprKApNCE.**

(d) **CpCqKNqNCAprKANpNAqrp.**

(e) **pNqANpqNprpKANACpEqAK.**

(f) **rqNANpCqNEprqNENAC.**

CHAPTER SEVEN

Graph Theory and Tree

The knowledge of graph theory has found immense application in various fields of study today. Computer network is an important example to cite here. We may take the requirements usually imposed on computer networks that they must be reliable, even in the face of failure of some nodes and communication lines. To achieve a fare degree of reliability with known unreliable components, the network must be redundant. A sufficiently redundant network can lose a small number of components (nodes or communication line or both) and still may function properly, albeit with lower performance. A systematic method is needed to analyze the degree of redundancy in a computer network. The graph theory plays an important role here. Other areas of application of graph theory are Operation Research, Knowledge representation in Artificial Intelligence, Algorithm representation in programming, Data Flow Diagram in System Requirement Studies and so on.

The theory of graph basically deals with the analysis of structures called ***graph***. These are not the graph of a functions e.g. graph of sin(x), cos(x) etc, but rather are structures such as the edges and vertices of a polyhedron, or the flowchart for a program. It implies that a graph of our concern, here, is made up of points joined by line segments.

A graph may be classified *as **directed, undirected, finite, infinite, labeled, unlabeled, weighted, unweighted, cyclic, acyclic, bipartite, complete, linear*** and so on. A ***tree*** is a special type of graph with some special characteristics. The concept of tree is used in many searching and sorting algorithms as well as in parsing of sentences in a language. All these algorithms are basically based on traversal of tree. We shall be discussing about this in the latter half of this chapter.

7.1 Graphs Introduction

According to the terminology of the theory of graphs, a 'graph' consists of a set of junction points called '***nodes***' (or ***vertices***) with certain pairs of the nodes joined by lines called '***arcs***' (or ***branches***, or ***links***, or ***edges*** or ***incident***). The pair of nodes joined by an arc is called ***end points*** of the arc. Thus, figure 7.1.1(a) and 7.1.1(b) are examples of graphs, where the circles are the nodes and the lines connecting them are the arcs.

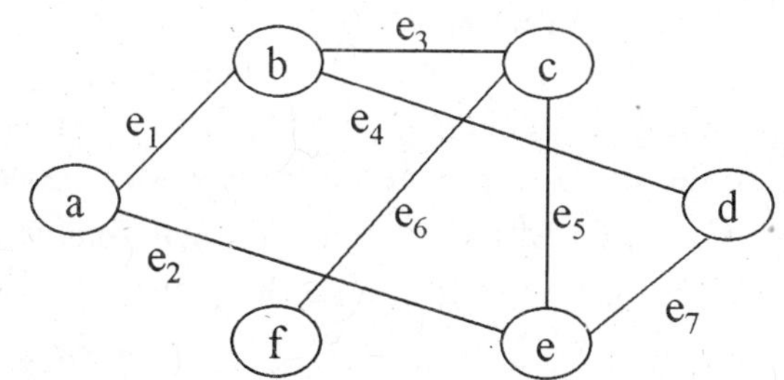

Figure 7.1.1(a)

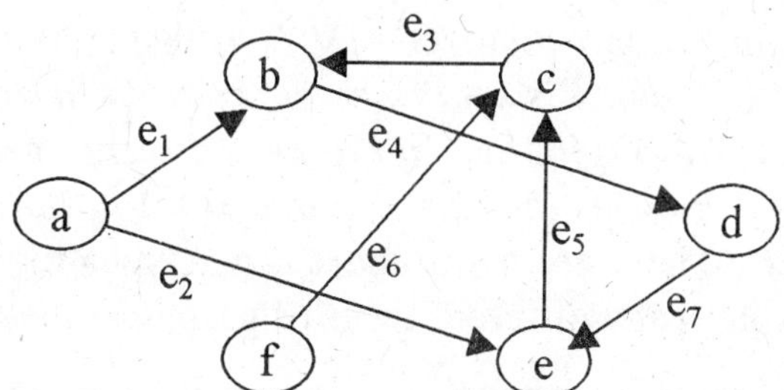

Figure 7.1.1(b)

All arcs in the graph of the figure 7.1.1(a) are *undirected* whereas those are directed in the graph of the figure 7.1.1(b). Let V be the set of all nodes and E be the set of all edges in the graph shown in the figure 7.1.1(a). Then, we have

$$V = \{a, b, c, d, e, f\} \text{ and } E = \{\mathbf{e_1}, \mathbf{e_2}, \mathbf{e_3}, \mathbf{e_4}, \mathbf{e_5}, \mathbf{e_6}, \mathbf{e_7}\}$$

Let us define a function γ: $E \rightarrow V \times V$ as:

$f(\mathbf{e_1})= (a, b)$ or (b, a), $f(\mathbf{e_2}) = (a, e)$ or (e, a), $f(\mathbf{e_3}) = (b, c)$ or (c, b),
$f(\mathbf{e_4}) = (b, d)$ or (d, b), $f(\mathbf{e_5}) = (c, e)$ or (e, c), $f(\mathbf{e_6}) = (c, f)$ or (f, c) and
$f(\mathbf{e_7}) = (d, e)$ or (e, d).

Then the graph of the figure 7.1.1(a) can be thought of as a mathematical structure (V, E, γ). In fact a graph is defined as

"A mathematical structure (V, E, γ), where V is a nonempty set of vertices, E is a set of edges and γ is a function from set E to $V \times V$ i.e. it returns an ordered pair from $V \times V$ for an edge in E."

In an undirected graph an edge between two nodes say, x and y, may be represented either by an ordered pair (x, y) or by an ordered pair (y, x). However, in the case of a directed graph, if the direction of edge is from x to y then the edge is represented by (x, y) whereas an ordered pair (y, x) denotes the directed arc from y to x. A directed graph may have both the arcs at the same time between a pair of nodes x and y. Therefore, in a directed graph, arcs (x, y) and (y, x) are distinct and called ***anti-parallel*** arcs whereas in an undirected graph both are treated as equivalent. In other words, in an undirected graph G = (V, E, γ), E is a set of unordered pairs whereas in a directed graph, E is a set of ordered pairs. If (V, E, γ) be the graph of the figure 7.1.1(b) then γ can be defined as:

$f(\mathbf{e_1}) = (a, b)$, $f(\mathbf{e_2}) = (a, e)$, $f(\mathbf{e_3}) = (c, b)$, $f(\mathbf{e_4}) = (b, d)$,
$f(\mathbf{e_5}) = (e, c)$, $f(\mathbf{e_6}) = (f, c)$ and $f(\mathbf{e_7}) = (d, e)$

It is the set E and the function γ of any graph G = (V, E, γ) which determines what type of graph we are dealing with. *A graph G is said to be an* ***undirected graph*** *if all the edges in G are undirected.* Similarly, *a graph G is said to be a* ***directed graph*** *if all the edges in G are directed. If some of the edges are directed and some are undirected then the graph is called a* ***mixed graph***. There are many more other types of graph. Let us now discuss some basic definitions here and different types of graphs in the next section.

In a directed graph the number of incoming arcs on a node is called ***indegree*** of the node and number of outgoing arcs is called ***outdegree*** of the node. The sum of indegree and outdegree of a node is called ***degree*** or ***valence*** of the node. A directed graph is said to be an ***isograph*** if indegree of every node is equal to its outdegree. It implies that an isograph is always traversable and it plays important role in Information flow network design. In a directed graph, a node with indegree zero and outdegree ≥ 1 is called a ***source node.*** Similarly, a node with outdegree zero and indegree ≥ 1 is called ***sink node***.

In an undirected graph, degree of a node is defined as the number of arcs arriving at the node. The highest degree of all nodes is called ***degree the graph***. A graph is said to be a ***regular graph*** if all nodes of the graph have equal degree. For a regular graph, degree of graph is equal to the degree of any of the node in the graph. If the degree of a regular graph is k, it is also called ***k-regular*** graph. In the graph of figure 7.1.1(a), the degree of nodes a, b, c, d, e and f are 2, 3, 3, 2, 3 and 1 respectively. Degree of the graph is 3. It is not a regular graph.

Similarly in the graph of figure 7.1.1(b), the indegree of nodes a, b, c, d, e and f are 0, 2, 2, 1, 2 and 0 respectively and outdegree are 2, 1, 1, 1, 1 and 1 respectively. Therefore, degrees of nodes in graph of 7.1.1(b) are 2, 3, 3, 2, 3 and 1 respectively. Here too, the degree of the graph is 3 and it is not regular.

In both the cases, graph has 7 edge and sum of degrees of all nodes is 14, twice of the number of edges in the graph. It is not by chance that it has happened. It is always true as proved in the following theorems.

Theorem 1 Sum of degrees of all nodes in a graph is equal to twice the number of edges in the graph.

Proof: Let G = (V, E, γ) be a graph. Let |E| = e. Since each edge add degree one to two nodes of G, sum of degrees of all nodes ∈ V is equal to the 2*e.

Proved.

Theorem 2 No regular graph of odd degree exists with an odd number of nodes.

Proof: Let G = (V, E, γ) be a k-regular graph. Let us assume that there are n nodes in G. Since G is k-regular, all n nodes must have degree k. Thus the sum of degrees of all nodes in G is n×k. If both n and k are odd then n×k is also an odd number. Therefore, number of edge in G is n×k/2, a fractional number, which cannot be a count of edges in a graph. Thus both n and k cannot be odd.

Proved.

In a graph G, a node having degree zero is called ***isolated node***. A node is called a ***dangle node*** if its degree is 1. Two nodes are said to be ***adjacent*** if there is an arc between them. An arc is called a ***loop*** if its starting and end vertex is the same. In the graph of figure 7.1.1(a) and (b), node f is a dangle node. There is no isolated node in these graphs. Nodes a and b are adjacent nodes where as a and f are not adjacent. Find other adjacent nodes in these graphs.

In the graph 7.1.2(a) node E is an isolated node. There is no dangle node in the graph. Node B is adjacent to nodes A, C and D, A is adjacent to B and C. Node D is adjacent to B and C. C is then adjacent to A, B and D. E is isolated so it cannot be adjacent to any of the node.

In the graph of the figure 7.1.2(b), E is an isolated node and D is a dangle node. Reader may determine which node is adjacent to which node.

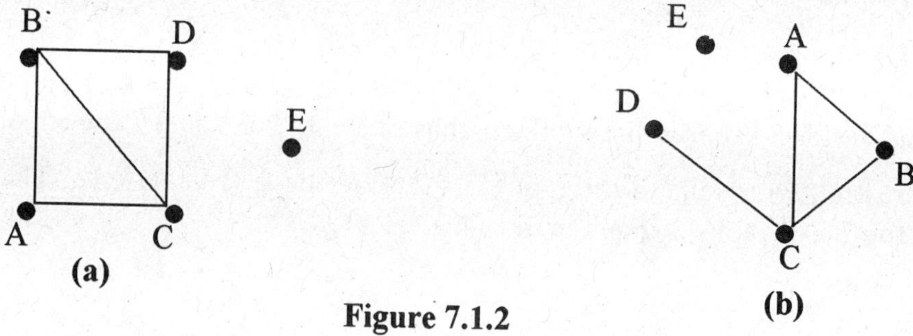

Figure 7.1.2

In the graphs of figure 7.1.2, we have used a thick dot to denote a node and its name just around it. We will be using this notation as well as the notation used in graph of 7.1.1, which is a hollow circle and name inside it, to denote a node in a graph. A graph G = (V, E, γ) is said to be an ***infinite graph*** if either V or E or both are infinite set. However, in real life application, we always deal with a finite graph. A graph G is said to be ***finite*** if both V and E are finite set. A finite graph can be represented in many ways. The most commonly used representations for graph manipulation through any programming language are list representation and matrix representation.

List Representation of Graph

We have already used an edge list representation of graph G = (V, E, γ). Where? The set E is just list of edges in a graph. If we have this list we can always draw a graph. The other list representation is *vertex adjacency* list. In this representation, a list of vertices adjacent to each vertex is maintained. For example, the graph of figure 7.1.1(a) can be represented using vertex adjacency list as show in table 7.1.1.

Vertex	Adjacent Vertices
a	b, e
b	a, c, d
c	b, e, f
d	b, e
e	a, c, d
f	c

Table 7.1.1

Example 1 Find vertex adjacency list for the graphs of figure 7.1.2.

Solution: There are two graphs in the figure 7.1.2. Both have five vertices. The vertex adjacency list for these graphs is given in the table 7.1.2.

Graph 7.1.2(a)		Graph 7.1.2(b)	
Vertex	Adjacent Vertices	Vertex	Adjacent Vertices
A	B, C	A	B, C
B	A, C, D	B	A, C
C	A, B, D	C	A, B, D
D	B, C	D	C
E	Nil	E	Nil

Table 7.1.2

Ans.

We have shown how an undirected graph can be represented using list of adjacent vertices. In the case of a directed graph, a care has to be taken, according to the direction of an edge, while placing a node in the adjacent vertex list of other node. For example, see the vertex adjacency list in the table 7.1.3 for the directed graph 7.1.1(b).

Vertex	Adjacent Vertex
a	b, e
b	d
c	b
d	e
e	c
f	c

Table 7.1.3

Matrix Representation of Graph

Many matrix representations for a graph are possible. But in practice, only two forms of representation are used: ***adjacency matrix*** and ***incidence matrix***. Let G = (V, E, γ) be a graph, with no loops or multiple edges. Let |V| = m and |E| = n. An incidence matrix for G is an m×n matrix $M_I = (\mathbf{a}_{ij})$, such that

$$\mathbf{a_{ij}} = \begin{cases} 1 & \text{if } v_i \text{ is an incident to } e_j \\ 0 & \text{otherwise} \end{cases}$$

Where $v_1, v_2, v_3, \ldots, v_m$ are m vertices and $e_1, e_2, e_3, \ldots, e_n$ are n incidents of G. An incidence matrix is also called *edge incidence matrix*. Similarly, an adjacency matrix, also called *vertex adjacency matrix*, for G is an m×m matrix $M_A = (\mathbf{a_{ij}})$, such that

$$\mathbf{a_{ij}} = \begin{cases} 1 & \text{if } (v_i, v_j) \text{ is an edge of } G \\ 0 & \text{otherwise} \end{cases}$$

Example 2 Show the adjacency matrix and incidence matrix representation for the graphs of figure 7.1.1(a).

Solution: The given graph has 6 vertices and 7 edges. The incidence matrix M_I for this graph is of order 6×7 and is given in figure 7.1.3(a).

$$M_I = \begin{array}{c} \\ a \\ b \\ c \\ d \\ e \\ f \end{array} \begin{array}{c} \begin{array}{ccccccc} e_1 & e_2 & e_3 & e_4 & e_5 & e_6 & e_7 \end{array} \\ \begin{bmatrix} 1 & 1 & 0 & 0 & 0 & 0 & 0 \\ 1 & 0 & 1 & 1 & 0 & 0 & 0 \\ 0 & 0 & 1 & 0 & 1 & 1 & 0 \\ 0 & 0 & 0 & 1 & 0 & 0 & 1 \\ 0 & 1 & 0 & 0 & 1 & 0 & 1 \\ 0 & 0 & 0 & 0 & 0 & 1 & 0 \end{bmatrix} \end{array}$$

Figure 7.1.3(a)

Similarly, the adjacency matrix M_A for this graph is of order 6×6 and is given in figure 7.1.3(b).

$$M_A = \begin{array}{c} \\ a \\ b \\ c \\ d \\ e \\ f \end{array} \begin{array}{c} \begin{array}{cccccc} a & b & c & d & e & f \end{array} \\ \begin{bmatrix} 0 & 1 & 0 & 0 & 1 & 0 \\ 1 & 0 & 1 & 1 & 0 & 0 \\ 0 & 1 & 0 & 0 & 1 & 1 \\ 0 & 1 & 0 & 0 & 1 & 0 \\ 1 & 0 & 1 & 1 & 0 & 0 \\ 0 & 0 & 1 & 0 & 0 & 0 \end{bmatrix} \end{array}$$

Figure 7.1.3(b)

Ans.

The representation of graph plays an important role while implementing any graph algorithms using a programming language. Readers having basic knowledge of data structure may realize the importance. To represent any graph in the form just discussed, we have required data structures available in many programming languages.

7.2 Different Types of Graphs

An undirected graph G = (V, E, γ) is said to be a ***simple graph*** if γ is one to one function from E to V×V i.e. there is either no or one arc between a pair of nodes in G. If γ is many to one mapping from E to V×V, then the graph is called a ***multi graph***. In a *multi graph, set E is a multi set*. As an example, the graph of the figure 7.1.1(a) is an *undirected simple graph* whereas the graph of the figure 7.2.1 is an *undirected multi graph*. There are two arcs between node b and c i.e. γ returns the same ordered pair (b, c) for the edges e_2 and e_3.

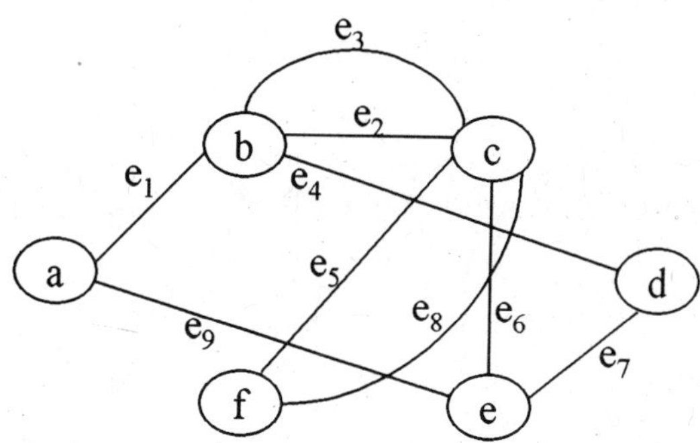

Figure 7.2.1

Now let us consider the directed graph of the figure 7.2.2. In this graph, there are two arcs: e_2, e_3, between nodes b and c, and another two arcs: e_6, e_9, between nodes a and e. The two arcs e_2 and e_3 have different directions whereas e_6 and e_9 have same direction. Multiple arcs between a pair of nodes are said to be ***parallel*** if they have same direction or said to be ***anti parallel*** if they have opposite direction. A directed graph is said to be *multi graph* if it has some parallel multiple arcs. The graph shown in the figure 7.2.2 is an example of a directed *multi graph* because of the presence of arcs e_6 and e_9.

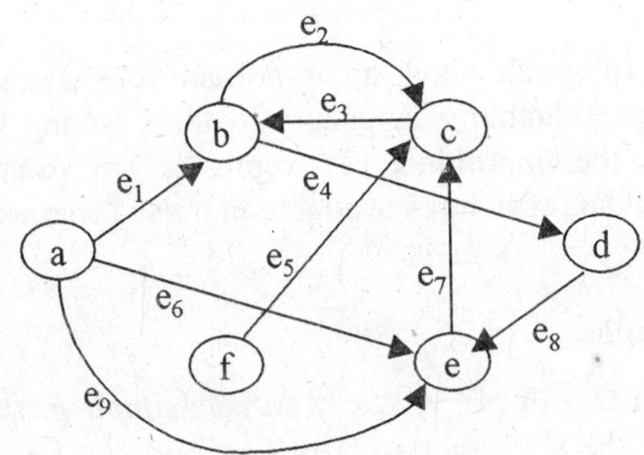

Figure 7.2.2

In this text unless and otherwise stated by a graph we mean a simple graph. Redundant arc has its own importance in the application of graph theory. Wherever required, we shall be referring it by the qualifying the graph with the word *'multi'*. A graph having one or more loops is called a ***pseudo graph.*** A graph of single node and no edge is called a ***trivial graph***.

Discrete Graph

A graph G = (V, E, γ) is said to be a ***discrete graph*** if $|E| = \phi$ i.e. there is no arc in G. All the nodes are isolated in a discrete graph. The name D_n is used to represent a discrete graph, where n is the number of nodes in G. Graphs D_1, D_2, D_3 and D_4 are shown in the figure 7.2.3. It is obvious that a discrete graph is always 0-regular graph. D_1 is a trivial graph.

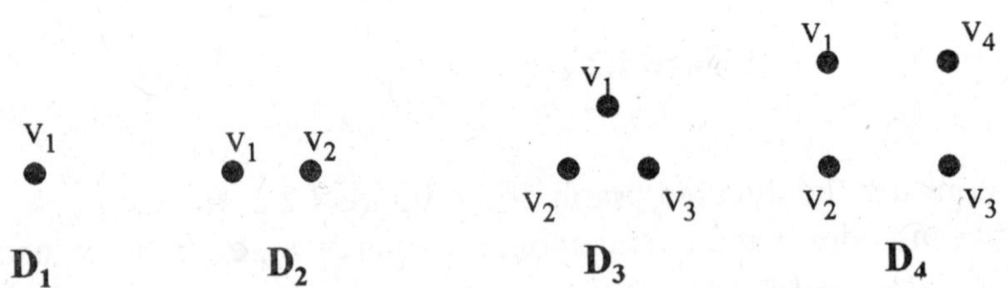

Figure 7.2.3

Linear Graph

Let G = (V, E, γ) be a graph. If it is possible to arrange the nodes of G in sequence, say $v_1, v_2, v_3, \ldots, v_n$, such that (v_i, v_j) is an edge for $1 \le i < j \le n$, then G is called a ***Linear graph***. Obviously, $|E| = |V| - 1$. The first and last nodes in $L_{n\geq 2}$ are dangle nodes and all other nodes have degree 2. The symbol L_n is used to represent a linear graph of n vertices. Graphs L_1, L_2, L_3 and L_4 are shown in the figure 7.2.4. A word of caution for the readers; in many books and papers linear graph and simple graph are used synonymously. A Simple graph G can be called a linear graph only if G can be stretched to form a line without any edge and node overlapping.

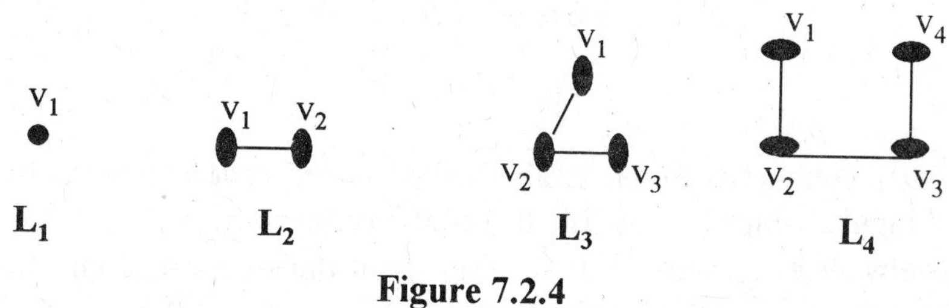

Figure 7.2.4

A linear graph L_n becomes ***n-cycle*** if $n \geq 3$ and arc (v_n, v_1) is present in the graph. An n-cycle graph is denoted as C_n. Graphs C_3, C_5 and C_6 are shown in the figure 7.2.5.

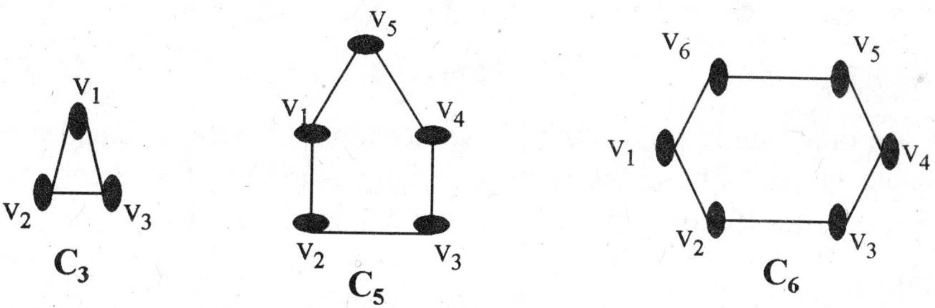

Figure 7.2.5

Complete Graph

A graph $G = (V, E, \gamma)$ is said to be a complete graph if every node v is adjacent to every other node w. We use the notation K_n to represent a complete graph of n nodes. If $|V| = n$ then $|E| = n(n-1)/2$. A complete graph is a regular graph of degree $(n-1)$, where n is the number of nodes in the graph. Graphs K_1, K_2, K_3 and K_4 are shown in the figure 7.2.6.

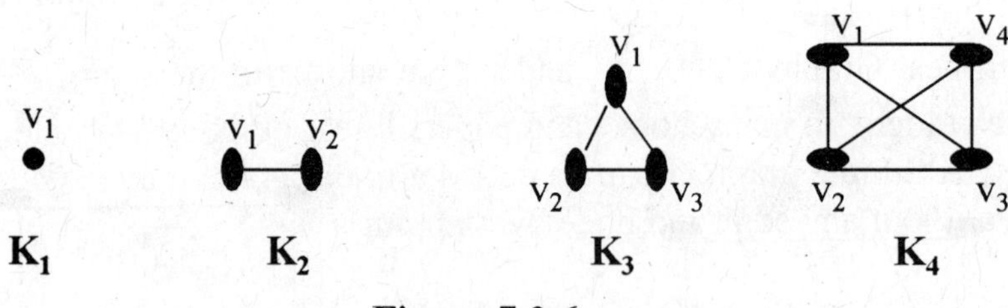

Figure 7.2.6

Cubic Graph

A 3-regular connected graph is also called ***cubic graph***. It constitutes the first nontrivial set of regular graphs. From the theorem 2 of the section 7.1, it is obvious that a cubic graph has always even number of vertices and it must be ≥ 4. Some cubic graph on 4, 6 and 8 vertices are given below in the figure 7.2.7.

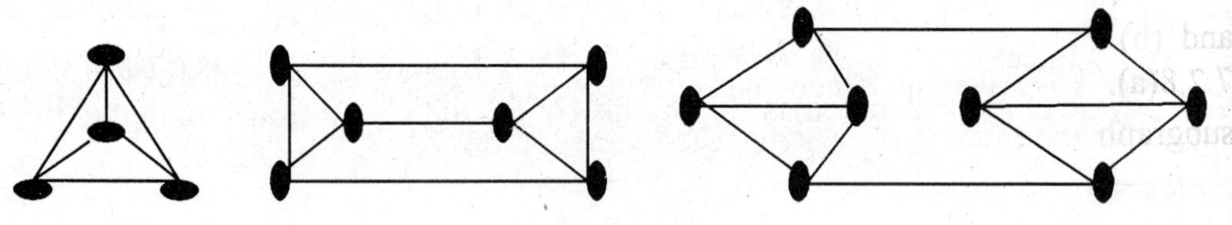

Figure 7.2.7

There are other cubic graphs with 6 vertices and 8 vertices. Reader may try to draw those cubic graphs. The cubic graphs of large number of vertices and Regular graphs of higher degrees are too complex a topic to be discussed here.

N Cube Graph

We have seen in the chapter six that how to draw Hasse diagram for a finite Boolean algebra B_n. There are 2^n vertices and each vertex represents an n-bit string. Two vertices are connected if the corresponding bit strings differ at exactly one bit position. See the figure 6.3.1, which contains the Hasse diagrams for B_1, B_2 and B_3. These diagrams are also called ***N Cube graph*** and denoted as Q_n. A lot about this type of graph has already been discussed in the chapter six.

Sub-graph

Let $G = (V, E, \gamma)$ be a graph. A graph $H = (V_1, E_1, \gamma)$ is said to be a subgraph of G if $E_1 \subseteq E$, $V_1 \subseteq V$ and V_1 contains all the end points of edges in E_1. A subgraph H of a graph G is called a ***spanning subgraph*** if H contains all vertices of G i.e. $|V| = |V_1|$. If H contains all edges of G that joins vertices in V_1 then H is called an ***induced subgraph*** of G.

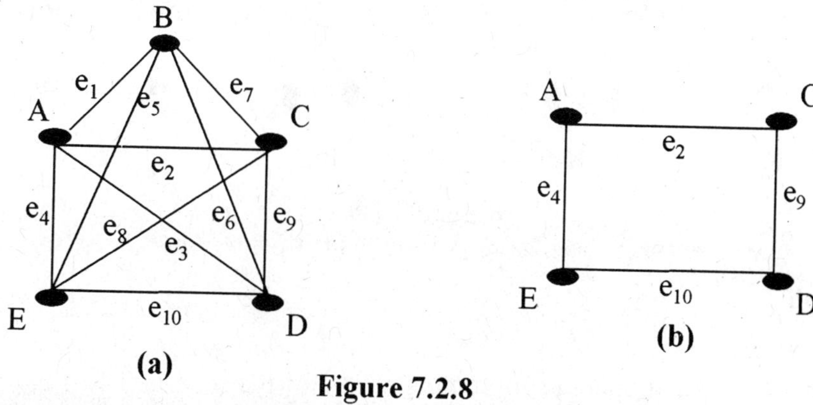

Figure 7.2.8

In the graphs of figure 7.2.8, graph (b) is a subgraph of graph (a). The graph (a) and (b) of figure 7.2.9 are spanning subgraph and induced subgraph respectively of 7.2.8(a). To understand these definitions, reader may form set V_1 and set E_1 for each subgraph and then verify whether the figure conforms to the definition.

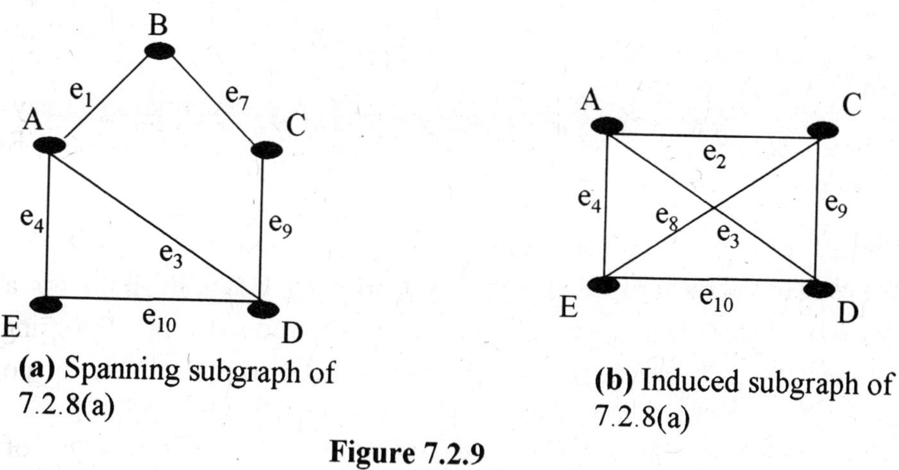

(a) Spanning subgraph of 7.2.8(a)

(b) Induced subgraph of 7.2.8(a)

Figure 7.2.9

For a given graph G, there can be many subgraphs. Let $|V| = m$ and $|E| = n$. The otal non-empty subsets of V is $2^m - 1$ and total subsets of E is 2^n. Thus, maximum umber of subgraphs is equal to $(2^m - 1) \times 2^n$. The number of spanning subgraphs is qual to 2^n because all vertices are to be included in a spanning subgraph.

Example 1 Find the total number of subgraphs and spanning subgraphs in K_6, L_5 and Q_3.

Solution: In graph K_6, we have $|V| = 6$ and $|E| = 15$. Thus, maximum number of subgraph is

$$(2^6 - 1) \times 2^{15} = 63 \times 32768 = 2064384$$

And the total number of spanning subgraphs is

$$2^{15} = 32768$$

In the linear graph L_5, we have $|V| = 5$ and $|E| = 4$. Thus, maximum number of subgraph is

$$(2^5 - 1) \times 2^4 = 31 \times 16 = 496$$

And the total number of spanning subgraphs is

$$2^4 = 16$$

In the 3-Cube graph Q_3, we have $|V| = 8$ and $|E| = 12$. Thus, maximum number of subgraph is

$$(2^8 - 1) \times 2^{12} = 127 \times 4096 = 520192$$

And the total number of spanning subgraphs is

$$2^{12} = 4096$$

Ans.

Complement of a Graph

Let H be a subgraph of a graph G. A ***complement of H*** with respect to G, denoted as G – H, is a graph that contains all the edges of G that are not in H and vertices of G – H are end points of the edges in G – H. In particular, if number of vertices in G is n, the complement of graph G, $\bar{G}$ is the complement of G with respect to K_n. The graph of the figure 8.2.10 is the complement of subgraph 8.2.8(b) with respect to the graph 8.2.8(a).

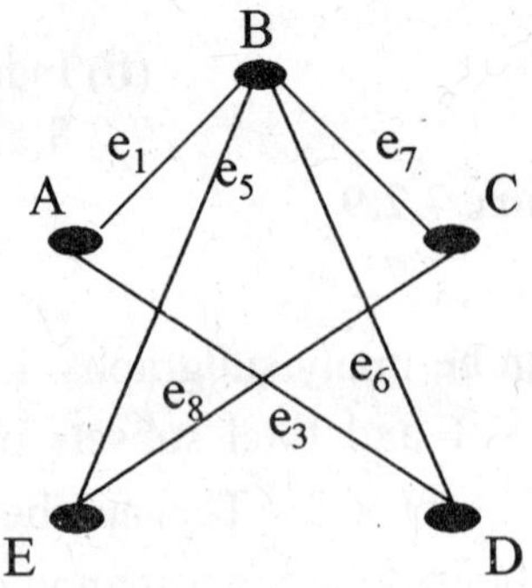

Figure 8.2.10

Quotient Graph

Let G = (V, E, γ) be a graph and R be an equivalence relation defined on set V. Then set $V/_R$ is the set of equivalence classes of V by R. Let $E/_R$ be the set of edges ($[v_i]$, $[v_j]$) for some $[v_i]$, $[v_j]$ in $V/_R$, such that there is an edge between a vertex of class $[v_i]$ to a vertex of equivalence class $[v_j]$ in the graph G. If γ^R be a function from $E/_R$ to $V/_R \times V/_R$ then the graph $G_Q = (V/_R, E/_R, \gamma^R)$ is called a ***quotient graph*** of G by R. To make the thing comprehensible to you, let us consider an example.

Let us take the graph G of figure 7.2.8(a). Let R be an equivalence relation defined on set V of G such that it gives the partition {{A, C}, {B, E}, {D}} i.e.

$$V/_R = \{\{A, C\}, \{B, E\}, \{D\}\}=\{[A], [B], [D]\}$$

Now there will be an edge from [A] to [B] because there is an age from node A to node B in G. Similarly ([A], [D]) and ([B], [D]) will be edge in the quotient graph. Let $[v_i]$ contains m vertices and $[v_j]$ contains n vertices. Possibly there can be $x = m \times n$ non-loop edges between vertices of $[v_i]$ and $[v_j]$. Whenever $x \geq 1$, there will be an edge between $[v_i]$ and $[v_j]$. The graph of the figure 7.2.11 is the quotient graph $G_Q = (V/_R, E/_R, \gamma^R)$ of G obtained by the relation R.

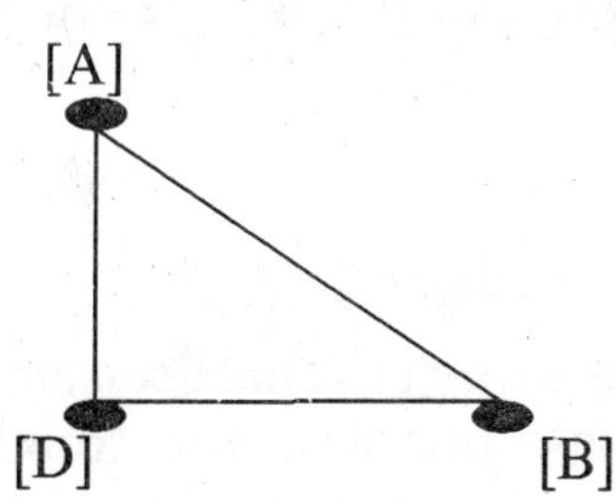

Figure 7.2.11

Bipartite Graph

A graph G = (V, E, γ) is called a ***bipartite graph*** if vertex set V can be partitioned into two non empty disjoint subsets V_1 and V_2 in such a way that every edge in E joins a vertex in V_1 to a vertex in V_2. That means a node in V_1 (or in V_2) is not adjacent to another node in V_1 (or in V_2) but some nodes in V_1 are adjacent to some node in V_2.

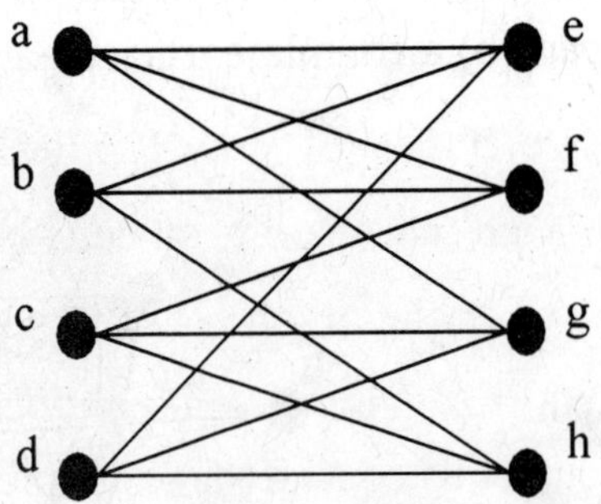

Figure 7.2.12

Incidentally, the bipartite graph of the figure 7.2.12 is also a cubic graph. A bipartite graph G with V_1 and V_2 as above is called a ***complete bipartite graph*** if every vertex in V_1 is adjacent to every vertex in V_2. If $|V_1| = r$ and $|V_2| = s$, then the complete bipartite graph is written as $K_{r, s}$. The $K_{3, 4}$ graph is shown in the figure 7.2.13.

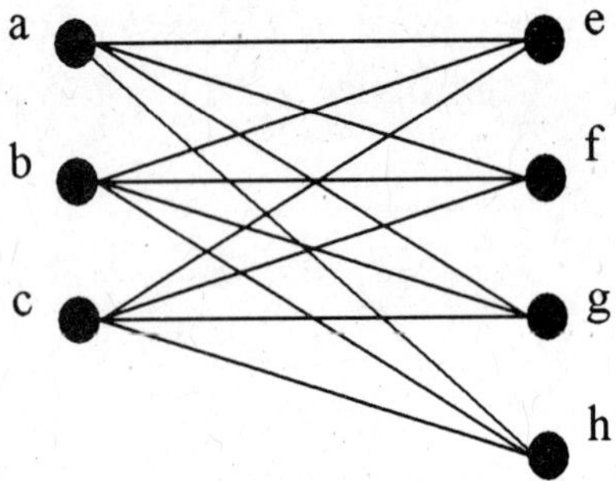

Figure 7.2.13

It is obvious that a bipartite graph cannot contain a triangle as its subgraph. Out of the three nodes, to form a triangle, two must come from one of the subset V_1 or V_2. Since these two nodes are not adjacent, a triangle cannot form. Other observation is that if $K_{r, s}$ is regular, then r and s must be equal.

Tripartite Graph

The definition of a tripartite graph is an extension of the definition of a bipartite graph. A graph $G = (V, E, \gamma)$ is called a ***tripartite graph*** if the vertex set V can be divided into three non empty disjoint subsets V_1, V_2 and V_3 so that vertices in the same subset are not adjacent. A tripartite graph is called a ***complete tripartite graph*** if every pair of vertices that are not in the same subset is adjacent. We write a complete tripartite graph as $K_{r, s, t}$, where $|V_1| = r$, $|V_2| = s$ and $|V_3| = t$. In the figure 7.2.14, a tripartite of seven vertices is shown in part (a) and in part (b) a complete tripartite graph $K_{2, 3, 4}$ is shown.

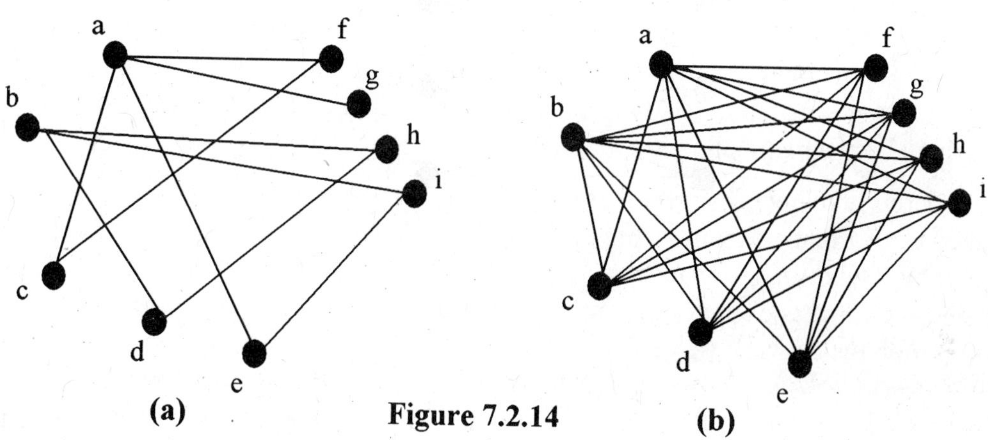

Figure 7.2.14

It is possible that that a tripartite graph may contain a triangle. It is easy to determine that the number of edges in $K_{r, s, t}$ is equal to $r\times(s + t) + s\times t$. If a complete tripartite graph $K_{r, s, t}$ is regular, then r, s and t must be equal.

Isomorphic Graph

Two graphs $G_1 = (V_1, E_1, \gamma_1)$ and $G_2 = (V_2, E_2, \gamma_2)$ are said to be isomorphic graphs iff $\exists$ functions $f: V_1 \rightarrow V_2$ and $g: E_1 \rightarrow E_2$ such that both f and g are bijection and the incidences are preserved i.e. if there is an edge **e** between nodes u and v of V_1 then g(**e**) should be an edge between f(u) and f(v) of V_2. The two graphs of the figure 7.2.15 are isomorphic graphs.

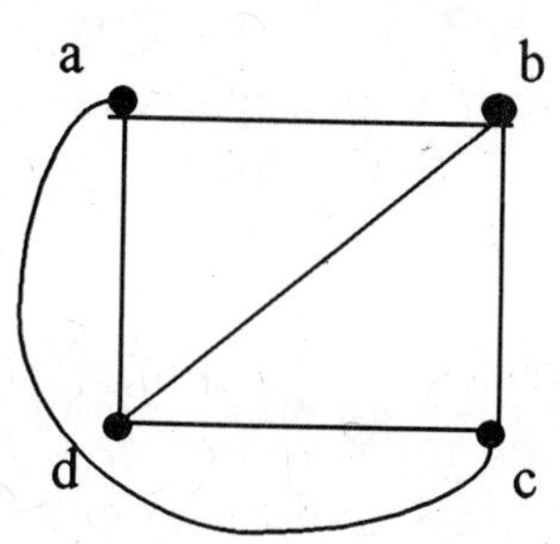

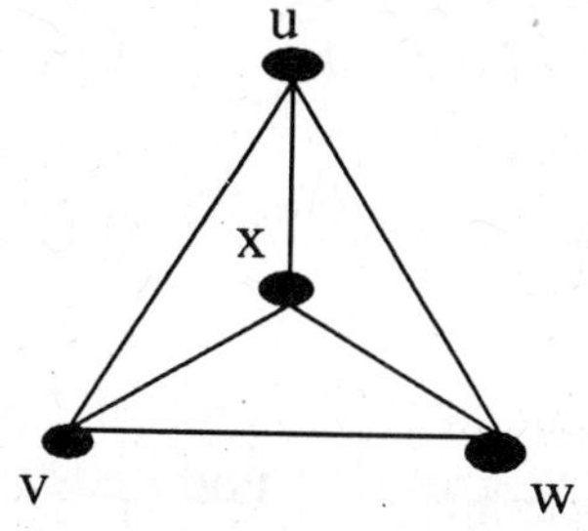

Figure 7.2.15

Testing of graph isomorphism, in most of the time, is quite simple contrary to the theoretical difficulties involved. It is practically not possible to check for all possible functions f: $V_1 \rightarrow V_2$. For example, if there are 20 vertices in a graph then there are 20! such functions. The easiest way is to look for the fixed parameters like number of nodes, number of edges and ***degree spectrum*** of graphs. A degree spectrum is defined as the list of degrees of all nodes in a graph. Two graphs cannot be isomorphic if any of these parameters are not matching. However, it cannot be said that two graphs will always be isomorphic if these parameters are matching.

There are many other types of graph. One of the important types of graph that has not got enough attention from students of discrete mathematics is ***hypergraph***. As the use of hypertext, hyperlinks and others related hyper-things in the web page designing is increasing the topic is bound to be interesting. Details can be found about it in any good book on graph theory.

7.3 Path and Circuits

Let $v_1, v_2, v_3, \ldots, v_n$ be vertices in a graph G = (V, E, γ). A ***path***, π, is defined as a sequence of vertices $v_1, v_2, v_3, \ldots, v_k$ such that v_i is adjacent to v_{i+1}. A path is said to be ***elementary*** if no node appears twice. If no edge appears twice on the path then the path is called a ***simple*** path. In the graph of the figure 7.3.1(a), path π: A B C D A C E, is a simple path but not an elementary path. Path π: A B C E, is an elementary path. It is a simple path also. Is it possible for a path to be elementary and not simple?

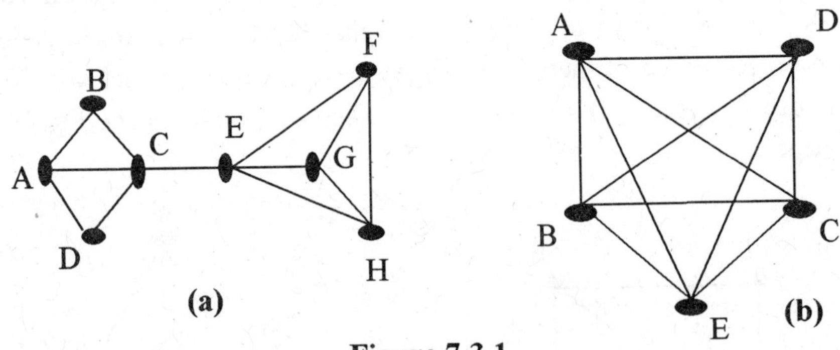

Figure 7.3.1

The number of edges in a path is called ***length of path.*** In an unweighted graph a path between a pair of node is called ***shortest*** if its has minimum number of edges. In weighted graph a path has an associated weight that equals the sum of weights of all the edges in the path. A path between two vertices of such graph is called a ***shortest path*** if it has minimum weight. The shortest path between any pair of nodes is called ***geodesic***. The largest geodesic is called ***diameter*** of the graph. Finding shortest path in a graph is an interesting area of study. We shall discuss a few algorithms related to this in the next section.

A path is said to be a ***circuit,*** if its first node and the last node are the same i.e. start node of the first edge and the terminating node of the last edge is the same. A circuit is also called ***cycle***. The number of edges in the cycle is called ***length of the cycle.*** A loop is a cycle of length one. A circuit corresponding to an elementary path is ***elementary circuit*** and corresponding to a simple path is a ***simple circuit***. In the graph of the figure 7.3.1(b) π: A B C D A, is a simple circuit as well as an elementary circuit. The path π: A B D C E D A, is a simple circuit but not an elementary circuit. The circuit BADEADB is not a simple circuit.

Theorem 1 In a graph (directed or undirected) with n vertices, if there is a path from vertex u to vertex v then the path cannot be of length greater than (n – 1).

Proof: Let π: u, v_1, v_2, v_3, ..., v_k, v be the sequence of vertices in a path from u to v. If there are m edges in the path then there are (m + 1) vertices in the sequence. If m < n, then the theorem is proved by default. Otherwise, if $m \geq n$ then there exists a vertex v_j in the path such that it appears more than once in the sequence (u, v_1, ..., v_j ..., v_j ..., v_k, v). Deleting the sequence of vertices that leads back to the node v_j, all the cycles in the path can be removed. The process when completed yields a path with all distinct nodes. Since there are n nodes in the graph, there cannot be more than n distinct nodes and hence n – 1 edges.

Proved.

An undirected graph is called ***connected*** if there is a path for every pair of nodes in the graph. Among the different types of graph discussed in the previous section, L_n, K_n, Q_n, C_n etc are always a connected graph whereas D_n is always a *disconnected* graph. A graph is called ***disconnected,*** if there are at least two non-empty subsets V_1 and V_2 of V such that there is no path from any vertex of V_1 to any vertex of V_2. Every maximal connected subgraph H of a graph G is called a component of G. A disconnected graph has always more than one component. The graph of the figure 7.3.2 is an example of disconnected graph having two components.

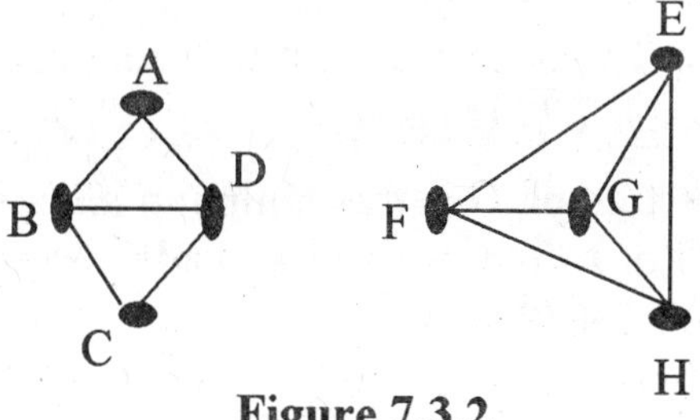

Figure 7.3.2

See the directed graph of the figure 7.3.3. There is no path from node E to the node D in the graph. However this graph is one-component graph. If we consider this graph as undirected then obviously it is connected.

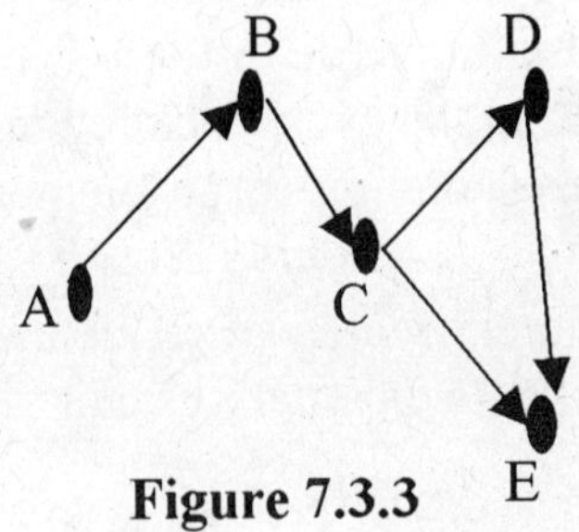

Figure 7.3.3

We cannot use the same criterion to determine the connectivity of a directed graph as that for an undirected graph. A directed graph is said to be a ***weekly connected*** if the corresponding undirected graph is connected. A directed graph is called ***unilaterally connected*** if given any two vertices of the graph, there exists a path from one vertex to another although a reverse path does not necessarily exists. Finally, a directed graph is called ***strongly connected*** if for any pair of vertices u & v, there exists a path from u to v as well as from v to u. For example see the graphs of figure 7.3.4.

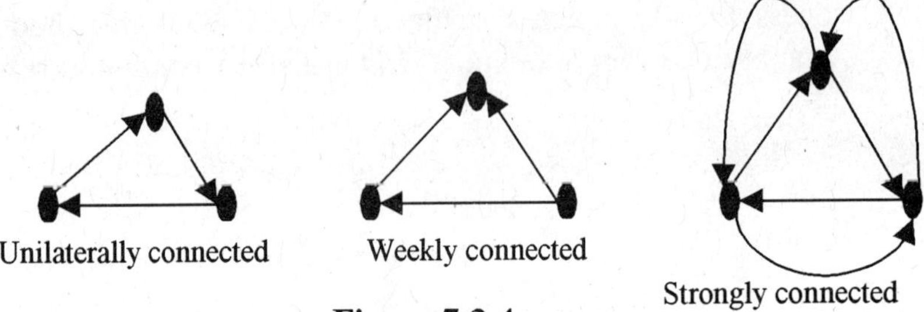

Figure 7.3.4

A connected graph may become disconnected once either one or more edge (or node) is removed from the graph. Larger the number of nodes (or arcs) needed to be removed for a graph to be disconnected, more strongly the graph is connected. Connectivity of a graph is measured in terms of vertex or edge.

Vertex connectivity, V(G), of a graph G is the minimum number of vertices that must be removed to either disconnect G or to reduce it to a single vertex graph. V(G) of the graph 7.3.1(a) is 1 and of the graph 7.3.1(b) is 4.

Edge connectivity, E(G), of a graph G is the minimum number of edges that must be removed to either disconnect G or to reduce it to a single vertex graph. E(G) of the graph 7.3.1(a) is 1 and of the graph 7.3.1(b) is 4.

Example 1 Find the V(G), E(G) and deg(G) for the graph of the figure 7.3.5.

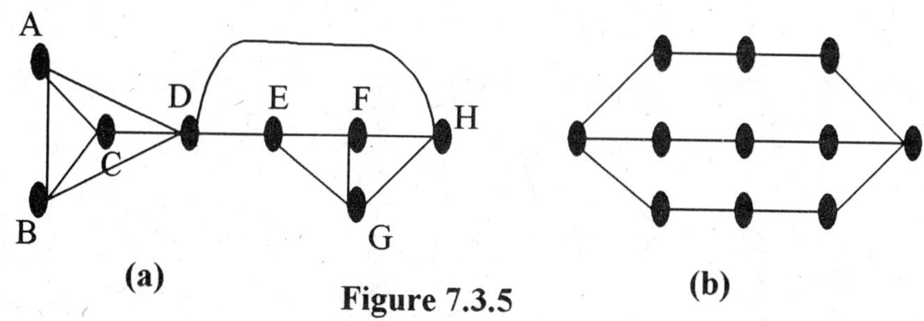

Figure 7.3.5

Solution: (a) The degree of the graph G, deg(G) = 5. If we remove node D from the graph then graph becomes two components graph. Thus, V(G) = 1. By the removal of arcs (D, H) and (D, E) the graph G turns into two components graph hence E(G) = 2.

(b) Here, deg(G) = 3, V(G) = 2 and E(G) = 2.

Ans.

One of the ways to compute E(G) and V(G) of a graph is to find the number of *arc disjoint path* and *node disjoints paths* respectively between all pair of nodes. Two paths are said to be ***arc disjoint*** if no arc is common in them. Similarly, two paths are said to be ***node disjoint*** if no node is common in them. Now E(G) and V(G) can be defined as:

E(G) = min{Number of *arc disjoint paths* for all pairs of node in the graph}
V(G) = min{Number of *node disjoint paths* for all pairs of node in the graph}

Example 2 Find the E(G) and V(G) of the graph shown in figure 7.3.6.

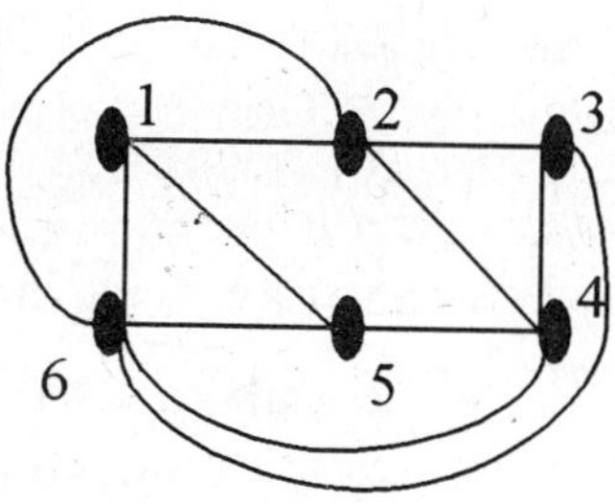

Figure 7.3.6

Solution: To calculate number of arc disjoint paths between any pair of nodes, maximum flow between that pair of node is calculated. The procedure is shown in the table 7.3.1. It is assumed that

- an arc can carry only one unit of flow and
- a node has infinite capacity.

SL.No.	Node Pair	Maximum Flow	Remark
1.	(1, 2)	3	Three arcs from node 1 can carry at the most 3 units of flow and node 2 can receive all of them.
2.	(1, 3)	3	Same as above
3.	(1, 4)	3	Same as above
4.	(1, 5)	3	Same as above
5.	(1, 6)	3	Same as above
6.	(2, 3)	3	Though node 2 can send 4 units of flow, node 3 can accept only 3 units.
7.	(2, 4)	4	Node 2 can sent 4 units and node 4 can accept all of them.
8.	(2, 5)	3	Same as in sl. no. 6
9.	(2, 6)	4	Same as in sl. no. 7
10.	(3, 4)	3	Same as in sl. no. 1
11.	(3, 5)	3	Same as in sl. no. 1
12.	(3, 6)	3	Same as in sl. no. 1
13.	(4, 5)	3	Same as in sl.no. 6
14.	(4, 6)	4	Same as in sl. no. 7
15.	(5, 6)	3	Same as in sl. no. 1

Table 7.3.1

The minimum value of maximum flow between any pair of node is 3. This the count of minimum number of arc disjoint path between any pair of nodes in G. Hence E(G) =3. Similarly, we can compute V(G) of the graph. The following assumptions are made to compute node disjoint path between one node to another.

- Arc has infinite capacity so it can carry any amount of flow
- Any intermediate node in the path can accept I units of flow along any one incoming arc and can pass only one unit at a time along any one outgoing arc. If an intermediate node b has 5 incoming arcs from a node a then b can accept only one unit of flow from a. Similarly if b has 4 outgoing arcs, it can pass only one unit of flow along any one out of four arcs.
- If nodes are adjacent then it can sustain loss of all other nodes, so maximum flow is assumed to be n - 1, where n is |V|.

The calculation is shown in the table 7.3.2. From the table it is clear that V(G) = 3.

SL.No.	Node Pair	Maximum Flow	Remark
1.	(1, 2)	n -1	Both nodes 1 and 2 are adjacent.
2.	(1, 3)	3	Node disjoint paths are (1, 2, 3), (1, 5, 4, 3) and (1, 6, 3)
3.	(1, 4)	3	Node disjoint paths are (1, 2, 4), (1, 5, 4) and (1, 6, 4)
4.	(1, 5)	n - 1	Both nodes 1 and 5 are adjacent.
5.	(1, 6)	n - 1	Both nodes 1 and 6 are adjacent.
6.	(2, 3)	n - 1	Both nodes 2 and 3 are adjacent.
7.	(2, 4)	n - 1	Both nodes 2 and 4 are adjacent.
8.	(2, 5)	3	Node disjoint paths are: (2, 1, 5), (2, 4, 5) and (2, 6, 5)
9.	(2, 6)	n - 1	Both nodes 2 and 6 are adjacent.
10.	(3, 4)	n - 1	Both nodes 3 and 4 are adjacent.
11.	(3, 5)	3	Node disjoint paths are: (3, 4, 5), (3, 2, 1, 5) and (3, 6, 5)
12.	(3, 6)	n - 1	Both nodes 1 and 2 are adjacent.
13.	(4, 5)	n - 1	Both nodes 1 and 2 are adjacent.
14.	(4, 6)	n - 1	Both nodes 1 and 2 are adjacent.
15.	(5, 6)	n - 1	Both nodes 1 and 2 are adjacent.

Table 7.3.2

Ans.

Theorem 2 In any graph G, $V(G) \leq E(G) \leq deg(G)$.

Proof: Let deg(G) = n. Then there exists a node v in G such that degree of v is n. If we drop all arcs for which v is an incidence (*a node is called an incidence of an arc if the node is either a start or an end point of the arc*), the graph becomes disconnected. Thus, E(G) cannot exceed n otherwise there exists a node which is incidence of $m > n$ number of arcs. That is in contradiction with the assumption that deg(G) = n. Thus,

$$E(G) \leq deg(G) \quad \ldots\ldots\ldots\ldots(1)$$

Next, let E(G) = r. Then there exist a pair of nodes such that there are r arc disjoint paths between them. These r paths may cross through $s \leq r$ number of nodes. If we remove these s nodes from the graph, the r arcs get deleted from the graph making the graph a disconnected. That means V(G) cannot exceed r. Thus,

$$V(G) \leq E(G) \quad \ldots\ldots\ldots\ldots(2)$$

Combining result (1) and (2), we have

$$V(G) \le E(G) \le \deg(G)$$

Proved.

If V(G) = K then the graph G is called ***k-connected***. A 1-connected graph is called ***separable***. If G is separable then any vertex that separates the graph into two or more components is called a ***cut point*** or ***articulation point***. The graph of figure 7.3.1(a) is separable and its cut point is E.

Theorem 3 Let v is a cut point of a connected graph G = (V, E, γ). The remaining set of vertex V – {v} can be partitioned into two non empty disjoint subsets U and W such that for any node u ∈ U and w ∈ W, the node v lies on every u-w path.

Proof: When cut point v is removed from G it becomes disconnected. Let U be a set of vertices of the largest connected subgraph of G and W = V – {v} – U. Let v is not on every u–w path. This implies that a path from u to w exists even after removal of v from G. That means U is not the set of vertices of largest connected subgraph of G after removal of v. This is contrary to the assumption that U is the largest component. Hence v lies on every u-w path.

Proved.

Euler Path and Circuit

In a graph G, a path is called an ***Euler path*** if it contains every edge of the graph exactly once. An Euler path that is circuit is called an ***Euler circuit***. For example, graph of the figure 7.3.7(a) has an Euler path but no Euler circuit. The graph of figure 7.3.7(b) has both Euler circuit and Euler path whereas no Euler circuit or path is possible in the graph of 7.3.7(c).

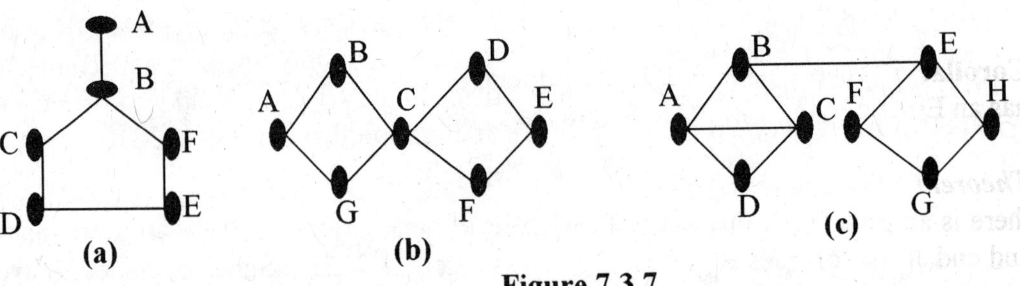

Figure 7.3.7

There are two nodes A and B of odd degrees 1 and 3 respectively in the graph 7.3.7(a). That means arc AB can be used to either arrive at node A or leave node A but not for both the actions. Thus, an Euler path can be found if we start either from node A or from B. Starting from any other node an Euler circuit can never be found in this graph. All nodes in the graph 7.3.7(b) are of even degrees whereas four nodes of graph 7.3.7(c), are of odd degrees. The existence of an Euler path and Euler circuit in a graph depends on the degrees of vertices. We have the following theorems in support of these observations.

Theorem 4 If a graph G = (V, E, γ) has a vertex of odd degree then there can be no Euler circuits in G.

Proof: Let v be a vertex of odd degree, say 2n + 1, in G. For G to have an Euler circuit, each of the 2n + 1 edges must be traveled once. We may **travel in** to node v along n arcs and **travel out** from v along another n arcs, leaving one arc not traveled. This last edge can be used to either **travel in** or to **travel out** from node v leaving the circuit incomplete.

Proved.

Theorem 5 In a connected graph G = (V, E, γ), if every vertex has even degree then there is an Euler circuit in G.

Proof: This theorem will be proved by the method of construction. It is assumed that graph G is undirected. It is given that every node in G is of even degree. Let us take half of edges emanating from a node as incoming and half as outgoing arcs. Starting from a node u, visit a node v adjacent to it and remove the edge. This will decrement the degree of both u and v by one. Move out from v to another node w adjacent to v and remove the used arc from G. Once the degree of a node becomes zero, remove that node from G. It should be understood that degree of a node v becomes zero if even number of arcs, to which v is an incidence, are removed from G. Since every node is of even degree, eventually every node will get removed after finally reaching at the start node. This will complete the circuit.

Proved.

Corollary: If every node in a directed graph G has indegree equal to its outdegree then G has an Euler circuit in it. It can be proved in the same way as the main theorem.

Theorem 6 If a connected graph G = (V, E, γ) has exactly two vertices of odd degree then there is an Euler path in G. Any Euler path in G must begin at one vertex of odd degree and end at the other.

Proof: Let u and v be the two vertices of G with odd degree. Let us add one arc (u, v) to the graph G and say this new graph G'. This new graph G' has all the nodes of even degree and hence has an Euler circuit in which u and v are adjacent. By removing this arc (u, v) from G' we get graph G and an Euler path that begins at u (or v) and ends in v (or u).

Proved.

Theorem *7* If a graph G = (V, E, γ) has more than two vertices of odd degrees then there can be no Euler path in G.

Proof: If G is not connected then result is obvious. Let us suppose that G is a connected graph and there are three vertices a, b and c of odd degrees. We have seen in the theorem 6 that if G has exactly two vertices of odd degree then G has an Euler path in which one of the odd degree node is starting node and other is end node. Let us assume that these two nodes are a nd b. Let c has degree 2n + 1. Then starting from a or b, c can be reached in along n arcs and reached out along another n arcs. Next either the $(2n + 1)^{th}$ arc is left not traveled or in the process of travelling this arc, path is trapped at c. In either case no Euler path is possible.

Proved.

To determine whether a graph G has an Euler path or Euler circuit, we first find the *degree spectrum* of the graph. If any value in the spectrum is zero, the graph is not connected and hence it cannot have an Euler path or Euler circuit in it. If all values in the spectrum are even, then G has both Euler path and Euler circuit. If exactly two values are odd then G has Euler path but no Euler circuit. In all other cases G does not have either an Euler path or an Euler circuit. Once known that a graph G has an Euler circuit or Euler path, an algorithm due to Fleury can be applied to find them.

Fleury's algorithm

Let G = (V, E, γ) be a connected graph such that deg(v) = 2n, n ≥ 1 $\forall v \in V$. To find an Euler circuit in G proceed according to the following steps.

Step 1: Select any node v from V as the starting node of Euler circuit π. Initialize π to v.

Step 2: Select an edge e = (v, w). If there are many such edges, select one that is not a bridge. A ***bridge*** is an arc when removed makes a graph disconnected. Extend the path π to πw and set E = E- {e}. If e is a bridge then set V = V - {v}. Now from node w proceed further.

Step 3: Repeat step 2 until E = ϕ.

The same algorithm can be used to find an Euler path in a graph. Only modification needed is in step 1. The choice of selection of starting node is limited to one of the two odd degree nodes. If all the nodes are of even degree then no modification is required. The working of algorithm is shown on the graph of 7.3.8 in the following example.

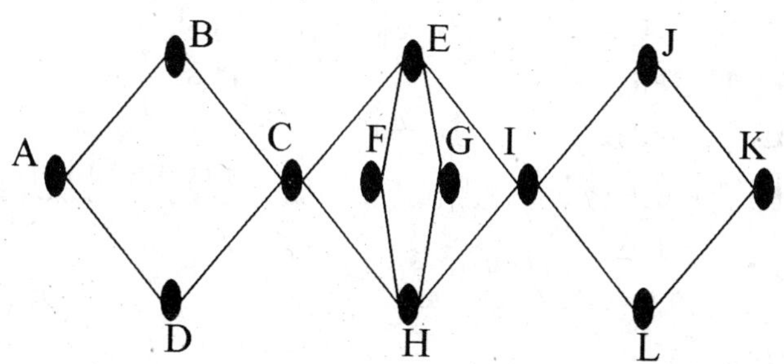

Figure 7.3.8

Example 3 Using Fleury's algorithm, find Euler circuit in the graph of figure 7.3.8.

Solution: The degree spectrum of the graph is (2, 2, 4, 2, 4, 2, 2, 4, 4, 2, 2, 2) considering the node from A to L in alphabetical order. Since all values are even there exists an Euler circuit in it. The process is summarized in the following table 7.3.3. The start node is A.

Sl No	*Current Path*	*Next Edge considered*	*Remark*
1.	π:A	(A, B)	We select (A, B). Add B to π and remove (A, B) from E.
2.	π:AB	(B, C)	It is the only option. Remove (B, C) from E and B from V. Add C to π.
3.	π:ABC	(C, E)	(C, D) cannot be selected, as it is a ***bridge***. Add E to π and remove (C, E) from E.
4.	π:ABCE	(E, F)	Other options are there..
5.	π:ABCEF	(F, H)	As in sl. no. 2
6.	π:ABCEFH	(H, G)	Other option is (H, I). We cannot select (H, C), as it is a ***bridge.***
7.	π:ABCEFHG	(G, E)	As in sl. no. 2
8.	π:ABCEFHGE	(E, I)	As in sl. no. 2
9.	π:ABCEFHGEI	(I, J)	Other options are also there. Edge (I, H) is a ***bridge***.
10.	π:ABCEFHGEIJ	(J, K)	As in sl. no. 2
11.	π:ABCEFHGEIJK	(K, L)	As in sl. no. 2

12.	π:ABCEFHGEIJKL	(L, I)	As in sl. no. 2
13.	π:ABCEFHGEIJKLI	(I, H)	As in sl. no. 2
14.	π:ABCEFHGEIJKLIH	(H, C)	As in sl. no. 2
15.	π:ABCEFHGEIJKLIHC	(C, D)	As in sl. no. 2
16.	π:ABCEFHGEIJKLIHCD	(D, A)	As in sl. no. 2
17.	π:ABCEFHGEIJKLIHCDA		This is the Euler cycle.

Table7.3.3

Ans.

Hamiltonian Path and Circuit

In this book I have tried to avoid giving any emphasis on historical evolution of any terminology rather emphasizing the concept involved and its applicability in computer industry. We will continue with the trend here also. We shall begin with the definition of a Hamiltonian path ignoring the history of its inception. Interested reader may find it elsewhere.

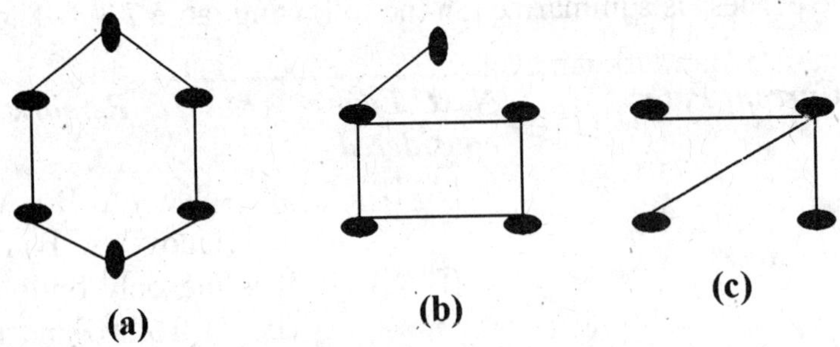

Figure 7.3.9

A path is called a ***Hamiltonian path*** if it contains every vertex of the graph exactly once. If a Hamiltonian path is a circuit, it is called a ***Hamiltonian circuit*** . In the graphs of the figure 7.3.9, (a) has a Hamiltonian circuit, (b) has a Hamiltonian path but no cycle and (c) has none.

As we have seen in the above figure, loops and multiple edges are of no use while finding Hamiltonian circuit in a graph. One edge between any two nodes is sufficient to cross over from one node to its adjacent node in a Hamiltonian path in a graph. Ironically, there is no theorem or formula, like that for Euler path and circuit, which can be used to determine whether a graph contains a Hamiltonian circuit or not. There are a few sufficient conditions for a graph to have a Hamiltonian circuit, but no necessary condition has yet been determined. Reader may try to find one! Here are some observations, which reader may find useful in finding Hamiltonian path in a graph.

1. A Linear graph has Hamiltonian path but no circuit.
2. A circular graph has both Hamiltonian path and circuit.
3. A complete graph $K_{n \geq 3}$ has always a Hamiltonian path and circuit.
4. A graph of n vertices must contain at least n edges to have a Hamiltonian circuit in it.
5. If a graph G contains a dangle node then it cannot have a Hamiltonian circuit. If there is any Hamiltonian path in such a graph then it must begin from any of the dangle node.

The following algorithm can be used on any connected graph to find a Hamiltonian path (circuit) if it is in the graph. This algorithm is based on the concept that if x is any intermediate node in a Hamiltonian path (or circuit) then once its neighbor has been visited from x, x is no longer required. (Every node is visited exactly once in a Hamiltonian path or circuit.) This algorithm is tested on a number of different types of graph by the author. Reader may try to find a counter example to falsify the claim.

Algorithm for Hamiltonian Circuit

Let $G = (V, E, \gamma)$ be a connected graph such that $|v| = n$ and $|E| \geq n$. Otherwise a graph cannot have a Hamiltonian cycle.

Step 1: Select a vertex v in G such that while coming from left (or lower) part of the graph to the right (or upper) part and vice versa, we need not to cross the same node v. If there is any dangle node, then select a dangle node to start with. Initialize the path π to v i.e. π: v. Set

Current_node ← v
Prev_Cuurent_node ← NULL

Step 2: Visit a neighbor w of the Current_node so that a cycle of length less than n does not form in the graph. If Prev_Current_node is set to a node, drop the Current_node from the graph together with all arcs to which Current_node is an incidence. Set

Prev_Cuurent_node ← Current_node
Current_node ← w

Step 3: Repeat Step 2 as long as visit of a neighbor is possible.

In the step 2, if a move to a neighbor results into a cycle of length < n, alternative path from previous nodes can be explored if it is available. At the end if only one node of degree zero is left in the graph, that must be the first one i.e. v. Therefore, we get a Hamiltonian cycle. If two nodes of degree zero are left in the graph, there is a

Hamiltonian path in G but no circuit. Otherwise neither of them is available in G. See the following examples.

Example 4 Determine whether a Hamiltonian path or circuit exists in the graph of Figure 7.3.10.

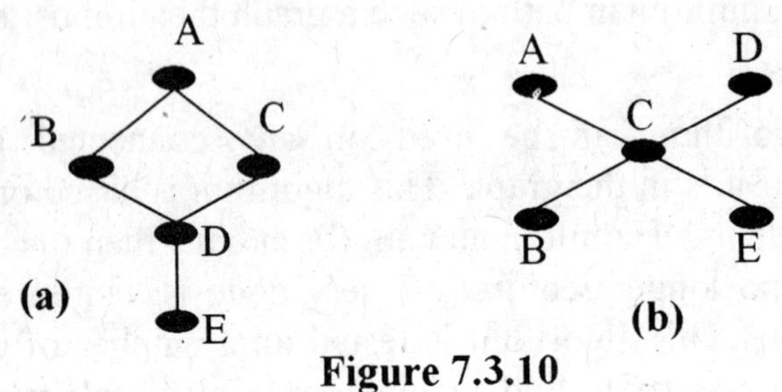

Figure 7.3.10

Solution: In Graph (a), there is a dangle node E. This graph cannot have a Hamiltonian circuit. Let us find whether a path exists or not. We will start with node E. So π: E, initially. Only neighbor of node E is the node D. Visit node D and extend path π. So we have π: E D. Now visit neighbor of D, which is either B or C. We can select any of them, say B. Extend the path up to B i.e. π: E D B and drop D from the graph together with edges (E, D) and (D, B) and (D, C). Next visit A and after extending the path drop node B and edge (B, A) from graph. Path is now π: E D B A. Next, visit node C and do the rest. By this time our path is π: E D B A C, and only two node of degree zero is left in the graph. The path is obviously a Hamiltonian path.

In the graph (b), there are four dangle nodes. Thus, this graph cannot have a Hamiltonian circuit. Select any of the dangle nodes, say A, to start with. Initialize the path π: A. Visit a neighbor of node A, which is C. Extend the path up to C i.e. π: A C. Now visit the neighbor of node C, which could any of the nodes: B, D and E. Select one, say B. Extend the path up to B i.e. π: A C B. Now drop node C from the graph and all edges: (C, A), (C, B), (C, D) and (C, E). We get then four nodes with degree zero. This shows that the graph does not have even a Hamiltonian path.

Ans.

Example 5 Determine whether a Hamiltonian path or circuit exists in the graph of Figure 7.3.11.

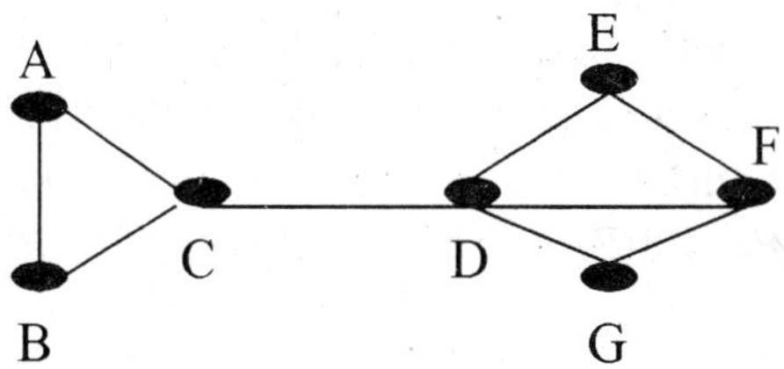

Figure 7.3.11

Solution: Let us start with the node A. We can select any one but node C and D. Initialize the path π: A. Next move to the node B. We cannot move to C from A. Because any move to D from B and to B from D need node C. Extend the path up to B. Then move to node C, extend the path up to C and drop node B together with edges (B, A) and (B, C). We have now the path π: A B C. Now move to D, extend the path up to D and drop node C together with arcs (C, A) and (C, D). Then move to either node G or E but not to F. Extend the path and do the rest. Finally, proceeding in this way, we get π:A B C D E F G. And two nodes A and G, with degree zero, are left. Thus, this graph has a Hamiltonian path π but no Hamiltonian circuit.

Ans.

Example 6 Determine whether a Hamiltonian path or circuit exists in the graph of Figure 7.3.12.

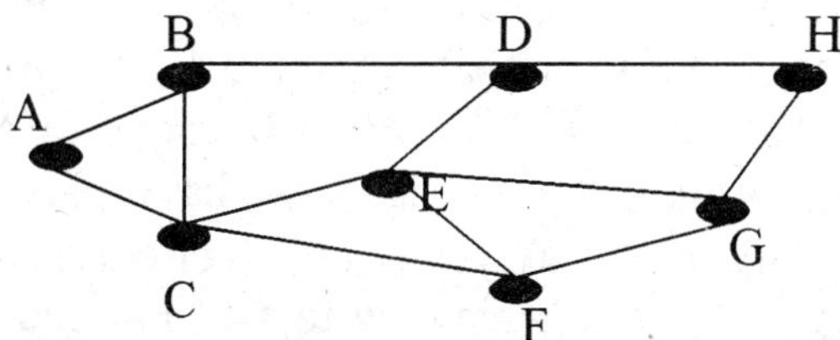

Figure 7.3.12

Solution: Let us take node A to start with. Next, move to either B or C, say B. Extend the path up to B. Next move to D and not to C as a cycle of length 3 could be formed here. Extend the path up to D and drop node B and edges (B, A), (B, C) and (B, D). Then move to H. Why we cannot move to E? Drop D and edges from it. Then move to G, then to F (or E) then to E (or F), then to C and finally to A dropping the nodes and edged from them on the way. At the end, only one node A is left with degree zero and π is A B D H G F E C A. This is a Hamiltonian cycle.

Ans.

In a weighted graph, a Hamiltonian path is called ***minimum Hamiltonian path (circuit)*** if the sum of weights of all edges in the path is minimum. To determine a minimum Hamiltonian path in a weighted graph, we used NNM (Nearest Neighborhood Method). We use the same algorithm, but a neighbor is selected on the basis of its closeness with respect to the weight of an arc. Less the weight of the arc, more closure is the neighbor.

Example 7 Find the minimum Hamiltonian circuit starting from node E in the graph of the figure 7.3.13.

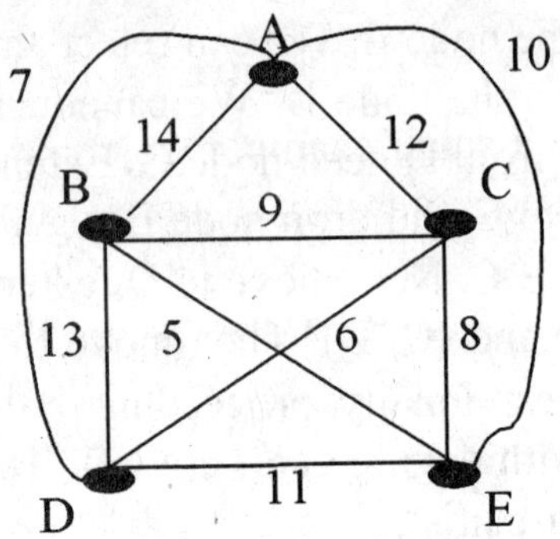

Figure 7.3.13

Solution: We have to start with the node E. Closest node to E is the node B. Move to B. Now closest node to B is C. Move to C, extend path up to C and drop node B and all edges from it, from the graph. From C move to D. Drop C and all edges from it. Extend the path up to D. Now, we have π: E B C D. From D, move to A and then to E back. Finally, we have only node E left in the graph. Thus, we have a Hamiltonian circuit in the graph, which is π: E B C D A E. The total weight of this circuit is (EB + BC + CD + DA + EA = 5 + 9 + 6 + 7 + 10) 37. Reader may try to find minimum Hamiltonian circuit starting from other nodes and verify whether all have the same minimum weight or not.

Ans.

7.4 Shortest Path Algorithms

In many areas like transportation, cartoon motion planning, communication network topology design etc, problems related to finding shortest path arise. The shortest path problem is concerned with finding the least cost (that costs minimum) path from an originating node in a weighted graph to a destination node in that graph. Let us consider a graph shown in the figure 7.4.1. in which number associated with each arc represents the weight of (or the cost associated with) the arc. Of the possible paths from vertex 1 to vertex 5 (e.g. 1-2-4-5, 1-3-4-5, etc) how can we find the shortest path from vertex 1 to vertex 5?

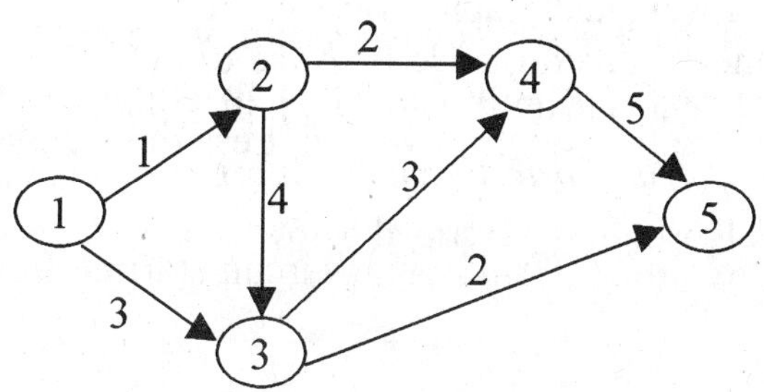

Figure 7.4.1

Clearly with such a small example we could find the shortest path from vertex 1 to vertex 5 simply by inspection, or by enumerating all possible paths between vertex 1 and vertex 5. However for a much larger graph, this approach would not be feasible.

There exist many algorithms for finding the shortest path in a weighted graph. One such algorithm is developed by Dijsktra in the early 1960's for finding the shortest path in a graph with non-negative weight associated with edge/arc without explicitly enumerating all possible paths. This algorithm is based upon a technique known as *dynamic programming*. We shall discuss this algorithm here. This algorithm determines shortest path between a pair of nodes in a graph. If there are n nodes in a graph, we need to run the algorithm $^nC_2$ times. In a network of 100 or more nodes, the time taken to compute the shortest path for all possible pair of nodes can be any body's guess. To overcome this we shall discuss a modification of Dijsktra's algorithm to find shortest distance between one node to all other nodes in a graph and Floyd Warshall's algorithm to compute all pair shortest path.

In an unweighted graph, the cost of a path is just the number of edges on the shortest path. This can be found using Breadth First Search (BFS). We shall learn about BFS in the section 7.6 of this chapter. In a weighted graph, the weight of a path between two nodes is the sum of the weights of the edges on a path. BFS does not work on a weighted graph because sometimes visiting more edges can lead to shorter distance.

Dijsktra's algorithm

As stated above this algorithm works for a weighted graph in which weight (cost) associated with each arc is non-negative. If the graph is unweighted, we can assume that all edges have equal weight of **one**. If a directed arc in a graph has negative weight, this algorithm can be applied after making the weight positive by reversing the direction of the edge. What about the case when an undirected edge of a graph has negative weight? Reader may explore.

Let us consider a weighted graph shown in the figure 7.4.2. Let we have to find shortest path between the originating node A to the destination node D. It works on the *principle of optimality.* It states that "*if there is a path from* v_1 *to* v_n *and if* v_j *is a node on the path such that* $\pi: v_1 \ldots v_j$ *is shortest then the shortest path from* v_1 *to* v_n *must contain* π." The working of the algorithm is described below.

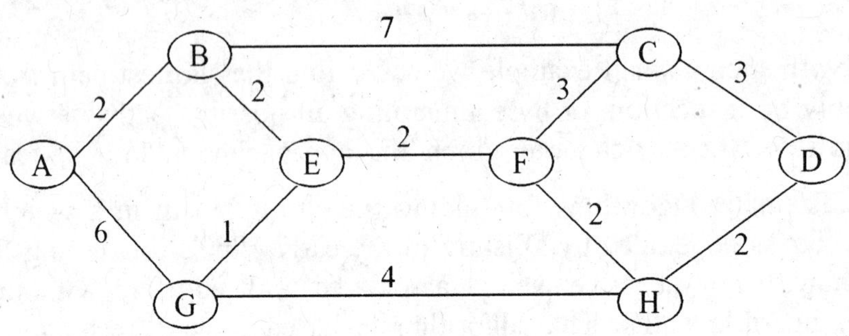

Figure 7.4.2

A node is marked as ***permanent*** if no further updation for the distance from the originating node is possible. However if there is scope for the updation, the node is marked as ***tentative***. We use a hollow circle (o) to mark a tentative node and thick circle (•) to mark a permanent node. We use an ordered pair ***(<distance>, <arrived from node>)*** for every node to represent shortest distance and route. Since we have to find distance from the node A to the node D, we mark the node A as permanent and all other node as tentative. We take node A as the ***current working node***. The *current working node* is shown with an arrow pointing to it. For every node in the graph other than the starting node, ordered pair ***(<distance>, <arrived from node>)*** is assigned a value. For adjacent nodes like B and G we write (2, A) and (6, A) respectively because distance of B from A is 2 and that of G from A is 6. For all other nodes we use ordered pair ($\propto$, –). This indicates that we have not yet arrived on the node from the originating node. A tentatively marked node indicates that the shortest route is yet to be ascertained. The starting graph is shown in the figure 7.4.2a.

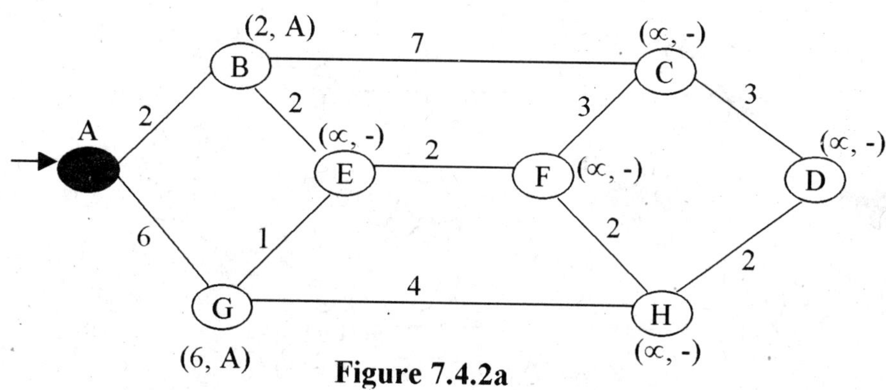

Figure 7.4.2a

Now we select from the tentatively marked nodes, the node having shortest distance. Here B is the node that satisfies the criteria. We select B and mark it as permanent and make it our current working node. The ordered pair ***(<distance>, <arrived from node>)*** is now updated for all tentatively marked node that are adjacent to the current working node i.e. B here. E and C are adjacent to B. Arc BC has length 7. So the distance of C from A through B is 9. Similarly E is updated with ordered pair (4, B). The updated graph is shown in the figure 7.4.2b.

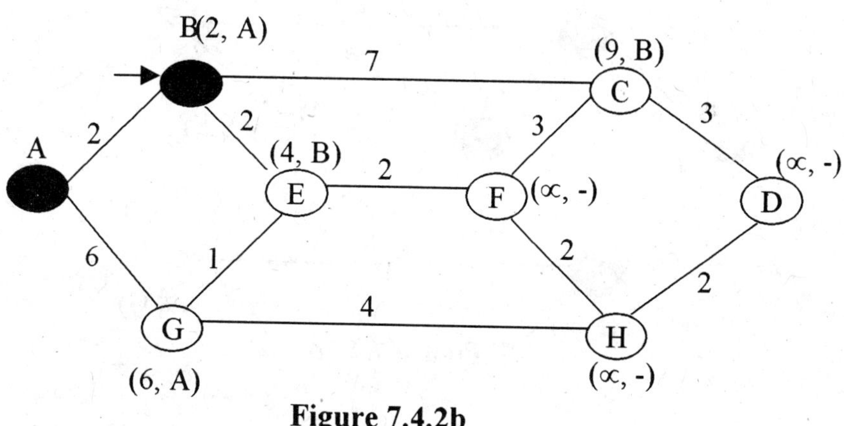

Figure 7.4.2b

Again select the node having shortest distance. It is now the node E. Mark it permanent and make it the current working node. Update the ordered pair ***(<distance>, <arrived from node>)*** for adjacent nodes G and F. For the node G, direct distance from A is 6 whereas distance through E is 4 +1 = 5. Thus shortest distance is 5 through E. The updated graph is shown in the figure 7.4.2c.

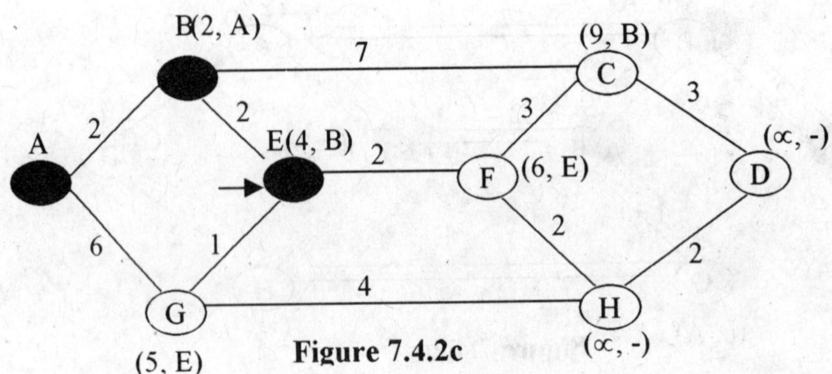

Figure 7.4.2c

Following the procedure, node G is now selected and marked permanent. This node becomes working node. The only adjacent node is H. The ordered pair ***(<distance>, <arrived from node>)*** for the node E is updated which is (9, G). The graph thus obtained is shown in the figure 7.4.2d.

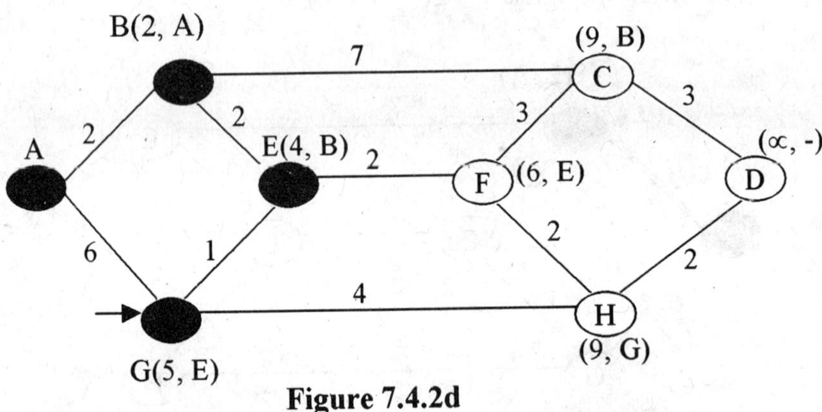

Figure 7.4.2d

Now node F is eligible to be marked as permanent and to work as working node. Sot we select it. It has two adjacent nodes: C and H. Distance of H through F is now 8 so it is updated. However, distance of C from F is 9, that is equal to the earlier found distance. It needs no updation. After assignment and modification, we get the graph as shown in the figure 7.4.2e.

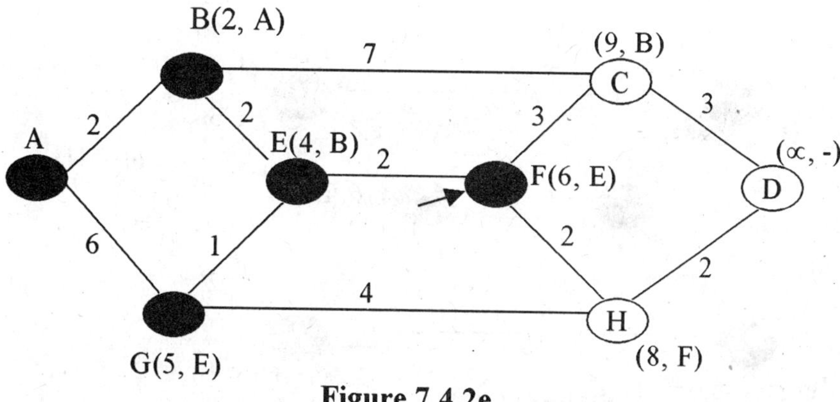

Figure 7.4.2e

Now among the tentatively marked nodes C, H and D, H has the shortest distance. Thus H is selected, marked permanent and designated as current working node. The only adjacent node is D. Thus the ordered pair ***(<distance>, <arrived from node>)*** for the node D is updated which is now (10, H). The graph showing this result is in the figure 7.4.2f.

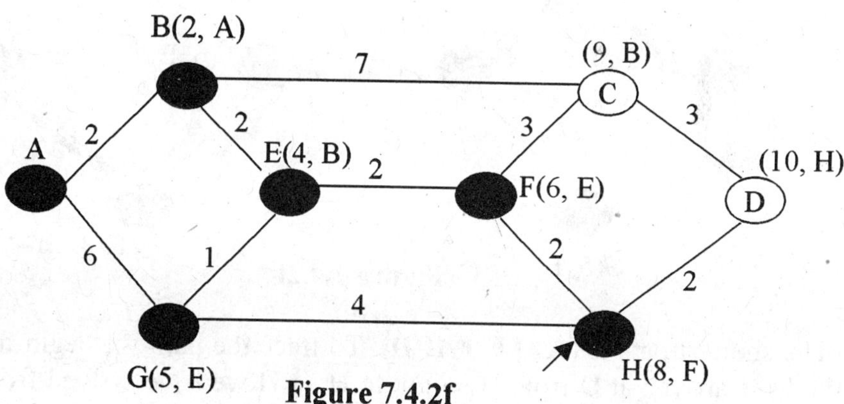

Figure 7.4.2f

Now node C is selected, marked permanent and designated as current working node. The only adjacent node is D. Thus the ordered pair ***(<distance>, <arrived from node>)*** for the node D is to be updated. If we go through C the distance is 11 that is greater than the earlier distance of 10. Thus distance for D is not updated. The updated graph is shown in the figure 7.4.2h.

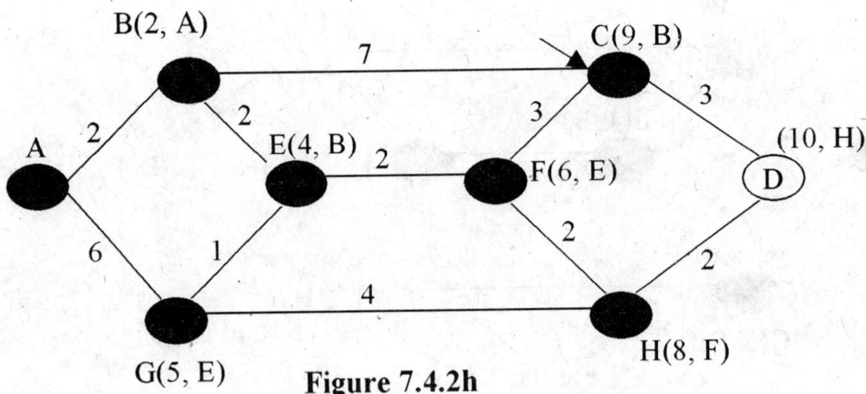

Figure 7.4.2h

Finally we have reached to the destination node D. The procedure of finding the shortest path terminates when we reach to the destination node. See the graph shown in the figure 7.4.2i that also shows the traced back shortest path.

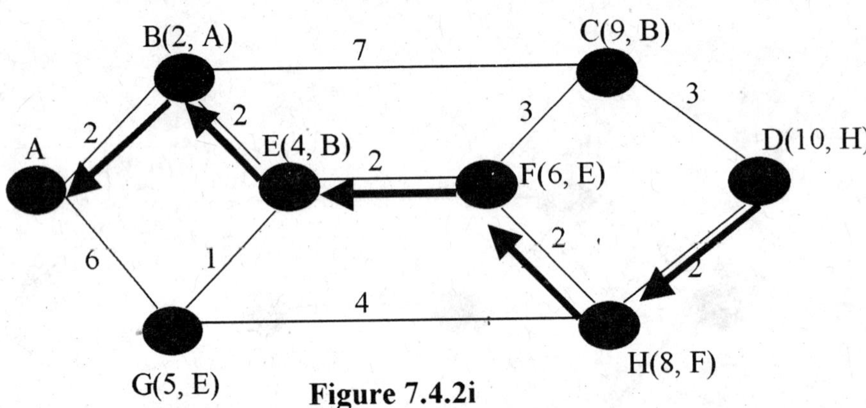

Figure 7.4.2i

The shortest path is (**A B E F H D**). To trace the path we begin from destination node. We have arrived at D from H so got to H, at H we have arrived from F and so on. Finally we reach at A –the originating node. The shortest path is shown in the graph 7.4.2i by thick directed line. The Dijsktra's algorithm can be summarized in stepwise procedure as:

Step 1: Start with the given originating node x. Mark x as permanent and all other nodes as tentative. Treat the node x as working node.

Step 2: Update the adjacent node of the current working node (node marked permanent in previous step) with distance from source node x to the node through current node in the form of ordered pair ***(<distance>, <arrived from node>)***.

Step 3: Select a node from tentatively marked node having minimum distance, Say it is y. Mark it permanent and consider it as current working node.

Step 4: Repeat the step 2 and the step 3 in sequence until destination node is marked permanent.

Here we have not only found the shortest distance between the node A and the node D but between A and any other node in the graph. This has happened because before marking the destination node as permanent, every other node was already marked permanent. Had the destination node marked permanent before marking any node in the graph, we would not have got that. To find the shortest path from one node to every other nodes proceed with the procedure discussed above until all nodes have been marked permanent. How to solve the problem of finding shortest path between all pair of nodes? One solution is to run Dijsktra's algorithm from every node treating it as originating node. This is tedious! We shall discuss now an algorithm called Floyd Warshall's algorithm that finds all pair shortest path at one run of the algorithm.

Floyd Warshall's algorithm

Recall the algorithm learnt in the chapter one to find transitive closure of a relation. This algorithm is exactly the same with exception that matrix representation of the graph contains weight of arc rather than 1 or 0. Let N be the number of nodes in a graph. If the graph is directed, there will be ${}^{n}P_2$ distinct pair of nodes. If the graph is undirected, there will be ${}^{n}C_2$ distinct pair of nodes. In a directed graph pair (A B) and (B A) are different. In a mixed graph, distinct number of pairs can be calculated accordingly. If we include pairs (A, A) for all nodes A of a graph, there will be N^2 pairs. For simplicity, we always take N^2 pairs and represent the graph by its adjacency matrix of order N×N.

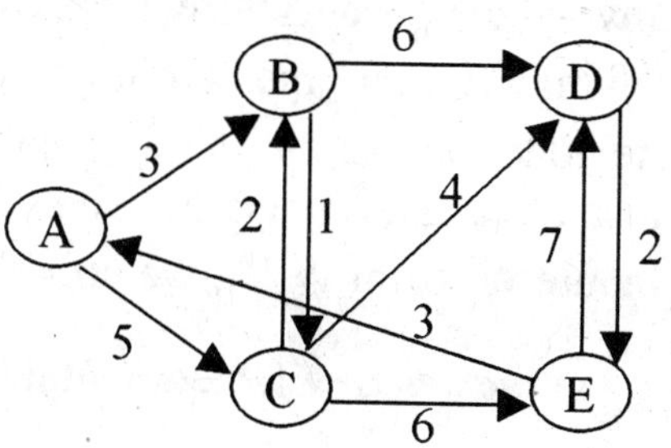

Figure 7.4.3

Let us consider the weighted directed graph shown in the figure 7.4.3. In this graph, let $d(v_i, v_j)^k$ represents the minimum distance between node v_i and v_j through k numbers of intermediate nodes. Let $d(v_i, v_j)^0$ denotes the weight of arc, if any, between two nodes v_i and v_j. Whenever, there is no direct arc between two nodes v_i and v_j, we write $d(v_i, v_j)^0 = 0$ for $i = j$ and $d(v_i, v_j)^0 = \infty$ for $i \neq j$. The adjacency matrix of the graph is then given by a matrix that contains $d(v_i, v_j)^0$ as its element at the location (i, j). This matrix is given below. It is called the initial Warshall's matrix and is denoted by W_0.

$$W_0 = \begin{bmatrix} \mathbf{0} & \mathbf{3} & \mathbf{5} & \infty & \infty \\ \infty & 0 & 1 & 6 & \infty \\ \infty & 2 & 0 & 4 & 6 \\ \infty & \infty & \infty & 0 & 2 \\ \mathbf{3} & \infty & \infty & 7 & 0 \end{bmatrix}$$

Since the shortest path between any two nodes must use at the most N edges (unless we have negative cost cycles), we must repeat the procedure N times, where N is the number of nodes in the graph. The algorithm is given as below.

```
For k = 1 to N
    For i = 1 to N
        For j = 1 to N
            d(i, j)^k = min(d(i, j)^(k-1), d(i, k)^(k-1) + d(k, j)^(k-1))
```

Algorithm

Wow! At first look, it seems very complicated. Let us find a working procedure. We take the graph of the figure 7.4.3. We shall start with the matrix W_0. This matrix represents all pair shortest path involving **Zero** intermediate nodes. Similarly, W_1 represents all pair shortest path involving **One** intermediate node, W_k represents all pair shortest path involving **k** intermediate nodes and so on. In the graph of our consideration, if we find W_5, our job is done, because there cannot be a path between any pair of nodes involving more than 5 nodes. To compute W_k from W_{k-1}, we proceed as follows:

- Consider k^{th} row of W_{k-1} as ***master row*** for computation of the elements of W_k.

- Take $(1, k)^{th}$ element of W_{k-1} as ***lead element*** for 1st row, $(2, k)^{th}$ element of W_{k-1} as lead element for 2nd row and so on.
- Let $(i, j)^{th}$ element of W_{k-1} be x. Compute the sum of ***lead element*** for i^{th} row (i.e. $(i, k)^{th}$ element of W_{k-1}) and j^{th} element of ***master row*** (i.e. $(k, j)^{th}$ element of W_{k-1}). Let the sum be y. Take minimum of x and y. This is the $(i, j)^{th}$ element of W_k).

In our example, at first, we have to find W_1 from W_0. Thus we K = 1. This implies that 1st row of W_0 is ***master row*** for the computation of the elements of W_1 and element of 1st column of W_0 will act as ***lead element*** for the corresponding rows. The master row and lead elements are shown in bold italic style in W_0. See how elements of 5th row of W_1 are calculated. For 5th row lead element is 3 and for 1st element of master row is 0. Thus sum is 3 + 0 = 3. The minimum of 3 and 3 is 3. Therefore, $(5, 1)^{th}$ element of W_1 will be 3. The 2nd element of master row is 3. Its sum with lead element is 6 (3 + 3). Since 6 is less than ∝, $(5, 2)^{th}$ element of W_1 will be 6. . The 3rd element of master row is 5. Its sum with lead element is 8 (3 + 5). Since 8 is less than ∝, $(5, 3)^{th}$ element of W_1 will be 8. The 4th element of master row is ∝. Its sum with lead element is ∝ (3 + ∝). Since 7 is already less than ∝, $(5, 4)^{th}$ element of W_1 will be 7 i.e. $(5, 4)^{th}$ element of W_0. The W_1 so computed is shown below.

$$W_1 = \begin{bmatrix} 0 & \boldsymbol{3} & 5 & \propto & \propto \\ \boldsymbol{\propto} & \boldsymbol{0} & \boldsymbol{1} & \boldsymbol{6} & \boldsymbol{\propto} \\ \propto & \boldsymbol{2} & 0 & 4 & 6 \\ \propto & \boldsymbol{\propto} & \propto & 0 & 2 \\ 3 & \boldsymbol{6} & 8 & 7 & 0 \end{bmatrix}$$

In W_1, 2nd row will be master row and elements of 2nd column will be lead elements for corresponding rows for the computation of elements of W_2. To make the thing visible, 2nd row and 2nd column is shown in bold italic font in W_1. Following the above procedure, we have computed W_2 as shown below.

$$W_2 = \begin{bmatrix} 0 & 3 & \boldsymbol{4} & 9 & \infty \\ \infty & 0 & \boldsymbol{1} & 6 & \infty \\ \boldsymbol{\infty} & \boldsymbol{2} & \boldsymbol{0} & \boldsymbol{4} & \boldsymbol{6} \\ \infty & \infty & \boldsymbol{\infty} & 0 & 2 \\ 3 & 6 & \boldsymbol{7} & 7 & 0 \end{bmatrix}$$

See the $(1, 3)^{th}$ element of W_2 and compare it with the corresponding element in W_1. The lead element for 1st row in W_1 is 3 and 3rd element of master row in W_1 is 1, thus sum is $(3 + 1) = 4$. It is smaller than 5 –the $(1, 3)^{th}$ element of W_1. Therefore we have 4 as element of W_2 at $(1, 3)^{th}$ location. Proceeding in the same way, we get W_3, W_4 and W_5 as shown below.

$$W_3 = \begin{bmatrix} 0 & 3 & 4 & \boldsymbol{8} & 10 \\ \infty & 0 & 1 & \boldsymbol{5} & 7 \\ \infty & 2 & 0 & \boldsymbol{4} & 6 \\ \boldsymbol{\infty} & \boldsymbol{\infty} & \boldsymbol{\infty} & \boldsymbol{0} & \boldsymbol{2} \\ 3 & 6 & 7 & \boldsymbol{7} & 0 \end{bmatrix}, W_4 = \begin{bmatrix} 0 & 3 & 4 & 8 & \boldsymbol{10} \\ \infty & 0 & 1 & 5 & \boldsymbol{7} \\ \infty & 2 & 0 & 4 & \boldsymbol{6} \\ \infty & \infty & \infty & 0 & \boldsymbol{2} \\ \boldsymbol{3} & \boldsymbol{6} & \boldsymbol{7} & \boldsymbol{7} & \boldsymbol{0} \end{bmatrix} \text{ and } W_5 = \begin{bmatrix} 0 & 3 & 4 & 8 & 10 \\ 10 & 0 & 1 & 5 & 7 \\ 9 & 2 & 0 & 4 & 6 \\ 5 & 8 & 9 & 0 & 2 \\ 3 & 6 & 7 & 7 & 0 \end{bmatrix}$$

In W_2, 3rd row will be master row and elements of 3rd column will be lead elements for corresponding rows for the computation of elements of W_3. Similarly, in W_3, 4th row will be master row and elements of 4th column will be lead elements for corresponding rows for the computation of elements of W_4. Finally to compute W_5, 5th row 5th column of W_4 will be taken into consideration. The master rows and leading elements column have been shown in in bold italic font in respective matrices W_i. Let nodes A, B, C, D and E of the graph be numbered 1, 2, 3, 4 and 5 respectively. Then $(i, j)^{th}$ element of W_5 is the shortest distance between node i and node j. For example, element at location (5, 3) in W_5 is the shortest distance between node E and node C of the graph. Reader may verify the other entries.

Unlike the Dijsktra's algorithm, this algorithm takes care of negative weight of an edge automatically. It is important to note that in data transmission **zero** weight edge between two nodes simply implies that two nodes are not connected, if weight denotes the bandwidth of line.

7.5 Coloring of Graphs

In real life we come across various types of maps like city map, railway map, road map, maps of buildings in an area, country map, state boundary map, district boundary map etc. A map may be of line segments like rail network map, road network map etc. Some of the map may be of ***regions*** like district map, state map, country map building map etc. A ***region*** is defined as a ***polygon***. A ***polygon*** is a shape surrounded by a number of lines (arcs/edges). A polygon covers an area. If the covered area by a polygon is finite, the polygon is called ***closed polygon*** otherwise it is called ***open polygon***. In the graph of figure 7.5.1, we have four closed polygons r_1, r_2, r_3, r_4. To understand the concept of region, draw this graph on the page of your notebook and cut the graph along the edges. You shall have five pieces of paper; four corresponding to r_1, r_2, r_3, r_4 and one left over part of the page. Therefore, there are five regions in the graph of the figure 7.5.1.

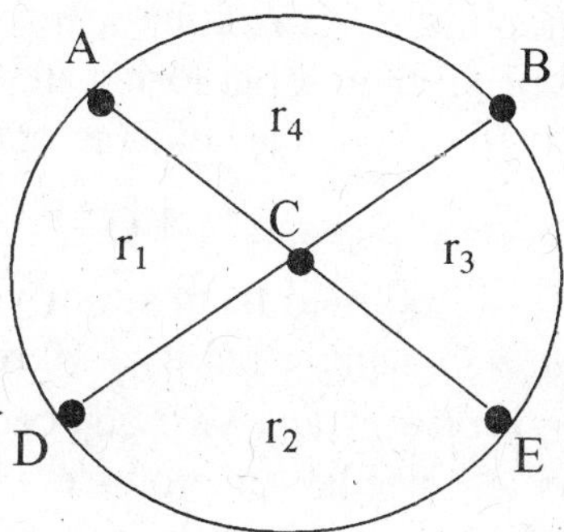

Figure 7.5.1

Coloring of graph plays a very important role in the presentation of maps. It will be more visible and appealing if adjacent regions are colored using distinct colors. The problem of finding the least number of distinct colors that must be required to color a map in such a way that no adjacent region should have same color, has given the birth of topics like 'coloring of graph'. A map can be converted into a graph. See the map of the figure 7.5.2. Let every closed polygon be denoted by a vertex. Draw an edge between two nodes if the corresponding polygons have common boundary. The graph thus obtained is shown in the figure 7.5.3.

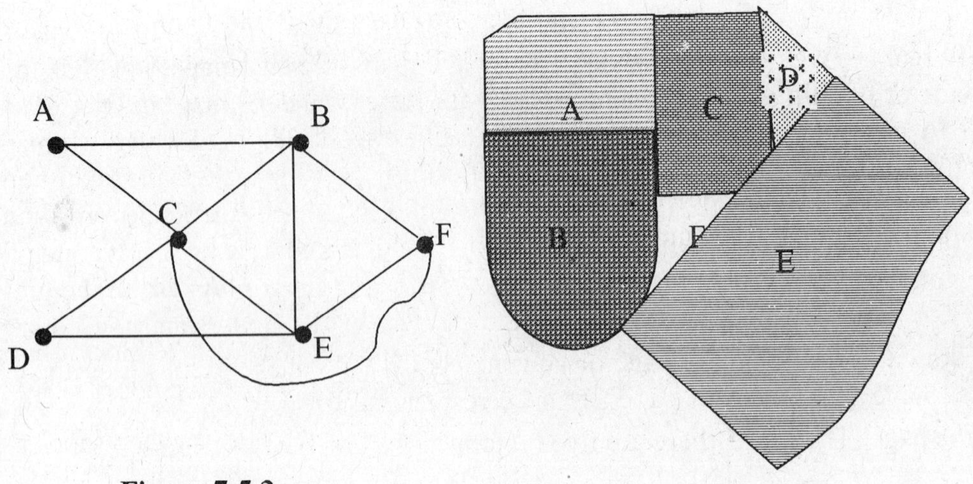

Figure 7.5.3

Figure 7.5.2

Now if we color the graph of 7.5.3 in such a way that two adjacent nodes should not have the same color then our problem of coloring the map is solved. This implies that problem of coloring a map has reduced to the coloring of graph. Let us define this now.

Let G = (V, E, γ) be a simple graph and C = {c_1, c_2, c_3, ..., c_n} be the set of n distinct colors. A function f: V→C defined from set of vertices V to set the set of colors C is called ***coloring of the graph*** G using |C| colors. Here f(v) gives the color of a node v ∈ V. A coloring is said to be ***proper*** if any two adjacent nodes v and w have different colors. The least number of colors needed to produce a proper coloring of a graph G is called ***chromatic number*** of G and is generally denoted as X(G). Let us consider the graph of the figure 7.5.4. It has four vertices A, B, C and D. Let there be x number of colors i.e. |C| = x.

If we start coloring from vertex A, we have x choices, thus vertex A can be colored in x number of ways. Once chosen a color, say red, node B must not be colored with red. Thus we are left with (x – 1) choices. Since D is adjacent to both A and B, the colors used for A and B cannot be used for D. Thus we have only (x – 2) options left.

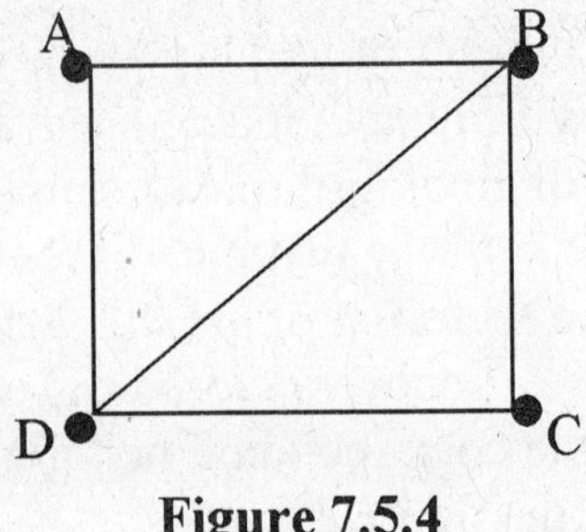

Figure 7.5.4

For the node C, we have to avoid the color used for B and D only i.e. we can use the color used for the node A as node A is not adjacent to node C. Thus, we have (x – 2) options. Using the multiplication principle of counting, we have x(x – 1)(x – 2) (x – 2) number of ways to color the graph. Obviously, this expression is a polynomial in x. *The polynomial which gives the number of ways a graph G can be properly colored using x number of colors is called* ***chromatic polynomial*** *of graph G.* It is denoted as P(G). Thus chromatic polynomial of the graph of the figure 7.5.4 is given by

$$P(G) = x\,(x-1)(x-2)\,(x-2)$$

The minimum value of x for which P(G) is positive is 3. Thus at least 3 colors are needed to properly color the graph of the figure 7.5.4 and hence its chromatic number is 3.

Example 1 Find the chromatic polynomial and chromatic number for the graphs K_3, K_4 and $K_{3,3}$.

Solution: The graphs (a), (b) and (c) of the figure 7.5.5 are for K_3, K_4 and $K_{3,3}$ respectively. Now let us take the graphs one by one.

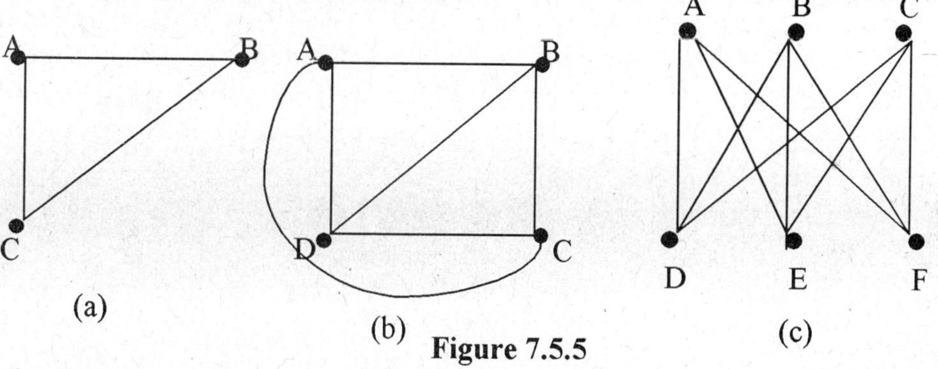

Figure 7.5.5

(a) Chromatic polynomial for K_3 is given by x(x – 1)(x – 2). Thus chromatic number of this graph is 3.

(b) Chromatic polynomial for K_4 is given by x(x – 1)(x – 2)(x – 3). Thus chromatic number of this graph is 4.

(c) Chromatic polynomial for $K_{3,3}$ is given by $x(x-1)^5$. Thus chromatic number of this graph is 2. It is interesting to note here that only two distinct colors are required to color $K_{3,3}$. Intuitively, we can infer that nodes A, B and C may have one color, as they are not adjacent. Similarly, nodes D, E and F can be colored in proper way using one color. But a node from {A, B, C} and a node from {D, E, F} both cannot have the same color. In fact every bipartite graph has chromatic number 2.

Ans.

It is not always possible to find chromatic polynomial for a given graph. This is still an open area. Some results have been observed and some theorems have been formulated to simplify the problem of finding chromatic polynomial for a graph. We shall discuss a few of them in this text.

Theorem 1 Let G_1, G_2, G_3, ..., G_n be n connected component of a disconnected simple graph G. Let P_1, P_2, P_3, ..., P_n be the chromatic polynomial in x colors for G_1, G_2, G_3, ..., G_n respectively. Then the product $P_1P_2P_3 \ldots P_n$ is chromatic polynomial for the graph G.

Proof: It is given that connected component G_1 can be properly colored in P_1 ways and G_2 can be properly colored in P_2 ways. Thus, by the multiplication principle of counting, a graph having components G_1 and G_2 can be can be colored in P_1P_2 ways because no nodes of G_1 is adjacent to any node in G_2. Extending this formulation for G_3, for G_4 and up to G_n, we get

$$P_G(x) = P_1P_2P_3 \ldots P_n$$

Proved.

Now let e = {x, y} be any edge in a graph G = (V, E, γ). If the edge e is removed from G, a subgraph of G is obtained. Let us say this subgraph G_e. And if nodes x and y are merged, a quotient graph of G is obtained. Let us call it G_Q. A formula based on this concept and proved in the following theorem may be helpful in finding the chromatic polynomial of a given connected graph.

Theorem 2 Let G = (V, E, γ) be a connected graph and e = {x, y} be any edge in G. Let G_e and G_Q be subgraph and quotient graph of G obtained respectively by removing the edge e from G and merging nodes x and y, then

$$P_G = P_S - P_Q$$

Where, P_S and P_Q are chromatic polynomials for G_e and G_Q respectively.

Proof: Let us consider the subgraph G_e. It has two nodes x and y that are not adjacent to each other. While coloring the graph G_e, these two nodes may have either the same color or distinct color. If the nodes have same color, then the graph is colored as if both the nodes are merged into one. The chromatic polynomial according to this coloring is the same as that for quotient graph G_Q i.e. P_Q. On the other hand, if the two nodes have different color, the graph is colored as if edge e is present in the graph. Thus the chromatics polynomial so obtained is that for graph G itself. Now, by the addition principle of counting, we have

$$P_S = P_G + P_Q$$

$$\text{Or, } P_G = P_S - P_Q$$

Proved.

Example 2 Find the chromatic polynomial and hence the chromatic number for the graph of the figure 7.5.6(a).

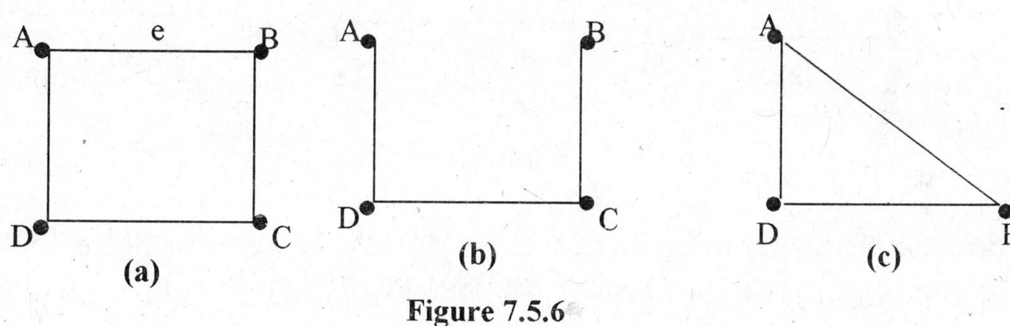

Figure 7.5.6

Solution: Let us remove the edge e to get subgraph G_e as shown in the figure (b) of 7.5.6(b). Let the number of colors be x. Then the chromatic polynomial P_S for G_e is given by

$$P_S = x(x-1)(x-1)(x-1).$$

Now, if we merge the nodes A and B, we get quotient graph G_Q as shown in the figure 7.5.6(c). Then, we have

$$P_Q = x(x-1)(x-2)$$

Using the result $P_G = P_S - P_Q$, we get

$$\begin{aligned} P_G &= x(x-1)^3 - x(x-1)(x-2) \\ &= x(x-1)\{(x-1)^2 - (x-2)\} \\ &= x(x-1)(x^2 - 3x + 3). \end{aligned}$$

From the polynomial, we have X(G) = 2.

Ans.

Example 3 Find the chromatic polynomial and hence the chromatic number for the graph shown in the figure 7.5.7(a).

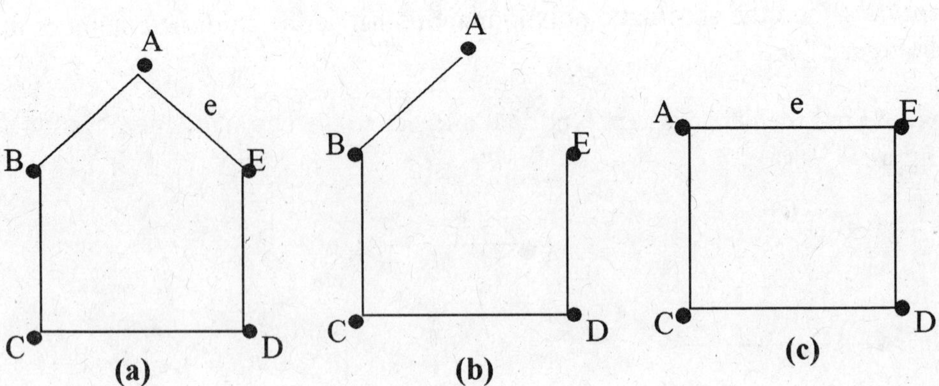

Figure 7.5.7

Solution: Removing the edge e from G, we get subgraph G_e of the figure 7.5.7(b) If the number of colors be x, the chromatic polynomial P_S for G_e is given by

$$P_S = x(x-1)(x-1)(x-1)(x-1)$$

And, if we merge the nodes A and E, we get quotient graph G_Q of the figure 7.5.7(c), which is same as the graph of figure 7.5.6(a). Thus, we have

$$P_Q = x(x-1)(x^2 - 3x + 3).$$

Therefore, we have

$$\begin{aligned} P_G &= x(x-1)^4 - x(x-1)(x^2 - 3x + 3). \\ &= x(x-1)\{(x-1)^3 - (x^2 - 3x + 3)\} \\ &= x(x-1)(x-2)(x^2 - 2x + 2) \end{aligned}$$

From the polynomial, we have X(G) = 3.

Ans.

The above theorem is very useful in finding the chromatic polynomial and chromatic number of any graph. However, reader may find the following heuristic approach more interesting. Let $v_1, v_2, v_3, \ldots, v_n$ be n vertices in a graph G. Arrange these vertices in the descending order of their degrees. Start coloring from the node having highest degree, say v_k. Let there are x colors. Then v_k can be colored in x ways. Then proceed to color the adjacent node of v_k having highest degree. This can be colored in (x −1) ways. Proceed in the same way until all the nodes have been colored properly. Finally use the multiplication principle of counting to find P_G. See the following example for the illustration.

Example 4 Find the chromatic polynomial and hence the chromatic number for the graph of the figure 7.5.8.

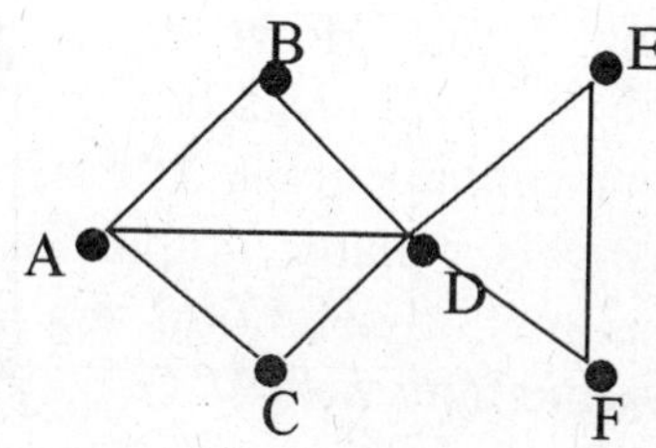

Figure 7.5.8

Solution: The degree of node D is 5, of node A is 3 and of the remaining nodes is 2. Let there are x colors. Since D has the highest degree, we start coloring from D. D can be colored in x ways. The adjacent nodes of D having highest degree is A. It will be colored next, and there are (x – 1) ways to do this. B and C are adjacent to A and both have equal degree. The color used for A and D cannot be used for either B or C, but both B and C may have, possibly, the same color. Thus B and C each can be colored in (x – 2) ways. For the node E we have (x – 1) options to choose from. Finally, to color the node F, color used for the node D and E cannot be used. Hence we are left with (x – 2) options. Therefore, we have

$$P_G = x(x-1)(x-2)(x-2)(x-1)(x-2) = x(x-1)^2(x-2)^3$$

Ans.

It is important to note that chromatic number of a subgraph cannot exceed the chromatic number of its parent graph. There are many such observations mentioned below that may be found very useful in proving theorems and corollary related to a planar graph.

1. Chromatic number of any discrete graph D_n is always one.
2. Chromatic number of any linear graph L_n is always 2 for $n > 1$.
3. Chromatic number of any bipartite graph is always 2.
4. Chromatic number of any tripartite graph is always 3.
5. Chromatic number of a complete graph K_n is n.
6. A cyclic graph C_n has chromatic number 3 for $n > 2$.
7. An n-ary tree having vertex ≥ 2, has always a chromatic number 2.

Planar Graph

A graph corresponding to a map is ***planar*** i.e. if it can be drawn on a 2-dimentional plane so that there is no false crossing of edges. If there is a map, we can always draw a corresponding graph. However, it is not possible to draw a map corresponding to every graph. It is ironical but true. A graph that does not have a corresponding map is called ***non-planar*** graph. If it is possible to draw a graph such that it has no overlapping regions, it is a planar graph. This is another interpretation of the above fact that can be taken as another definition of a planar graph. This concept is used tremendously in the currently emerging field of Geographical Information System (GIS). Researchers and graph suave Mathematicians have developed quite a few methods to determine whether a graph is planar or non-planar.

Example 5 Show that K_n is a planar graph for $n \leq 4$ and non-planar for $n \geq 5$.

Solution: A K_4 graph can be drawn in the way as shown in the figure 7.5.5(b). This does not contain any false crossing of edges. Thus, it is a planar graph. Graphs K_1, K_2 and K_3 are by construction a planar graph, since they do not contain a false crossing of edges. K_5 is shown below in the figure 7.5.9.

It is not possible to draw this graph on a 2-dimentional plane without false crossing of edges. Whatever way we adopt, at least one of the edges, say **e**, must cross the other for graph to be completed. Hence K_5 is not a planar graph. For any $n > 5$, K_n must contain a subgraph isomorphic to K_5. Since K_5 is not planar, any graph containing K_5 as its one of the subgraph cannot be planar.

Ans.

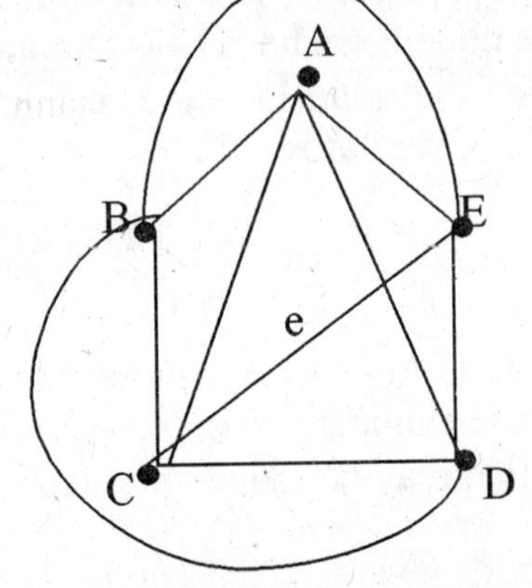

Figure 7.5.9

A theorem called "***four color theorem***" has been proved by exhaustive method, taking almost 2000 graphs of different types. According to this theorem, if a graph is planar, its chromatic number cannot exceed 4 i.e. any graph having chromatic number greater than 4, is a non-planar graph. However, the theorem does not ensure that a graph having chromatic number less than or equal to 4, is always a planar graph. Below is an example that shows that a graph has chromatic number 2 and yet it is a non-planar graph.

Example 6 Show that $K_{3,3}$ is a non-planar graph.

Solution: Graph $K_{3,3}$ is shown in the figure 7.5.5(c). It is not possible to draw this graph such that there is no false crossing of edges. This is classic problem of designing direct lanes without intersection between any two houses, for three houses on each side of a road.

In this graph there exists an edge, say **e,** that cannot be drawn without crossing another edge. Hence $K_{3,3}$ is a non-planar graph. It is easy to determine that the chromatic number of this graph is 2.

Ans.

The graph K_5 and $K_{3,3}$ are called **Kuratowski's Graphs**. Any graph obtained by adding one or more vertices or edges or both to either K_5 or $K_{3,3}$ will again be non-planar i.e. any ***super graph*** of K_5 or $K_{3,3}$ is non-planar. This fact can be stated as "*Any graph G is planar iff it does not contain a subgraph isomorphic to either* K_5 *or* $K_{3,3}$." This statement is also known as **Kuratowski's Theorem**. Again this is not a simple task to determine all possible subgraph of a graph G and then find whether any of them is isomorphic to K_5 or $K_{3,3}$.

To simplify the process of determining whether a graph is planar or non-planar a formula has been developed by Euler in terms of number of vertices(v), number of edges(e) and number of regions(r) in a graph. Before discussing this formula, we shall establish a relationship between e and r.

Theorem 3 Sum of the degrees of all regions in a map is equal to twice the number of edges in the corresponding graph.

Proof: As discussed earlier, a map can be drawn as a graph, where regions of the map is denoted by vertices in the graph and adjoining regions are connected by edges. Degree of a region in a map is defined as the number of adjoining region. Thus, degree of a region in a map is equal to the degree of the corresponding vertices in the graph. We know that the sum of the degrees of all vertices in a graph is equal to the twice the number of edges in the graph. Therefore, we have

$$2e = \sum \deg(R_i)$$

Proved.

Theorem 4 **(Euler Theorem)** Let G = (V, E, γ) be any connected planar graph such that $|V| = v$, $|E| = e$ and r is the number of regions in the graph. Then

$$r = e - v + 2$$

Proof: Let G_n be a graph having n edge. We shall prove this theorem by mathematical induction on number of edges. As a **basis step** of induction let us take a graph of single edge. A graph of single edge may have either v = 1 or v = 2. When v = 1, number of region = 2: one inside the loop (closed polygon) and other outside the loop (open polygon). And, when v = 2 and e = 1, we have r = 1. This is shown in the figure 7.5.10. The formula r = e – v + 2 is satisfied in both the cases.

Figure 7.5.10

Inductive Step: Let us suppose that the formula holds good for G_{k-1}. Let G_k be any graph with k. edges. If there is any dangle node in G_k, removal of it together with the arc terminating at it, reduces the graph to G_{k-1}. The formula is satisfied for G_{k-1}. Now putting back the removed arc and node in the graph increases e and v by one leaving r unchanged. Thus formula is again satisfied.

If there is no dangle node in G_k then all edges are parts of some regions. Removing an edge from the graph reduces r by one and leaves v unchanged. Also G_k becomes G_{k-1} for which formula is satisfied. Now we place back the removed arc into the graph. It increments both e and r by 1. The formula is again satisfied. Therefore the Euler formula is true for all $k \geq 1$.

Proved.

Corollary: In any connected simple planar graph without loop and $v \geq 3$

$$e \leq 3v - 6.$$

Proof: In a simple graph without loop (every edge is assumed to be straight-line segment and no curved line segment), a region is formed by at least 3 edges. If there are r regions in the graph then number of edges ≥ 3r. Also a single edge can be on common boundary of at the most two regions. This implies that the sum of the number of edges constituting boundary of all regions cannot exceed the sum of the degrees of all regions. And this value is 2e. Therefore,

$$2e \geq 3r \quad or, \quad \frac{2e}{3} \geq r$$

Using this relation in Euler formula, we get

$$\frac{2e}{3} \geq e - v + 2 \quad or, \quad \frac{2e}{3} - e \geq -v + 2 \quad or, \quad -\frac{e}{3} \geq -v + 2 \quad or, \quad -e \geq -3v + 6$$

$$or, \quad e \leq 3v - 6$$

Proved.

Example 7 Prove that K_4 and $K_{2,2}$ are planar.

Solution: In K_4, we have v = 4 and e = 6. Obviously, 6 ≤ 3*4 – 6 = 6. Thus this relation is satisfied for K_4. For $K_{2,2}$, we have v = 4 and e = 4. Again in this case, the relation e ≤ 3v – 6 i.e. 4 ≤ 3*4 – 6 = 6 is satisfied. Hence both K_4 and $K_{2,2}$ are planar.

Ans.

It is important to note that the application of this formula depends upon drawing of the graph. A false crossing may give incorrect number of regions in a graph and hence wrong interpretation.

7.6 Definitions and Types of Tree

See the graph of figure 7.6.1. There are 12 vertices and 11 edges. The graph is connected. It is acyclic. *A graph that contains no cycle (circuit or loop) is called an **acyclic** graph.* There is unique path from node A to every other node in this graph. This graph is of the form of an inverted plant (root at the top and leaves at the bottom). A graph of this type, resembling an inverted plant, is known as ***tree*** in our computer science terminology. Thus, we can say that a tree is a particular type of graph such that it is connected, it has no loop or circuit of any length and number of edges in it is one less than the number of vertices.

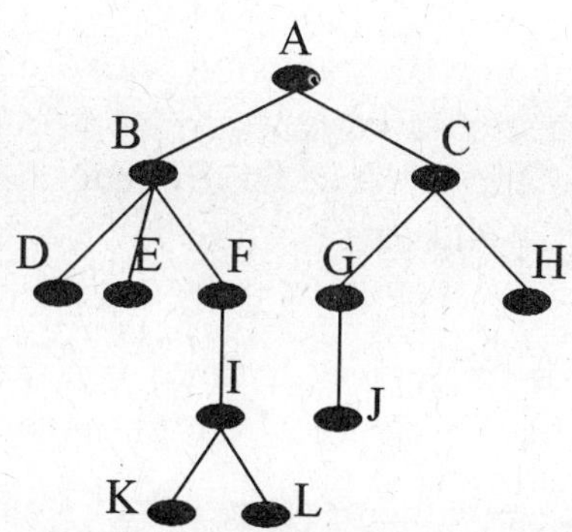

Figure 7.6.1

If the downward slop of an arc is taken as the direction of the arc then the above graph can be treated as a directed graph. The corresponding tree is called a ***directed tree***. A directed tree is also called ***arborescence***. The indegree of node A is zero and of all other nodes is one. There are nodes like D, E, K, L, J and H with outdegree zero and all other nodes have outdegree > 0. A node with indegree zero in a tree is called ***root*** of the tree. *Every tree has one and only one root and that is why a directed tree is also called a* ***rooted tree***. A node having outdegree zero is called ***leaf*** of the tree. A node with indegree one and outdegree greater than zero is called ***internal*** node of the tree. In the above graph, A is the root, D, E, K, L, J, H are leaves and B, C, F, G, I are internal nodes.

In a tree every node is reachable through a unique path. While descending from root towards leaf node, we may encounter many internal nodes. A node that can be reached from root through a path of one arc is called ***first level node*** of the tree. The node that can be reached through a path of two arcs is called ***second level node*** of the tree and so on. A node that can be reached through a path of n arcs is called ***n^{th} level node*** of the tree. In the tree of figure 7.6.1, B and C are first level nodes whereas nodes K and L are of 4th level node. The longest path from root to any leaf is called ***depth or height of the tree***. The depth of the above tree is 4. All nodes that can be reached from a node while descending from it are called ***descendent*** of that node. A first level descendents of a node are called its ***children*** and node itself is called ***parent*** of its children. In the above tree, A is parent of B and C and B and C are children of A. Similarly, K and L are children of I and I is parent of K and L.

Theorem 1 A tree with n vertices has (n – 1) edges.

Proof: From the definition of a tree a root has indegree zero and all other nodes have indegree one. There must be (n – 1) incoming arcs to the (n – 1) non-root nodes. If there is any other arc, this arc must be terminating at any of the nodes. If the node is root, then its indegree will become one and that is in contradiction with the fact that root always has indegree zero. If the end point of this extra edge is any non-root node then its indegree will be two, which is again a contradiction. Hence there cannot be more arcs. Therefore, a tree of n vertices will have exactly (n – 1) edges.

Proved.

Example 1 A tree has five vertices of degree 2, three vertices of degree 3 and four vertices of degree 4. How many vertices of degree 1 does it have?

Solution: Let x be the number of nodes of degree one. Thus, total number of vertices = 5 + 3 + 4 + x = 12 + x. The total degree of the tree = 5×2 + 3×3 + 4×4 + x = 35 + x. Therefore number of edges in the three is half of the total degree of the tree. If G = (V, E, γ) be the tree then, we have

$$|V| = 12 + x \quad and \quad |E| = \frac{35 + x}{2}$$

In any tree, |E| = |V| –1. Therefore, we have

$$\frac{35 + x}{2} = 12 + x - 1$$

$$or \quad 35 + x = 24 + 2x - 2$$

$$or \quad x = 13$$

Thus, there are 13 nodes of degree one in the tree.

Ans.

Example 2 A tree has 2n vertices of degree 1, 3n vertices of degree 2 and n vertices of degree 3. Determine the number of vertices and edges in the tree.

Solution: It is given that total number of vertices in the tree is 2n + 3n + n = 6n. The total degree of the tree is 2n×1 + 3n×2 + n×3 = 11n. The number of edges in the tree will be half of 11n. If G = (V, E, γ) be the tree then, we have

$$|V| = 6n \quad and \quad |E| = \frac{11n}{2}$$

In any tree, |E| = |V| –1. Therefore, we have

$$\frac{11n}{2} = 6n - 1$$

$$or \quad 11n = 12n - 2$$

$$or \quad n = 2$$

Thus, there are 6×2 = 12 nodes and 11 edges in the tree.

Ans.

In a tree, the root or any internal node may have more than one child. A tree is called n-ary tree if the maximum number of descendents for any node is n. As a special case, if the maximum number of children for any node in a tree is two, the tree is called a ***binary tree***. If every non-leaf node has 2 children, the tree is called ***complete binary tree***. Similarly, a tree is called ***complete n-ary tree*** if every non-leaf node has exactly n descendents.

Theorem 2 A complete n-ary tree with m non-leaf nodes contains $n \times m + 1$ nodes.

Solution: Since there are m internal nodes, and each internal node has n descendents, there are $n \times m$ nodes in tree other than root node. Since there is one and only one root node in a tree, the total number of nodes in the tree will $n \times m+1$.

Proved.

Theorem 3 There are at the most n^h leaves in an n-ary tree of height h.

Solution: Let us prove this theorem by mathematical induction on the height of the tree. As basis step take h = 0 i.e. tree consists of root node only. Since $n^0 = 1$, the basis step is true. Now let us assume that the above statement is true for h = k i.e. an n-ary tree of height k has at the most n^k leaves. If we add n nodes to each of the leaf node of n-ary tree of height k, the total number of leaf nodes will be at the most $n^h \times n = n^{h+1}$. Hence inductive step is also true. This proves that above statement is true for all $h \geq 0$.

Proved.

Theorem 4 In a complete n-ary tree with m non-leaf nodes, the number of leaf node, l, is given by the formula.

$$l = \frac{(n-1)(x-1)}{n}$$

Where, x is the total number of nodes in the tree.

Solution: It is given that the tree has m non-leaf nodes and it is complete n-ary, so total number of nodes $x = n \times m + 1$. Thus, we have

$$m = \frac{(x-1)}{n}$$

It is also given that l is the number of leaf nodes in the tree. Thus, we have

$$x = m + l + 1.$$

Substituting the value of m in this equation, we get

$$x = \frac{(x-1)}{n} + l + 1$$

$$or, \quad l = \frac{(n-1)(x-1)}{n}$$

Proved.

A tree is always separable. Deletion of any one internal node or root node disconnects the tree. Thus any internal node or root node is a cut point in a tree. Each component of a disconnected tree is a subtree of the tree. A forest consists of many trees.

A disconnected tree also presents multiple trees if each subtree is treated as a tree. A collection of trees is called ***forest.*** A disconnected tree is an example of forest. The following theorems are based on the connectivity of a tree.

Theorem 5 If T = (V, E, γ) be a rooted tree with v_0 as its root then

1. T is a acyclic.
2. v_0 is the only root in T.
3. Each node other than root in T has indegree 1 and v_0 has indegree zero.

Proof: We prove the theorem by the method of contradiction

1. Let there is a cycle π in T that begins and end at a node v. Since the indegree of root is zero, $v \neq v_0$. Also by the definition of tree, there must be a path from v_0 to v, let it be p. Then πp is also a path, distinct from p, from v_0 to v. This contradicts the definition of a tree that there is unique path from root to every other node. Hence T cannot have a cycle in it i.e. a tree is always acyclic.
2. Let v_1 is another root in T. By the definition of a tree, every node is reachable from root. Thus, v_0 is reachable from v_1 and v_1 is reachable from v_0 and the paths are π_1 and π_2 respectively. Then, $\pi_1\pi_2$, combination of these two paths is a cycle from v_0 to v_0. Since a tree is always acyclic, v_0 and v_1 cannot be different. Thus, v_0 is a unique root.
3. Let w be any non-root node in T. Thus, $\exists$ a path π: $v_0, v_1 \ldots v_k$ w from v_0 to w in T. Now let us suppose that indegree of w is two. Then $\exists$ two nodes w_1 and w_2 in T such that edges (w_1, w) and (w_2, w) are in E. Let π_1 and π_2 be paths from v_0 to w_1 and w_2 respectively. Then, π_1:$v_0v_1\ldots v_k\, w_1w$ and π_2:$v_0v_1\ldots v_k\, w_2w$ are two possible paths from v_0 to w. This is in contradiction with the fact that there is unique path from root to every other nodes in a tree. Thus indegree of w cannot be greater than 1.

Next, let indegree of $v_0 > 0$. Then $\exists$ a node v in T such that $(v, v_0) \in E$. Let π be a path from v_0 to v, thus $\pi(v, v_0)$ is a path from v_0 to v_0 that is a cycle. This is again a contradiction with the fact that any tree is acyclic. Thus indegree of root node v_0 cannot be greater than zero.

Proved.

Example 3 Let T = (V, E, γ) be a rooted tree. Obviously E is a relation on set V. Show that

- E is irreflexive.
- E is asymmetric
- If $(a, b) \in E$ and $(b, c) \in E$ then $(a, c) \notin E$, $\forall a, b, c \in V$.

Solution: Since a tree is acyclic, there is no cycle of any length in a tree. This implies that there is no loop in T. Thus, $(v, v) \notin E \; \forall a \in V$. Thus E is an irreflexive relation on V.

Let $(x, y) \in E$. If $(y, x) \in E$, then there will be cycle at node x as well as on node y. Since no cycle is permissible in a tree, either pair (x, y) or (y, x) can be in E but never both. This implies that presence of (x, y) excludes the presence of (y, x) in E and vice versa. Thus E is a asymmetric relation on V.

Let $(a, c) \in E$. Thus presence of pairs (b, c) and (a, c) in E implies that c has indegree two. This in contradiction with the fact that in a tree no non-root node can have indegree > 1. Hence $(a, c) \notin E$.

Ans.

Example 4 Prove that a tree T is always separable.

Solution: Let w be any internal node in T and node v is the parent of w. By the definition of a tree, in degree of w is one. If w is dropped from the tree T, the incoming edge from v to w is also removed. Therefore all children of w will be unreachable from root and tree T will become disconnected. See the forest of the figure 7.6.2, which has been obtained after removal of node F from the tree of figure 7.6.1.

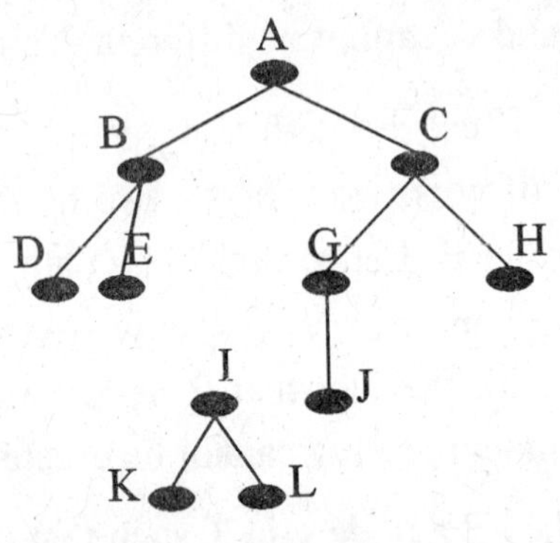

Figure 7.6.2

Ans.

7.7 Tree and Graph Traversals

We have studied in basic programming techniques that while searching for an item in a list, the list has to be traversed from one end to the other end until the desired item is found or list is exhausted. Traversal of tree or graph means exactly the same i.e. tree searching or graph searching. We traverse a graph or tree for a purpose. The purpose may be for searching or sorting of the items contained in a tree. A tree may contain an item at its node as a label. A tree in which every node is labeled is called a ***labeled tree***.

The tree of the figure 7.6.1 is an example of a labeled tree. The tree shown in the figure 7.7.1 is not a labeled tree.

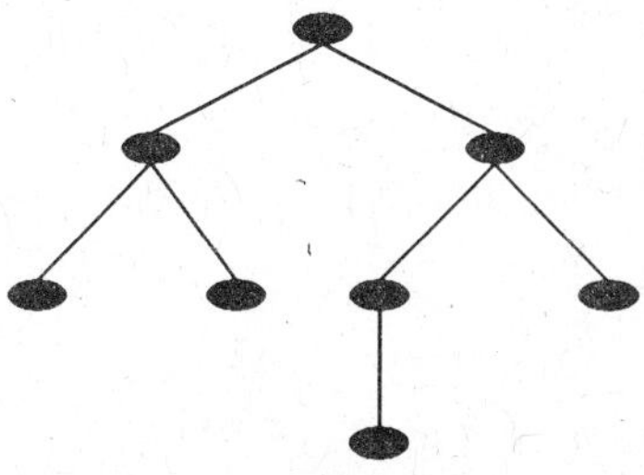

Figure 7.7.1

If we traverse this tree, it is of no use. It is just like roaming in the street without purpose. Thus whenever we talk about traversal of a tree or graph, we talk about a labeled tree or graph. Now consider the following tree of the figure 7.7.2.

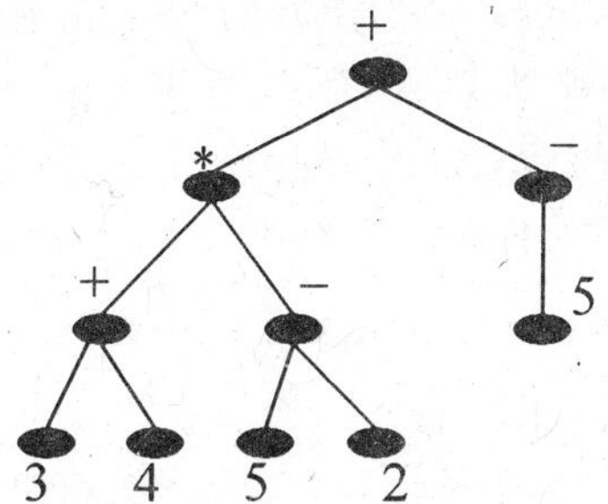

Figure 7.7.2

This tree corresponds to the mathematical expression (3+4)*(5–2)+(-5). As we have studied in our basic data structures course, a mathematical expression can be expressed in prefix, postfix or infix notations. The expression written above is in infix form. The tree of the figure 7.7.2 is also called an expression tree. A labeled tree corresponding to a mathematical, boolean or logical expression is also called an ***expression tree***. In an expression tree, position of every node is fixed because any change in the position of a node will change the expression itself. A labeled tree in which position of every node is fixed is called ***positional tree***. The tree of the figure 7.7.2 is a positional tree. The same expression written above can be written in prefix form as + * + 3 4 – 5 2 –5 and in postfix form as 3 4 + 5 2 – * 5 – +. The three forms of expressions can very well be obtained by just traversal of the tree. Traversing a tree is a recursive procedure. To implement this a tree is considered to have three components: root, left subtree and right subtree. These three components can be arranged in six different ways: *(left, root, right), (root, left, right), (left, right, root), (right, left, root), (right, root, left)* *and (root, right, left)*. The first three are used whereas the last three combinations are of no use as it alters the positions of a node in a positional tree.

Inorder Traversal

In this form of traversal a tree is traversed in the sequence

- Left subtree
- Root
- Right subtree

Consider the expression tree of figure 7.7.2. To traverse this in inorder, we begin at the root node marked, +. Since first we have to traverse its left subtree, so move to the root of left subtree i.e. node marked, *. Again it has a left subtree with root node marked, +. Visit it. This subtree has a node labeled 3, which has no left subtree, so out put 3. Then root of this subtree i.e. '+' and then right subtree which is again a node labeled with 4, so output it. Thus we have expression obtained till here is 3 + 4. Proceeding this way we get (3+4)*(5–2)+(-5). Parentheses signify both precedence and portion of the subtree to which this sub-expression corresponds.

Preorder Traversal

In this form of traversal a tree is traversed in the sequence

- Root
- Left subtree
- Right subtree

Again consider the expression tree of figure 7.7.2. To traverse this in preorder, we begin at the root, output its label i.e. +. Then move to its left subtree. Output the label of the root of the left subtree i.e. *. Then move to the left subtree of the current subtree and output its root. Continue till we get a leaf node. Out put the label of the leaf node. Them move to right subtree of the current subtree and proceed as above. Apply the algorithm recursively till all nodes have been visited. Proceeding this way we get + * + 3 4 – 5 2 –5. Parentheses are not needed here as the sequence of operators and operands take care of the precedence of operators in a preorder expression.

Postorder Traversal

In this form of traversal a tree is traversed in the sequence

- Left subtree
- Right subtree
- Root

We shall continue with the same expression tree to make the concept more clear to the reader. To traverse this in postorder, we begin at the root node marked, +. Since first we have to traverse its left subtree, so move to the root of left subtree i.e. node marked, *. Again it has a left subtree with root node marked, +. Visit it. This subtree has a node labeled 3, which has no left subtree, so output 3. Then right subtree, which is again a node labeled with 4, so output it. Then finally root of this subtree i.e. '+'. Thus we have expression obtained till here is 3 4 +. Proceeding this way we get 3 4 + 5 2 – * 5 – +. Parentheses are not needed here too as the sequence of operators and operands take care of the precedence of operators in a postorder expression as in a preorder expression.

It is obvious that an infix notation is obtained by traversing the expression tree in inorder, prefix notation is obtained by traversing the tree in preorder and postfix notation is obtained if the tree is traversed in postorder. For a graph that is not a tree, we cannot traverse a graph in any of the three way discussed above. There are other two important types of tree traversal: Breadth First Search (BFS) and Depth First Search (DFS). These traversal methods we can use for both tree and graph. Consider the labeled tree of the figure 7.6.1 and directed graph of the figure 7.7.3 for the illustration of DFS and BFS.

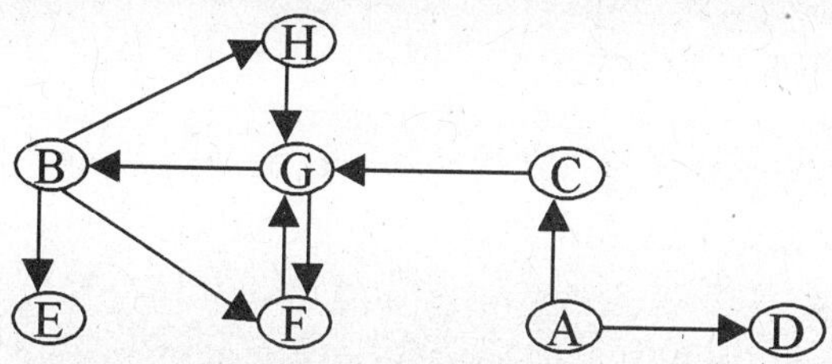

Figure 7.7.3

Breadth First Search (Traversal)

In this traversal a tree is traversed in the order of level of nodes. At a particular level, nodes are traversed from left to right. First visit the root node. Then go to the first level nodes and visit all nodes from left to right. Then go to the second level and visit all nodes from left to right and so on until all nodes have been visited. The algorithm is stated as *"visit all successors of a visited node before visiting any successors of any of those successors"*. If we visit the tree of the figure 7.6.1, we get the sequence of node as **A B C D E F G H I J K L**.

To traverse a graph using this algorithm, first select a node to start with because there is no node in a graph which we can designate as root node and start traversing from there. Select a node having no indegree, if any, to start with. If there is no such node, select a node that has minimum indegree. Then follow the above visiting criteria. In the given graph of 7.7.3, node A is of indegree 0. We start traversing this graph from this node. Then we visit all of its successors. A has C and D as its successors. Then visit the successors of C and D. There is no successor of D. We move to the successors of C. There is only one successor of C that is G. G has two successors B and F. visit them. F has only one successor that is G, which has already been visited. So move to the successor of B. B has two unvisited successors E and H. Visit them. H has one visited successor, so leave it. Thus path traced in the graph by BFS is A C D G B F E H.

Depth First Search (Traversal)

A depth first traversal of a tree is similar to the preorder traversal of a tree. Thus we are not discussing this for a tree. In DFS, a graph is traversed along a single path of the graph as far as it can go until a node with no successor is visited or a node all of which successors already have been visited. Next, resume the traversal of the graph at the last node on the path just traversed that has an unvisited successor. Continue traversing the graph until all nodes in the graph have been visited.

To traverse a graph using this algorithm, again selection of a node to start with is required. Select a node having no indegree, if any, to start with. If there is no such node, select a node that has minimum indegree. Then follow the above visiting criteria. In the given graph of 7.7.3, node A is of indegree 0. We start traversing this graph from this node. Then go to node C. From the node C go to the node G, then to the node B then to E. Node E has no successor, so backtrack to the node B. Then visit H. H has visited successor G, so backtrack to B. Now visit node F. Again node F has a visited successor so backtrack to B then to G then to C and finally to A. A has one unvisited successor D. Visits it and complete the traversal of the graph. Thus path traced in the graph by DFS is A C G B E H F D.

7.8 Minimum Spanning Tree

Let G = (V, E, γ) be a connected graph. Any connected spanning subgraph H = (V, E_1, γ_1) of graph G is called ***spanning tree*** if H is a tree i.e. $|E_1| = |V| - 1$. The complement of a spanning tree with respect to the graph is called ***cotree*** of the graph. Let $|V| = n$ and $|E| = m$. Since a spanning tree of G must contain exactly (n – 1) edges, the remaining $(m - (n - 1)) = m - n + 1$ edges will be in its cotree. Number of edges in a cotree of a graph G is called ***Cyclomatic number*** of the graph. The concept of a spanning tree is used basically in communication network design for remote switch installation. Where to place all the required remote switches so that total cables requirement (cost) should be minimum. This requires finding of minimum spanning tree in a weighted graph. A spanning tree of a weighted graph is called ***minimum spanning tree*** if the sum of edges in the spanning tree is minimum.

See the graph of the figure 7.8.1. There are six edges and four nodes in the graph. A spanning tree of the graph will have 3 edges. Which three edges out of the six should be taken. In this graph we can find it easily because it is a simple case. A spanning subgraph with edges e_1, e_4 and e_5 is a spanning tree of the graph as shown in 7.8.2.

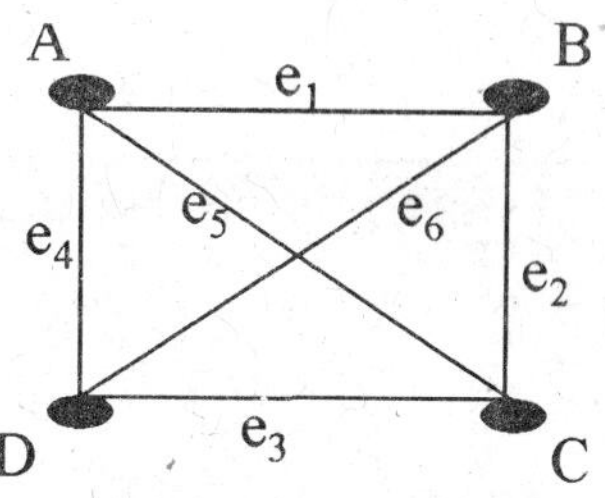

Figure 7.8.1

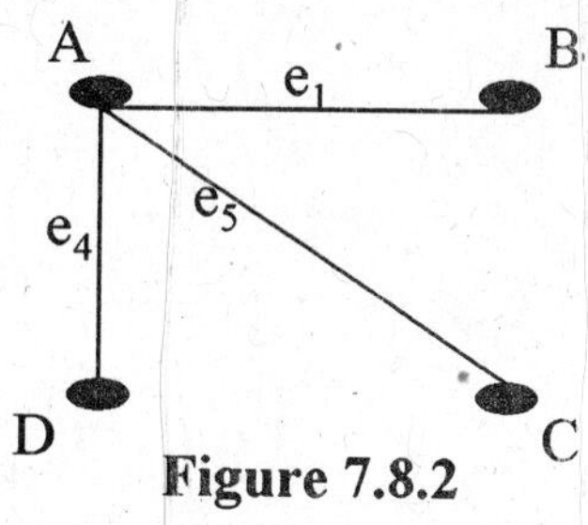

Figure 7.8.2

There are other possible spanning trees for this graph. User may use the formula discussed in section 7.2 under heading subgraph to find the maximum possible number of spanning tree in any connected graph. Now, we shall discuss an algorithm to find a spanning tree for a given connected graph. The algorithm is called Prim's algorithm.

Prim's Algorithm

Let $G = (V, E, \gamma)$ be a connected graph with $|V| = n$. Find the adjacency matrix for G. Now proceed according to the following steps.

Step 1: Select a vertex $v_1 \in V$ and arrange the adjacency matrix of the graph in such a way that the first row and first column of the matrix corresponds to v_1.

Step 2: Choose a vertex v_2 of V such that $(v_1, v_2) \in E$. Merge v_1 and v_2 into a new vertex, call it $v_m^{\,i}$ and drop v_2. Replace v_1 by $v_m^{\,i}$ in the graph. Find the new adjacency matrix corresponding to this new quotient graph.

Step 3: While merging select an edge from those edges which are going to be removed (or merged with other edge) from the graph. Keep a record of it.

Step 4: Repeat steps 1, 2 and 3 until all vertices have been merged into one vertex.

Step 5: Now construct a tree from the edges, collected at different iterations of the algorithm.

Example 1 Find a minimum spanning tree from the graph of the figure 7.8.1.

Solution: The working of the procedure is shown in the following table. We have selected node A to start with. The first row of the table contains the adjacency matrix of the given graph. If there are n nodes in a graph then merging process will continue up to nth iterations.

Matrix	**Nodes merged**	**Next node**	**Edge Kept**
$\begin{array}{c} \\ A \\ B \\ C \\ D \end{array}\begin{array}{c} \begin{matrix} A & B & C & D \end{matrix} \\ \begin{bmatrix} 0 & 1 & 1 & 1 \\ 1 & 0 & 1 & 1 \\ 1 & 1 & 0 & 1 \\ 1 & 1 & 1 & 0 \end{bmatrix} \end{array}$		B	
$\begin{array}{c} \\ A_m^1 \\ C \\ D \end{array}\begin{array}{c} \begin{matrix} A_m^1 & C & D \end{matrix} \\ \begin{bmatrix} 0 & 1 & 1 \\ 1 & 0 & 1 \\ 1 & 1 & 0 \end{bmatrix} \end{array}$	$A_m^1 = \{A, B\}$	C	(A, B)
$\begin{array}{c} \\ A_m^2 \\ D \end{array}\begin{array}{c} \begin{matrix} A_m^2 & D \end{matrix} \\ \begin{bmatrix} 0 & 1 \\ 1 & 0 \end{bmatrix} \end{array}$	$A_m^2 = \{A, B, C\}$	D	(A, C)
$\begin{array}{c} \\ A_m^3 \end{array}\begin{array}{c} A_m^3 \\ [0] \end{array}$	$A_m^3 = \{A, B, C, D\}$		(A, D)

Therefore, we get three arcs in the process. The tree constructed with these arcs is shown in the figure 7.8.2.

Ans.

The same Prim's algorithm with little modification can be used to find a minimum spanning tree for a weighted graph. The stepwise algorithm is given below. Let $G = (V, E, \gamma)$ be graph and $S = (V_S, E_S, \gamma_S)$ be the spanning tree to be found from G.

Step 1: Select a vertex v_1 of V and initialize

$$V_S = \{v_1\} \text{ and}$$
$$E_S = \{\}$$

Step 2: Select a nearest neighbor of v_i from V that is adjacent to some $v_j \in V_S$ and that edge (v_i, v_j) does not form a cycle with members edge of E_S. Set

$$V_S = V_S \cup \{v_i\} \text{ and}$$
$$E_S = E_S \cup \{(v_i, v_j)\}$$

Step 3: Repeat step2 until $|E_S| = |V| - 1$.

Example 2 Find the minimum spanning tree for the weighted graph of the figure 7.8.3.

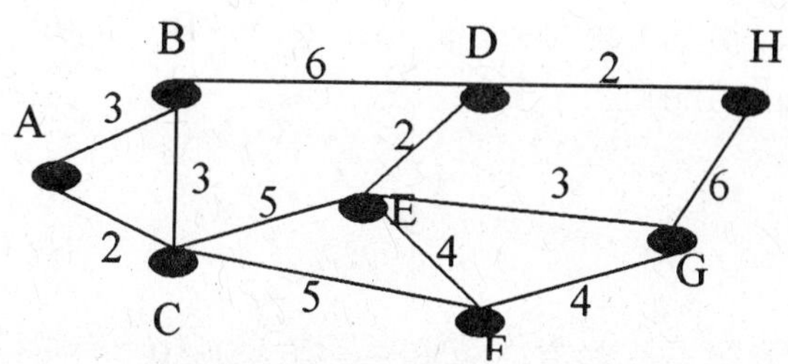

Figure 7.8.3

Solution: Let us begin with the node A of the graph. Let $S = (V_S, E_S, \gamma_S)$ be the spanning tree to be found from G. Initialize $V_S = \{A\}$ and $E_S = \{\}$. There are eight nodes so the spanning tree will have seven arcs. The iterations of algorithm applied on the graph are given below. The number indicates iteration number.

1. Nodes B and C are neighbors of A. Since node C is nearest to the node A we select C. Thus, we have $V_S = \{A, C\}$ and $E_S = \{(A, C)\}$.

2. Now node B is neighbor of both A and C and C has nodes E and F as its neighbor. We have AB = 3, CB = 3, CE = 5 and CF = 5. Thus, the nearest neighbor is B. We can select either AB or CB. We select CB. Therefore, $V_S = \{A, C, B\}$ and $E_S = \{(A, C), (C, B)\}$.

3. Now D, E, F are neighbor of nodes in V_S. An arc AB is still to be considered. This arc forms cycle with arcs AC and CB already in E_S so it cannot be selected. Thus we have to select from BD = 6, CE = 5, CF = 5. We may take either CE or CF. We select CF. Therefore, $V_S = \{A, C, B, F\}$ and $E_S = \{(A, C), (C, B), (C, F)\}$.

4. Now we have to select an arc from BD = 6, CE = 5, FE = 4, FG = 4. We select FE. Therefore, $V_S = \{A, C, B, F, E\}$ and $E_S = \{(A, C), (C, B), (C, F), (F, E)\}$.

5. The selection of arc CE is ruled out as it forms a cycle with the edges CF and FE. Thus, we have to select an arc from BD = 6, ED = 2, FG = 4. We select ED. Therefore, $V_S = \{A, C, B, F, E, D\}$ and $E_S = \{(A, C), (C, B), (C, F), (F, E), (E, D)\}$.

6. Now BD is ruled out as it forms cycle with CB, CF, FE and ED. Thus we have to consider DH = 2, EG = 3, FG = 4. We select DH. Therefore, V_S = {A, C, B, F, E, D. H} and E_S = {(A, C), (C, B), (C, F), (F, E), (E, D), (D, H)}.
7. Now left over arcs are EG = 3, HG = 6, FG = 4. We select EG. Therefore, V_S = {A, C, B, F, E, D, H, G} and E_S = {(A, C), (C, B), (C, F), (F, E), (E, D), (D, H), (E, G)}.

Since number of edges in E_S is seven process terminates here. The spanning tree so obtained is shown in the figure 7.8.4.

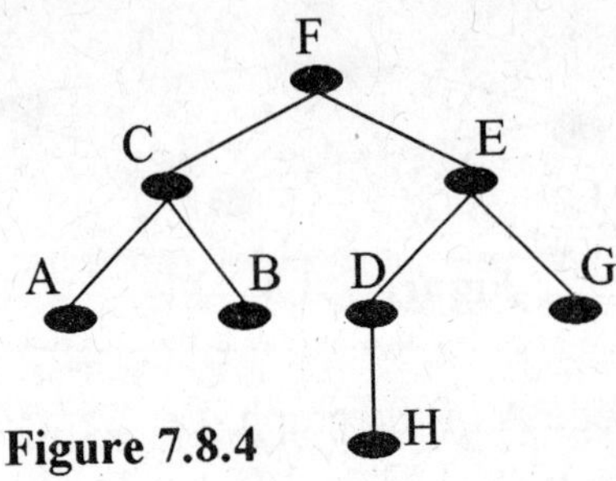

Figure 7.8.4

Ans.

It is obvious that Prim's algorithm is a **greedy algorithm** as it confines to the local conditions while determining for the optimal arcs. Unlike the many other greedy algorithms, Prim's algorithm always finds an optimal spanning tree. This has been proved mathematically as a theorem. But there is a problem regarding its efficiency. If there are n nodes in a graph, then order of this algorithm is $O(n^2)$. However, for a given n , if number of edges in a graph is relatively less than possible number n(n – 1)/2, then an algorithm called Kruskal's algorithm solves the same problem in more efficient way. This algorithm is also greedy algorithm.

Kruskal's Algorithm

Let G = (V, E, γ) be graph and S = (V_S, E_S, γ_S) be the spanning tree to be found from G. Let |V| = n and E = {e_1, e_2, e_3, …, e_m}. The stepwise algorithm is given below.

Step 1: Select an edge e_1 from E such that e_1 has least weight. Replace

$$E = E - \{e_1\} \text{ and}$$
$$E_S = \{e_1\}$$

Step 2: Select an edge e_i from E such that e_i has least weight and that it does not form a cycle with members of E_S. Set

$$E = E - \{e_i\} \text{ and}$$

$$E_S = E_S \cup \{e_i\}$$

Step 3: Repeat step2 until $|E_S| = |V| - 1$.

Example 3 Using Kruskal's algorithm, find the minimum spanning tree for the weighted graph of the figure 7.8.3.

Solution: Let $S = (V_S, E_S, \gamma_S)$ be the spanning tree to be found from G. There are eight nodes so the spanning tree will have seven arcs. The iterations of algorithm applied on the graph are given below and it runs at the most seven times. The number indicates iteration number.

1. Since arcs AC, ED and DH have minimum weight 2. Since they do not form a cycle, we select all of them and set E_S = {(A, C), (E, D), (D, H)} and E = E − {(A, C), (E, D), (D, H)}.
2. Next arcs with minimum weights 3 are AB, BC and EG. We can select only one of the AB and BC. Also we can select EG. Therefore, E_S = {(A, C), (E, D), (D, H), (B, C), (E, G)} and E = E − {(B, C), (E, G)}.
3. Next arcs with minimum weights 4 are EF and FG. We can select only one of them. Therefore, E_S = {(A, C), (E, D), (D, H), (B, C), (E, G), (F, G)} and E = E − {(F, G)}.
4. Next arcs with minimum weights 5 are CE and CF. We can select only one of them. Therefore, E_S = {(A, C), (E, D), (D, H), (B, C), (E, G), (F, G),(C, E)} and E = E − {(C, E)}.

Since number of edges in E_S is seven process terminates here. The spanning tree so obtained is shown in the figure 7.8.5.

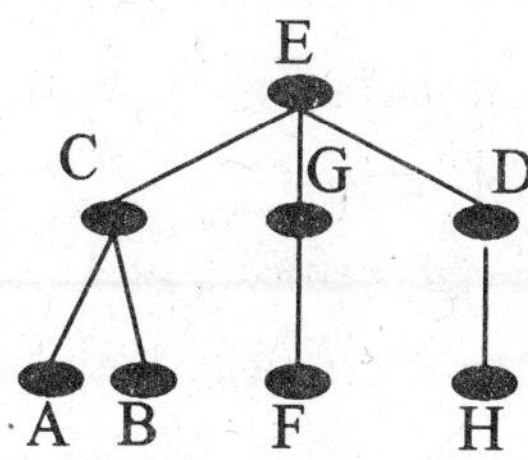

Figure 7.8.5

Ans.

The above algorithm on the same graph 7.8.3 terminated after four iterations instead of seven iterations in Prim's algorithm. If the number of edges is not too large the efficiency of this algorithm is even more noticeable. We have seen in this chapter the various application of graph theory e.g. connectivity analysis of computer network topology, minimum spanning tree for hierarchical computer network design, shortest path algorithm etc. These are the areas of computer science where the knowledge of graph theory is immensely used. Readers, by now, must have realized the importance of the topics covered in this chapter. That is why almost every book on discrete mathematics contains one or more chapters on graph theory. A collection of problems is provided in exercise below for hands on to the reader. An honest attempt to solve them will certainly help the reader in conceptualizing the basics of graph theory.

Exercise

1. In how many ways a finite graph can be represented? Write a program to converts between these representations.
2. Draw connected regular graphs of degree 0, 1, 2 and 3.
3. Use matrix multiplication method to determine the connected components of the graph whose vertex adjacency list is

1:	6, 8	7:	3, 5
2:	4, 10, 12	8:	1, 6
3:	5, 7, 9	9:	3
4:	2, 10	10:	2, 4, 12
5:	3, 7, 11	11:	5
6:	1, 8	12:	2, 10

4. Draw a complete graph K_n for n = 1, 2, 3, 4, 5 and 6.
5. Draw complete bipartite graph $K_{m, n}$ for m = 1, 2, 3 and n = m, m +1, m + 2.
6. Show that T is a tree iff it has no cycles and the addition of any edge forms a cycle.
7. Show that a tree T is a connected graph with n vertices and exactly n - 1 edges.
8. Show that sum of the degrees of the vertices of a graph is twice the number of edges of the graph.
9. Show that each cubic graph has an even number of vertices.
10. Show that there are exactly five cubic graphs on eight vertices.
11. Determine the degrees of the vertices of $K_{m, n}$. Hence determine the number of edges.
12. Show that a graph T on n vertices is a tree iff one of these properties holds:
    (a) T is connected, but the deletion of any one edge disconnects T.
    (b) For any two vertices u and v of T, there is one simple path from u to v;
    (c) T has exactly n – 1 edges and no cycles.
13. A forest is a collection of one or more trees. Show that any sub-graph of a tree is a forest.
14. Find the minimum and maximum possible degrees for a tree on n vertices.
15. If v(G) = e(G) = 1, G is necessarily a tree? Either prove it or give a counter example to disprove it.

16. Find all spanning trees for the graph of the figure 7.16ex.

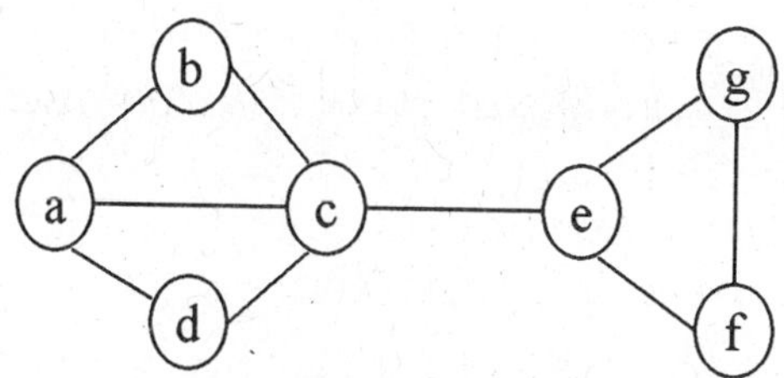

Figure 7.16ex

17. Prove that G is connected iff it has a spanning tree.

18. Does a connected graph G necessarily have a spanning tree that is a simple path?

19. How many spanning trees are there for

(a) A tree (b) K_m (c) $K_{m,n}$ (d) a connected regular graph of degree 2?

20. Show that a connected **k-regular** graph on n vertices has at the most

$$^{\frac{nk}{2}}C_{n-1} = \frac{\frac{nk}{2}\left(\frac{nk}{2}-1\right)\left(\frac{nk}{2}-2\right)\cdots\left(\frac{nk}{2}-n+2\right)}{1\cdot 2\cdot 3\cdot\cdots\cdot(n-1)}$$

spanning trees. Find a k-regular graph having fewer number of spanning trees. You may take a 3-regular graph on 8 vertices to start with.

21. Show that a graph is bipartite iff it has no cycles of odd length.

22. Find all cubic bipartite graphs on six eight and ten vertices.

23. Show that if a finite graph has no vertices of odd degree and no isolated vertices, then it must contain a cycle.

24. If a digraph has a spanning arborescence, what can be said about its connectivity type?

25. Let A be an adjacency matrix for a directed graph. If the graph is arborescence, what can be said about A and its different powers? Do their properties characterize arborescence?

26. Use the concept of topological sorting discussed in the chapter 6 to write out all the topological sorts of the graph shown in the figure 7.26ex.

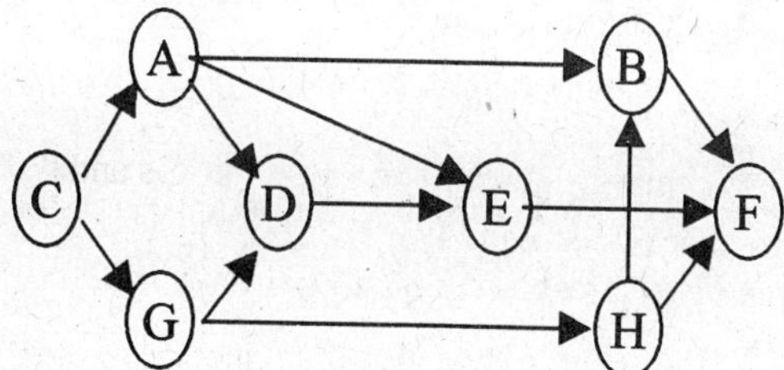

Figure 7.26ex

27. Find the tree corresponding to the following preorder expressions:
(a) + * 3 + a * b c + * 7 a * b + 3 c.
(b) + + * 3 * 4 a c + + 7 + * a b c 2.
(c) * + * + * + 1 a 2 b 3 c + * + + * + * c 3 b 2 a 1 a.
(d) . + ? * :: a b ? c d a . b c a + a c d
The stratification is
0: a, b, c, d
1: .
2: +, *
3. ?
4: ::

28. Construct the tree corresponding to the following post order expressions:
(a) a c 3 + b * c d c a + + c b / * b * + a + / *
(b) b b b * b * c b a a a / a + c c + * b + / a * + * c / + +

29. Construct the tree corresponding to the following level expressions:
(a) + a * + / b * c * c / b + d * a * c + b c b a.
(b) + * b + * a + c a / d + + * / b c a + d a b * c d.

30. Given that the string GDHBAECJIKF is the in-order representation of a binary tree and the string ABDGHCEFIJK is the preorder representation of the same tree, draw the tree.

31. (a) which of the platonic graph is bipartite?
(b) Is it possible for a bipartite graph to contain a triangle as a subgraph? (You may use chromatic number concept of a triangle)
(c) What is the maximum number of edges that a complete bipartite graph $K_{r, s}$ with a total of n vertices i.e. r + s = n, could have?
(d) If a complete bipartite graph $K_{r, s}$ is regular, what can you say about r and s?

32. A bipartite graph is used in a certain college to model the relationship between students and courses. Let the vertices in one part (say V) represent the students and the vertices in the other part (say U) represent the courses. What do the following represent?
(a) The degree of a vertex s in V
(b) The degree of a vertex c in U
(c) The fact that two vertices x and y are adjacent to the same vertex z in U.

33. Look at the adjacency matrices of a few complete bipartite graphs and describe the pattern you observe.

34. (a) Which complete bipartite graph is trivalent?
(b) Let S_n be star graph on n vertices. Find the chromatic number of S_5, S_8 and S_{10}
(c) Is star graph bipartite?
(d) Which complete bipartite graph is isomorphic to S_n?

35. (a) How many edges does $K_{r, s ,t}$ have?
(b) If a complete tripartite graph $K_{r, s ,t}$ is regular, what can you say about r, s and t?
(c) Which complete tripartite graph is regular of degree m?
(d) Why can't a complete tripartite graph be trivalent?
(e) Write down the adjacency matrix of $K_{2,3,4}$.
(f) What is the chromatic number of a complete tripartite graph?

36. If $G = (V, E, \gamma)$ be an undirected graph with k components and $|V| = n$, $|E| = m$ then prove that

$$m \geq n - k$$

37. n cities are connected by a road network of k highways. It is assumed that a highway is a road between two cities that does not go through any intermediate cities. Show that one can always travel between any two cities through connecting highways if

$$k > \frac{1}{2}(n-1)(n-2)$$

38. An ordered n-tuple $(d_1, d_2, d_3, \ldots, d_n)$ of nonnegative integers is said to be *graphical* if there exists a simple graph with no self loops that has n vertices with the degrees of the vertices being $d_1, d_2, d_3, \ldots, d_n$.
    (a) Show that (4, 3, 2, 2, 1) is graphical.
    (b) Show that (3, 3, 3, 1) is not graphical.
    (c) Without loss of generality, let $d_1 \geq d_2 \geq d_3 \geq \ldots \geq d_n$. Show that $(d_1, d_2, d_3, \ldots, d_n)$ is graphical iff $(d_2 - 1, d_3 - 1, \ldots, d_n - 1)$ is graphical.

39. (a) Show that the sum of the in-degrees over all vertices is equal to the sum of out-degrees over all vertices in any directed graph.
    (b) Show that the sum of the squares of the in-degrees over all vertices is equal to the sum of the squares of the out-degrees over all vertices, in any directed complete graph.

40. A graph is said to be *self-complementary* if it is isomorphic to its complement. Now answer the followings:
    (a) Show a self-complementary graph with four vertices.
    (b) Show a self-complementary graph with five vertices.
    (c) Is there a self-complementary graph with three vertices?
    (d) Show that a self-complementary must have either 4n or 4n + 1 vertices.

41. A set of vertices in an undirected graph is said to be a *dominating* set if every vertex not in the set is adjacent to one or more vertices in the set. A *minimal* dominating set is a dominating set such that no proper subset of it is also a dominating set. Now answer the followings:
    (a) For the graph shown in the figure 7.41ex, find two minimal dominating sets of different sizes.
    (b) Let the vertices of the graph represent cities and the edges denote road links between cities. Give a physical interpretation of the notion of a dominating set in this case.

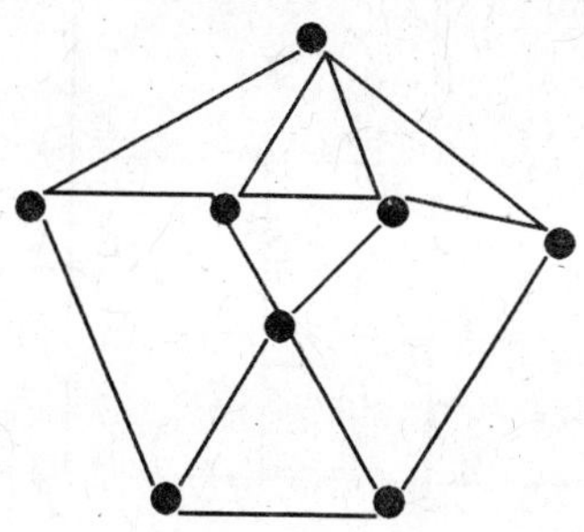

Figure 7.41ex

42. A set of vertices in an undirected graph is said to be an *independent set* if no two vertices in it are adjacent. A *maximal independent set* is an independent set if no proper super set of it is an independent set.
   (a) For the graph shown in the figure 7.41ex, find two maximal independent sets of different sizes.
   (b) How can the problem of placing eight queens on a chessboard so that no one captures another be stated in graph theoretic terms?

43. Show that an undirected graph can be properly colored with two colors iff it contains no circuit of odd length.

44. Apply Dijsktra's algorithm to find shortest path between node A and node Z in the graph shown in the figure 7.44ex (a) and 7.44ex (b) below.

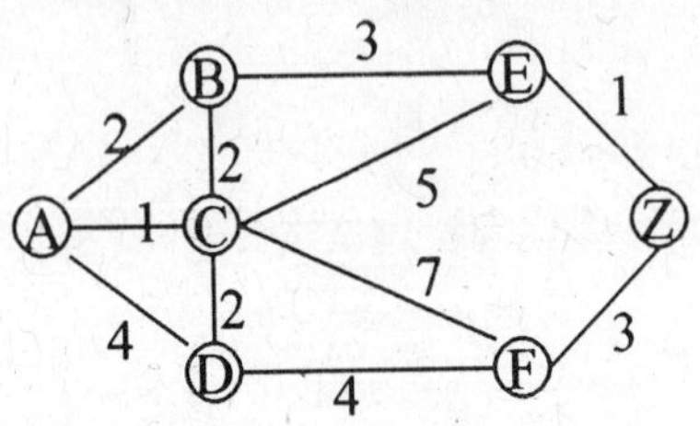

Figure 7.44ex(a)

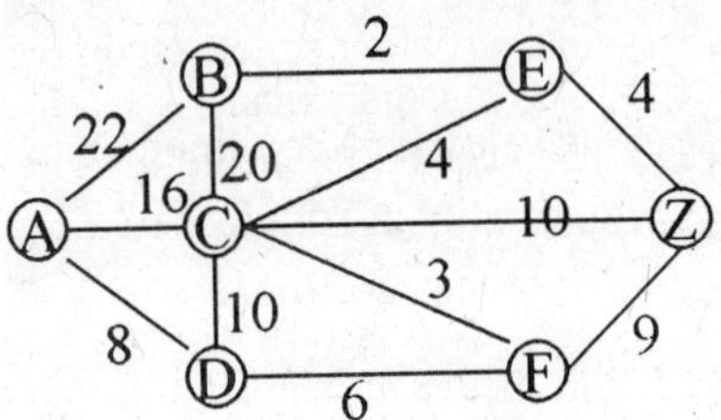

Figure 7.44ex(b)

45. Apply Floyd Warshall's algorithm to compute all pair shortest distance in the graphs shown in the figure 7.44ex (a) and 7.44ex (b) above.

46. Apply Dijsktra's algorithm to find shortest path from node A to all other nodes in the graph shown in the figure 7.44ex (a) and 7.44ex (b) above.

47. Using Dijsktra's algorithm find the shortest path between node A and node Z in the mesh graph of the figure 7.47ex.

Figure 7.47ex

48. (a) Draw a graph that has both Eulerian and Hamiltonian circuit.

    (b) Draw a graph that has an Eulerian but has no Hamiltonian circuit.

    (c) Draw a graph that has no Eulerian but has a Hamiltonian circuit

    (d) Draw a graph that has neither an Eulerian nor a Hamiltonian circuit

49. (a) Does K_{13} have an Eulerian circuit? A Hamiltonian circuit?

    (b) Repeat problem (a) for K_{14}.

50. (a) Do $K_{4,4}$, $K_{4,5}$ and $K_{4,6}$ have an Hamiltonian circuit?

    (b) Do $K_{4,4}$, $K_{4,5}$ and $K_{4,6}$ have a Hamiltonian path?

    (c) State a necessary and sufficient condition on the existence of a Hamiltonian circuit in $K_{m,n}$.

    (d) State a necessary and sufficient condition on the existence of a Hamiltonian path in $K_{m,n}$.

51. Let G be a complete directed graph. A nonempty subset of the vertices of G is said to be an ***outclassed group*** if any edge joining a vertex in the subset and a vertex not in the subset is always directed from the latter to the former. Show that G has a directed circuit containing all the vertices if there is no outclassed group of vertices. .

52. An undirected graph G is said to be ***orientable*** if direction can be assigned to each of the edges of the graph so that the resultant graph is strongly connected. Now answer the following:

    (a) Show that the graph shown in the figure 7.52ex is orientable.

    (b) Show that any graph with an Eulerian circuit is orientable.

    (c) Show that any graph with a Hamiltonian circuit is orientable.

    (d) Show that a connected undirected graph is orientable iff each edge of the graph is contained in at least one circuit.

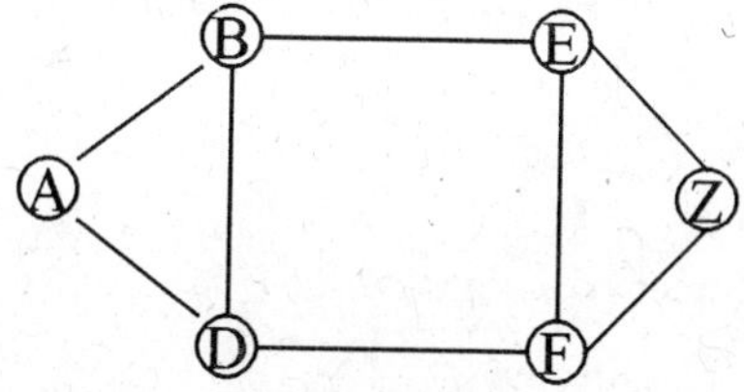

Figure 7.52ex

53. (a) Use the nearest-neighbor method to determine a hamiltonian circuit for the graph in the figure 7.53ex starting at the vertex a..

    (b) Repeat part (a) starting at vertex d.

    (c) Determine a minimum Hamiltonian circuit for the graph.

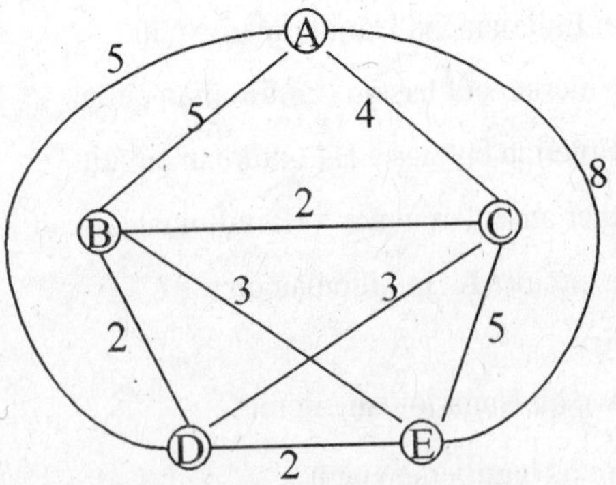

Figure 7.53ex

54. (a) Show that a linear planar graph has a vertex of degree 5 or less.
    (b) Show that a linear planar graph with less than 30 edges has a vertex of degree 4 or less.

55. A tree has 2n vertices of degree 1, 3n vertices of degree 2, and n vertices of degree 3. Determine the number of vertices and edges in the tree.

56. (a) A tree has two vertices of degree 2, one vertex of degree 3 and three vertices of degree 4. How many vertices of degree 1 does it have?
    (b) A tree has n_2 vertices of degree 2, n_3 vertex of degree 3, ... and n_k vertices of degree k. How many vertices of degree 1 does it have?

57. Let T be a tree with 50 edges. The removal of a certain edge from T yields two disjoint trees T_1 and T_2. Given that the number of vertices in T_1 equals the number of edges in T_2, determine the number of vertices and the number of edges in T_1 and T_2.

58. (a) Show that the sum of the degrees of the vertices of a tree with n vertices is 2n – 2.
    (b) For $n \geq 2$, let $d_1, d_2, d_3, \ldots, d_n$ be n positive integers such that

$$\sum_{i=1}^{n} d_i = 2n - 2$$

Show that there exists a tree whose vertices have degrees $d_1, d_2, d_3, \ldots, d_n$.

59. Prove that the complement of a spanning tree does not contain a cut-set and that the complement of a cut-set does not contain a spanning tree.

60. Let L be a circuit in a graph G. Let e_1 and e_2 be any two edges in L. Prove that there exists a cut set C such that $L \cap C = \{e_1, e_2\}$.

61. (a) Let L_1 and L_2 be two circuits in a graph G. Let e_1 be an edge that is in both L_1 and L_2, and let e_2 be an edge that is in L_1 but not in L_2. Prove that there exists a circuit L_3 which is such that $L_3 \subseteq (L_1 \cap L_2)$ and $b \in L_3$.
    (b) Repeat part (a) when the term 'circuit' is replaced by the term 'cut-set'.

62. For the graphs of figures 7.62ex, give the degree of each vertex and hence degree of graph. Which of the graph is regular?

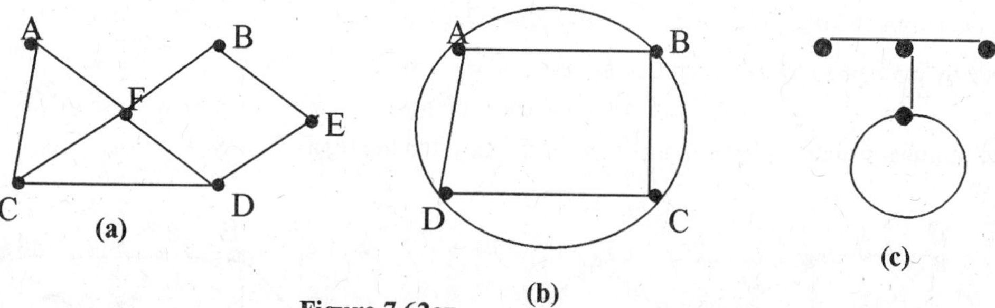

Figure 7.62ex

63. Give an example of a regular connected graph on six vertices that is not a complete graph.

64. For the graph of the figure 7.62ex(a) find quotient graph where relation R on V is defined by the partition
(a) {{a, f}, {e, b, d}, {c}}
(b) {{a, b}, {c}, {d}, {f, c}}

65. For the graph of the figure 7.65ex find quotient graph where relation R on V is defined by the partition
(a) {{1, 2}, {3, 4}, {5, 6}, {7, 8}, {9, 16}, {10, 11}, {12, 13}, {14, 15}}
(b) {{1, 10}, {3, 12}, {5, 14}, {8, 9}, {7, 16}, {2, 11}, {4, 13}, {6, 15}}

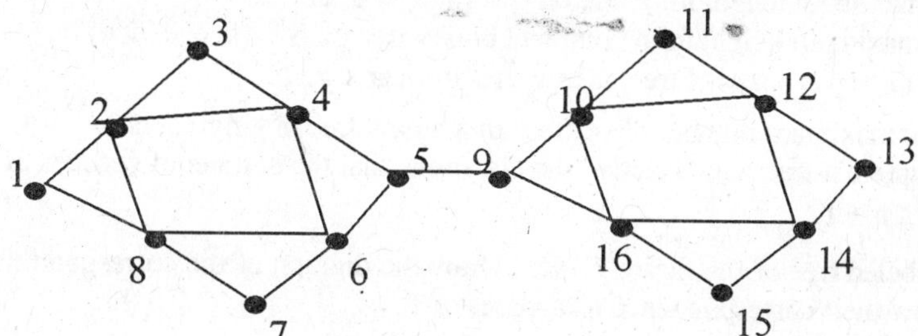

Figure 7.65ex

66. Which of the graphs of 7.62ex and 7.65ex have an Euler circuit and (or) Euler path.
67. Modify the graph of figure 7.65ex so that it can have an Euler circuit. Use Fleury's algorithm to find Euler circuit in them.
68. Draw graphs corresponding to the following maps of figure 7.68ex. Verify that sum of degrees of all regions in them is equal to twice of number of edges in the corresponding graphs.

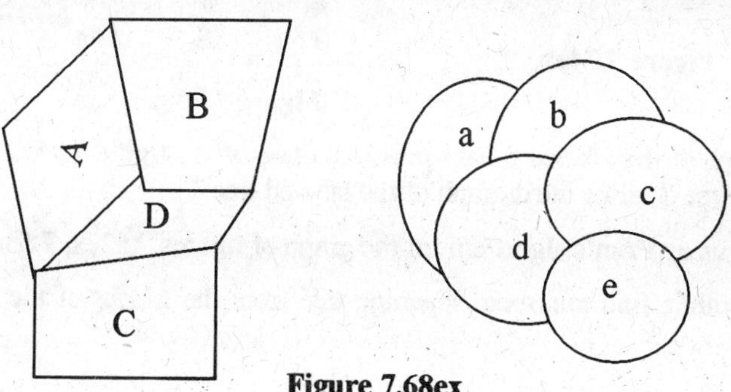

Figure 7.68ex

69. For the graph obtained from maps of 7.68ex, find chromatic polynomial and chromatic number.

70. Prove by mathematical induction that for any linear graph L_n,

$$P(L_n) = x(x-1)^n \text{ for } n \geq 1.$$

71. Find minimal weight Hamiltonian circuit for the graph in the figure 7.71ex.

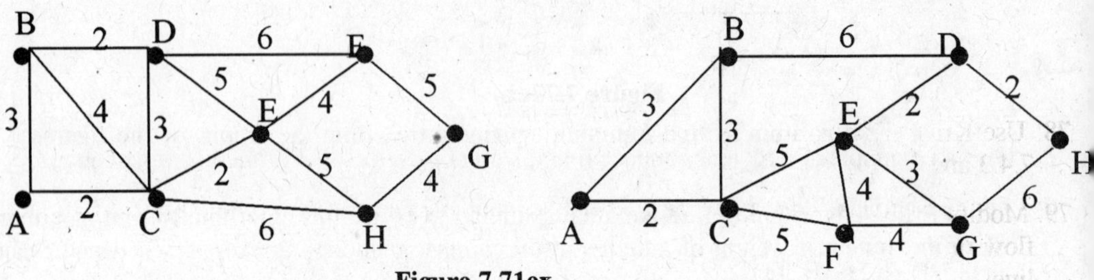

Figure 7.71ex

72. Determine whether a Hamiltonian circuit exists in the graphs of figures 7.62ex and 7.65ex.

73. (a) Let T be a tree. Suppose that T has r vertices and s edges. Find a formula relating r to s.
   (b) Draw all possible unordered trees on the set S = {a, b, c}.
   (c) What is the maximum height for a tree on S = {a, b, c, d, e}?
   (d) What is the maximum height for a complete binary tree on S = {a, b, c, d, e}?
   (e) Show that if (T, v_0) is a rooted tree, then v_0 has in-degree zero.
   (f) Show that the maximum number of vertices in a binary tree of height n is $2^{n+1} - 1$.
   (g) If T is a complete n-tree with exactly 3 levels, prove that the number of vertices of T must be 1 + kn, where $2 \leq k \leq n + 1$.

74. Consider the labeled tree of the figure 7.74ex. Draw the digraph of the corresponding binary positional tree B(T) to show their correspondence to vertices of T.

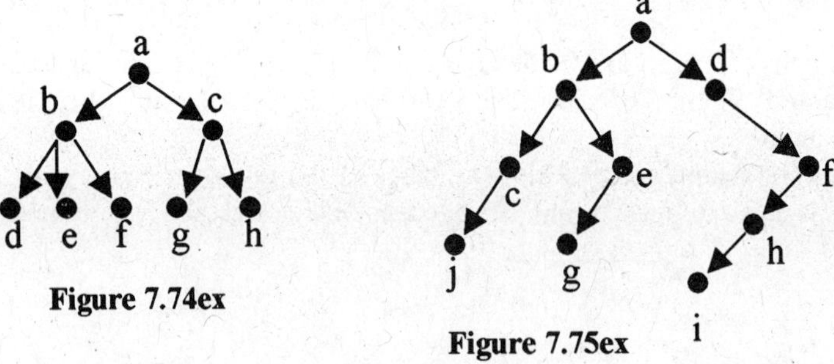

Figure 7.74ex

Figure 7.75ex

75. Consider the digraph of the labeled binary positional tree of the figure 7.75ex. If this tree is the binary form B(T) of some tree T, draw the digraph of the labeled tree T.

76. Find spanning tree using Prim's algorithm for the graph of figures 7.52ex, 7.53ex and 7.62ex.

77. Using Prim's algorithm, find minimum spanning tree from the graphs of the figure 7.4.1, 7.4.2, 7.4.3 and 7.77ex.

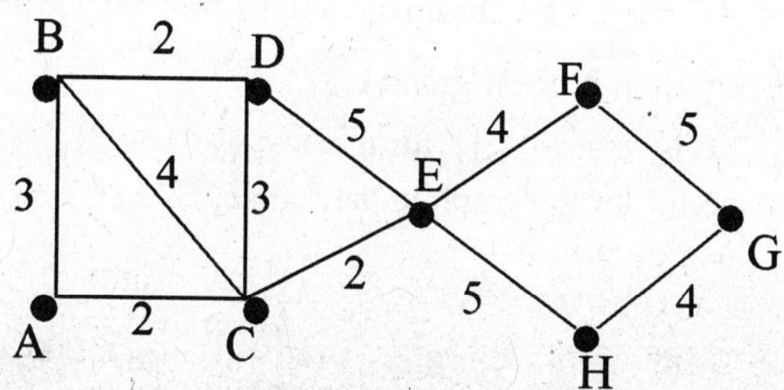

Figure 7.77ex

78. Use Kruskal's algorithm to find minimum spanning tree from the graphs of the figures 7.4.1, 7.4.2, 7.4.3 and 7.77ex.

79. Modify Prim's algorithm to find maximal spanning tree. Use this algorithm to find maximum possible flow in the system if weight of an edge in the graphs mentioned in exercise 77 denotes capacity of a line.

CHAPTER EIGHT

Finite Automata

Since the days of realization that an electronic device can compute any complex mathematical expression, substantial effort has been made to define a computing device that will be general enough to compute every "computable" function. In 1936 ***Turing*** suggested the use of a machine. Since then it is known as the ***Turing machine***, which is considered to be the most general computing device. The study of the Turing machine is beyond the scope of this course. It is however, important to mention here that any computation that can be described by means of a ***Turing machine*** can be mechanically carried out. Conversely, any computation that can be performed on a modern day digital computer can be described by means of a ***Turing machine***. This is according to the definition of a ***Turing machine***.

The ***Turing machine is the most general possible computing device.*** There are different classes of ***Turing machines***. We devote this chapter to discuss the simplest possible class of computing device called ***finite automata*** or ***finite states machine.*** We shall study a finite state machine as

- an acceptor of a *language*
- a generator of a language
- an algorithm to test whether a sentence is valid ?

Any language is suitable for communication provided the syntax and semantic of the language is known to the participating sides. It is made possible by forcing a standard on the way to make sentences from words of that language. This standard is forced through a set of rules. This set of rules is called ***grammar*** of the language. We shall study in this text the most general type of grammars called **Chomsky grammar** and the languages defined by it.

8.1 Grammar and Languages

Let A be a set of symbols. Let A^* be a set of all finite sequences of symbols of A. The set A is referred as **alphabets** and a sequence in A^* is called **word**. The number of symbols in a word is called **length of the word**. A word of length zero is called an **empty string**. A^* may contain an empty string. An empty string is also called **NULL** string and is denoted by $\wedge$ or by ε (pronounced as epsilon). For example, let

A = {a, b, c, …, z} then

$A^*=\{x \mid x$ is an ordinary word of dictionary in case-insensitive form including $\varepsilon\}$

Grammar is probably the most important class of generator of languages. A grammar is a mathematical system for defining a language. As well, it is a device for giving the sentences in the language a useful structure.

A grammar for a language L uses two finite disjoint sets of symbols. These are the set of **non-terminal** symbols, denoted by N in this text, and the set of **terminal** symbols, denoted by Σ. The set of terminal symbols is the set of alphabets over which the language is defined. Non-terminal symbols are used in the generation of words in the language. The heart of a grammar is a finite set P - production rules, which describe how the sentences of the language are to be generated. A production rule is of the form

$$\alpha \rightarrow \beta$$

where $\alpha \in (N \cup \Sigma)^* N(N \cup \Sigma)^*$ and $\beta \in (N \cup \Sigma)^*$. That is, α is any string containing at least one non-terminal and β is any string including a null string. The union $N \cup \Sigma$ of sets of non-terminal and of terminal symbols is also called **total vocabulary** and is denoted by V. Now, we can give a formal definition of a grammar as below.

A grammar is defined as a 4-tuple mathematical structure G (N, Σ, P, S) where

1. N is a non-empty, finite set of non-terminal symbols (sometimes called variables or syntactic categories).
2. Σ is a non-empty, finite set of terminal symbols. The set N is disjoint from Σ i.e. *a symbol cannot be both terminal and non-terminal in the same grammar*.
3. P is a non-empty, finite set of production rules of the form $\alpha \rightarrow \beta$. Notice that P is a non-empty finite subset of Cartesian product of sets $(N \cup \Sigma)^* N(N \cup \Sigma)^*$ and $(N \cup \Sigma)^*$
4. S is the distinguished symbol in N called the ***sentence*** (or **start** symbol).

Example 1 An example of a grammar is G({A, S}, {1,0}, P, S) where, P consists of

1. $S \rightarrow 0A1$ 2. $0A \rightarrow 00A1$ 3. $A \rightarrow \varepsilon$

The non-terminal symbols are A, S and terminal symbols are 1, 0. The symbol ε is used to denote an empty string.

A grammar defines a language in a recursive manner. A ***sentential form*** of a grammar G (N, Σ, P, S), is defined recursively as

1. S is a ***sentential form***
2. If α β γ is a sentential form and $\beta \rightarrow \delta$ is in P, then α δ γ is also a sentential form.

A sentential form of G containing no non-terminal symbols is called a ***sentence*** generated by G. A ***language*** generated by a grammar G is the set of sentences generated by G and is denoted by L(G).

Example 2 Find the language of the grammar G given in the example 1 of this section 8.1.

Solution: We apply the production rules one by one. From rule 1, we have

$S \rightarrow 0A1$	[From rule 1]
$\rightarrow 00A11$	[From rule 2 $0A \rightarrow 00A1$]
$\rightarrow 000A111$	[Applying rule 2 2nd time]
$\rightarrow 0000A1111$	[Applying rule 2 3rd time]
...	
...	
$\rightarrow 0^n A\, 1^n$	[Applying rule 2 (n – 1) times]
$\rightarrow 0^n\, 1^n$	[Applying rule 3 $A \rightarrow \varepsilon$]

Optionally, when rule 3 is applied just after rule 1, it gives $S \rightarrow 01$. This shows that rule 2 can be applied 0 or more times giving the string $0^n 1^n$ for $n > 0$. Therefore, L(G) $= \{0^n\, 1^n \mid n > 0\}$.

Ans.

A sentence in a language can be derived from its grammar. Before proceeding further, let us get acquainted with some terminology. Let (N, Σ, P, S) be a grammar G. We define a relation $\Rightarrow$ (read as *directly derives*) on V^* as:

If αβγ is a string in V^* and $\beta \rightarrow \delta$ is a production rule in P, then

$$\alpha\beta\gamma \Rightarrow \alpha\, \delta\, \gamma$$

The n–steps transitive closure of $\Rightarrow$ is denoted as $\overset{n}{\Rightarrow}$. That is to say $\alpha \overset{n}{\Rightarrow} \beta$ if there is a sequence $\alpha_0, \alpha_1, \alpha_2, \ldots, \alpha_k, \ldots \alpha_n$ of n+1 strings (not necessarily distinct) such that

$$\alpha = \alpha_0;$$

$\alpha_{k-1} \rightarrow \alpha_k$ **for** $1 \leq k \leq n$ and

$\alpha_n = \beta$.

This sequence of strings is called a derivation of length n of β from α in G. In general, the transitive closure of ⇒ is denoted as $\overset{+}{\Rightarrow}$ (read as *derives in a nontrivial way*). It implies that $\alpha \overset{+}{\Rightarrow} \beta$ if and only if β has been derived from α by applying at least one production from P. Similarly, the reflexive and transitive closure of ⇒ is denoted as $\Rightarrow^*$ (read as *derives*). It implies that $\alpha \Rightarrow^* \beta$, if and only if β has been derived from α by applying zero or more productions from P.

Example 3 Let G = (N, Σ, P, S) be a grammar, where

N = {S, L, D, W},

Σ = {a, b, c, 0, 1, 2, 3, 4, 5, 6, 7, 8, 9} and

P is given by:

1. S → L 2. S → LW 3. W → LW 4. W → DW 5. W → L

6. W → D 7. L → a 8. L → b 9. L → c 10. D → 0

11. D → 1 .... 19 D → 9

Which of the following statements are true for this grammar.

(a) ab092 ∈ L(G) (b) 2a3b ∈ L(G) (c) aaaa ∈ L(G) (d) S ⇒ a

(e) $S \Rightarrow^*$ ab (f) DW ⇒ 2 (g) $DW \Rightarrow^*$ 2 (h) $W \Rightarrow^*$2abc

(i) $W \Rightarrow^*$ ba2c

Solution: (a) ab092 ∈ L(G) is true if ab092 is a valid sentence i.e. it is derived from starting symbol S using one or more productions in P. We have the following derivation.

S → LW	[From rule 2]
→ LLW	[From rule 3, W→ LW]
→ LLDW	[From rule 4, W→ DW]
→ LLDDW	[From rule 4, W→ DW]
→ LLDDD	[From rule 6, W→ D]
→ LLDD2	[From rule 12, D→ 2]
→ LLD92	[From rule 19, D→ 9]
→ LL092	[From rule 10, D→ 0]
→ Lb092	[From rule 8, L→ b]

$\rightarrow$ ab092 [From rule 7, L$\rightarrow$ a]

As shown above, ab092 is a sentence in L(G). Therefore, this statement is true. The above derivation can be better shown as the following **derivation tree**.

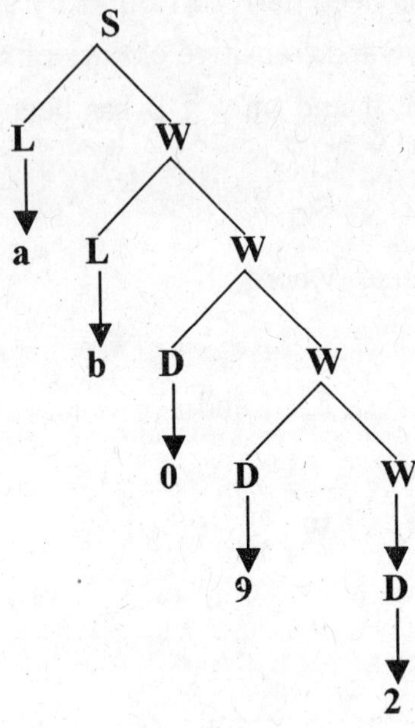

Figure 8.1.1

In the above derivation tree, the starting symbol S is at the root, all the intermediate nodes are non-terminal symbols and leaves are terminal symbols from G. *For a string to be a valid sentence of a language L(G), all the leaves in the derivation tree must be terminal symbols. Otherwise the sentence does not belong to the language.* We read symbols at the leaves from left to right to get the sentence derived and in this case it is **ab092.**

(b) Here again, 2a3b $\in$ L(G) is true if 2a3b is a valid sentence. See the 19 productions in P. As usual, we have to begin with S to derive any sentence or to test validity of any string. While proceeding from S, we can have either L or LW in the next step. This means, the left most non-terminal will always be an L. The non-terminal L can be replaced either by **a** or **b** or **c**. Therefore, any string acceptable (derivable) by this grammar must have either **a** or **b** or **c** as its left most symbol. The given string 2a3b does not satisfy this condition. Hence, 2a3b does not belong to L(G).

(c) The derivation of the strings is shown in the following derivation tree.

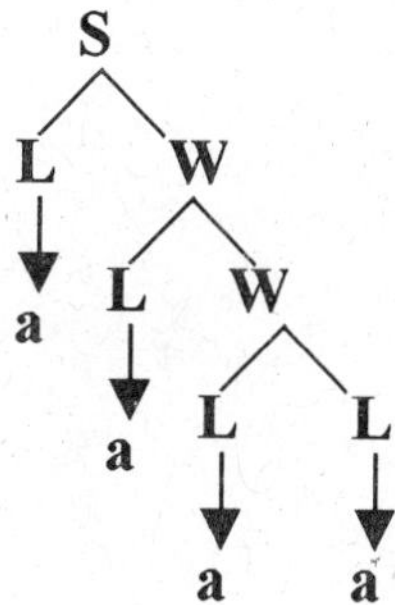

Figure 8.1.2

From the derivation tree, it is obvious that string **aaaa** is a valid sentence of the language L(G). Hence the statement is true.

(d) As $\Rightarrow$ stands for directly derives, "S $\Rightarrow$ a" will be valid if there is any production in P of the form S $\rightarrow$ a which directly provides for substituting S by a. Since there is no production of the form S $\rightarrow$ a in P, the statement "S $\Rightarrow$ a" is not true.

(e) As $\Rightarrow^*$ stands for derives in zero or more steps (reflexive and transitive closure), "S $\Rightarrow^*$ ab" will be valid if the string **ab** belongs to L(G). Let us derive this as follows:

S $\Rightarrow^1$ LW	[From rule 2 S $\rightarrow$LW]
$\Rightarrow^2$ aW	[From rule 7 L $\rightarrow$a]
$\Rightarrow^3$ aL	[From rule 5 W $\rightarrow$L]
$\Rightarrow^4$ ab	[From rule 8 L $\rightarrow$b]

This shows that S $\Rightarrow^*$ ab is valid.

The statements in (f) and (g) are not valid. Reader may verify this.

(h) Here, both W and 2abc are in V^*. The statement W $\Rightarrow^*$ 2abc will be valid if we can derive 2abc from W using the productions in P. See the following derivation tree which starts from W as it's root.

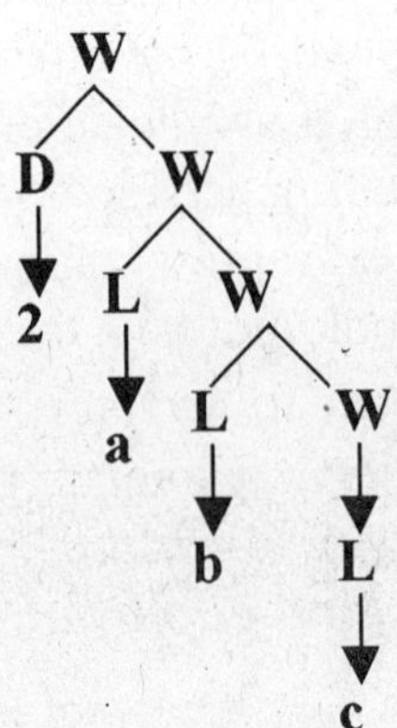

Figure 8.1.3

In the above derivation tree all the leaves are terminal symbols. This shows that starting from W we can derive 2abc using productions of P. Therefore, it is a valid statement.

The statement in (i) is also true. Reader may verify this as an exercise.

Ans.

From examples 2 and 3, it should be clear, how to find language of a grammar and test whether a given sentence is valid according to the grammar. The process of testing whether a sentence is valid begins from sentence itself. Derivation tree is an important tool to test the validity of a sentence under some types of grammar. Let us now summarize the definition of a ***derivation tree***.

Derivation Tree

A derivation tree is also called a *parse* tree. A tree is a derivation tree of a valid sentence in a grammar $G = (N, \Sigma, P, S)$ if

1. Every vertex is a symbol of the set $N \cup \Sigma \cup \{\varepsilon\}$.
2. The root is the start symbol S.
3. If A is an interior vertex in the tree, then A must be in N.
4. If a vertex A has n siblings $X_1, X_2, X_3, \ldots, X_n$ in order from left to right, then $A \rightarrow X_1X_2X_3\ldots X_n$ must be a production in P.
5. If ε is a vertex, then ε must be a leaf and it must be the only offspring of its parent.

There are certain grammars in which derivations of sentences cannot be expressed as trees. Construction of a derivation tree works only if the left hand side of every production in G contains a single, non-terminal symbol. See the grammar of ***example 1*** above. In the grammar of this example, the left hand side does not have this simple form. Although it is possible to construct a graphical representation of derivations (***example 2***) under this grammar, the resulting digraph would not be a tree. Many other problems arise in an unrestricted grammar. To simplify the situation, some restrictions are imposed on α in the format of production $\alpha \to \beta$ of a general grammar. The subsequent restrictions yield different types of grammar. The general format of the grammar we have been discussing is also known as ***phrase structure grammar, or Chomsky grammar***. A grammar $G = (N, \Sigma, P, S)$ is said to be of

Type 0: if there is no restriction on the production rules i.e., in $\alpha \to \beta$, where $\alpha, \beta \in (N \cup \Sigma)^*$. This type of grammar is also called an ***unrestricted grammar***.

Type 1: if in every production $\alpha \to \beta$ of P, $\alpha, \beta \in (N \cup \Sigma)^*$ and $|\alpha| \leq |\beta|$. Here $|\alpha|$ and $|\beta|$ represent number of symbols in string α and β respectively. This type of grammar is also called a ***context sensitive*** grammar (or ***CSG***).

Type 2: if in every production $\alpha \to \beta$ of P, $\alpha \in N$ and $\beta \in (N \cup \Sigma)^*$. Here α is a single non-terminal symbol. This type of grammar is also called a ***context free*** grammar (or ***CFG***).

Type 3: if in every production $\alpha \to \beta$ of P, $\alpha \in N$ and $\beta \in (N \cup \Sigma)^*$. Here α is a single non-terminal symbol and β may consist of at the most one non-terminal symbol and one or more terminal symbols. The non-terminal symbol appearing in β must be the extreme right symbol. This type of grammar is also called a ***right linear*** grammar or ***regular*** grammar (or ***RG***).

This classification is widely known as ***Chomsky hierarchy***. The hierarchy can be better shown diagrammatically through the adjoining figure 8.1.4. It is obvious that type-I grammar is of type-J whenever I > J.

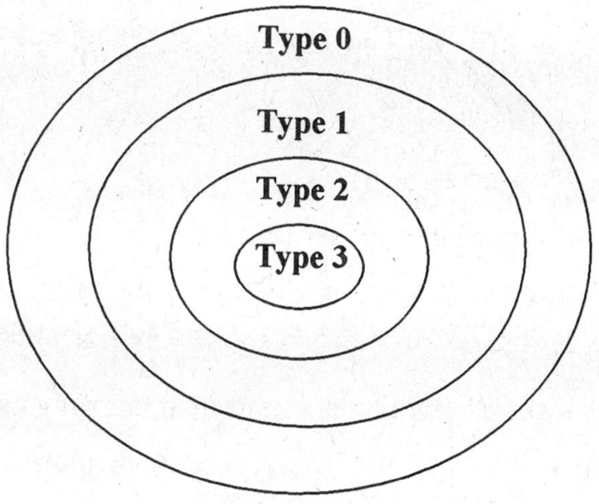

Figure 8.1.4

Now we shall learn how to find the language generated by a given grammar. The process involves a number of steps starting from the production with start symbol S. See the following examples.

Example 4 Find the Language L(G), where $G = (N, \Sigma, P, S)$ is given as: $N = \{S, W\}$; $\Sigma = \{a, b, c\}$ and $P = \{S \to aW, W \to bbW, W \to c\}$.

Solution: The given grammar is of type-3. To deduce the language for this grammar we proceed step by step as follows.

Step 1: Start from one of the production $\alpha \to \beta$ in which α = S (start symbol). If β contains start symbol, eliminate the start symbol from the strings so obtained by using the appropriate production. In this example, we have only one production beginning with S. So, we have

$$S \to aW$$

The string "aW" does not contain S. So its elimination is not relevant here. Go to the step 2.

Step 2: Use other productions successively to eliminate all non-terminals present in the string so obtained. If the production is recursive, use it **n** times.

$$\begin{aligned} S &\to aW \\ &\to a(bbW) && \text{[Apply production 2]} \\ &\to a(bb)^n W && \text{[Apply production 2, n times]} \\ &\to a(bb)^n c && \text{[Apply production 3]} \end{aligned}$$

Step 3: If there are more than one production $\alpha \to \beta$ having the same α, apply them to get all possible alternatives. Combine all these alternatives with **OR** ($\vee$). In this case, we have two productions for W: one we have used in the step 2 & the other is $W \to c$. This we can apply without applying production 2 even once. This gives a string

$$\begin{aligned} S &\to aW \\ &\to ac && \text{[Apply rule 3 once]} \end{aligned}$$

Step 4: Combine all the alternatives so obtained. In this case combining the two alternatives, we get

$$\begin{aligned} S &\to ac \vee a(bb)^n c \\ &= a(\varepsilon \vee (bb)^n)c \\ &= a(bb)^* c \end{aligned}$$

Therefore, the language of this grammar can be written in set theoretic notation as

$$L(G) = \{x \mid x = a(bb)^n c \text{ for } n \geq 0\}$$

Ans.

Example 5 Find the Language L(G) where, G = (N, Σ, P, S) is given as: N = {S, W}; Σ = {a, b, c} and P = {S → aSb, Sb → bW, abW → c}.

Solution: The given grammar is of type-0. To deduce the language for this grammar we proceed step by step as follows.

Step 1: Let us start with the production starting with start symbol S. In this example, there is one such production which is recursive. From this production, we have

$$S \rightarrow aSb$$
$$\rightarrow a^n S b^n \quad \text{[Apply the production 1, n times]}$$

Now our immediate goal is to eliminate the non-terminal S from the string $a^n S b^n$ We have a production Sb → bW in P. Applying this, we get

$$S \rightarrow a^n bWb^{n-1}$$

Step 2: Now eliminate W from right hand side of the above derivation. Using the production 3 abW → c, we get

$$S \rightarrow a^{n-1} c\, b^{n-1}$$

Step 3: There is an alternative to the substitution in the step 1. We could have proceeded as below

$$S \rightarrow aSb$$
$$\rightarrow abW \quad \text{[Apply the production 2 once]}$$
$$\rightarrow c \quad \text{[Apply the production 2 once]}$$

Step 4: Combining the alternatives, we conclude that this grammar generates the language of the form $a^m c\, b^m$ where, m ≥ 0. In set theoretic notation the language can be expressed as

$$L(G) = \{x \mid x = a^m c\, b^m \text{ for } m \geq 0\}$$

Ans.

Example 6 Find the Language L(G) where, G = (N, Σ, P, S) is given as: N = {S, A, B}; Σ = {(,), a, +} and P = {S → (S), S → a + A, A → a + B, B → a + B, B→a }.

Solution: The given grammar is of type-2. To deduce the language for this grammar, we proceed from start symbol. There are two productions beginning with S. We may start optionally with S → (S). This is a recursive production. Therefore, we have

$$S \rightarrow (S) \quad \text{[Apply production 1]}$$
$$\rightarrow (^n S)^n \quad \text{[Apply production 1, n times]}$$

Now symbol S is to be eliminated from the string obtained above. We apply production 2 to achieve this. It is important to note that the production 1 can be used any number of times, optionally. Thus, $n \geq 0$. We now have

$$S \rightarrow (^n a + A)^n,\ n \geq 0 \qquad \text{[Apply production 2]}$$
$$\rightarrow (^n a + a\ + B)^n \qquad \text{[Apply production 3]}$$

One non-terminal B is yet to be eliminated from the string. There are two productions for B: one is recursive and the other ending in terminal symbol **a**. Combining these two for B, we can have a production for B as below

$$B \rightarrow [a+]^m B,\ m \geq 0 \qquad \text{[Apply production 4 m times]}$$
$$B \rightarrow [a+]^m a,\ m \geq 0 \qquad \text{[Apply production 5]}$$

Using this result for B in the derivation $S \rightarrow (^n a + a\ + B)^n$, we get

$$S \rightarrow (^n a + a\ + [a+]^m a)^n$$

In the above derivation, we may replace m with * but n cannot be replaced with *, because n stands for count of left and right parentheses. Therefore, L(G) can be expressed in set theoretic notation as

$$L(G) = \{x \mid x = (^n a + a\ + [a+]^* a)^n \text{ for } n \geq 0\}$$

Ans.

We have seen some examples to learn how to find language for a given grammar. It is true that for a given grammar there can be only one language. The same argument is not valid for the converse. For a given language we may find different grammars producing the same language. Therefore finding a grammar for a given language is a somewhat complicated process; though it is always possible to do that. See the following examples to know how to do it.

***Example** 7* Find a grammar for the language $L = \{x \mid x = a^n b^n \text{ for } n \geq 1\}$.

Solution: Here, language is defined over the alphabet $\Sigma = \{a, b\}$. This language contains n ($\geq$ 1) number of a and b & a's come first and all b's follow. Let S be the start symbol. Then, we have to think for a production such that whenever an "a" is added to the left side a "b" must be added to the right side. This can be achieved by the production

$$S \rightarrow aSb.$$

Next, for a sentence to be valid, it must not contain any non-terminal. This requires that there must be a production such that right side contains only terminal symbol(s). This is achieved by the production

$$S \rightarrow ab.$$

Therefore, $G = (N, \Sigma, P, S)$ is a grammar for the given language where, $N = \{S\}$, $\Sigma = \{a, b\}$, $P = \{S \rightarrow aSb, S \rightarrow ab\}$. This is a type –2 grammar.

Ans.

Example 8 Find a grammar for the language $L = \{x \mid x$ is a string of 0's and 1's with equal number n $(\geq 0)\}$.

Solution: The given language is defined over the alphabet $\Sigma = \{1, 0\}$. A string in this language contains equal number of 0 and 1. Strings 0011, 0101, 1100, 1010, 1001 and 0110 are all valid strings of length 4 in this language. Here, order of appearance of 0's and 1's is not fixed. Let S be the start symbol. A string may start with either 0 or 1. So, we should have productions $S \rightarrow 0S1$, $S \rightarrow 1S0$, $S \rightarrow 0A0$ and $S \rightarrow 1B1$ to implement this. Next A and B should be replaced in such a way that it maintains the count of 0's and 1's to be equal in a string. This is achieved by the productions $A \rightarrow 1S1$ and $B \rightarrow 0S0$. The **null** string is also an acceptable string, as it contains **zero** number of 0's and 1's. Thus $S \rightarrow \varepsilon$ should also be a production. Therefore, the required grammar G can be given by (N, Σ, P, S), where $N = \{S, A, B\}$, $\Sigma = (1, 0\}$ and $P = \{S \rightarrow 0S1, S \rightarrow 1S0, S \rightarrow 0A0, S \rightarrow 1B1, A \rightarrow 1S1, B \rightarrow 0S0, S \rightarrow \varepsilon\}$.

Ans.

Example 9 Construct a phrase structure grammar g for the language

$$L = \{a^m b^n \mid m, n \geq 1, m \neq n\}$$

Solution: This language is defined over the alphabet $\Sigma = (a, b\}$. It is important to note that

$$L = L_1 \cup L_2 \text{ where,}$$
$$L_1 = \{a^m b^n \mid m, n \geq 1, m > n\}$$
$$L_2 = \{a^m b^n \mid m, n \geq 1, m < n\}$$

A production $P_1 = \{A \rightarrow aA, A \rightarrow aB, B \rightarrow aBb, B \rightarrow ab\}$ generates the language L_1. Similarly, production $P_2 = \{C \rightarrow Cb, C \rightarrow Db, D \rightarrow aDb, D \rightarrow ab\}$ generates the language L_2. Now, if we add two productions: $S \rightarrow A$ and $S \rightarrow C$ to the set $P_1 \cup P_2$, they together generate the language L. Therefore, the required grammar G can be given, in simplified way, by (N, Σ, P, S) where, $N = \{S, A, B, C\}$, $\Sigma = \{a, b\}$ and $P = \{S \rightarrow A, S \rightarrow C, A \rightarrow aA, A \rightarrow aB, B \rightarrow aBb, B \rightarrow ab, C \rightarrow Cb, C \rightarrow Bb\}$. Here, reader may note that non-terminal D has been combined with B.

Ans.

8.2 Regular Expression and Regular Sets

We have studied phrase structure grammar in the previous section. In type 2 grammar (which includes type 3 grammars), all productions contain a single, non-terminal symbol on its left-hand side. Because of this special characteristics, type 2 and type 3 grammars can be represented by some useful alternative methods. One of them is BNF (Backus-Naur form) notation. The following steps are followed to represent productions in type 2 or type 3 grammar in BNF form.

Step 1: All non-terminals, wherever they occur, are enclosed in angle brackets (< >). For example S, A, B etc are represented as <S>, <A>, <B>.

Step 2: If there is more than one production corresponding to same non-terminal, then they are listed together separated with |. ('|' stands for alternative option).

Step 3: The → symbol in production is replaced with the symbol "::= "

To make the things more clear, let us see the following examples.

Example 1 Let G = (N, Σ, P, S) be a grammar, where N = {S, A}, Σ = {x, y, z} and P = {S → xS, S → yA, A → yA, A→z}. Give BNF notation for the productions of G.

Solution: The BNF representation for the productions of G can be given as:

<S> ::= x <S> | y <A>
<A> ::= y <A> | z

Ans.

Example 2 Let G = (N, Σ, P, S) be a grammar, where N = {S, A}, Σ = {0, 1} and P = {S→0A, A→11A, A→010A, A→1}. Give BNF notation for the productions of G.

Solution: The BNF representation for the productions of G can be given as:

<S> ::= 0 <A>
<A> ::= 11 <A> | 010 <A> | 1

Ans.

Note that the symbol on the left-hand side of a production may also appear in one of the strings on the right-hand side. For example, in **<S> ::= x <S>,** <S> appears on both the sides. Similarly, in **<A> ::= y <A>,** <A> appears on both the sides. When this happens, we say that the corresponding production is **recursive**. A **recursive** production is said to be **normal** if the non-terminal symbol appearing on the right-hand side is the **rightmost** symbol in the production and it appears **only once**. *Note that a recursive production that appears in a type 3 grammar is a normal production by definition.* A BNF notation can also be shown as a **syntax diagram**. To draw a syntax diagram we use two geometrical shapes: **circle** for terminal symbol and **rectangle** for non-terminal. The step by step method to draw a syntax diagram is outlined below.

Step 1: Identify the terminals and non-terminals in a production. Represent a non-terminal by □ and a terminal by ○. Put the symbols inside the respective geometric shapes.

Step 2: Connect all these shapes in the sequence in which the corresponding symbol appears in the production. If the production is normal, a loop is drawn from rightmost side to the initial state.

Step 3: If there are more than one production for a non-terminal, all of them are connected as alternatives.

A syntax diagram for the productions of the grammar in example 1 can be given as in figure 8.2.1. The non-terminal <S> has two alternatives: one is a normal recursive and other is a linear one. Both these alternatives have been combined and shown as a syntax diagram for <S>. Similarly we can do for <A>.

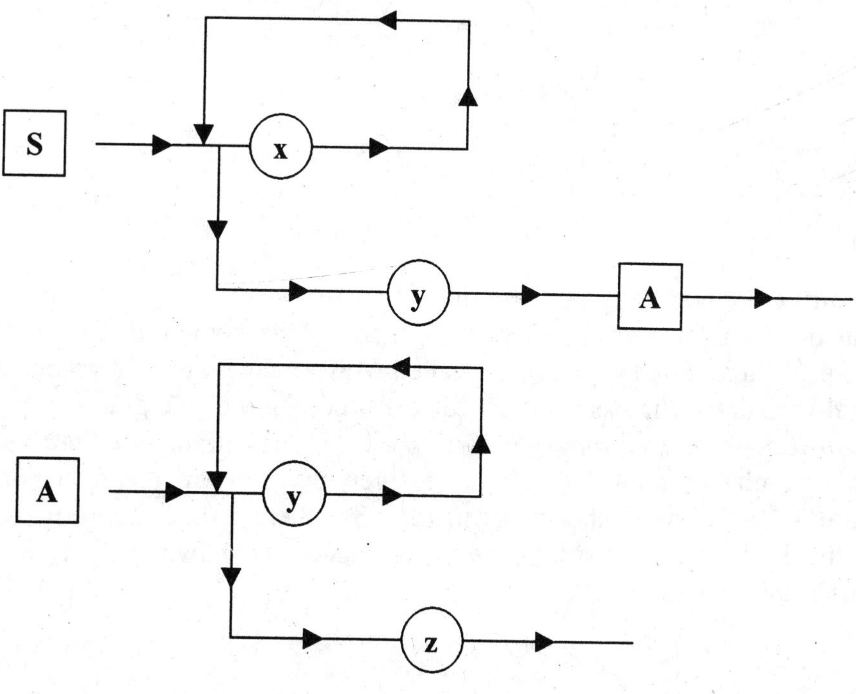

Figure 8.2.1

In example 2, <S> has only one production and that is linear in <A>. This is shown in a line. The non-terminal <A> has three alternatives. These are connected in such a way that while moving from <A>, one can select any one of the three paths available at a time. The syntax diagram is shown in figure 8.2.2.

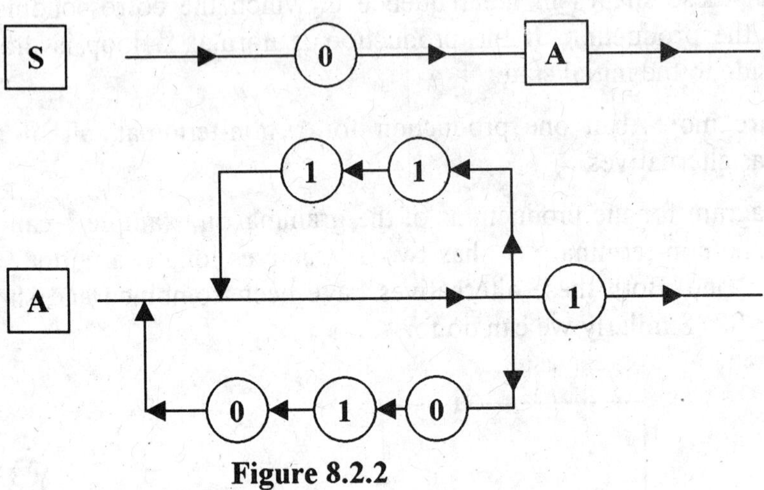

Figure 8.2.2

Further, syntax diagrams for individual productions can be combined in such a way that the diagram contains the start symbol in the beginning and only the terminal symbols elsewhere. Such a diagram is called **Master diagram**. If a grammar is of type 3, we can always draw a maser diagram for the production of the grammar. *In case of type 2, we can draw a master diagram if all the recursive productions are normal.* This is obtained by replacing non-terminals successively by its syntax diagram. For example, see the figure 8.2.1. In the syntax diagram for <S>, there is one non-terminal <A>. If we replace this by its syntax diagram, we get a diagram as shown in figure 8.2.3. This does not contain any non-terminals.

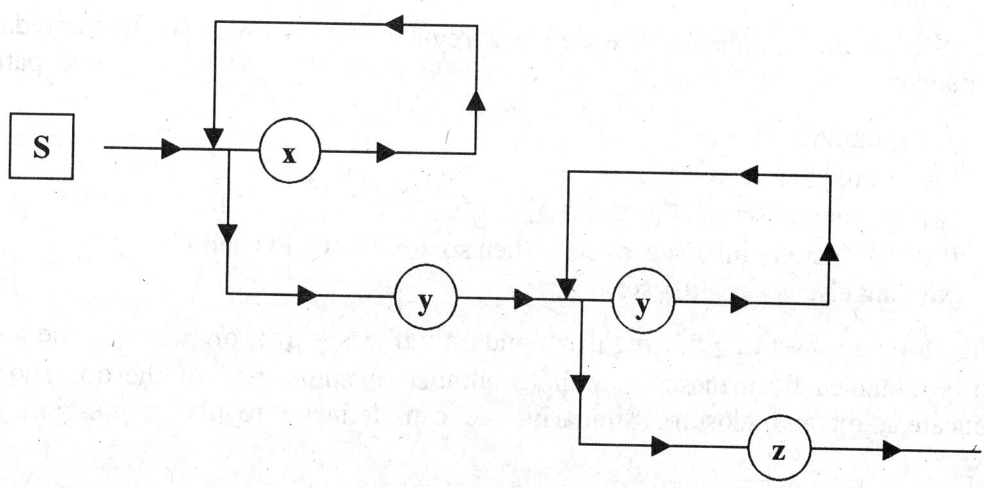

Figure 8.2.3

Similarly, in example 2, a master diagram for the production of grammar can be given by combining the syntax diagrams in figure 8.2.2. The diagram so obtained is shown in the figure 8.2.4.

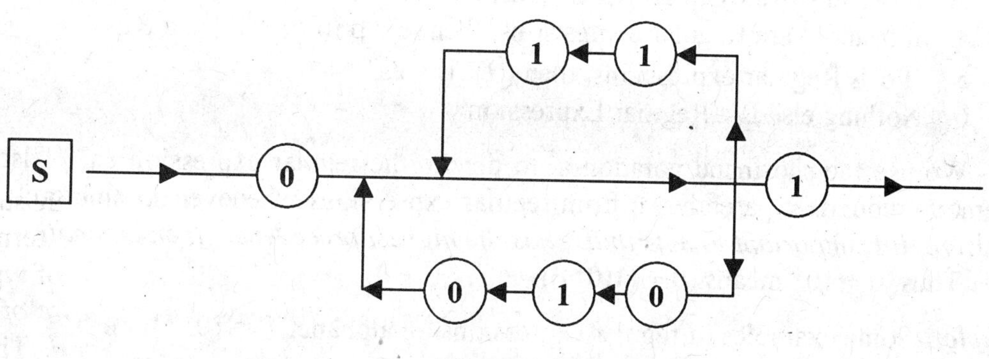

Figure 8.2.4

We have studied that a language generated by a regular grammar (type 3) is a regular language. From a master diagram of a regular grammar we can deduce the regular language generated by the grammar. The master diagram of figure 8.2.4 generates the expression 0 (11 ∨ 010)*1 and the master diagram of figure 8.2.3 generates the expression x*yy*z. Reader may wonder that how this has been done? Before responding to the query in your mind, let us first define a regular set and then a regular expression.

Let Σ be a finite alphabet. We define a regular set over Σ recursively in the following manner:

1. ϕ is a regular set over Σ.
2. $\{\varepsilon\}$ is a regular set over Σ.
3. $\{a\}$ is a regular set over Σ $\forall$ $a \in \Sigma$.
4. If P and Q are regular sets over Σ, then so are $P \cup Q$, PQ and P*.
5. Nothing else is a regular set.

Therefore, a subset of Σ^* is regular if and only if it is ϕ, $\{\varepsilon\}$, or $\{a\}$ for some a in Σ, or can be obtained from these by a finite number of application of the operations union, concatenation and closure. Similarly, we can define a regular expression as follows.

A ***regular expression*** (RE) over an alphabet Σ (set of symbols) is a word (string) constructed from elements of Σ and symbols • (catenation), $\vee$ (or +), ε(null) and * (closure) according to the following definition.

1. The symbol ε is a RE.
2. If "a" is any symbols of alphabet Σ, then a is RE.
3. If α and β are Regular expressions, then $\alpha \bullet \beta$ (or $\alpha\beta$) is RE.
4. If α and β are Regular expressions, then $\alpha \vee \beta$ (or $\alpha + \beta$) is a RE.
5. If α is Regular expressions, then $(\alpha)^*$ is a RE.
6. Nothing else is a Regular Expression.

We use the shorthand notation a^+ to denote the regular expression aa*. Also we can remove redundant parentheses from regular expressions whenever no ambiguity can arise. *Here, it is important to note that* * *has the highest precedence, then catenation, and then* +. Thus, 0 + 10* means (0 + (1(0*))).

Example 1 Some examples of regular expressions on alphabet $\Sigma = \{0, 1\}$ are

(a) 01, denoting regular set {01}.

(b) 0*, denoting regular set {0}*.

(c) (0 + 1)*, denoting regular set {0,1}*.

(d) (0 + 1)*011, denoting the regular set of all strings of 0's and 1's ending in 011.

(e) (00 +11)*((01 + 10)(00 + 11)*(01 + 10)(00 + 11)*)*, denoting the set of all strings of 0's and 1's containing an even number of both 0's and 1's.

(f) $0^*(0 \vee 1)^*$, denoting the set of any string in 0's and 1's.

(g) $00^*(0 \vee 1)^*1$, denoting the set of all strings in 0's and 1's that begin with 0 and ends in 1.

Example 2 (a + b)(a + b + 0 + 1)*, denoting the set of all strings in {0, 1, a, b}* beginning with a or b.

It should be quite clear that for each regular expression we can construct the regular set denoted by that regular expression. Similarly, for each regular set, we can find at least one regular expression denoting that regular set. *Unfortunately, for each regular set, there is infinity of regular expressions denoting that set.* **Two regular expressions are equal if they denote the same regular set.** Some basic algebraic properties of regular expressions are given in the following theorem.

Theorem 1 Let α, β and γ be regular expressions, then

(a) $\alpha + \beta = \beta + \alpha$
(b) $\phi^* = \varepsilon$
(c) $\alpha + (\beta + \gamma) = (\alpha + \beta) + \gamma$
(d) $\alpha(\beta\gamma) = (\alpha\beta)\gamma$
(e) $\alpha(\beta + \gamma) = \alpha\beta + \alpha\gamma$
(f) $(\alpha + \beta)\gamma = \alpha\gamma + \beta\gamma$
(g) $\varepsilon\alpha = \alpha\varepsilon = \alpha$
(h) $\phi\alpha = \alpha\phi = \phi$
(i) $\alpha^* = \alpha + \alpha^*$
(j) $(\alpha^*)^* = \alpha^*$
(k) $\alpha + \alpha = \alpha$
(l) $\alpha + \phi = \alpha$

Proof: (a) Let α and β denote the regular sets L_1 and L_2 respectively. Then $\alpha + \beta$ denotes the regular set $L_1 \cup L_2$ and $\beta + \alpha$ denotes the regular set $L_2 \cup L_1$. But $L_1 \cup L_2 = L_1 \cup L_2$ by the commutative property of union operation on sets. Therefore, $\alpha + \beta = \beta + \alpha$ because both the regular expressions denote the same regular set.

Proved.

Proofs of the other parts of the theorem are left as an exercise for the reader.

Now, it is the time to answer the query, which might have come to your head while reading the regular expressions corresponding to the master diagram of figure 8.2.3 and 8.2.4. Each such diagram has sections corresponding to terminal symbols. These sections may be in ***sequence*** or in ***parallel*** or in ***loop*** or may be ***combination of two or more of these kinds***. The technique to find the expression lies in detecting these sections and then combining them according to the following rules.

1. Each terminal symbol constitutes a segment and the symbol itself is the RE corresponding to the segment.
2. If a segment D is composed of two segments D_1 and D_2, in sequence; and D_1, D_2 correspond to the RE's α_1 and α_2, respectively, then D corresponds to the regular expression $\alpha_1\alpha_2$ (concatenation).

3. If a segment D is composed of alternative segments D_1 and D_2, in parallel; and D_1, D_2 correspond to the RE's α_1 and α_2, respectively, then, D corresponds to the regular expression $\alpha_1 \vee \alpha_2$ (union).
4. If a segment D of master diagram is in a loop through segment D_1, and D_1 represents the RE α, then D corresponds to the regular expression α^* (closure).

Now, let us apply the above procedure on the master diagram of the figure 8.2.4. This diagram has three segments D_1, D_2 and D_3, in sequence. They are marked with dotted rectangles.

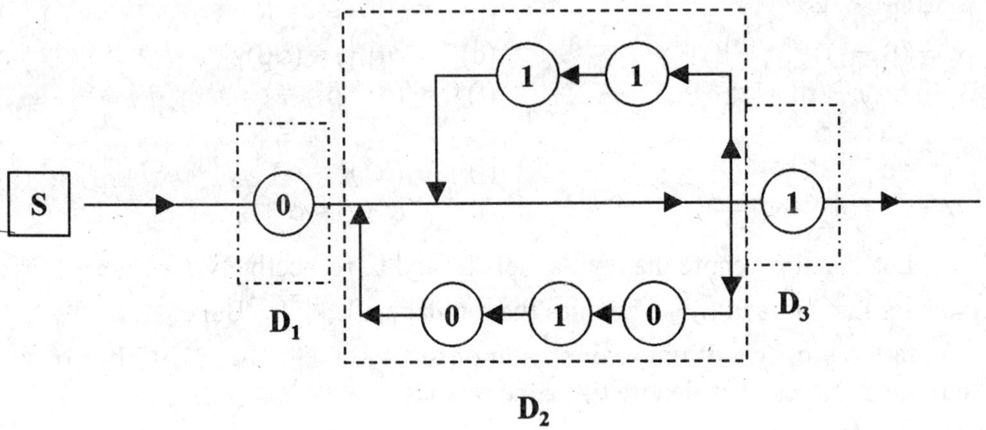

Figure 8.2.5

The segment D_1 represent the RE 0 and D_3 represents the RE 1. D_2 contains two segments in parallel and is in loop. The top one represents the RE 11 and bottom one corresponds to 010. Thus, segment D_2, without loop, represents the RE $(11 \vee 010)$. Therefore, D_2 corresponds to the RE $(11 \vee 010)^*$. Then, the complete master diagram represents the RE corresponding to $D_1D_2D_3$, and this corresponds to $0(11 \vee 010)^*1$. I hope this must have satisfied your curiosity. Similarly, you can try to find the RE from the diagram shown in the figure 8.2.3.

A regular grammar can also be expressed as a set of equations. For example, G defined by productions

$$\langle S \rangle ::= 0\langle A \rangle \mid 1\langle S \rangle \mid \varepsilon$$
$$\langle A \rangle ::= 0\langle B \rangle \mid 1\langle A \rangle$$
$$\langle B \rangle ::= 0\langle S \rangle \mid 1\langle B \rangle$$

can also be written as

$$S = 0A + 1S + \varepsilon$$
$$A = 0B + 1A$$
$$B = 0S + 1B$$

In dealing with languages, it is often convenient to use equations in which variables (non-terminals) and coefficients represent regular expressions. A set of equations whose coefficients are regular expressions are called ***regular expression equations***. Solving these equations mean finding the RE's for each variable satisfying the equations. By drawing master diagram for each variable and then finding RE representing the diagram a regular expression equation can be solved. We can also use Gaussian elimination method to do the same.

We shall now, show that a language is a regular expression if and only if it is defined by a right linear grammar (regular grammar). A few observations are needed to show that every regular expression has a right linear grammar.

Theorem 2 Let Σ be a finite alphabet. Then, (a) ϕ, (b) $\{\varepsilon\}$ and (c) $\{a\}$ for all a in Σ, are right linear language.

Proof: We know that a language generated by a right linear grammar is right linear. Thus, our task is show that there exist a right linear grammar for language in (a), (b) and (c) each.

(a) Let $G = (\{S\}, \Sigma, \phi, S)$ is a right linear grammar such that $L(G) = \phi$.

(b) Let $G = (\{S\}, \Sigma, \{S \rightarrow \varepsilon\}, S)$ is a right linear grammar such that $L(G) = \{\varepsilon\}$.

(c) Let $G = (\{S\}, \Sigma, \{S \rightarrow a\}, S)$ is a right linear grammar such that $L(G) = \{a\}$.

Proved.

Theorem 3 If L_1 and L_2 are right linear languages then, so are (a) $L_1 \cup L_2$ (b) L_1L_2 and (c) L_1*.

Proof: Since L_1 and L_2 are right linear languages, we can assume that there exist right linear grammars $G_1 = (N_1, \Sigma, P_1, S_1)$ and $G_2 = (N_2, \Sigma, P_2, S_2)$ such that $L(G_1) = L_1$ and $L(G_2) = L_2$. We may also assume that N_1 and N_2 are disjoint. This can be achieved by renaming the non-terminals arbitrarily. So, this assumption is without any loss of generality.

(a) Let G_3 be a right linear grammar given by

$$(\mathbf{N_1 \cup N_2 \cup \{S_3\}, \Sigma, P_1 \cup P_2 \cup \{S_3 \rightarrow S_1 \mid S_2\}, S_3})$$

Where, S_3 is a new non-terminal neither in N_1 nor in N_2. It should be clear that $L(G_3) = L(G_1) \cup L(G_2)$ because for each derivation $S_3 \overset{+}{\Rightarrow} w$ in G_3, there is either a derivation S_1

$\stackrel{+}{\Rightarrow}$ w in G_1 or $S_2 \stackrel{+}{\Rightarrow}$ w in G_2 and conversely. Since G_3 is a right linear grammar, $L(G_3)$ is a right linear language.

(b) Let G_4 be a right linear grammar given by **($N_1 \cup N_2$, Σ, P_4, S_1)** where, P_4 is defined as follows:

1. If $A \to xB$ is in P_1, then $A \to xB$ is in P_4,
2. If $A \to x$ is in P_1, then $A \to x\ S_2$ is in P_4,
3. All productions in P_2 are in P_4.

Note that, if $S_1 \stackrel{+}{\Rightarrow}$ w in G_1 then $S_1 \stackrel{+}{\Rightarrow}$ w S_2 in G_4. And, if $S_2 \stackrel{+}{\Rightarrow}$ x in G_2 then $S_2 \stackrel{+}{\Rightarrow}$ x in G_4. Thus, $L(G_1)L(G_2) \subseteq L(G_4)$. Now suppose that $S_1 \stackrel{+}{\Rightarrow}$ w in G_4. There is no production of the form $A \to x$ in P_4 that came out of P_4. Thus, we can write a derivation in the form $S_1 \stackrel{+}{\Rightarrow} xS_2 \stackrel{+}{\Rightarrow}$ xy where, w = xy and all productions used in the derivation $S_1 \stackrel{+}{\Rightarrow} xS_2$ arose from rules (1) and (2) of the construction of P_4. Thus, we must have the derivations $S_1 \stackrel{+}{\Rightarrow}$ x in G_1 and $S_2 \stackrel{+}{\Rightarrow}$ y in G_2. Therefore, $L(G_4) \subseteq L(G_1)L(G_2)$. It is therefore concluded that $L(G_4) = L(G_1)L(G_2)$.

(c) Let G_5 be a right linear grammar given by **($N_1 \cup \{S_5\}$, Σ, P_5, S_5)** such that S_5 is not in N_1 and P_5 is constructed as follows:

1. If $A \to xB$ is in P_1 then $A \to xB$ is in P_5,
2. If $A \to x$ is in P_1 then $A \to x\ S_5$ and $A \to x$ are in P_5,
3. $S_5 \to xS_1 | \varepsilon$ are in P_5.

Now, to prove that $L(G_5) = (L(G_1))^*$, it is enough to prove that

$S_5 \stackrel{+}{\Rightarrow} x_1S_5 \stackrel{+}{\Rightarrow} x_1x_2S_5 \stackrel{+}{\Rightarrow} x_1x_2x_3S_5 \ldots \stackrel{+}{\Rightarrow} x_1x_2x_3\ldots x_{n-1}S_5 \stackrel{+}{\Rightarrow} x_1x_2x_3\ldots x_{n-1}x_n$, if, and only if $S_1 \stackrel{+}{\Rightarrow} x_1$, $S_1 \stackrel{+}{\Rightarrow} x_2$, $S_1 \stackrel{+}{\Rightarrow} x_3$, …, $S_1 \stackrel{+}{\Rightarrow} x_n$ are productions in P_1. This is easy to prove and is left as an exercise for the reader.

Proved.

Theorem 4 A language is a regular expression if, and only if it is a right linear language.

Proof: **Only if part**: Here, we have to prove that if a language is a regular expression then the language is a right linear language. It can be proved that a right linear grammar exists, which generates the given regular expression. From theorem 2 and 3, it is obvious that for any regular expression, we can find a right linear grammar and hence it is a right linear language.

If part: In this part, we have to show that if a language is a right linear language then there exists a regular expression representing this language. We know that there exists a right linear grammar for every right linear language. Let this be G = (N, Σ, P, S). A master diagram representing P can be drawn for this grammar. From this master diagram, a regular expression can always be found. This regular expression is the representation of the language L(G).

Proved.

We have seen that a regular expression can be generated (represented) by a right linear grammar. There exists a mechanism that recognizes a regular expression. Having known this, we can now construct, now, a mathematical model in such a way that set of non-terminals in a grammar G is taken as a set of **states** and set of terminals is taken as set of **input** symbols. Any productions A → xB is interpreted as "*while scanning a string a machine goes to state B from A after reading input symbol x*". This arrangement is shown as below:

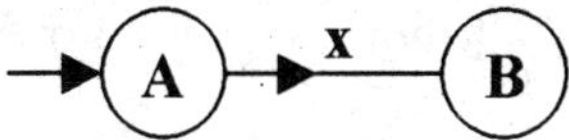

Based on this concept, we can define a finite state machine, which is the simplest recognizer of a regular expression. The following section deals with Finite States Machines (FSM).

8.3 Finite State Machine

We take a Finite State Machine (FSM) as a recognizer of a RE. Consider the directed graph shown in the figure 8.3.1 and the adjoining table 8.3.1.

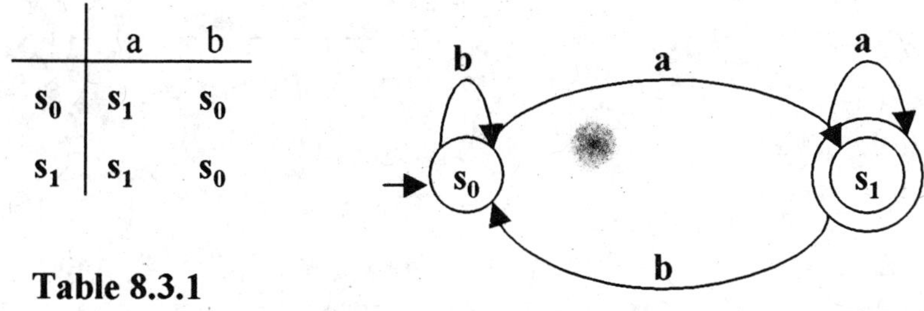

	a	b
s_0	s_1	s_0
s_1	s_1	s_0

Table 8.3.1

Figure 8.3.1

The entry in table 8.3.1 corresponding to row s_0 and column "a" is interpreted as: "*If a recognizer of a string is in state* s_0 *and reads an input symbol a, then it changes its state to* s_1." Similarly if it is on state s_1 and input symbol is b, then the recognizer goes to state s_0. Thus, entries in the table 8.3.1 are listing of possible states to which a machine may go depending upon the input symbol it reads while scanning a given string. A table like this is called **state transition table** of a machine. The same transition can be shown diagrammatically as in figure 8.3.1. If the number of states in a machine is finite it is called a finite state machine.

We define a **Finite State Machine** (FSM) as a mathematical model, (S, I, $\Im$, S_0, T). Where, S is a non-empty, finite set of states, I is a non-empty, finite set of input symbols, S_0 is a start state, T is set of final states (also called acceptance states and T $\subseteq$ S) and $\Im$ is a set of *state transition functions,* defined as:

$$\Im : S \times I \rightarrow S$$

The **FSM** shown in the figure 8.3.1 can be formalized as M = (S, I, $\Im$, S_0, T) where, S = {s_0, s_1}, I = {a, b}, $S_0 = s_0$, T = {s_1} and $\Im$ is given by the table 8.3.1.

Let us consider another diagram shown in the figure 8.3.2 below. This is a machine M = (S, I, $\Im$, S_0, T).

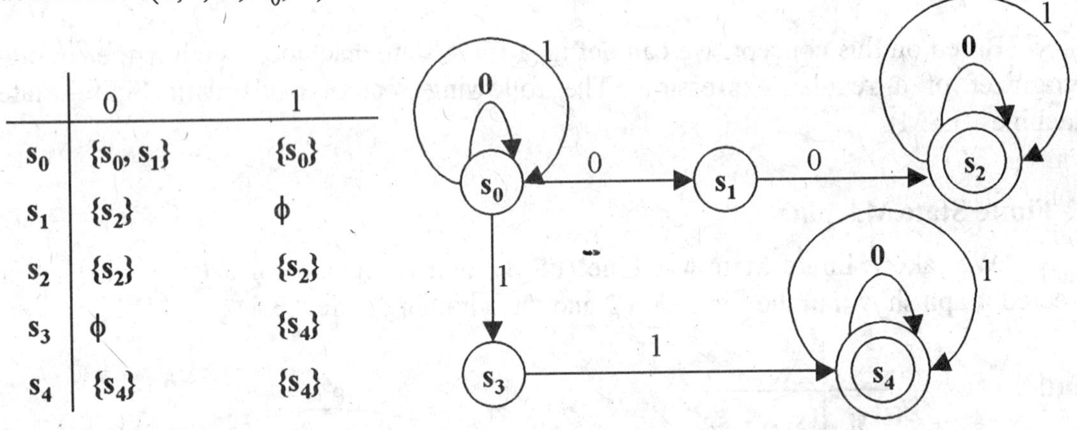

	0	1
s_0	{s_0, s_1}	{s_0}
s_1	{s_2}	ϕ
s_2	{s_2}	{s_2}
s_3	ϕ	{s_4}
s_4	{s_4}	{s_4}

Table 8.3.2 **Figure 8.3.2**

In this machine M, S = {s_0, s_1, s_2, s_3, s_4}, I = {0, 1}, $S_0 = s_0$, T = {s_2, s_4} and $\Im$ is given by the table 8.3.2. In this table, entries corresponding to (s_1, 1) and (s_3, 0) are null set i.e., once the machine M is in state s_1 and reads input symbol 1, it hangs. No transition is defined. Similarly, M hangs when it reads input symbol 0 while in state s_3. Now, notice the entry corresponding to (s_0, 0). It is a set {s_0, s_1}. This implies that if the machine M is in state s_0 and reads input symbol 0, it has two states to move to. It can move to only one of them. Such a general machine is called **Non-deterministic Finite State Machine**

(NFSM), or **Non-deterministic (NFA) Finite Automata**. If every entry in the transition table is a set containing **exactly one state**, the machine is called **Deterministic Finite State Machine (DFSM),** or **Deterministic Finite Automata (DFA)**. Thus Machine of figure 8.3.1 is a deterministic finite state machine. In this text we shall study only **DFSM.** Hence onward in this text, FSM implies DFSM only unless otherwise stated.

Any $M = (S, I, \mathfrak{I}, S_0, T)$ is a FSM, if

- S is a non-empty, finite set of states,
- I is non-empty, finite set of symbols over which RE is defined,
- There is one, and only state that is marked as initial (or starting) state with arrow sign $\rightarrow$,
- T is a non-empty subset of S. Possibly F may be equal to S i.e. all states may be a final state, and
- The transition table $\mathfrak{I}$ must have a singleton as entry at all places. No entry should be a null set or a set having cardinality ≥ 2. Here, we represent $\{s_n\}$ by s_n without loss of any generality. Thus, we can write $\mathfrak{I} = \{f_x \mid f_x : S \rightarrow S \; \forall \; x \in I\}$

Sometimes, some memory is provided to every state in a machine so that it can remember the input symbol it reads. In our model of consideration, a machine does not have any memory. It simply reads an input symbol, changes its state and throws the symbol. It recognizes a string as valid if the machine is in one of the final state when the string is completely scanned. Otherwise, the string is an invalid sentence. A machine scans any input strings starting from its initial state. Thus, any string $w \in I^*$ is valid for a machine $M = (S, I, \mathfrak{I}, S_0, T)$, if $f_w(S_0) \in T$.

Finding Language of FSM

An FSM is a recognizer of a language. The language of M is denoted as L(M). We can write

$$L(M) = \{w \mid w \in I^* \text{ and } f_w(S_0) \in T\}$$

Given a FSM we can always find its language. Finding language of a machine M means finding all those w that satisfies the above conditions. For a given machine, there may be infinite many such strings. It is not possible to test and list all such strings. Thus, clue lies in identifying the pattern which all such strings must have. Once the pattern is identified, express the language in a set theoretic notation. Let us now see some examples.

Example 1 Find the language of FSM shown in the figure 8.3.1.

Solution: Language of the machine is defined in input symbol I = {a, b}. The I is the alphabet of L(M). A string containing single a is acceptable as $f_a(s_0) = s_1 \in T$. If a string contains b, then M will remain at s_0 if it is at s_0 or it returns to s_0 if it is at s_1. This shows that any string to be accepted as valid, it must end in "a". Thus, we have

$$L(M) = \{w \mid w \in \{a, b\}^* \text{ and } w \text{ ends in "a"}\}$$

Ans.

Example 2 Find the language of FSM shown in the following figure 8.3.3.

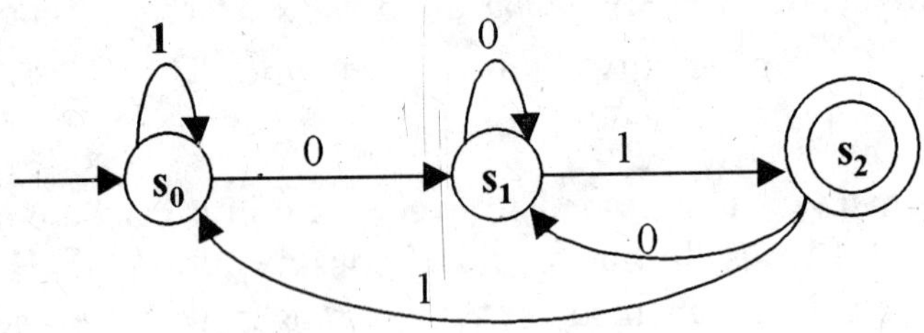

Figure 8.3.3

Solution: Language of the machine is defined in input symbol I = {0, 1}. The I is the alphabet of L(M). A string "01" is acceptable, because we have $f_{01}(s_0) = f_1(f_0(s_0)) = f_1(s_1) = s_2 \in T$. If machine M is at state s_2 and string is not completely scanned, M changes its state to s_0 or s_1 depending upon inputs 1 and 0 respectively. After a little bit of reasoning, it can be found that any string terminating in "01" is acceptable. And a string not terminating in "01" is invalid. Thus, we have

$$L(M) = \{w \mid w \in \{0, 1\}^* \text{ and } w \text{ ends in "01"}\}$$

Ans.

Example 3 Find the language of FSM shown in the following figure 8.3.4.

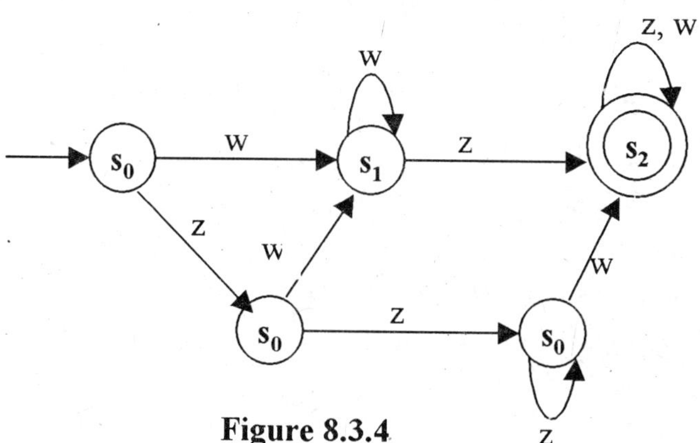

Figure 8.3.4

Solution: Language of the machine is defined in input symbol I = {w, z}. The I is the alphabet of L(M). A string "wz" is acceptable. Similarly, "zzw" is acceptable. There is a loop at the final state on input symbol w, z. This implies that, once reached at the final state, machine is not going to change its state. Once at the final state, it successfully scans the remaining symbols. Thus, the clue lies in identifying the patterns, which transfer the machine from the initial state to a final state. Certainly, any string containing "wz" or "zzw" can transfer the state of the machine from its initial to final state. Hence, we have

$$L(M) = \{w \mid w \in \{w, z\}^* \text{ and } w \text{ contains either "wz" or "zzw"}\}.$$

Ans.

Finding a Grammar from FSM

Every FSM language is a Regular language. For every regular language there exists a regular grammar. Thus, we can find a regular grammar G from a given FSM such that L(G) = L(M). To obtain a grammar from FSM, we apply the following rules.

(a) Let A and B be two states in FSM such that machine transfers its state from A to

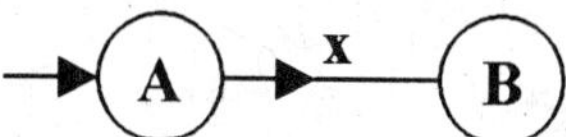

B on input symbol x then, we have a production A→ xB

(b) In the above diagram, if B is a final state then it yields two productions: A→ xB and A→ x.

(c) A loop is interpreted accordingly.

See the following examples to learn the procedure of finding a Regular grammar from a given FSM.

Example 4 Find a regular grammar corresponding to the FSM shown in the figure 8.3.3.

Solution: Let G = {N, Σ, P, S) be the corresponding RG. Here Σ = I = {0, 1}. The alphabets remain the same. The set S of M becomes N of G. Thus, we have N = {s_0, s_1, s_2}. The states corresponding to the initial state of M, becomes the start symbol of G. Thus, S = s_0. The production P has to be found. And it is given as

1. $s_0 \rightarrow 1s_0$
2. $s_0 \rightarrow 0s_1$
3. $s_1 \rightarrow 1s_2$
4. $s_1 \rightarrow 0s_1$
5. $s_1 \rightarrow 1$ [Since on 1 machine goes to final state from s_1]
6. $s_2 \rightarrow 0s_1$
7. $s_2 \rightarrow 1s_0$

Ans.

Example 5 Find a regular grammar corresponding to the FSM shown in the figure 8.3.1.

Solution: Let G = {N, Σ, P, S) be the corresponding RG. Here Σ = I = {a, b}, N = {s_0, s_1} and S = s_0. The production P has to be found. And it is given as

1. $s_0 \rightarrow bs_0$
2. $s_0 \rightarrow as_1$
3. $s_0 \rightarrow a$
4. $s_1 \rightarrow a$ [Since on a machine goes to final state from s_0.]
5. $s_1 \rightarrow bs_0$
6. $s_1 \rightarrow as_1$

Ans.

Drawing a FSM for a given language

Given a FSM, we can find L(M). We can also find a RG G from FSM such that L(G) = L(M). In this section we shall learn that given a language, we can draw a FSM. Every language is not a FSM language. A language is said to be a FSM language if there exist a FSM to recognize this language and rejects everything else. In brief, we can say that if the language is a regular language, there exist a FSM to recognize it. Conversely, if there is a FSM M, then the L(M) is a regular language. There is a theorem by "Kleen" stating exactly the same. Proof of this theorem is beyond the scope of the intended course.

In this part of the section, we gradually learn how to design a machine that accepts the given language and rejects all strings not belonging to the language.

Example 6 Draw a finite state machine that can accept any string of {0, 1} that terminates with the symbol 1 i.e. 1 must be the last symbol in a string.

Solution: Let the FSM be M and the given language be L. Any string like 1, 11, 111, 101, 10001110001 etc must be acceptable to M. However, strings like 10, 0, 11110, 1101000000 etc must not be acceptable. Let S_0 be the initial state of M. Since "1" is a valid string in L, M must transfer its own state to a final state, say S_1, from S_0 on input symbol 1. This arrangement can be shown as below.

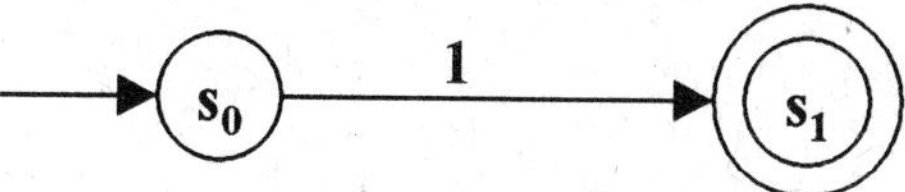

This is not a complete FSM. To complete this we have to define move from S_0 on input symbol 0 and all moves from final state S_1. Once at final state, M can scan all the following 1's in the string without changing its state. Thus, a loop on input symbol 1 will be sufficient at S_1. If M gets an input 0 at state S_1, it must change its state to ensure that a valid string must terminate in 1. If string contains any number of 0's in the beginning, it may be scanned without transferring state. A final arrangement is shown in figure 8.3.5.

Ans.

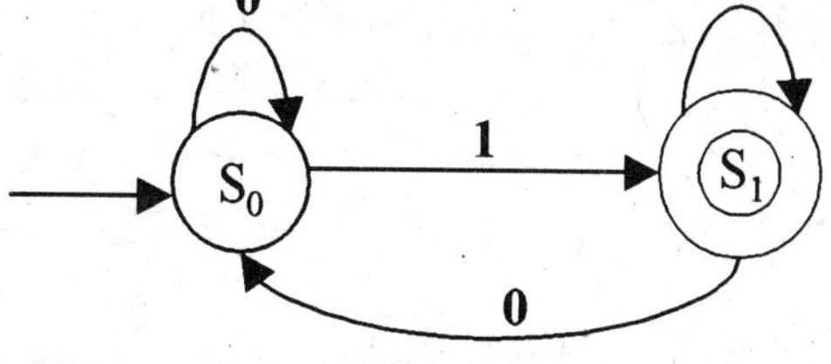

Figure 8.3.5

Example 7 Draw a finite state machine that can recognize only string 0120 on the input symbols {0, 1, 2}.

Solution: Let M is the required FSM. This accepts only string "0120". Let S_0 be the initial state of M. If a string contains any thing other that 0 as its first symbol, M must not move toward its final state after reading those symbols. The skeleton of the proposed FSM can be drawn as below.

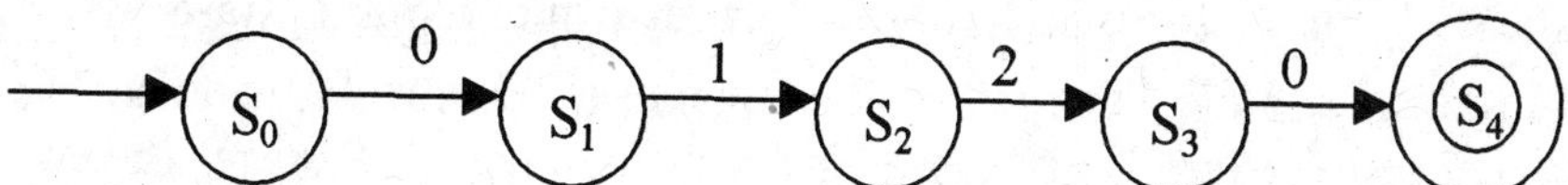

This is not a complete FSM. To complete this we have to define all the remaining moves from all states. If M is at S_0 and reads symbol 1 or 2 it must go to a **state of no return**, as those strings are not valid. Similarly, we can complete other moves. The final FSM is shown in the figure 8.3.6.

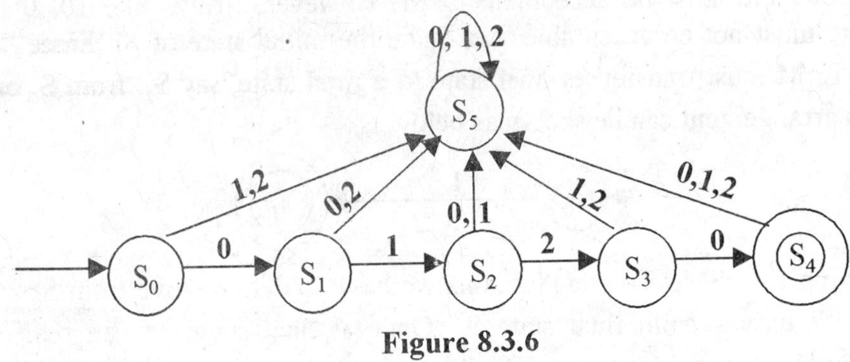

Figure 8.3.6

Ans.

In the machine of figure 8.3.6, state S_5 has a peculiar property. Once M has reached to this state from anywhere, there is no escape route from here. Such state is called **Trapping State.** Sometimes, it becomes necessary to introduce a trapping state in a FSM, for a language L, to ensure that a string having particular pattern must be rejected if it does not belong to L. A state is said to be **inaccessible** if there is no incoming arc to this state from any other state. A state is said to be **redundant**, if a FSM can accept the same language even without this state. More about this, we shall learn in next section. Now, let us see some more examples on drawing a FSM for a given language.

Example 8 Draw a FSM that can recognize a binary representation of any non-negative integer divisible by 5.

Solution: Let M is the required FSM and L be the language containing all bit string, the decimal equivalent of which is divisible by 5. Thus "0", "00", "00…0" all belong to L as they are divisible by 5. "101", "1010", "1111", "10100" etc are divisible by 5, thus is a valid string in L. 11, 110 etc does not belong to L. While designing a machine these patterns are very helpful. The skeleton of the proposed FSM can be drawn successively as below. Since a null string can be interpreted as decimal 0, a null string must also be acceptable to M. In order to ensure this, we have to mark initial state S_0 as final state. Skeleton in Figure 8.3.8(a) scans "000…0", that in (b) scans 101 and 1010, that in (c) scans all previous strings and 1111 also. This way, we can finally get the required FSM M, as shown in the figure 8.3.8.

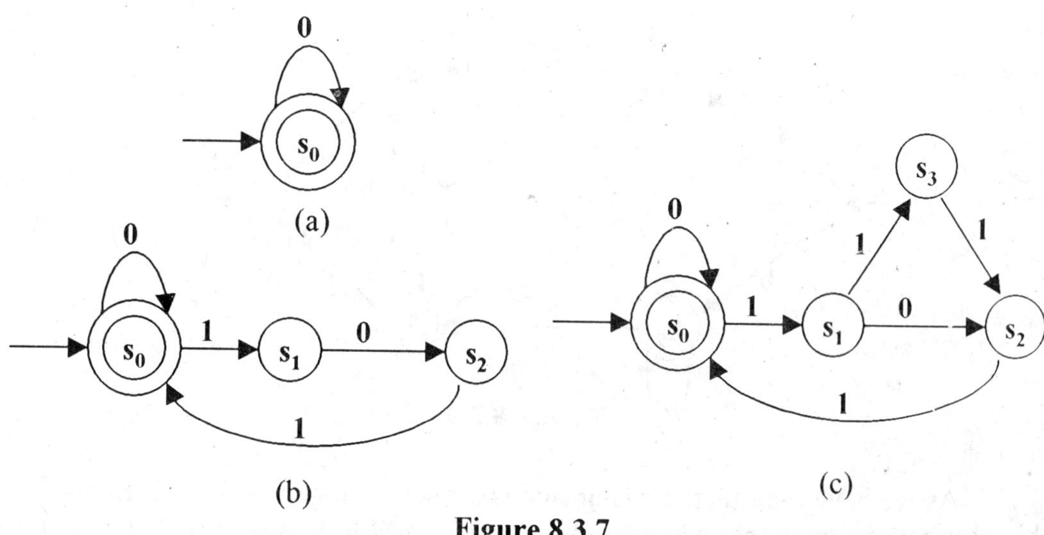

Figure 8.3.7

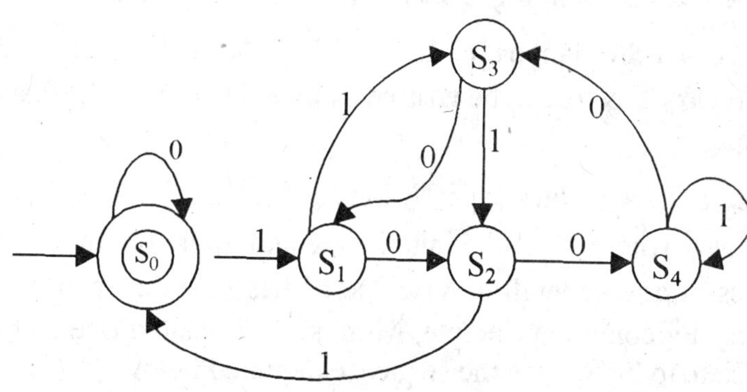

Figure 8.3.8

Example 9 Draw a finite state machine that can accept any string in a, b that contains '**abb**' or '**baa**'.

Solution: A FSM can be drawn as shown in the figure 8.3.9. How it has evolved to this stage, reader can very well reason it out. As a volunteer guide, I may suggest to reader that you never mug up the thing, which you are not able to comprehend. Try it again and again. Once you have got it, it is yours.

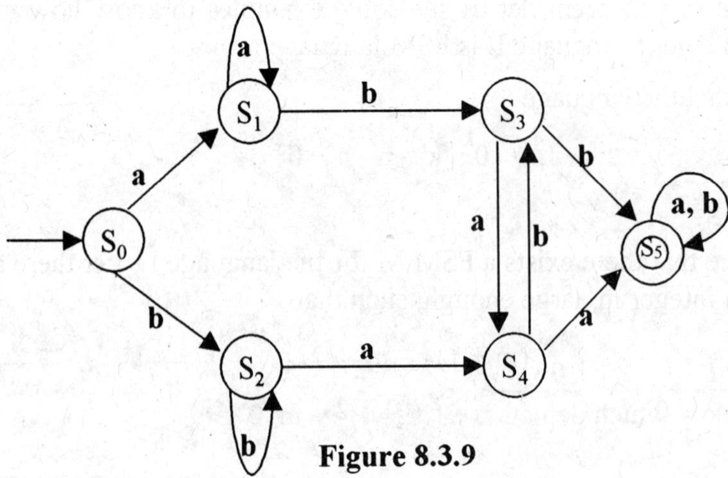

Figure 8.3.9

As we have seen that if a language is a regular language, we can find a FSM for this. But how to know that whether a language is FSM language or not? Trying to find a FSM for a language that is not regular, will be an effort in vain. Thus, it is better to ensure that given L is a regular language before proceeding to find a FSM for it. The following theorem, known as *pumping Lemma,* helps in this regard.

Theorem 1 Let L be a FSM language. Let FSM be M and it has N states. For any sequence $\alpha \in L$ where, $|\alpha| \geq N$, α can be written as **uvw** such that $|v| \neq 0$ and $\mathbf{uv^kw}$ for $k \geq 0$, is also acceptable.

Proof: Let $\alpha = x_1\, x_2\, x_3 \ldots x_N$. Thus, $|\alpha| = N$. Let $S_0, S_1, S_2, S_3, \ldots, S_N$ be states visited by the machine to accept α where, S_0 is the initial state and S_N is the final state. This we can assumes without loss of any generality. Machine M has N states, and it has visited N+1 states to accept α. By Pigeonhole principle, M must have visited one state at least twice. Let us assume that state to be S_k. See the sequence depicted below.

$$\overbrace{a \quad a \quad a \quad \cdots \quad a}^{u} \quad \overbrace{a \quad a \quad a \quad \cdots \quad a}^{v} \quad \overbrace{a \quad a \quad a \quad \cdots \quad a}^{w}$$

$$S_0\ S_1\ S_2 \cdots\cdots S_k \quad S_k \cdots\cdots\cdots\cdots S_k \quad S_{k+1} \cdots\cdots\cdots\cdots S_k$$

The depicted sequence can be divided into three segments: u, v and w where, v can be in loop and can be scanned $k \geq 0$ times. Therefore, $\mathbf{uv^kw}$ for $k \geq 0$, is a language accepted by M.

Proved.

Having done this theorem, let us see some examples to know how to apply this theorem to know whether a language L is FSM language or not.

Example 10 Show that the language

$$L = \{0^k \mid k = n^2, n > 0\}$$

is not a FSM language.

Proof: Let us assume that there exists a FSM M for the language L. Let there are N states in M. Let us take an integer m, large enough, such that

$$(m + 1)2 - m^2 > N$$

See the sequence below which depicts $p = (m + 1)2 - m^2$ 0's.

$$0\ 0\ 0\ 0\ \cdots\ \underset{m^2}{0}\ \overbrace{0\ 0\ 0\ 0\ \cdots\ \underset{p}{0}}^{>N}$$

Clearly, there are 0's > N between m^2 and $(m + 1)2$. If the machine M accepts the string $0^k \mid k = (m+1)^2$, then a certain state S_k is visited at least twice. Removal of 0's between these two visits of S_k, yields a sequence of 0's having q number of 0's such that $0 < q < p$ and q lies between m^2 and $(m + 1)^2$ i.e.,

$$(m + 1)2 > q > m^2$$

This implies that M accepts a string that does not belong to L. Hence L is not a FSM language.

Proved.

Example 11 Show that the language

$$L = \{a^n b^n \mid n \geq 1\}$$

is not a FSM language.

Proof: Let us assume to the contrary that the L is a FSM language. Thus, there exists a FSM, say M. Let M has N states. Let $a^N b^N$ be a string in L. To scan N a's, M changes states N + 1 times. Let S_0 be the initial state of M. Let us number the states as M visits it. Machine M has N states and it has visited N+1 states to accept N 0's. By Pigeonhole principle, M must have visited one state at least twice. Let us assume that state to be S_k. See the sequence depicted below.

$$\overbrace{a \quad a \quad a \cdots a \quad a \cdots a \quad a \cdots a}^{N} \quad \overbrace{b \quad b \quad b \cdots b}^{N}$$
$$S_0 \quad S_1 \quad S_2 \; S_K \qquad S_K \qquad\qquad\qquad S_{2N}$$

This shows that at S_k, an optional loop is created, that can be avoided. This implies that a string $x = a^{N-m} b^N$ may also be accepted for $m > 0$. Since $x \notin L$, FSM, so assumed, is not a FSM for this language. Therefore, L is not a FSM language.

Proved.

This section can be summarized by stating the following equivalence in form of a theorem without proof.

Theorem 2 The following statements are equivalent.

1. L is a regular set.
2. L is a right linear language
3. L is a FSM language
4. L is denoted by regular expression.

Important points to remember while drawing a FSM for a language.

- If a null string is in language, the initial state must be a final state.
- If a string terminates in a fixed sequence of symbols (e.g. 011, 001, az, aa, etc) then, there must not be a loop on final states on symbols not mentioned in the string.
- If a string terminates in a symbol, then there must be a loop at the final state on the same symbol, but not on the other symbol $\in$ I.
- If a string starts with a particular symbol. a trapping state is required in the FSM.

These are just some tips that may help you in drawing a FSM for a given language. The best way to develop a concept is to do a rigorous practice. Try to solve problems given in exercise at the end of this chapter.

8.4 Simplification of Finite State Machine

For a given FSM M we can find the smallest FSM equivalent to M by eliminating all ***inaccessible*** states in M and then ***merging*** all redundant states in M. The redundant states are determined by partitioning the set of all accessible states into equivalence classes such that each equivalence class contains indistinguishable states and is as large as possible. We then choose one representative from each equivalence class as a state for the reduced FSM. Thus we can reduce the size of M if M contains inaccessible states or two or more indistinguishable states. We shall show that this reduced machine is the smallest and the most efficient finite automation that recognizes the language defined by the original FSM M. We use the number of states in a FSM as a measure of its efficiency. Let us first learn some basic concepts needed to understand the simplification process.

Reader may recall the definition of a free semi-group and of congruence relation. For quick reference it is reproduced here. Let (S, *) be a semi-group, then an equivalence relation R defined on S is said to be ***congruence relation*** if

$$aRb \text{ and } cRd \Rightarrow (a * c)\, R\, (b * d)$$

Example 1 Let (I, +) be a mathematical structure & R be a relation defined on I as aRb iff $a \equiv_2 b$. The relation R is a congruence relation on I.

Example 2 Let A = {1, 0} and consider the free semi-group (A*, •), where • is concatenation operation. Define a relation R on A* as $\alpha R \beta$ iff α and β have same number of 1's. The relation R is a congruence relation on A*.

Let M = (S, I, $\mathfrak{I}$, s_0, T) be a FSM and suppose R is an equivalence relation on S, we say that R is a ***machine congruence*** on M if for any s, t $\in$ S

$$sRt \Rightarrow f_x(s)\, R\, f_x(t) \text{ for all x in I}$$

That is R-equivalent pair of states are always taken into R-equivalent pair of states by every input symbols. Since R is an equivalence relation on S, it partitions S into equivalence classes S/R = {[s] | s $\in$ S}.

$$\text{Since } \{s] = [t] \Rightarrow sRt$$

$$\Rightarrow f_x(s)\, R\, f_x(t)$$

$$\Rightarrow [f_x(s)] = [f_x(t)]$$

Thus we can define a new FSM M/R = (S/R, I, $\mathfrak{I}$/R, [s_0], T/R), where

$$\mathfrak{I}/R = \{g_x |\, g_x : S/R \rightarrow S/R \text{ for all x in I and } g_x([s]) = [f_x(s)]\},$$

[s_0] is the class of states equivalent to initial states in M,

T/R = Set of equivalence classes containing one or more states of T of M.

This shows that this FSM is again a machine containing less number of states than that in M. This FSM M/R is called ***quotient finite state machine (QFSM)*** corresponding to relation R. Generally a QFSM is simpler and efficient than the original FSM.

Example 3 Let M = (S, I, $\mathfrak{I}$, s_0, T) be a finite state machine whose state transition table is according to the table 8.4.1. Let R be an equivalence relation on S and it is defined according to the matrix M_R of table 8.4.2. It is also given that T = {S_2} and s_0= S_0. Show that R is machine congruence on M and find the corresponding QFSM.

$$M_R = \begin{pmatrix} 1 & 0 & 0 & 0 & 1 & 0 \\ 0 & 1 & 0 & 0 & 0 & 1 \\ 0 & 0 & 1 & 1 & 0 & 0 \\ 0 & 0 & 1 & 1 & 0 & 0 \\ 1 & 0 & 0 & 0 & 1 & 0 \\ 0 & 1 & 0 & 0 & 0 & 1 \end{pmatrix}$$

Table 8.4.2

$\mathfrak{I}$ =

	0	1
S_0	S_0	S_1
S_1	S_3	S_5
S_2	S_2	S_4
S_3	S_3	S_0
S_4	S_4	S_5
S_5	S_2	S_1

Table 8.4.1

Solution: In this FSM M, S = {S_0, S_1, S_2, S_3, S_4, S_5}, I = {0, 1} and other things are already mentioned. Based on the given states transition table, diagrammatic representation of FSM can be given as in figure 8.4.1.

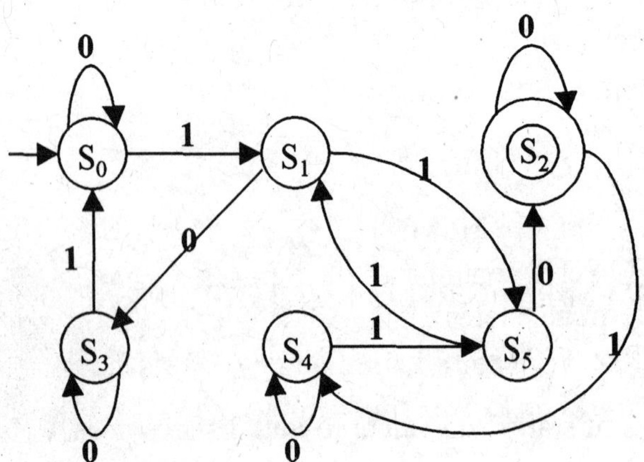

Figure 8.4.1

The given relation on S is an equivalence relation, which can be easily verified. This relation partitions S into three equivalence classes:

$$[S_0] = \{S_0, S_4\},$$
$$[S_1] = \{S_1, S_5\},$$
$$[S_2] = \{S_2, S_3\}$$

Notice that (S_0 & S_4) are R equivalent states. Similarly, (S_1 & S_5) and (S_2 & S_3) are R equivalent pairs of states. Here, we have

$$f_0(S_0) = S_0,\ f_0(S_4) = S_4;\ f_1(S_0) = S_1,\ f_1(S_4) = S_5$$

This implies that on input symbol 0, equivalence class $[S_0]$ is mapped to the equivalence class $[S_0]$ and on input symbol 1 $[S_0]$ is mapped to the equivalence class $[S_1]$. Similarly we can find transitions for other equivalence classes. Therefore R is a machine congruence to M. In the QFSM, S/R = $\{[S_0], [S_1], [S_2]\}$, I = {0,1}, $\Im$/R is given by the table 8.4.3. The class that contains initial states of original FSM, becomes initial state in QFSM and all the classes containing at least one final state of original FSM become final states in QFSM. Therefore, in QFSM $s_0 = [S_0]$ and T/R = $\{[S_2]\}$. The diagrammatic representation of QFSM is shown in the figure 8.4.2.

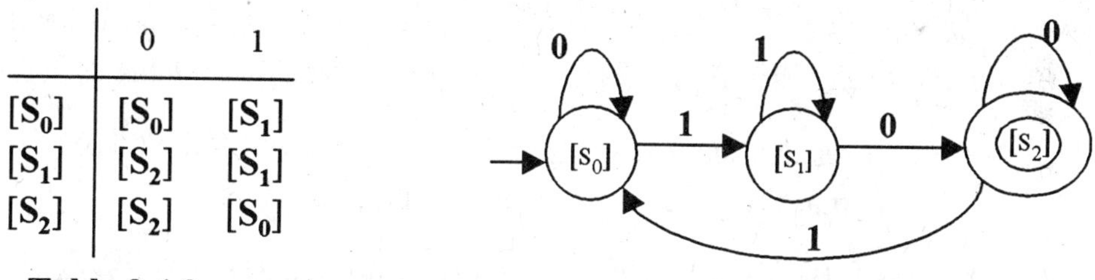

	0	1
$[S_0]$	$[S_0]$	$[S_1]$
$[S_1]$	$[S_2]$	$[S_1]$
$[S_2]$	$[S_2]$	$[S_0]$

Table 8.4.3

Figure 8.4.2

Ans.

Let M = (S, I, $\Im$, s_0, T) be a FSM. Let $T^C = S - T$. Then set S is decomposed into two disjoint subsets T and T^C. Set T contains all the terminal (acceptance) states of the M and T^C contains all the non-terminal states of M. Let $w \in I^*$ be any string. For any state $s \in S$, $f_w(s)$ belongs to either T or T^C. *Any two states: s and t are said to be* ***w-compatible*** *if both* $f_w(s)$ *and* $f_w(t)$ *either belongs to T or to* T^C *for all* $w \in I^*$.

Theorem 1 Let M = (S, I, $\Im$, s_0, T) be a FSM. Let R be a relation defined on S as sRt if and only if both s and t are w-compatible. Show that

(a) R is an equivalence relation.

(b) R is machine congruence to M.

Proof: (a) Let s be any state in S and w be any string in I*. Then $f_w(s)$ is either in T or in T^C. This implies that sRs $\forall$ s$\in$ S. Hence R is reflexive on S. Let s and t be any two states in S such that sRt. Therefore, by definition of R both s and t w-compatible states. This implies that tRs. Thus, R is symmetric also. Finally, let s, t and u be any three states in S such that sRt and tRu. This implies that s t and u are w-compatible states i.e. s and u are w-compatible. Hence sRu. Thus, R is a transitive relation. Since R is reflexive, symmetric and transitive, it is an equivalence relation.

Proved.

(b) To prove that R is machine congruence to M, we prove that every R-equivalent pair of states is taken into R-equivalent pair of states by R i.e.,

if s, t$\in$ S and **sRt** then $\mathbf{f_x(s)\ R\ f_x(t)}$ $\forall$ x$\in$ I

Let $w_1 \in$ I* and w= w_1•x, where • is a concatenation operation and x is any input symbol from I.

Since s & t are R related, s & t are w-compatible i.e. $f_w(s)$ and $f_w(t)$ both are either in T or in T^C. But w= w_1•x, thus

$$f_w(s) = f_{w_1 \bullet x}(s) = f_{w_1}(f_x(s)) \text{ and}$$

$$f_w(t) = f_{w_1 \bullet x}(t) = f_{w_1}(f_x(t))$$

The above two results show that $f_x(s)$ and $f_x(t)$ are w_1-compatible. Since w_1 is any string in I*, we say that $f_x(s)$ and $f_x(t)$ are w-compatible. Thus,

$f_x(s)\ R\ f_x(t)$ $\forall$ x$\in$ I [Since x is any symbol of I]

Thus we have proved that sRt $\Rightarrow$ $f_x(s)\ R\ f_x(t)$ $\forall$ x $\in$ I. Therefore R is machine congruence to M.

Proved.

We have discussed a quotient machine, which is obtained from a given FSM. In the following theorem, we shall show that a language of a quotient machine is same as that of original FSM. Therefore, the process of simplification of FSM only makes the machine more efficient without losing any generality of the nature of language acceptable to the machine.

Theorem 2 Let M = (S, I, $\Im$, s_0, T) be a FSM and R be a relation of w-compatibility on S. If M/R = (S/R, I, $\Im$/R, [s_0], T/R) be the corresponding QFSM then L(M) = L(M/R).

Proof: Let $\mathfrak{I}/R = \{g_x | g_x: S/R \to S/R \text{ for all x in I}\}$ and $\mathfrak{I} = \{f_x | f_x: S \to S \text{ for all x in I}\}$. Let $w \in L(M)$ then $f_w(s_0) \in T$. The equivalence class $[s_0]$ contains all states of S that are R-equivalent to s_0. Since R is machine congruence to M (as proved in theorem 1(b)), all R-equivalent states are mapped to r-equivalent pair of states. Thus $f_w(s_0) \in T \Rightarrow [f_w(s_0)] \in T/R$. Also $g_w([s_0]) = [f_w(s_0)]$ because M/R is QFSM. This shows that $g_w([s_0]) \in T/R$. Therefore, $w \in L(M/R)$. Hence from the result

$$w \in L(M) \Rightarrow w \in L(M/R)$$

we conclude that $L(M) \subseteq L(M/R)$. ———————————— (A)

Conversely, let $w \in L(M/R)$. Then $g_w([s_0]) \in T/R$. Since $g_w([s_0]) = [f_w(s_0)]$, $[f_w(s_0)] \in T/R$. This implies that $\exists$ some state $t \in T$ such that $t \, R \, f_w(s_0)$. That is t and $f_w(s_0)$ are w-compatible states $\forall \, w \in I^*$. If we take $w_1 = \varepsilon$ then $f_{w_1}(t)$ and $f_{w_1}(f_w(s_0))$ both either belong to T or to T^C. Since $f_{w_1}(t) = t \in T$, $f_{w_1}(f_w(s_0)) = f_w(s_0)$ also belongs to T. Thus w is acceptable by M and hence $w \in L(M)$. This shows that

$$L(M/R) \subseteq L(M). \quad \text{———————————— (B)}$$

From (A) and (B), we have $L(M) = L(M/R)$.

Proved.

Simplification of a FSM is an evolutionary process. It is achieved through step-by step approximation. Let us define a relation R_k on S of $M = (S, I, \mathfrak{I}, s_0, T)$ as for any two states s and t, sRt iff s and t are w-compatible for all $w \in I^*$ with $|w| \leq k$. The following theorems give the basic concepts involved in simplifying a given FSM.

Theorem 3 Let R_k be relation on S of $M = (S, I, \mathfrak{I}, s_0, T)$ as defined above. Then

(a) $R_{k+1} \subseteq R_k \;\; \forall \, k \geq 0$.

(b) Each R_k is an equivalence relation.

(c) $R \subseteq R_k \;\; \forall \, k \geq 0$.

Proof: (a) Let $(s, t) \in R_{k+1} \Rightarrow s \, R_{k+1} \, t$

$\Rightarrow$ s and t are w-compatible $\forall \, w \in I^*$ with $|w| \leq k+1$

$\Rightarrow$ s and t are w-compatible $\forall \, w \in I^*$ with $|w| \leq k$

$\Rightarrow s \, R_k \, t$

$\Rightarrow (s, t) \in R_k$

Therefore, $R_{k+1} \subseteq R_k$

Proved.

(b) It can be proved according to the proof of the theorem 1(a).

(c) Prove as in part (a).

Theorem 4 For any FSM M = (S, I, $\Im$, s_0, T), and relation R_k as defined above the following are true.

(a) $S/R_0 = \{T, T^C\}$

(b) For $k \geq 0$ R s, $t \in S$, $s\, R_{k+1}\, t$ if and only if

(i) $s\, R_k\, t$ and

(ii) $f_x(s)\, R_k\, f_w(t)\ \forall\ x \in I$.

Proof: Since $|\varepsilon| = 0$, no input symbol is there for state transition in M. Thus a state is either in T or in T^C. Therefore, $S/R_0 = \{T, T^C\}$.

(b) Let $w \in I^*$ and $|w| \leq k+1$. Then $w = vx$ for some $x \in I$ and $v \in I^*$ with $|v| \leq k$. Similarly, if $x \in I$ and $v \in I^*$ with $|v| \leq k$ then $w = vx \in I^*$ with $|w| \leq k+1$.

Now for any state s, $t \in S$, we have

$$f_w(s) = f_{vx}(s) = f_v(f_x(s))$$

and

$$f_w(t) = f_{vx}(t) = f_v(f_x(t))$$

Now s and t are w-compatible if and only if both $f_x(s)$ and $f_x(t)\ \forall\ x \in I$ are v-compatible i.e. $sR_{k+1}t$ **iff** $f_x(s)\, R_k\, f_x(t)\ \forall\ x \in I$

Next, since $R_{k+1} \subseteq R_k\ \forall\ k \geq 0$ (Theorem 3(a)), we have $sR_{k+1}t \Rightarrow sR_kt$.

Proved.

Theorem 5 If $R_k = R_{k+1}$ for any non-negative integer k, then $R_k = R$.

Proof: From the theorem 4, for any two states s and t of S

$$\begin{aligned} sR_{k+2}t \ \textbf{iff} \quad & f_x(s)\, R_{k+1}\, f_x(t) \quad \forall\ x \in I \\ \textbf{iff} \quad & f_x(s)\, R_k\, f_x(t) \quad \forall\ x \in I \quad [\text{Since } R_k = R_{k+1}] \\ \textbf{iff} \quad & s\, R_{k+1}\, t \end{aligned}$$

This implies that $R_{k+2} = R_{k+1} = R_k$. Similarly by induction, $R_k = R_n\ \forall\ n \geq k$.

Since $R_0 \supseteq R_1 \supseteq R_2 \ldots \supseteq R_n \ldots$, we have

$$R = \bigcap_{n=0}^{\infty} R_n = R_k$$

Therefore $R = R_k$.

Proved.

Having seen all the basic theorems, it is the time to learn the working method to find QFSM from a given FSM. The algorithm is outlined below.

Step 1: Start with $P_0 = \{T, T^C\}$ of S corresponding to relation R_0.

Step 2: Construct successive partitions $P_1, P_2, P_3, \ldots, P_n$ corresponding to the equivalence relations $R_1, R_2, R_3, \ldots, R_n$ respectively, using

2.1 Let $P_k = \{A_1, A_2, A_3, \ldots, A_m\}$, where each A_i is an equivalence class of S obtained by equivalence relation R_k. Examine each A_i and decompose it into further subclasses where two elements s & t of A_i fall into same subclasses if a input $x \in I$ takes both s & t into same class A_i.

2.2 Call this new partition of S P_{k+1}.

Step 3: Repeat Step 2 until $P_k = P_{k+1}$.

Step 4: The resulting $P_k = P$ corresponds to the relation R. The QFSM $(S/R, I, \Im/R, [s_0], T/R)$ of M obtained by R is the simplified machine.

We shall demonstrate the working of this algorithm in the following examples.

Example 4 Simplify the FSM shown in the figures 8.4.3.

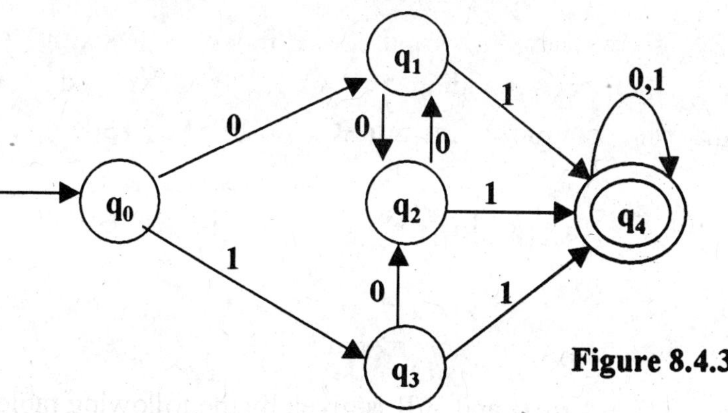

Figure 8.4.3

Solution: Here, in the given FSM M =(S, I, $\Im$, s_0, T), we have S = {q_0, q_1, q_2, q_3, q_4}, I = {0, 1}, $s_0 = q_0$, T = {q_4} and $\Im$ can be derived from the diagram of M. The initial partition P_0 is obtained on empty string ε i.e. without making a move. Thus, we have

P_0 = {T, T^C}, where T = {q_4 } and T^C = {q_0, q_1, q_2, q_3}. We can now write different partitions as collection of subsets of S only, like

$$P_0 = \{\{q_4\}, \{q_0, q_1, q_2, q_3\}\}$$

This completes the starting steps i.e. step 1. Now we have to decompose the subclasses so obtained. The subclass {q_4 } contains only one element, so it cannot be further decomposed. We, therefore, consider {q_0, q_1, q_2, q_3} for further decomposition, if possible.

On input symbol 0 all states in T^C goes to some states belonging to T^C only, but on input symbol 1, q_0 goes to q_1 in T^C and other states q_1, q_2 and q_3 goes to q_4 in T. This implies that on input symbol 0 q_1, q_2 and q_3 go to states in T^C and on input symbol 1 they go to a state in T, so they form a class different from the class to which q_0 belongs. Thus we have a new partition

$$P_1 = \{\{q_4\}, \{q_0\} \{q_1, q_2, q_3\}\}$$

We repeat the above procedure on P_1 to get P_2. Only subclass that can be, if at all, partitioned is {q_1, q_2, q_3}. On input symbol 0 all these states go to some state of {q_1, q_2, q_3} and on input symbol 1 they go to {q_4 }, so no further partition is possible. Therefore, we have

$$P_2 = \{\{q_4\}, \{q_0\} \{q_1, q_2, q_3\}\}$$

Now $P_1 = P_2$, so we stop here and it the P we are looking for. Thus P = {[q_0],[q_1],[q_4]}. The relation corresponding to P_0 is R_0, P_1 is R_1, and so on. Finally P corresponds to the machine congruence relation R. The QFSM = (S/R, I, $\Im$/R, [s_0], T/R), where

S/R = {[q_0],[q_1],[q_4]},

I = {0,1},

[s_0] = [q_0],

T/R = { [q_4]} and $\Im$/R is given by the following table 8.4.4.

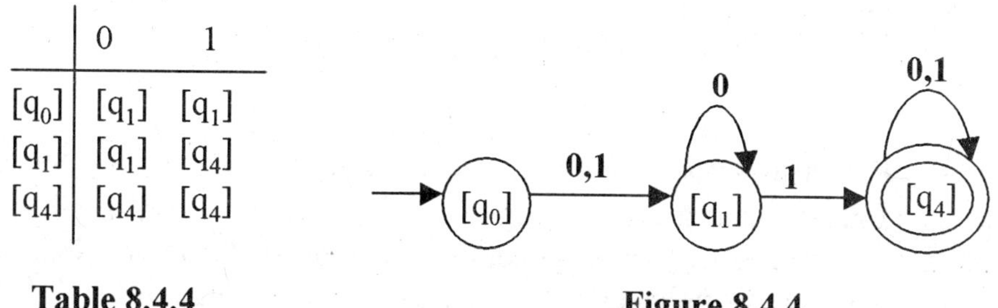

	0	1
$[q_0]$	$[q_1]$	$[q_1]$
$[q_1]$	$[q_1]$	$[q_4]$
$[q_4]$	$[q_4]$	$[q_4]$

Table 8.4.4 **Figure 8.4.4**

The final simplified QFSM is shown in the figure 8.4.4.

Ans.

Example 5 Simplify the FSM shown in the following figure.

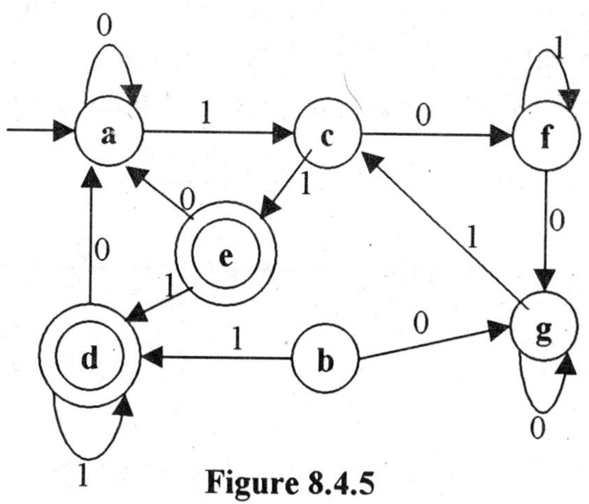

Figure 8.4.5

Solution: In the given FSM M = (S, I, $\Im$, s_0, T), we have S = {a, b, c, d, e, f, g}, I = {0, 1}, s_0 = a, T = {d, e} and $\Im$ is obvious from the diagram 8.4.5.

Here, T = {d,e}and T^C = {a, b, c, f, g}. Thus, the initial partition P_0 is given by

$$P_0 = \{\{d, e\}, \{a, b, c, f, g\}\}$$

Notice that, state b has no incoming arc and hence it is an inaccessible state. It can be eliminated from the machine. Now decomposition procedure is applied on T and T^C. The set {d, e} cannot be decomposed further as on input symbol 1, d and e go to states in {d, e} and on 0 it goes to T^C. In T^C, on input symbol 1, c goes to states in T where as a, g and f go to states in T^C. On input symbol 0, c goes to states in {a, g, f} and states in {a, g, f} get transferred to some states in {a, g, f}. Thus {a, c, f, g} has been decomposed into two subclasses: {a, g, f} and {c}. Hence, we have

$$P_1 = \{\{d, e\}, \{a, f, g\}, \{c\}\}$$

Now to find P_2, we consider each class of P_1. Again {d, e} cannot be decomposed further. In class {a, f, g}, states a and g go to states c in {c} on input symbol 1, whereas, state f goes to f itself. This implies that {a, f, g} can be decomposed into two subclasses {a, g} and {f}. Class {c} contains single state, no further decomposition is possible. Therefore, we have

$$P_2 = \{\{d, e\}, \{a, g\}, \{f\}, \{c\}\}$$

It can be easily verified that

$$P_3 = \{\{d, e\}, \{a, g\}, \{f\}, \{c\}\}$$

Since $P_2 = P_3$, we have $P = P_2 = \{[a], [c], [d], [f]\}$. Therefore, in the QFSM $M/R = (S/R, I, \Im/R, [s_0], T/R)$, we have $S/R = P$, $I = \{0, 1\}$, $s_0 = [a]$, $T = \{[d]\}$ and $\Im/R$ is given by the following table 8.4.5. The simplified machine is shown in the figure 8.4.6.

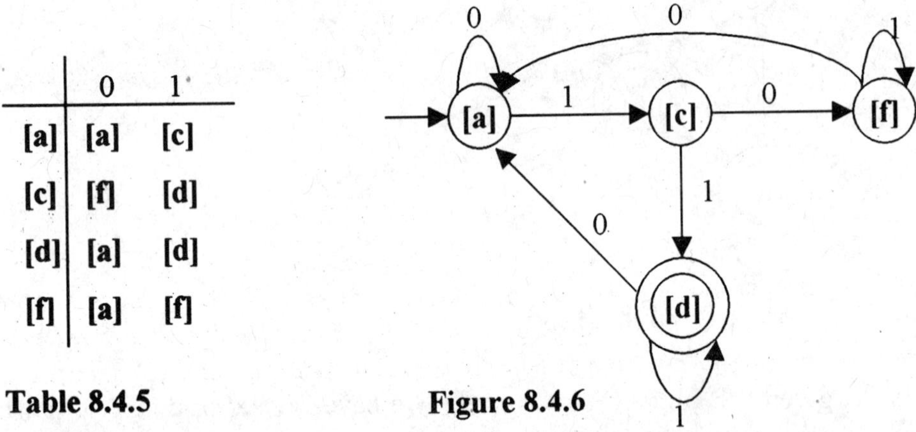

	0	1
[a]	[a]	[c]
[c]	[f]	[d]
[d]	[a]	[d]
[f]	[a]	[f]

Table 8.4.5

Figure 8.4.6

Example 6 Simplify the FSM shown in the following diagram.

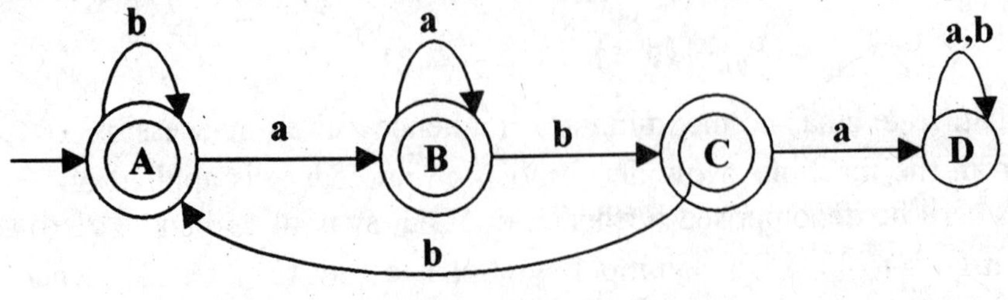

Figure 8.4.7

Solution: In the given FSM M = (S, I, $\mathfrak{I}$, s_0, T), we have S = {A, B, C, D}, I = {a, b}, s_0 = A, T = {A, B, C} and $\mathfrak{I}$ is obvious from the diagram 8.4.7.

Here, T = {A, B, C} and T^C = {D}. Thus, the initial partition P_0 is given by

$$P_0 = \{\{A, B, C\}, \{D\}\}$$

Notice that, there is no inaccessible state in this machine. Now decomposition procedure is applied on T and T^C. The set {D} cannot be decomposed further as it contains only one state. In T, on input symbol a, C goes to states in T^C whereas A and B go to states in T. Thus {A, B, C} has been decomposed into two subclasses: {A, B} and {C}. Hence, we have

$$P_1 = \{\{A, B\}, \{C\}, \{D\}\}$$

Now to find P_2, we consider each class of P_1. Again {C} and {D} cannot be decomposed further. In class {A, B}, state A goes to state A in {A, B} on input symbol b, whereas, state B goes to C on the same input symbol. This implies that {A, B} can be decomposed into two subclasses: {A} and {B}. Therefore, we have

$$P_2 = \{\{A\}, \{B\}, \{C\}, \{D\}\} = P$$

This shows that the given FSM is already in simplified form. Therefore, it cannot be simplified further.

Ans.

A programming language is based on certain syntax and semantics. A tool is needed to ensure that whatever is said to the computer in a language, it is said in exact way. It is the grammar that determines whether the said instructions are in exact way or not. In this chapter, an attempt has been made to acquaint the readers with basic concepts of language, generator of a language (grammar) and recognizer (machine). This basic foundation will help the reader in understanding the essentials of compiler design and Natural Language processing. Readers are advised to attempt the problems given in the exercise to have fare understanding of the topics covered in this chapter.

Exercise

1. Construct right-linear grammar for

   (a) Identifiers which can be of arbitrary length and must start with a letter and may contain any characters from roman alphabet and any digits.

   (b) Identifiers which can be one to six symbols in length and must start with I, J, K, L, M or N and may contain any lower/uppercase roman alphabet and digits.

(c) All strings of 0's and 1's having both an odd number of 0's and an odd number of 1's..

2. Construct context free grammars that generate

(a) All strings of 0's and 1's having equal number of 0's and 1's.

(b) Well formed statements in propositional calculus.

(c)$\{0^i1^j \mid i \neq j \text{ and } i,j \geq 0\}$.

(d) All possible sequences of balanced parentheses.

3. Describe the language generated by the productions $S \to bSS \mid a$. Observe that it is not always easy to describe what language a grammar generates.

4. What class of languages can be generated by grammars with only *left context*, i.e. grammar in which each production is of the form $\alpha A \to \alpha\beta$, where α and β belongs to $(N \cup \Sigma)^*$?

5. Show that every context free language can be generated by a grammar $G = (N, \Sigma, P, S)$ in which each production is of either the form $A \to \alpha$, α in N^* or $A \to w$, w in Σ^*.

6. Let G be a grammar defined by the productions:

1. $S \to aSBC \mid abC$ 2. $CB \to BC$ 3. $bB \to bb$ 4. $bC \to bc$ 5. $cC \to cc$.

Prove that $L(G) = \{a^nb^nc^n \mid n \geq 1\}$.

7. In an unrestricted grammar G there are many ways of deriving a given sentence that are essentially the same, differing only in order in which productions are applied. If G is context-free, then we can represent these essentially similar derivations by means of a derivation tree. However, if G is a context sensitive or unrestricted, we can define equivalence classes of derivations in the following manner.

Let $G = (N, \Sigma, P, S)$ be an unrestricted grammar. Let D be the set of all derivations of the form $S \Rightarrow^* w$. That means elements of D are sequences of the form $(\alpha_0, \alpha_1, \alpha_2, \ldots, \alpha_n)$ such that $\alpha_0 = S$, $\alpha_n \in \Sigma^*$ and $\alpha_{i-1} \to \alpha_i$ for $1 \leq i \leq n$.

Define a relation R_0 on D by $(\alpha_0, \alpha_1, \alpha_2, \ldots, \alpha_n)\ R_0\ (\beta_0, \beta_1, \beta_2, \ldots, \beta_n)$ if and only if there exists i between 1 and (n - 1) such that

1. $\alpha_j = \beta_j$ for $1 \leq j \leq n$ such that $j \neq I$.

2. We can write $\alpha_{i-1} = \gamma_1\gamma_2\gamma_3\gamma_4\gamma_5$ and $\alpha_{i+1} = \gamma_1\delta\gamma_3\varepsilon\gamma_5$ such that $\gamma_2 \to \delta$ and $\gamma_4 \to \varepsilon$ are in P, and either $\alpha_i = \gamma_1\delta\gamma_3\gamma_4\gamma_5$ and $\beta_i = \gamma_1\gamma_2\gamma_3\varepsilon\gamma_5$ or conversely. Let R be the equivalence closure of the relation R_0.

A grammar G is said to be ***unambiguous*** if each w in L(G) appears as the last component of a derivation in one and only one equivalence class under R as defined above. Show that every right-linear language has an ***unambiguous right linear grammar.***

8. Which of the following are regular sets? Give regular expressions for these regular sets.
(a) The set of words in $\{0,1\}^*$ having equal number of 0's and 1's.
(b) The set of words in $\{0,1\}^*$ with an even number of 0's and an odd number of 1's.
(c) The set of words in Σ^* whose length is divisible by 3.
(d) The set of words in $\{0, 1\}^*$ with no sub-string 101.

9. Show that if L is any regular set, then there is infinite number of regular expressions denoting L.
10. Show the following identities for regular expressions α, β and γ:
    (a) $\alpha(\beta + \gamma) = \alpha\beta + \alpha\gamma$
    (b) $\alpha + (\beta + \gamma) = (\alpha + \beta) + \gamma$
    (c) $\alpha(\beta\,\gamma) = (\alpha\beta)\gamma$
    (d) $\alpha\,\varepsilon = \varepsilon\alpha = \alpha$
    (e) $(\alpha + \beta)\,\gamma = \alpha\gamma + \beta\gamma$
    (f) $\alpha + \alpha = \alpha$
    (g) $\phi^* = \varepsilon$
    (h) $\alpha^* + \alpha = \alpha^*$
    (i) $(\alpha^*)^* = \alpha^*$
    (j) $\alpha + \phi = \alpha$
    (k) $(\alpha + \beta)^* = (\alpha^*\beta^*)^*$
11. Solve the following set of regular expressions:
    $A_1 = (01^* + 1)\,A_1 + A_2$
    $A_2 = 11 + 1A_1 + 00A_3$
    $A_3 = \varepsilon + A_1 + A_2$
12. A ***right linear grammar*** G = (N, Σ, P, S) is called a ***regular*** grammar when
    - All productions with the possible exception of $S \rightarrow \varepsilon$ are of the form $A \rightarrow aB$ or $A \rightarrow a$ where A and B are in N and a is in Σ.
    - If $S \rightarrow \varepsilon$ is in P, then S does not appear on the right of any production.

    Show that every regular set has a regular grammar. Next construct a regular grammar for the regular set generated by the right linear grammar

    $A \rightarrow B \mid C \qquad B \rightarrow 0B \mid 1B \mid 011 \qquad C \rightarrow 0D \mid 1C \mid \varepsilon \qquad D \rightarrow 0C \mid 1D$
13. Find the regular expressions accepted by the FSM shown in the figures a, b, c and d.

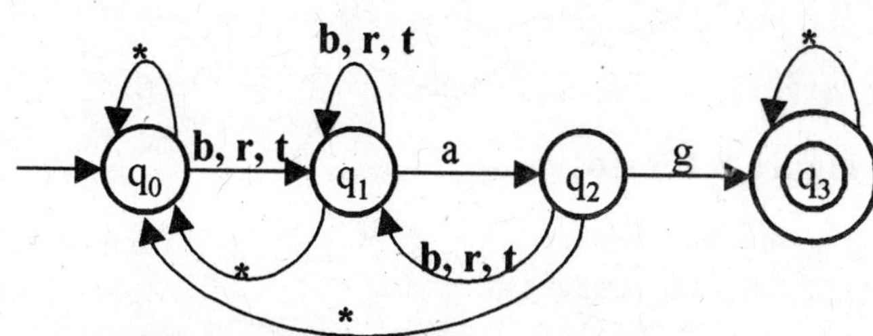

Figure 8.13ex(a)

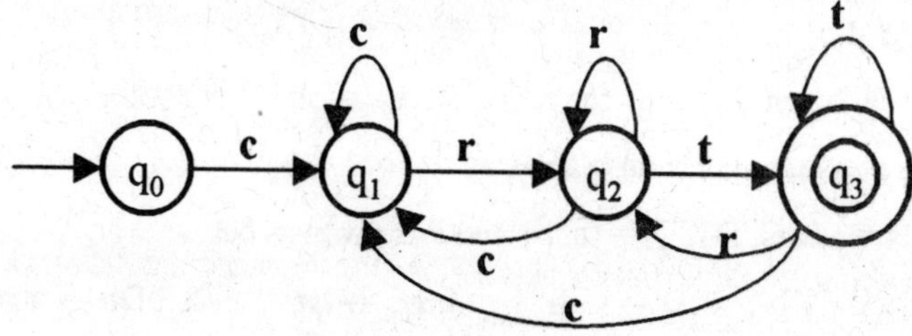

Figure 8.13ex(b)

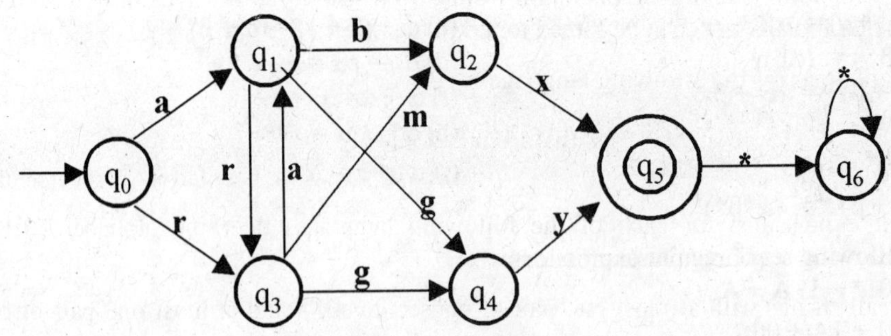

Figure 8.13ex(c)

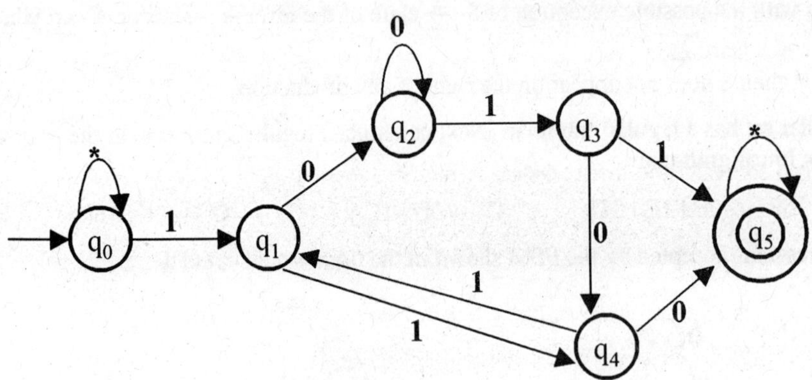

Figure 8.13ex(d)

14. Draw FSM corresponding to the following regular expressions:

(a) a + bb + bab*a (b) (ab*a + b)a*b

(c) a(ab* + ba)* (d) (b + a(a +bb)*ba)*(ε +(b(ba + ab)*a).

15. The regular expression a* represents the set {ε, a, aa, aaa, ...}. Find the regular expression representing the regular set {a, aa, aaa, ...}.

16. Find the set of strings on T = {a, b} produced by the regular expression b(a + b)*ab.

17. Draw a FSM to recognize the regular expression b(a+b)*ab.

18. Ler G = (N, Σ, P, S), where N = { S, C, D}, Σ = {a, b} and P is given as below:

1. S → aCDa 2. C → baCDb 3. D → Cab 4. aC → baa 5. bDb → abab.

(a) Show that baabbabaaabbaba ∈ L(G).

(b) Show that L(G) contains infinitely many elements (words).

(c) Show that if α is any string generated from S by P and includes some non-terminal symbols, then it is always possible to extend the generation from α to a string that involves only terminals. That is, any generation process in G can be forced to terminate.

19. Determine a grammar for the following languages.

(a) $\{a^{n^2} \mid n > 1\}$ (b) $\{a^n b^n a^m \mid n, m > 1\}$

(c) $\{ww \mid w \in \{a, b\}^*\}$ (d) $\{ww^R \mid w \in \{a, b\}^*$ and w^R is reverse of $w\}$.

20. Write regular expressions for each of the following languages over the alphabet {0,1}. Provide justification that your regular expression is correct.

(a) The set of all strings with at most one pair of consecutive 0's and at most one pair of consecutive 1's.

(b) The set of all strings in which every pair of adjacent 0's appears before any pair of adjacent 1's.

(c) The set of all strings not containing 101 as a sub-string.

21. Describe in statement the sets denoted by the following regular expressions.

(a) $(11 + 0)^*(00 + 1)^*$ (b) $(1 + 01 + 001)^*(\varepsilon + 0 + 00)$

(c) $[00 + 11 + (01 + 10)(00 + 11)^*(01 + 10)]^*$

22. Construct finite automata equivalent to the following regular expressions.

(a) $10 + (0 + 11)0^*1$ (b) $01[((10)^* + 111)^* + 0]^*1$

(c) $((0 + 1)(0 + 1))^* + ((0 + 1)() + 1)(0 + 1)]^*$

23. Construct regular expressions corresponding to the state diagram given in the following figures.

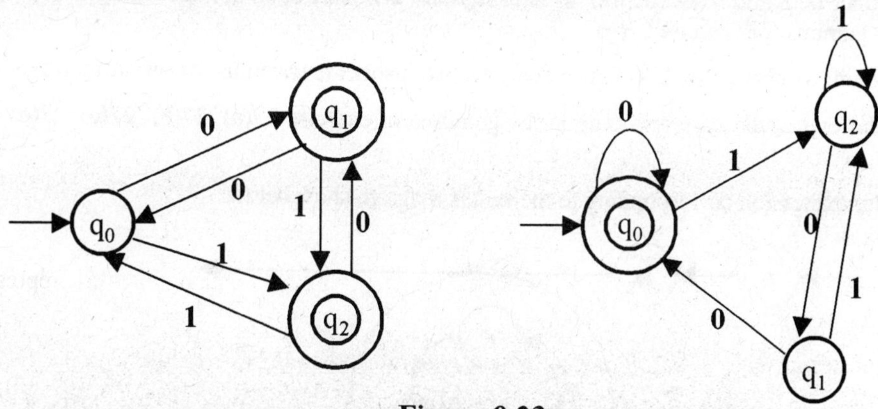

Figure 8.23ex

24. Consider the language L specified by the grammar G = (N, Σ, P, S), where N = {S, A, B}, Σ = {a, b, c} and P is set containing following productions:

1. $S \to AB$ 2. $A \to ab$ 3. $A \to aAb$ 4. $B \to c$ 5. $B \to Bc$

(a) Determine whether each of the following strings is a sentence in the language.

aabb aaabbc aaabbbccc ababcc

(b) Describe the language L in set-theoretic notations.

25. Give a grammar that specifies each of the following languages:

(a) $L = \{a^{2m} b^{2n} \mid m \geq 1, n \geq 1\}$ (b) $L = \{(ab)^m c^{2n} \mid m \geq 1, n \geq 1\}$
(c) $L = \{a^m b^n \mid 1 \leq m < n, n \geq 1\}$ (d) $L = \{a^m b^n \mid m \leq n \leq 2m, m \geq 1\}$
(e) $L = \{a^m b^m c^n \mid m \geq 1, n \geq 1\}$ (f) $L = \{a^m b^n c^q \mid m \geq 1, n \geq 1, m + n = q\}$

26. (a) Give a type-3 grammar that generates the language $L = \{x \mid x \in \{a, b\}^*$ and x does not contain two consecutive a's.}
(b) Give a type-2 grammar that generates the language $L = \{x \mid x \in \{a, b\}^*$ and x contains twice as many a's as b's.}

27. For each of the following grammar state whether it is type-1,2 or3. Find L(G) for each of them. Whenever grammar is type-2 or 3, give its BNF representation and syntax diagram.
(a) $G=(N, \Sigma, P, S)$, where $N=\{S, A\}$, $\Sigma=\{x, y, z\}$ and $P=\{S \to xS, S \to yA, A \to yA, A \to z\}$
(b) $G = (N, \Sigma, P, S)$, where $N = \{S\}$, $\Sigma = \{a\}$ and $P = \{S \to aaS, S \to aa\}$
(c) $G = (N, \Sigma, P, S)$, where $N = \{S\}$, $\Sigma = \{a, b\}$ and $P = \{S \to aaS, S \to a, S \to a\}$
(d) $G = (N, \Sigma, P, S)$, where $N = \{S\}$, $\Sigma = \{a, b, c\}$ and $P=\{S \to aS, S \to bS, S \to c\}$
(e) $G = (N, \Sigma, P, S)$, where $N = \{S, A, B\}$, $\Sigma = \{a, +, (,)\}$ and $P = \{S \to (S), S \to a + A, A \to a + B, B \to a + B, B \to a\}$
(f) $G = (N, \Sigma, P, S)$, where $N = \{S, A\}$, $\Sigma = \{a, b\}$ and $P = \{S \to aA, A \to bS\ A \to a\}$
(g) $G = (N, \Sigma, P, S)$, where $N = \{S, A, B\}$, $\Sigma = \{x, y, z\}$ and $P = \{S \to SA, SA \to BS, BS \to xy, B \to x, A \to z\}$

28. Give two distinct derivations (sequences of substitutions that start at S) for the string xyz of L(G), where G is the grammar of exercise 27(g).

29. Let G be the grammar of exercise 27(e). Can you give two distinct derivations for the string ((a+a+a))?

30. Find the regular expression corresponding to the grammar of exercise 27(a), 27(b), 27(c), 27(d) 27(e) and 27(f).

31. Find the regular expression corresponding to the syntax diagram show below.

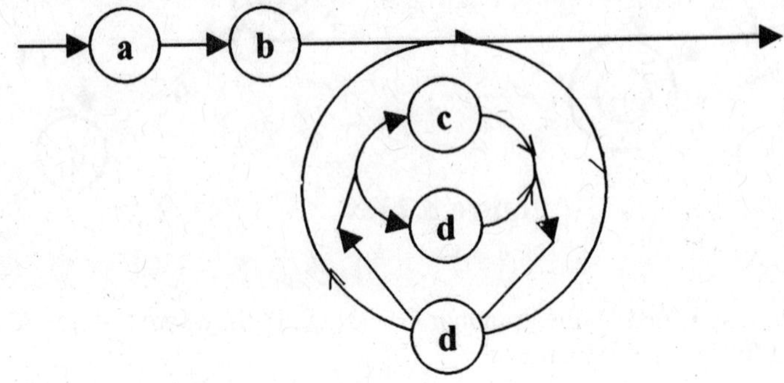

Figure 8.31ex

32. Draw the digraph of the machine whose state transition table is shown below.

	0	1	2
S_0	S_1	S_0	S_2
S_1	S_0	S_0	S_1
S_2	S_2	S_0	S_2

	a	b
S_0	S_1	S_0
S_1	S_2	S_1
S_2	S_3	S_2
S_3	S_3	S_3

Table 8.32ex

33. Construct the state transition table of the finite state machine whose digraph is shown in the figure 8.33ex.

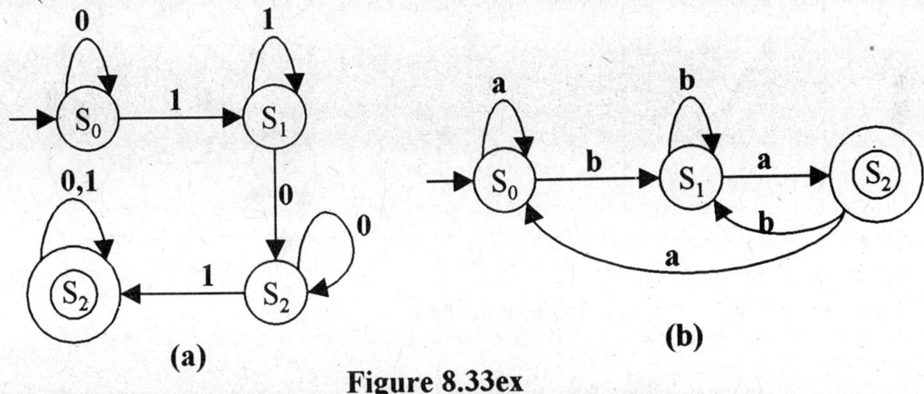

Figure 8.33ex

34. Let M =(S, I, $\Im$, s_0, T) be a finite state machine. Define a relation R on I as follows: $x_1 R\, x_2$ iff $f_{x_1}(s) = f_{x_2}(s)$ for every s in S. Show that R is an equivalence relation.

35. Let (S, *) be a finite semi group. Then we may consider the machine (S, S, $\Im$, s_0, T), where $\Im = \{ f_x \mid x \in S\}$, and $f_x(y) = x*y$ for all x, y $\in$ S. Define a relation R on S as follows: x R y iff there is some z $\in$ S such that $f_z(x) = y$. Show that R is transitive.

36. Consider the machine whose state transition diagram is given as below:

	0	1
1	1	4
2	3	2
3	2	3
4	4	1

Table 8.36ex

Here S = { 1, 2, 3, 4}. (a) Show that R = [(1, 1), (1, 4), (4, 1), (4, 4), (2, 2), (2, 3), (3, 2), (3, 3)} is a machine congruence. (B) Construct the state transition table for the corresponding quotient machine.

37. Let I = {1, 0} and S = {a, b}. Construct all possible state transition tables of finite state machines that have S as state set and I as input set. Let T = {b} and a is the start state of the machine.

38. Consider the finite state machine whose diagram is shown in the figure 8.36ex. Show that the relation R on S whose matrix M is as below, is a machine congruence. Draw the diagram of the corresponding quotient machine.

$$M_R = \begin{pmatrix} 1 & 0 & 0 & 0 & 1 & 0 \\ 0 & 1 & 0 & 0 & 0 & 1 \\ 0 & 0 & 1 & 1 & 0 & 0 \\ 0 & 0 & 1 & 1 & 0 & 0 \\ 1 & 0 & 0 & 0 & 1 & 0 \\ 0 & 1 & 0 & 0 & 0 & 1 \end{pmatrix}$$

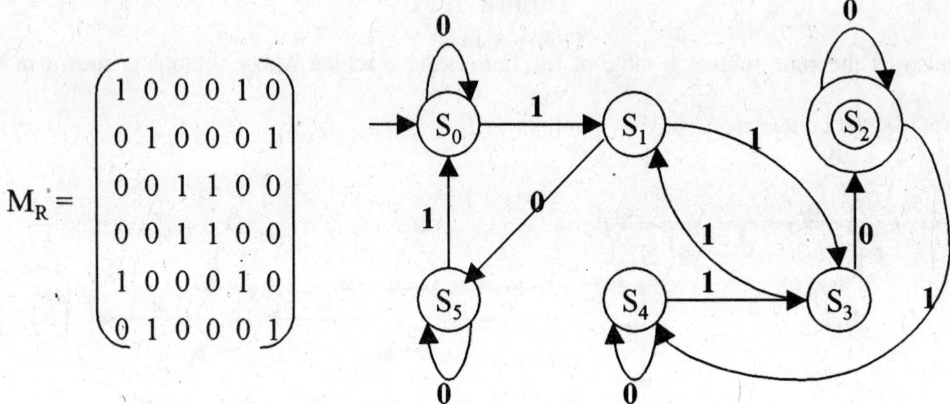

Figure 8.38ex

39. Draw the finite state machine whose state transition table is given as below and answer the following:

	0	1
S_0	S_0	S_1
S_1	S_1	S_2
S_2	S_2	S_3
S_3	S_3	S_0

Table 8.39ex

(a) List the values of transition function f_w for w = 01001.
(b) List the values of transition function f_w for w = 11100.
(c) Describe the set of binary words w having the property that $f_w = f_{010}$.
(d) Describe the set of binary words w having the property that $f_w(s_0) = s_2$.

40. See the finite state machine shown in the figure 8.40ex and answer the following questions.
(a) List the values of transition function f_w for w = abba.
(b) List the values of transition function f_w for w = babab.
(c) Describe the set of binary words w having the property that $f_w(s_0) = s_2$.
(d) Describe the set of binary words w having the property that $f_w(s_0) = s_0$.

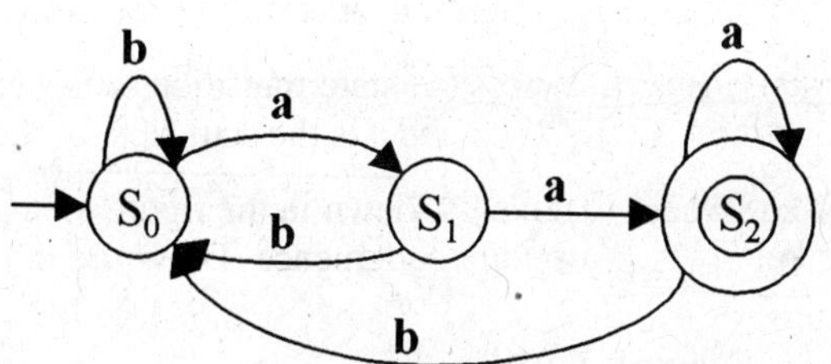

Figure 8.40ex

41. Describe the language accepted by the finite state machines shown in the figures 8.41ex.

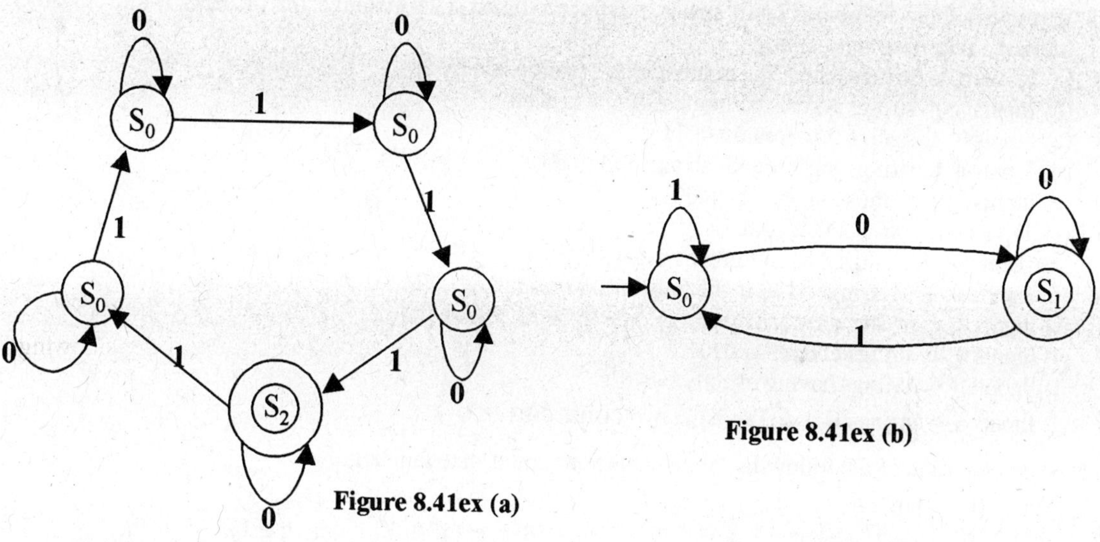

Figure 8.41ex (a)

Figure 8.41ex (b)

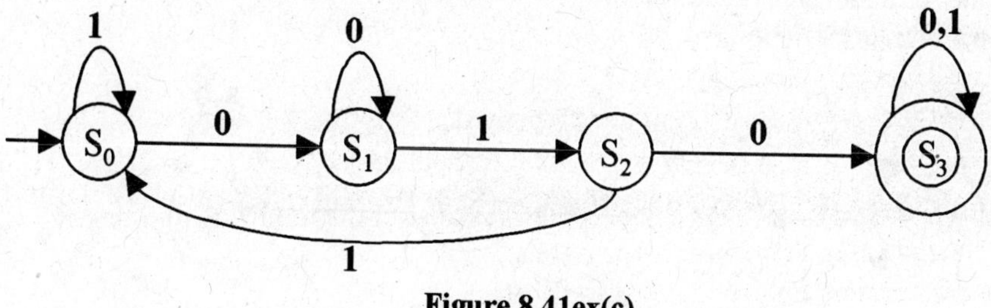

Figure 8.41ex(c)

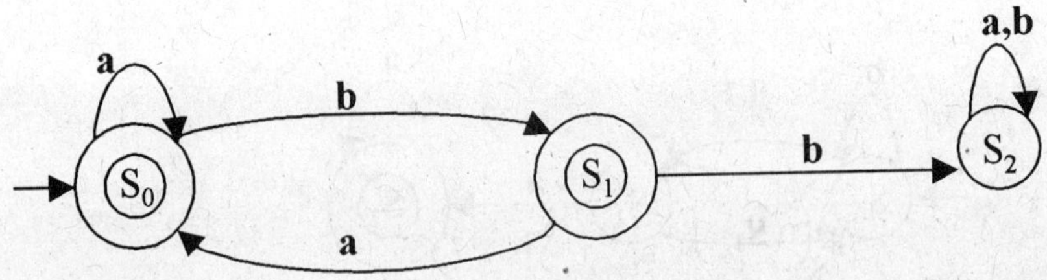

Figure 8.41ex(d)

42. Let M = {S, I, ℑ, s_0, T) be a finite state machine. Suppose that if s is in T and w in I*, then f_w (s) is in T. Prove that L(M) is a sub-semi group of (I*, •) , where • is catenation operation.

43. Construct type 3 grammar for the regular expression accepted by the finite state machines shown in exercise 41.

44. For each of the following input string construct a finite state machine that accepts it and no others. Mark the **trapping** state, if any.
(a) Inputs a, b; strings where the number of b's is divisible by 3.
(b) Inputs x, y; strings where the number of y's is even.
(c) Inputs 0, 1; strings that contain 0011.
(d) Inputs a, b; strings that contain ab and end in bbb.
(e) Inputs 0, 1; strings that end with 0011.
(f) Inputs w, z; strings that contain wz or zzw..
(g) Inputs w, z; strings that end in wz or zzw.
(h) Inputs 0, 1, 2; strings 0120 is the only string recognized.
(i) Inputs x, y, z.; strings xzx or yx or zyx are to be recognized.
(j) Inputs 0, 1; strings ending in 0101.
(k) Inputs x, y; strings having exactly two x's.
(l) Inputs a, b; strings that do not have two consecutive b's.

45. Show that each of the following language is not a finite state language.
(a) $L = \{0^m 1^n \mid m \geq n\}$
(b) $L = \{0^m 1^n \mid m \leq n\}$
(c) $L = \{0^m \mid m = 2^n$ such that $n \geq 1\}$
(d) $L = \{1^n 0^m 1^{n+m} \mid m, n \geq 1\}$
(e) L = {ww | w is a string of 0's and 1's}

46. Find the relation R defined on S -the set of states of the following finite state machines, in such a way that R partitions the set S into equivalent classes of w-compatible states. Then draw a simplified version of the FSMs shown in figure 8.46ex.

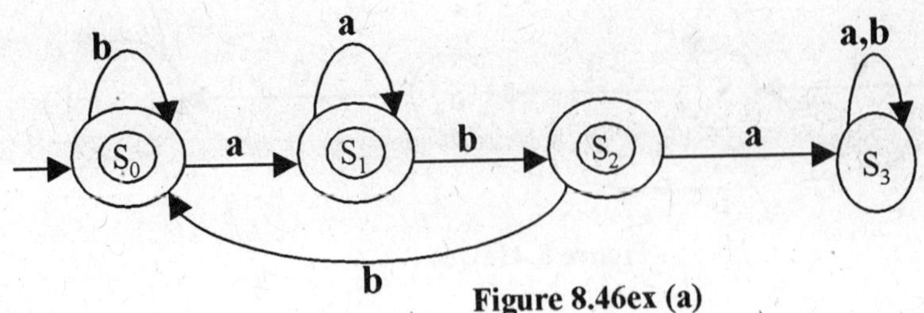

Figure 8.46ex (a)

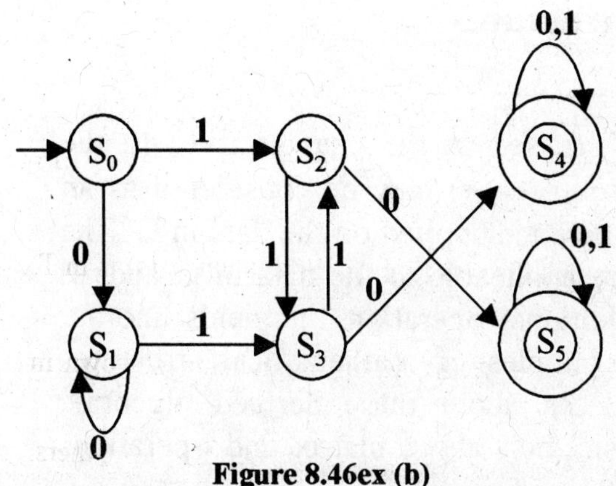

Figure 8.46ex (b)

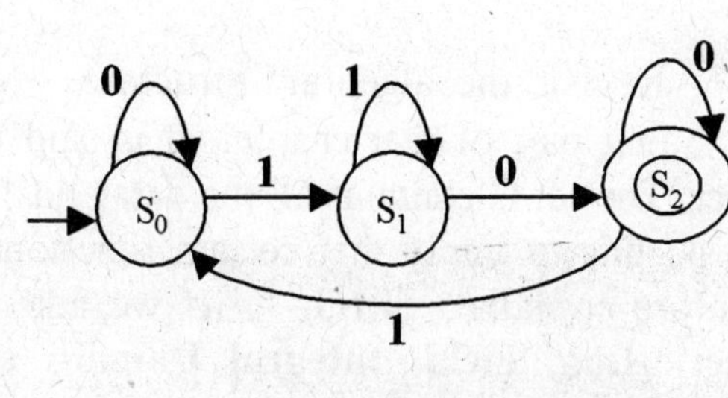

Figure 8.46ex (c)

CHAPTER NINE
Derived Algebraic Structures

Revisit the algebraic structure – group, discussed in the chapter five. In the concluding part of that chapter, it is said that a group (G, *) may be considered as an object. The set G contains all the data and * is the operation applied on the data in G. The four postulates imply that certain functionalities are applicable to the data in G and all these are contained in (G, *). If we add some additional operations, it yields another object. **Ring**, **Field**, **Integral Domain** etc are examples of mathematical structures derived from group. In this chapter, we shall learn about these derived algebraic structures. But before that let us discuss some basic facts about matrix and operations defined on it.

9.1 Matrix

The term **matrix** was first used to distinguish matrices from determinants. It seemed that matrix was intended to imply "*mother of determinants*". As with the vectors, matrices are arrays of real numbers unless stated otherwise. An **m** × **n** matrix A is a rectangular array of **mn** elements arranged in a definite order in **m** *rows* and **n** *columns*.

$$A = \begin{bmatrix} a_{11} & a_{12} & \cdots & a_{1n} \\ a_{21} & a_{22} & \cdots & a_{2n} \\ a_{31} & a_{32} & \cdots & a_{3n} \\ \vdots & & & \\ a_{m1} & a_{m2} & \cdots & a_{mn} \end{bmatrix}$$

The number a_{ij} appearing in the ith row and j^{th} column of A is called the ij^{th} component of A. For convenience, the matrix is written as $A = (a_{ij})$. Normally, matrices are represented by capital letters.

The algebra for symbolic operations on matrices is different from the algebra for operations on scalars, or single numbers. For example there is no division in matrix algebra, although there is an operation called "multiplying by an inverse". Although it is possible to express the exact equivalent of matrix algebra equations in terms of scalar algebra expressions, the results look rather messy.

It can be said that the matrix algebra notation is shorthand for the corresponding scalar longhand. A *vector* is a special kind of matrix. The n-component row vector

$$\mathbf{a} = [a_1, a_2, a_3, \ldots, a_n]$$

is a $1 \times n$ matrix, whereas n-component column vector

$$\mathbf{a} = \begin{bmatrix} a_1 \\ a_2 \\ a_3 \\ \vdots \\ a_n \end{bmatrix}$$

is a $n \times 1$ matrix. The scalars a_i are the elements of vector **a**. The *transpose* of **a**, denoted by **a'**, is the row arrangement of elements of the vector **a** as shown below.

$$\mathbf{a}' = [a_1 \ a_2 \ a_3 \ \ldots \ a_n]$$

Matrices like vectors can be added and multiplied by scalars. The sum of two matrices is the matrix of sums of corresponding elements. For example, let A and B be two $m \times n$ matrices given as

$$A = \begin{bmatrix} a_{11} & a_{12} & \cdots & a_{1n} \\ a_{21} & a_{22} & \cdots & a_{2n} \\ a_{31} & a_{32} & \cdots & a_{3n} \\ \vdots \\ a_{m1} & a_{m2} & \cdots & a_{mn} \end{bmatrix} \qquad B = \begin{bmatrix} b_{11} & b_{12} & \cdots & b_{1n} \\ b_{21} & b_{22} & \cdots & b_{2n} \\ b_{31} & b_{32} & \cdots & b_{3n} \\ \vdots \\ b_{m1} & b_{m2} & \cdots & b_{mn} \end{bmatrix}$$

then the sum A + B is given as

$$A + B = \begin{bmatrix} a_{11} + b_{11} & a_{12} + b_{12} & \cdots & a_{1n} + b_{1n} \\ a_{21} + b_{21} & a_{22} + b_{22} & \cdots & a_{2n} + b_{2n} \\ a_{31} + b_{31} & a_{32} + b_{32} & \cdots & a_{3n} + b_{3n} \\ \vdots \\ a_{m1} + b_{m1} & a_{m2} + b_{m2} & \cdots & a_{mn} + b_{mn} \end{bmatrix}$$

And the difference of two matrices is the matrix of differences of corresponding elements. For example, A – B is given as

$$A-B=\begin{bmatrix} a_{11}-b_{11} & a_{12}-b_{12} & \cdots & a_{1n}-b_{1n} \\ a_{21}-b_{21} & a_{22}-b_{22} & \cdots & a_{2n}-b_{2n} \\ a_{31}-b_{31} & a_{32}-b_{32} & \cdots & a_{3n}-b_{3n} \\ \vdots & & & \\ a_{m1}-b_{m1} & a_{m2}-b_{m2} & \cdots & a_{mn}-b_{mn} \end{bmatrix}$$

It is important to note here that addition and subtraction of two matrices is possible if and only if they are of same order. The number of row and column in matrix determines the order of a matrix. For example, matrices A and B are of order m×n.

The product of a scalar k, times a matrix A is k times each element of A. For example, kA is given as

$$kA=\begin{bmatrix} ka_{11} & ka_{12} & \cdots & ka_{1n} \\ ka_{21} & ka_{22} & \cdots & ka_{2n} \\ ka_{31} & ka_{32} & \cdots & ka_{3n} \\ \vdots & & & \\ ka_{m1} & ka_{m2} & \cdots & ka_{mn} \end{bmatrix}$$

Matrix multiplication involves the computation of the sum of the products of elements from a row of the first matrix (the pre-multiplier on the left) and a column of the second matrix (the post-multiplier on the right). This sum of products is computed for every combination of rows and columns. For example, if **A** is a 2 x 3 matrix and **B** is a 3 x 2 matrix, the product **AB** is

$$A=\begin{bmatrix} 2 & 4 & 6 \\ 6 & 5 & 8 \end{bmatrix} \qquad B=\begin{bmatrix} 2 & 5 \\ 4 & 1 \\ 0 & 9 \end{bmatrix} \qquad AB=\begin{bmatrix} 20 & 68 \\ 32 & 107 \end{bmatrix}$$

Thus, the product is a 2 x 2 matrix. This came as follows: The *number of columns* of **A** must be equal to the *number of rows of* ***B***. In this case this is 3. *If they are not equal, multiplication is impossible.* If they are equal, then the number of rows of the product **AB** is equal to the number of rows of **A** and the number of columns is equal to the number of columns of **B**. *Example of 3x2 matrix multiplied by a 2x3* is shown below. It follows that the result of the product **BA** is a 3 x 3 matrix.

$$BA = \begin{bmatrix} 34 & 33 & 52 \\ 14 & 21 & 32 \\ 54 & 45 & 72 \end{bmatrix}$$

In general, if **A** is a k x p matrix and **B** is a p x n matrix, the product **AB** is a k x n matrix. If $k = n$, then the product **BA** can also be formed. We say that matrices conform for the operations of addition, subtraction or multiplication when their respective orders (numbers of row and columns) are such as to permit the operations. Matrices that do not conform for addition or subtraction cannot be added or subtracted. Matrices that do not conform for multiplication cannot be multiplied.

It is obvious from the above result that multiplication of matrices is not a commutative operation. In fact it is not always possible to find AB if BA is possible and vice versa for any two matrices A and B. However if it is possible to find ABC for any three matrices A, B and C, then A(BC) = (AB)C i.e. multiplication of matrices is an associative operation. Similarly, addition of matrices is an associative operation. It is a commutative operation as well if the operation of addition on the elements of matrix is commutative.

A matrix can be classified on the basis of the types of element it possesses and on the equality and inequality of rows and columns of the matrix. A matrix m x n is, by definition, ***a rectangular matrix***. In an m x n matrix, if m is equal to n i.e. if number of rows in a matrix is equal to the number of columns in the matrix then it is called a ***square matrix***. The product matrix BA calculated above is a square matrix of order *3 x 3*. A n x n square matrix is also called a square matrix of *order n*. If the number of column is 1, the matrix is called a ***column matrix*** or column vector and if the number of row is 1, the matrix is called ***row matrix*** or row vector as mentioned above. The other types of matrices are given below with examples. The elements along diagonal in a square matrix are called **diagonal elements** of the matrix. A diagonal element is also called ***leading element***.

A square matrix in which all elements except the diagonal elements are zero, is called ***diagonal matrix***. For example,

$$A = \begin{bmatrix} 4 & 0 & 0 \\ 0 & 1 & 0 \\ 0 & 0 & 2 \end{bmatrix}$$

is a diagonal matrix. In a diagonal matrix if the diagonal elements are equal, the matrix is called ***scalar matrix***. For example,

$$A = \begin{bmatrix} 4 & 0 & 0 \\ 0 & 4 & 0 \\ 0 & 0 & 4 \end{bmatrix}$$

is a scalar matrix. In a scalar matrix if all the diagonal elements are equal to 1, it is called ***unit or identity matrix***. For example, The following matrices I_2, I_3 and I_4 are identity matrix of order 2, 3 and 4 respectively.

$$I_2 = \begin{bmatrix} 1 & 0 \\ 0 & 1 \end{bmatrix} \quad I_3 = \begin{bmatrix} 1 & 0 & 0 \\ 0 & 1 & 0 \\ 0 & 0 & 1 \end{bmatrix} \quad I_4 = \begin{bmatrix} 1 & 0 & 0 & 0 \\ 0 & 1 & 0 & 0 \\ 0 & 0 & 1 & 0 \\ 0 & 0 & 0 & 1 \end{bmatrix}$$

An identity matrix is generally represented by I. A matrix having all its elements zero is called ***zero matrix*** and is denoted by O. A zero matrix may be square or rectangular. A matrix obtained by omitting either some rows or columns from a matrix A is called ***sub matrix*** of A.

We used boolean matrix in the chapter one to represent a relation in matrix form and in the chapter seven to represent a graph by its adjacency matrix. A matrix is called boolean if all its elements are from set {0, 1} i.e. either 0 or 1. All the rules applied for addition and multiplication of matrices are very well applicable for boolean matrices also. However two boolean matrices cannot be subtracted as 1 – 1 (True – True), 1 – 0 (True – False) etc have no meaning in boolean algebra. *Reader must not confuse this with binary arithmetic.* Only thing to be taken care of is the boolean arithmetic on the elements of the boolean matrices. Here 1 + 1 is 1 because (T and T is true). Similarly, 1 + 0 = 0 + 1 = 1. 1*1 = 1. 0*1 = 1*0 = 0*0 = 0. And so on. See some example below.

Example 1 Let $A = \begin{bmatrix} 1 & 0 & 1 \\ 0 & 0 & 1 \\ 1 & 1 & 1 \end{bmatrix}$ be a boolean matrix of order 3×3 order. Find A + A, A^T and A^2.

Solution: Here $A + A = \begin{bmatrix} 1 & 0 & 1 \\ 0 & 0 & 1 \\ 1 & 1 & 1 \end{bmatrix} + \begin{bmatrix} 1 & 0 & 1 \\ 0 & 0 & 1 \\ 1 & 1 & 1 \end{bmatrix} = \begin{bmatrix} 1+1 & 0+0 & 1+1 \\ 0+0 & 0+0 & 1+1 \\ 1+1 & 1+1 & 1+1 \end{bmatrix} = \begin{bmatrix} 1 & 0 & 1 \\ 0 & 0 & 1 \\ 1 & 1 & 1 \end{bmatrix}$

$$A^T = \begin{bmatrix} 1 & 0 & 1 \\ 0 & 0 & 1 \\ 1 & 1 & 1 \end{bmatrix}^T = \begin{bmatrix} 1 & 0 & 1 \\ 0 & 0 & 1 \\ 1 & 1 & 1 \end{bmatrix}$$

So there is no difference. It is obvious that A is a symmetric matrix. The relation represented by this matrix will be a symmetric relation.

$$A^2 = \begin{bmatrix} 1 & 0 & 1 \\ 0 & 0 & 1 \\ 1 & 1 & 1 \end{bmatrix}\begin{bmatrix} 1 & 0 & 1 \\ 0 & 0 & 1 \\ 1 & 1 & 1 \end{bmatrix} = \begin{bmatrix} 1.1+0.0+1.1 & 1.0+0.0+1.1 & 1.1+0.1+1.1 \\ 0.1+0.0+1.1 & 0.0+0.0+1.1 & 0.1+0.1+1.1 \\ 1.1+0.1+1.1 & 1.0+1.0+1.1 & 1.1+1.1+1.1 \end{bmatrix}$$

$$= \begin{bmatrix} 1 & 1 & 1 \\ 1 & 1 & 1 \\ 1 & 1 & 1 \end{bmatrix}$$

Ans.

Two matrices $A=(a_{ij})$ and $B=(b_{ij})$ are said to be ***equal*** if both are of same order and the corresponding elements are equal in the matrices i.e. $a_{ij} = b_{ij} \ \forall$ i and j. For example, in the following matrices A, B and C, matrices A and B are equal whereas matrices A and C are not equal though both are of the same order. D is not equal to any of the matrices A, B and C as they are of different order.

$$A = \begin{bmatrix} 1 & 2 & 5 \\ 7 & 4 & 2 \\ 9 & 8 & 5 \end{bmatrix} \quad B = \begin{bmatrix} 1 & 2 & 5 \\ 7 & 4 & 2 \\ 9 & 8 & 5 \end{bmatrix} \quad C = \begin{bmatrix} 1 & 5 & 9 \\ 0 & 6 & 3 \\ 3 & 9 & 5 \end{bmatrix} \quad D = \begin{bmatrix} 1 & 4 & 0 \\ 0 & 1 & 2 \\ 0 & 3 & 1 \\ 0 & 0 & 6 \end{bmatrix}$$

Let $A=(a_{ij})$ be a m×n matrix. Then the ***transpose*** of A, denoted by A^T, is the n×m matrix obtained by interchanging the rows and columns of A. We can write then $A^T = (a_{ji})$. Transpose of a matrix is obtained simply by making the ith row of A as the ith column of A^T. For example see the following matrices.

$$\text{If } A = \begin{bmatrix} 1 & 5 & 9 \\ 0 & 6 & 3 \\ 3 & 9 & 5 \end{bmatrix} \text{ then } A^T = \begin{bmatrix} 1 & 0 & 3 \\ 5 & 6 & 9 \\ 9 & 3 & 5 \end{bmatrix}$$

A square matrix A is said to be a ***symmetric matrix*** if $A=A^T$. And if $A= -A^T$, the matrix A is called an ***anti-symmetric matrix*** or ***skew-symmetric matrix.*** A square matrix A is called ***upper triangular matrix*** if all its elements below diagonal are zero. It is called ***lower triangular matrix*** if all its elements above diagonal are zero. It can be mathematically summarized as " a square matrix $A = (a_{ij})$ is upper triangular if $a_{ij} = 0$ for $i>j$, lower triangular if $a_{ij} = 0$ for $i<j$, and diagonal if $a_{ij} = 0$ for $i\neq j$."

Determinant of a square matrix A, denoted by |A|, is obtained recursively by finding cofactors of A. Let A be an n×n square matrix and M_{ij} be the (n – 1) × (n – 1) matrix obtained from A by removing the ith row and the jth column of A. The sub matrix M_{ij} is called (i, j)th ***minor*** of A. And the factor $(-1)^{i+j}|M_{ij}| = A_{ij}$ is called (i, j)th ***cofactor*** of A. Then |A| is given by the following formula

$$|A| = a_{11}A_{11} + a_{12}A_{12} + a_{13}A_{13} + \cdots + a_{1n}A_{1n} = \sum_{k=1}^{n} a_{1k}A_{1k}$$

A square matrix A is called ***singular matrix*** if |A| = 0 otherwise it is called ***non-singular***. A non-singular matrix is also called an ***invertible matrix*** i.e. for a non-singular matrix A there exists a matrix B such that

$$AB = BA = I$$

The matrix B is called ***inverse matrix*** of A and is denoted as A^{-1}. Computing inverse of a matrix is a lengthy process. As we know, division is not a valid operation on matrices, so inverse of a matrix A cannot be obtained by simply dividing identity matrix I by A. To obtain A^{-1}, we first find adjoint matrix of A, denoted as adj(A). Then A^{-1} is given by the following formula.

$$A^{-1} = \frac{1}{|A|} adj(A)$$

Let A be an n×n matrix and B be the n×n matrix of cofactors A_{ij} of A. The transpose matrix B^T of the matrix B is called adjoint matrix of A.

Example 2 Find the inverse of the matrix $A = \begin{bmatrix} 2 & 4 & 3 \\ 0 & 1 & -1 \\ 3 & 5 & 7 \end{bmatrix}$.

Solution: Let us first determine whether A is invertible or not. According to the formula discussed above, we have

$$|A| = 2\begin{vmatrix}1 & -1\\5 & 7\end{vmatrix} - 4\begin{vmatrix}0 & -1\\3 & 7\end{vmatrix} + 3\begin{vmatrix}0 & 1\\3 & 5\end{vmatrix} = 2\times12 - 4\times3 + 3\times-3 = 24 - 21 = 3 \neq 0$$

Thus, A is non-singular and hence it is invertible. Now adj(A) = $\begin{bmatrix}12 & -13 & -7\\-3 & 5 & 2\\-3 & 2 & 2\end{bmatrix}$

Finally, the inverse of A is given by the matrix $A^{-1} = \frac{1}{3}\begin{bmatrix}12 & -13 & -7\\-3 & 5 & 2\\-3 & 2 & 2\end{bmatrix}$

Ans.

9.2 Ring

Suppose R is a non-empty set equipped with two binary operations called addition and multiplication denoted by '+' and '*' respectively. The mathematical structure (R, +, *) is called a ***ring*** if the following postulates are satisfied:

A. Laws of addition

1. Closure property: $a + b \in R \; \forall\, a, b \in R$.
2. Associative property: $a + (b + c) = (a + b) + c \; \forall a, b, c \in R$
3. Existence of Identity: There exists an element denoted by 0 in R such that $0 + a = a + 0 = a \; \forall\, a \in R$
4. Existence of Inverse: To each element a in R there exists an element –a in R such that $-a + a = a + (-a) = 0$
5. Commutative property: $a + b = b + a \; \forall a, b \in R$

B. Laws of multiplication

6. Closure property: $a * b \in R \; \forall\, a, b \in R$
7. Associative property: $a * (b * c) = (a * b) * c \; \forall a, b, c \in R$

C. Multiplication distributes over addition

8. Left Distributive law: $a*(b + c) = a*b + a*c \; \forall a, b, c \in R$
9. Right Distributive law: $(b + c)*a = b*a + c*a \; \forall a, b, c \in R$

In brief, we can say that a mathematical structure (R, +, *) is a ring, if (R, +) is an abelian group, (R, *) is a semigroup and * distributes over + in R. The additive identity of the ring R is called ***zero*** element of the ring and is denoted by **0** and it must not be confused with number 0. If the ring R possesses identity element for the second operation

*, then the ring R is called a ***ring with unity*** and the identity element is called unit element of the ring; It is denoted by **1**. A ring (R, +, *) is called a ***commutative ring*** if the second operator is commutative in R i.e., $a * b = b * a, \forall a, b \in R$. See the following basic properties of a ring.

Basic Properties of a Ring

If (R, +, *) be a ring and a, b, c be any elements of R then

(i) $a * \mathbf{0} = \mathbf{0} * a = \mathbf{0}$

(ii) $a * (-b) = -(a * b) = (-a) * b$

(iii) $(-a) * (-b) = a * b$

(iv) $a * (b - c) = a * b - a * c$

(v) $(b - c) * a = b * a - c * a$

Proof: (i) Here **0** is zero element of R, thus $a * \mathbf{0} = a * (\mathbf{0} + \mathbf{0}) = a * \mathbf{0} + a * \mathbf{0}$ using the left distribution law. This implies that, we have

$$a * \mathbf{0} = a * \mathbf{0} + a * \mathbf{0}$$

$$\Rightarrow \mathbf{0} + a * \mathbf{0} = a * \mathbf{0} + a * \mathbf{0}$$

$$\Rightarrow \mathbf{0} = a * \mathbf{0}$$

Similarly, we can prove that

$$\mathbf{0} = \mathbf{0} * a$$

Thus, we have $a * \mathbf{0} = \mathbf{0} = \mathbf{0} * a$

Proved.

(ii) We have $-b \in R$ for any $b \in R$. Also we have $-b + b = \mathbf{0}$ and then $a*(-b + b) = a * \mathbf{0}$. Thus, applying left distribution law, we get from

$a*(-b) + a * b = \mathbf{0}$ [since $a * \mathbf{0} = \mathbf{0}$]

$\Rightarrow a*(-b) = -(a * b)$ [since (a * b) is additive inverse of a*(–b)]

Similarly, we can prove that

$$(-a) * b = -(a * b)$$

Thus, we have $a*(-b) = -(a * b) = (-a) * b$

Proved.

Reader may prove the remaining properties.

Example 1 The set R consisting of a single element **0** with two binary operations: + and * defined by $\mathbf{0} + \mathbf{0} = \mathbf{0}$ and $\mathbf{0} * \mathbf{0} = \mathbf{0}$ is a ring. This ring is called **zero ring** or **null ring**. ♦

Example 2 The set I_E of even integers is a ring with respect to the addition and multiplication of integers as the two binary operations.

Solution: It is proved in the chapter five that (I_E, +) is an abelian group. The product of two even integers is even, so the structure is closed with respect to multiplication. It is also known that multiplication of integers distributes over addition of integers and multiplication of integers is an associative operation, Therefore, (I_E, +, *) is a ring. Further, this is a commutative ring without unity. ♦

Example 3 The set Q of all rational numbers is a commutative ring with unity under the binary operations of addition and multiplication of rational numbers. ♦

Example 4 Show that the set M of all square real matrix of order n is a non-commutative ring with unity with respect to the binary operations of addition and multiplication of matrices.

Solution: It can easily be proved that (M, +) is a commutative group. The product of two square real matrices of order n is a square real matrix of order n. Thus M is closed under the binary operation of multiplication. It is also known that multiplication of matrices is an associative operation. Multiplication of matrices distributes over the addition of matrices, thus (M, +, *) is a ring. Since identity matrix of order n is a real square matrix, it belongs to M. Hence M is a ring with unity. Since, multiplication of matrices is not a commutative operation, (M, +, *) is not a commutative ring.

Ans.

Example 5 Show that the set R = {0, 1, 2, 3, 4, 5} is a commutative ring under the binary operations of addition modulo 6 ($+_6$) and multiplication modulo 6 ($*_6$).

Solution: It is easy to prove that that (R, $+_6$) is an abelian group by constructing a composition table for $+_6$ as is shown in the example 14 of the 5.1 of the chapter 5. The following table 9.2.1 is the composition table for multiplication modulo 6 on the given R.

$*_6$	0	1	2	3	4	5
0	0	0	0	0	0	0
1	0	1	2	3	4	5
2	0	2	4	0	2	4
3	0	3	0	3	0	3
4	0	4	2	0	4	2
5	0	5	4	3	2	1

Table 9.2.1

From the composition table it is obvious that the set R is closed with respect to the second binary operation of the multiplication modulo 6. Also $*_6$ is an associative operation in R. Next we have to prove that $*_6$ distributes over $+_6$. Let a, b and c be any three elements in R, then

$$\begin{aligned} a *_6 (b +_6 c) &= a *_6 (b + c) \qquad [\text{Since } b +_6 c \equiv_6 b + c] \\ &= \text{The remainder when } a * (b + c) \text{ is divided by 6} \\ &= \text{The remainder when } a * b + a * c \text{ is divided by 6} \\ &= (a * b) +_6 (a * c) \\ &= (a *_6 b) +_6 (a *_6 c) \end{aligned}$$

Similarly, we can show that $(b +_6 c) *_6 a = (b *_6 a) +_6 (c *_6 a)$. Thus, $(R, +_6, *_6)$ is a ring. This ring is commutative ring with unity. Reader may verify from the composition table.

Ans.

The ring of the above example 5 is an example of a finite ring. It is also important to note that the zero element of the ring is 0 and the unit element is 1. The two non-zero element 2 and 3 yield result 0 when binary operation of multiplication modulo 6 is applied i.e. $2 *_6 3 = 0$. That means product of two non-zero elements is zero i.e. 2 and 3 are divisors of zero. A non-zero element $x \in (R, +, *)$ is called ***zero divisor*** if there exists a non-zero element y such that $x * y = \mathbf{0}$, where **0** is zero element in R The element y is also called a divisor of zero. A ring with zero divisors is called a ***ring with zero divisors*** otherwise it is called a ***ring without zero divisors***. The ring of example 5 is an example of a ring with zero divisors. The rings of examples 2 and 3 are examples of ring without zero divisors. Reader is asked to give two matrices to show that ring of example 4 is a ring with zero divisors.

Theorem 1 A ring R is without zero divisors iff the cancellation laws hold in R.

Proof: Let us first suppose that R has no zero divisors. Let x, y and $z \in R$ such that x is a non-zero element and $y * x = z * x$. Now

$$\begin{aligned} y * x = z * x &\Rightarrow y * x - z * x = 0 \\ &\Rightarrow (y - z) * x = 0 \\ &\Rightarrow y - z = 0 \text{ (since } x \neq 0 \text{ and R is a ring without zero divisors)} \\ &\Rightarrow y = z \end{aligned}$$

Hence right cancellation law holds good. Similarly we can show that left cancellation law holds good in R. Now let us suppose that cancellation laws hold in R. Then we have to show that R has no zero divisors. We proceed, contrarily, assuming that R has zero divisors i.e. $x * y =$ for $x \neq 0$ and $y \neq 0$. Thus, we have $x * y = x * 0 \Rightarrow y = 0$ by left cancellation law. This is contrary to the assumption that $y \neq 0$. Hence R does not have zero divisors.

Proved.

A commutative ring with unity without zero divisors is called ***integral domain***. If a ring with unity, having at least two elements, possesses multiplicative inverse for every non-zero element, it is called a ***division ring***. A division ring is also called a ***skew field***. More about field we shall discuss in the next section of this chapter. The set I of integers with the two binary operations of integer addition and integer multiplication is an example of integral domain. This ring (I, +, *) is not a division ring because the non-zero integers do not possess multiplicative inverse in I. Other examples of infinite integral domains are (Q, +, *) and (R, +, *), where Q and R are sets rational and real numbers respectively. If C be the set of all complex numbers of the form x + iy, where x and y are integers, then (C, +, *) is an integral domain. The binary operations + and * are addition and multiplication respectively of complex numbers. Reader may verify that (C, +, *) is not a division ring. However, if we take x and y, in x + iy, as rational numbers then (C, +, *) is a division ring. The ring ({0, 1, 2, 3, 4}, $+_5$, $*_5$) is an example of finite integral domain. Since this ring has unit element, contains 5 (> 2) elements and every non-zero elements possesses inverse (1 has 1, 2 has 3, 3 has 2, 4 has 4), this ring is a division ring of finite order.

Let (D, +, *) be an integral domain. D is said to be ***ordered*** if D contains a subset D_+ such that

- D_+ is closed with respect to addition and multiplication of D, and
- $\forall\, x \in D$, exactly one of the three situations: $x = 0$, $x \in D_+$, $-x \in D_+$ should be true. The elements of D_+ are called positive element and all other non-zero elements are called negative elements. This is known as principle of trichotomy.

For example, integral domain (I, +, *) of integers is an ordered integral domain, as I_+ is closed with respect to the integer addition and multiplication and an integer is either = 0, < 0 or > 0. However the integral domain (C, +, *) of complex numbers is not ordered. This can be proved by contradiction. Let C is an ordered integral domain, so $\exists$ C_+. Since $i \in C$, so either $i \in C_+$ or $-i \in C_+$. Let $i \in C_+$. Then $i * i = -1 \in C_+$ and so $i * -1 = -i \in C_+$. This cannot be possible in an ordered integral domain. Similarly we can show that if $-i \in C_+$ then $i \in C_+$. Thus C is not an ordered integral domain. Another example of an unordered integral domain is (I_p, $+_p$, $*_p$), where p is a prime number and other symbols have usual meaning.

Example 6 Do the following set form integral domains with respect to ordinary addition and multiplication? Whether they form division ring?
(a) the set of numbers of the form $b\sqrt{5}$, where b is a rational number.
(b) the set of even integers
(c) the positive integers.

Solution: (a) Let $A = \{x \mid x = b\sqrt{5}$, b is a rational number$\}$. We have $2\sqrt{5}$ and $3\sqrt{5}$ in A. The product of these two numbers is 30 which cannot be written in the form $b\sqrt{5}$. Thus A is not closed with respect to multiplication, hence the set A is not an integral domain. For the same reason this set does not form a division ring.

(b) We know that 1 is the identity element for the binary operation of integer multiplication and 1 is not an even integer so $1 \notin I_E$. Since the set does not possess unit element, it is not an integral domain. Since identity for multiplication does not exist, it is not a division ring either.

(c) It is known that the set of positive integer does not form a group with respect to the binary operation of integer addition. Thus this set cannot form integral domain or division ring.

Ans.

Example 7 Show that a division ring has no divisors of zero.

Solution: Let (R, +, *) be a division ring. By definition, R has a unit element and every non-zero element x has its multiplicative inverse in R. Let x and y be any two elements of R such that $x * y = 0$ and $x \neq 0$. Since x is non-zero its inverse $x^{-1} \in R$. Thus, we have

$$x * y = 0 \Rightarrow x^{-1} * (x * y) = x^{-1} * 0 \Rightarrow (x^{-1} * x) * y = 0 \Rightarrow y = 0$$

Similarly, we can prove that if $y \neq 0$ then $x * y = 0 \Rightarrow x = 0$. Thus R does not possess a zero divisor.

Ans.

Let $(R_1, +_1, *_1)$ and $(R_2, +_2, *_2)$ be two rings. They are said to be ***isomorphic*** to each other if there exists a function f: $R_1 \rightarrow R_2$ such that

1. f is one to one and onto, and
2. $f(x +_1 y) = f(x) +_2 f(y)$ and $f(x *_1 y) = f(x) *_2 f(y)$ $\forall$ x, y $\in R_1$

The function f is called ***isomorphism of*** R_1 ***onto*** R_2. If the function f is into function, f is called ***isomorphism of*** R_1 ***into*** R_2. If a ring R_1 is isomorphic to another ring R_2 then R_2 is called ***isomorphic image of*** R_1 and R_1 is called ***isomorphic pre-image of*** R_2 and they are denoted as $R_1 \cong R_2$. It is easy to show that relation of isomorphism is an equivalence relation on the set of all rings. Thus it will partition the set of all rings into equivalence classes such that any two rings of the same equivalence class are isomorphic to each other. Any two rings belonging to the same equivalence class are called ***abstractly identical***.

The definition of isomorphism can be extended to *transfer* the ring properties from one ring to another set. If f is a one to one function from a ring $(R_1, +_1, *_1)$ onto a set R_2 with two binary operations $+_2, *_2$ such that $f(x +_1 y) = f(x) +_2 f(y)$ and $f(x *_1 y) = f(x) *_2 f(y)\ \forall\ x, y \in R_1$, then $(R_2, +_2, *_2)$ is also a ring.

Theorem 2 If f is an isomorphism of $(R_1, +_1, *_1)$ onto $(R_2, +_2, *_2)$ then
(a) The image of zero element of R_1 is the zero element of R_2.
(b) The image of additive inverse of an element x in R_1 is additive inverse of f(x) in R_2.
(c) If $(R_1, +_1, *_1)$ is an integral domain then $(R_2, +_2, *_2)$ is also an integral domain.
(d) If $(R_1, +_1, *_1)$ is a division ring then $(R_2, +_2, *_2)$ is also a division ring.

Proof: (a) Let x be any element of R_1. Let $\mathbf{0_1}$ be the zero element of R_1 and $\mathbf{0_2}$ be the zero element of R_2. Thus f(x) and $f(\mathbf{0_1})$ are in R_2. Now we have to prove that $f(\mathbf{0_1}) = \mathbf{0_2}$. Since f is a homomorphism onto from R_1 to R_2, we have

$$f(\mathbf{0_1} + x) = f(\mathbf{0_1}) + f(x) \Rightarrow f(x) = f(\mathbf{0_1}) + f(x)$$
$$\Rightarrow \mathbf{0_2} + f(x) = f(\mathbf{0_1}) + f(x) \qquad [\mathbf{0_2} \text{ is zero element of } R_2]$$
$$\Rightarrow \mathbf{0_2} = f(\mathbf{0_1}) \qquad [\text{By left cancellation law}]$$

(b) Let x be any element of R_1. Then −x is additive inverse of x in R_1. Thus $x + (-x) = \mathbf{0_1}$. And for any element f(x) of R_2, we have

$$f(x) + f(-x) = f(x + (-x)) = f(\mathbf{0_1}) = \mathbf{0_2}$$
$$\Rightarrow f(x) + f(-x) = \mathbf{0_2}$$
$$\Rightarrow f(-x) = -f(x)$$

Hence, image of additive inverse of an element x in R_1 is additive inverse of f(x) in R_2.

(c) Here we have to show that if $(R_1, +_1, *_1)$ is an integral domain then $(R_2, +_2, *_2)$ is also an integral domain i.e.

- If $*_1$ is commutative in R_1 then $*_2$ is commutative in R_2
- If unit element exist in R_1 then unit element also exists in R_2 and
- If there is no zero divisors in R_1 then there is no zero divisors in R_2 also.

Reader should prove this as an exercise.

(d) To prove this we have to show that

- If unit element exist in R_1 then unit element also exists in R_2 and
- If R_1 has multiplicative inverse for every non-zero element then R_2 also has multiplicative inverse for every non-zero element.

Reader should prove this as an exercise.

Proved.

Let (R, +, *) be a ring and S be any non-empty subset of R. If (S, +, *) forms a ring, where + and * are the two binary operations of R, then S is called a ***subring*** of R. Using the alternative postulates for subgroup, we can say that (S, +, *) is a subring of (R, +, *) if and only if for any two elements x and y of S, we have

- $x - y \in S$, and
- $x * y \in S$.

If (R, +, *) be a ring then {**0**} and R itself are always subrings of R. These two subrings are called ***improper or trivial subrings*** of R. Other subrings are called ***proper subrings*** of R. We defined reflexive (symmetric, transitive) closure of a relation in the chapter one. Similarly, we can define closure of a subring. Let M be any subset of a ring R. A subring S is called ***subring closure*** of M if M is subset of S and any subring containing M also contains S.

Example 8 Show that the set of all integers is a subring of the ring of rational numbers with respect to the two binary operations of ordinary addition and multiplication.

Solution: Let I be the set of all integers and Q be the set of all rational numbers. It is already stated above that (Q, +, *) is a ring. It is also clear that $I \subseteq Q$. Let x and y be any two elements of I then x – y is also an integer so is in I. Also x * y is an integer and hence is in I. This proves that I is a subring of Q. It is a proper subring because set I is a proper subset of Q different from {**0**} and Q. ♦

Example 9 Show that set A of matrices $\begin{bmatrix} x & y \\ 0 & z \end{bmatrix}$ is a subring of the ring M of all real square matrices of order 2

Solution: Obviously, $A \subseteq M$. Let $\begin{bmatrix} x & y \\ 0 & z \end{bmatrix}$ and $\begin{bmatrix} a & b \\ 0 & c \end{bmatrix}$ be any two matrices of A. Then

$$\begin{bmatrix} x & y \\ 0 & z \end{bmatrix} - \begin{bmatrix} a & b \\ 0 & c \end{bmatrix} = \begin{bmatrix} x-a & y-b \\ 0 & z-c \end{bmatrix} \in A \quad \text{and} \quad \begin{bmatrix} x & y \\ 0 & z \end{bmatrix} * \begin{bmatrix} a & b \\ 0 & c \end{bmatrix} = \begin{bmatrix} xa & xb+yc \\ 0 & zc \end{bmatrix} \in A.$$

Therefore, A is a subring of M. ♦

Theorem 3 The intersection of any two subrings is a subring.

Proof: Let S and T be any two subrings of a ring R. Then We have to show that $S \cap T$ is also a subring of R. Obviously $S \cap T$ is not empty because, at least $\mathbf{0} \in S \cap T$. Now, let x and y be any two elements in $S \cap T$. Then, we have

$$\begin{aligned} x, y \in S \cap T &\Rightarrow x, y \in S \text{ and } x, y \in T \\ &\Rightarrow x - y \in S \text{ and } x - y \in T \\ &\Rightarrow x - y \in S \cap T \end{aligned}$$

Similarly,

$$\begin{aligned} x, y \in S \cap T &\Rightarrow x, y \in S \text{ and } x, y \in T \\ &\Rightarrow x * y \in S \text{ and } x * y \in T \\ &\Rightarrow x * y \in S \cap T \end{aligned}$$

Therefore $S \cap T$ is a subring of ring R whenever S ad T are subring of ring R.

Proved.

The result of the theorem 3 can be generalized to state that "*an arbitrary intersection of subrings is a subring*". Like the order of a group, we can also define order of a ring with respect to the binary operation of addition. Let (R, +, *) be a ring and x be any element of R. Suppose that $\exists$ a positive integer n such that x + x + x +...n times = **0** $\forall x \in R$. The smallest such positive integer n is called ***order of the ring R w. r. t.*** +. This smallest integer is also called ***characteristic*** of the ring R. If there exist no such integer then the ring is called of characteristic zero or infinite. Since an integral domain does not have a zero divisors, every non-zero element taken as member of the additive group is of the same order.

Example 10 The ring of integers with the binary operations of ordinary addition and multiplication is of characteristic zero. ♦

Example 11 The ring of rational numbers with the binary operations of ordinary addition and multiplication is of characteristic zero. ♦

Example 12 The ring of integers modulo 6 {0, 1, 2, 3, 4, 5} with the binary operations of addition modulo 6 and multiplication modulo 6 is of characteristic **six**. ♦

Theorem 4 The characteristic of an integral domain is either 0 or a prime number.

Proof: Let (D, +, *) be an integral domain and x be any element of D. If order of x is 0, then characteristic of D is zero and hence we have nothing to prove. Let the characteristic of D is a finite number p. Then o(x) = p. Let us suppose that p is not a prime number. Then we can write p = qr, where q < p and r < p. Now

$$O(x) = p \Rightarrow o(x^2) = p \qquad [\text{Since } x^2 \in D]$$
$$\Rightarrow x^2 + x^2 + x^2 + \ldots p \text{ times} = \mathbf{0}$$
$$\Rightarrow x^2 + x^2 + x^2 + \ldots p \text{ times} = \mathbf{0}$$
$$\Rightarrow x^2 + x^2 + x^2 + \ldots qr \text{ times} = \mathbf{0} \qquad [\text{Since } p = qr]$$
$$\Rightarrow (qr)x^2 = \mathbf{0}$$
$$\Rightarrow (qx)(rx) = \mathbf{0}$$
$$\Rightarrow \text{Either } qx = \mathbf{0} \text{ or } rx = \mathbf{0} \qquad [\text{Since D is integral domain}]$$

This leads to contradiction because p is the least positive integer such that $px = \mathbf{0}$. Hence p is a prime number.

Proved.

Having discussed the concept of subring and isomorphism of rings, we can define ***embedded ring***. Let $(R_1, +_1, *_1)$ be a ring and there exist a mapping f from R_1 ***into*** another ring $(R_2, +_2, *_2)$ such that f is one to one and

$$f(x +_1 y) = f(x) +_2 f(y) \text{ and } f(x *_1 y) = f(x) *_2 f(y)$$

Obviously, $(f(R_1), +_2, *_2)$ is a ring and because it is a subset of R_2, $f(R_1)$ is a subring of R_2. Since image of R_1 is embedded in R_2, R_1 is said to be an ***embedded ring*** in R_2. For example a non-commutative ring without unity can be embedded in a commutative ring, or in a ring with unit element or in a division ring or in an integral domain or in a field (to be discussed in the next section).

A nonempty subset S of a ring (R, +, *) is said to be ***left ideal*** of R if (S, +) is subgroup and $\mathbf{r * s} \in S \ \forall\ r \in R$ and $\forall\ s \in S$. Similarly, a nonempty subset S of a ring R is called a ***right ideal*** if (S, +) is subgroup and $\mathbf{s * r} \in S \ \forall\ r \in R$ and $\forall\ s \in S$. If S is both left and right ideals of R then it is called ***ideal of*** R. It can easily be inferred that *every ideal of a ring is a subring whereas the reverse is not true*. The reason is the requirement of the closure of ideals w. r. t. the second binary operation of multiplication. A subring is required to be closed w. r. t. the multiplication for all the elements of the subring, whereas in the case of an ideal, it is required to be closed w. r. t. the multiplication for all the elements of ideal with all the elements of ring. For example, the set of integer I is only a subring of the ring of rational numbers (Q, +, *) and not an ideal. Another example is set Q of rational numbers, which is a subring of the ring of real numbers (R, +, *) and not an ideal. However set P = {xm | m is an integer} is a subring as well as an ideal of the ring of integers (I, +, *).

Every ring R possesses at least two ideals: the ring R itself and the set {**0**} containing zero of R. These two ideal are improper ideals and are called ***unit ideal*** and ***null ideal*** respectively. All other ideals are called ***proper ideals***. A ring having no proper ideal is called ***simple ideals.*** In order to verify that whether a subset S of a ring R is ideal of R, it is enough to test the following two conditions:

- $\forall x, y \in S, x - y \in S$, and
- $\forall\, r \in R$ and $\forall s \in S$, $r * s \in S$ and $s * r \in S$.

If R is a commutative ring then it is enough to test that $\forall\, r \in R$ and $\forall s \in S$, $r * s \in S$. It can easily be shown, as it is shown in the case of subring, that intersection of two ideals is again a ideal of the ring. Similarly, arbitrary intersection of ideals of a ring is an ideal of the ring. An ***ideal closure*** of a subset M of a ring R is the smallest ideal S of R such that S contains M and S is contained in any ideal of R that contains M as its subset. If M is an ideal then it is closure of itself otherwise we can always add some elements to M such that M becomes an ideal of R. Recall the definition of reflexive closure, symmetric closure and transitive closure defined in the chapter 1 and subring closure defined in this section of this chapter. The ideal closure of a subset M is called an ideal generated by M and is written as (M). If M contains only one element, say x, then the ideal closure S can be written as (x).

An ideal S of a ring R is said to be a ***principal ideal*** if there exists an element x in S such that any ideal T containing x must also contains S i.e. S = (x). An integral domain D is called a ***principal ideal ring*** if every ideal S of D is a principal ideal. Finally, before discussing some examples and theorems, let us define polynomial on a ring and polynomial ring.

Let (R, +, *) be a ring. Let f(x) be a polynomial in a variable x given as

$$f(x) = a_0 + a_1x + a_2x^2 + a_3x^3 + \ldots\ldots = \sum_{n=0}^{\infty} a_n x^n$$

Where a_n (n = 0, 1, 2, 3, ...) are elements of R and only finite number of them are non-zero. x is any variable and it is not an element of R. The f(x) is called ***polynomial in x over ring R***. We can have many such polynomials over R. Collection of all such polynomials defined over the ring R is denoted as R[x]. If all the coefficients a_n (n = 0, 1, 2, 3, ...) in f(x) are zero, the f(x) is called ***zero polynomial***. We shall show in one of the examples to be discussed below that the set R[x] forms a ring with respect to the binary operations of polynomial addition and polynomial multiplication. Let $f(x) = a_0 + a_1x + a_2x^2 + a_3x^3 + \ldots\ldots$ and $g(x) = b_0 + b_1x + b_2x^2 + b_3x^3 + \ldots\ldots$ be any two polynomials in x over a ring R, then the polynomial addition is defined as:

$$f(x) + g(x) = (a_0 + b_0) + (a_1 + b_1)x + (a_2 + b_2)x^2 + (a_3 + b_3)x^3 + \ldots\ldots$$

and the polynomial multiplication is defined as:

$$f(x)*g(x) = c_0 + c_1x + c_2x^2 + c_3x^3 + \ldots\ldots$$

Where c_n is the convolution of two sequences $(a_0, a_1, a_2, a_3, \ldots\ldots)$ and $(b_0, b_1, b_2, b_3, \ldots\ldots)$ as defined in the chapter two of this book. It is obvious now that if f(x) and g(x) are polynomials in x over a ring R then f(x) + g(x) and f(x)*g(x) are polynomials in x over the same ring R. Thus R[x] is closed with respect to the binary operations of polynomials addition and multiplication.

Example 13 Let U be an ideal of a ring R with unity 1. If $1 \in U$ then prove that U = R.

Solution: Since U is an ideal of R, $U \subseteq R$. Now let x be any element of R. Since U is an ideal of R and $1 \in U$, we have $1*x \in U$ i.e. $x \in U$. Thus $R \subseteq U$. Therefore U = R.

Proved.

Example 14 Show that S is an ideal of S + T where S is any ideal of ring R and T any subring of R.

Solution: Let us first show that S + T is a subring of R. Let x and y are any two elements of S+T. Then x and y can be written as x = a + b and y = c + d, where a, $c \in S$ and b, $d \in T$. Now,

$$x - y = (a + b) - (c + d) = (a - c) + (b - d)$$

Since S and T both are subring of R, a, $c \in S \Rightarrow a - c \in S$ and b, $d \in T \Rightarrow b - d \in T$. Thus $(a - c) + (b - d) \in S + T \Rightarrow x - y \in S + T$.

Next,

$$x * y = (a + b) * (c + d) = (a * c + a * d + b * c) + b * d$$

Since S is an ideal of R, a, $c \in S$ and b, $d \in R \Rightarrow (a * c + a * d + b * c) \in S$. And b, $d \in T \Rightarrow b * d \in T$ because T is a subring of R. Thus,

$$(a * c + a * d + b * c) + b * d \in S + T \Rightarrow x * y \in S + T$$

Therefore, S + T is a subring of R. It is given that S is an ideal of R and $S \subseteq S + T$, thus S is an ideal of S + T.

Ans.

Example 15 For any given element a of the ring R let $Ra = \{xa : x \in R\}$. Show that Ra is a left ideal of R.

Solution: Let sa and ta be any two elements of Ra, where s, $t \in R$. Now $sa - ta = (s - t)a \in Ra$ because s, t in R implies s – t in R. Next let r be any element of R and sa be any element of Ra, then $r(sa) = (rs)a \in Ra$. Hence, Ra is a left ideal of R.

Proved.

Example 16 The set R[x] of all polynomials over an arbitrary ring R is a ring with respect to addition and multiplication of polynomials.

Solution: Let + and * denote the binary operations of polynomial addition and multiplication respectively. It is already explained above that R[x] is closed with respect to + and *. Now let us test for other postulates one by one.

Associativity of +: Let f(x) = $\sum_{n=0}^{\infty} a_n x^n$, g(x)= $\sum_{n=0}^{\infty} b_n x^n$ and h(x)= $\sum_{n=0}^{\infty} c_n x^n$ be any three polynomials in R[x]. Then

$$f(x) + (g(x) + h(x)) = \sum_{n=0}^{\infty} [a_n + (b_n + c_n)]x^n = \sum_{n=0}^{\infty} [(a_n + b_n) + c_n]x^n$$

$$=(f(x) + g(x)) + h(x)$$

Hence + is associative in R[x].

Commutativity of +: Let f(x) = $\sum_{n=0}^{\infty} a_n x^n$, g(x)= $\sum_{n=0}^{\infty} b_n x^n$ be any two polynomials in R[x]. Then

$$f(x) + g(x) = \sum_{n=0}^{\infty} [a_n + b_n]x^n = \sum_{n=0}^{\infty} [b_n + a_n]x^n$$

$$= g(x) + f(x)$$

Hence + is commutative in R[x].

Existence of additive identity: A zero polynomial **0**(x) = $\sum_{n=0}^{\infty} a_n x^n$ where a_n = 0 for n = 1, 2, 3, ..., is also a member of R[x]. If g(x) = $\sum_{n=0}^{\infty} b_n x^n$ be any polynomial in R[x] then

$$\mathbf{0}(x) + g(x) = \sum_{n=0}^{\infty} [a_n + b_n]x^n = \sum_{n=0}^{\infty} [b_n + a_n]x^n = \sum_{n=0}^{\infty} b_n x^n$$

$$= g(x) + \mathbf{0}(x) = g(x)$$

Thus **0**(x) is the identity of + and is in R[x].

Existence of additive inverse: Let f(x) = $\sum_{n=0}^{\infty} a_n x^n$ be any polynomial in R[x]. Since a_n is an element of the ring R, the additive inverse $-a_n$ of a_n exists in R and thus a polynomial

g(x) = $\sum_{n=0}^{\infty} -a_n x^n$ is in R[x] such that then

$$f(x) + g(x) = \sum_{n=0}^{\infty} [a_n - a_n]x^n = \sum_{n=0}^{\infty} 0x^n = \mathbf{0}(x) = g(x) + f(x)$$

Thus every polynomial f(x) of R[x] has an additive inverse in R[x].

Associativity of *: Let f(x) = $\sum_{n=0}^{\infty} a_n x^n$, g(x)= $\sum_{n=0}^{\infty} b_n x^n$ and h(x)= $\sum_{n=0}^{\infty} c_n x^n$ be any three polynomials in R[x]. Then

$$f(x) * (g(x) * h(x)) = \sum_{n=0}^{\infty} a_n x^n * \sum_{n=0}^{\infty} d_n x^n$$

where d_n is the convolution of the sequences b_n and c_n.

$$= \sum_{n=0}^{\infty} e_n x^n$$

Similarly, (f(x) * g(x)) * h(x) = $\sum_{n=0}^{\infty} d_n x^n * \sum_{n=0}^{\infty} c_n x^n$

where d_n is the convolution of the sequences a_n and b_n.

$$\sum_{n=0}^{\infty} e_n x^n$$

Hence * is associative in R[x].

Distribution of * over + in R[x]: Let f(x), g(x) and h(x) be three polynomials in R[x] as defined above. It is easy to test that

f(x) * [g(x) + h(x)] = f(x) * g(x) + f(x) * h(x), and
[g(x) + h(x)] * f(x) = g(x) * f(x) + h(x) * f(x),

Therefore, (R[x], +, *) is a ring. This ring is called ***polynomial ring***.

Proved.

Theorem 5 If x is an element in a commutative ring R with unity, then the set $S = \{rx \mid r \in R\}$ is a principal ideal of R generated by the element x i.e. S = (x).

Proof: Here, we have to prove

- $x \in S$
- S is an ideal of R
- If T be any ideal of R such that $x \in T$, then $S \subseteq T$.

Since it is given that R is a ring with unity, $1 \in R$. This implies that $1*x = x \in S$.

Next Let a and b be any two elements of S. Then there exists two elements r_1 and r_2 such that $a = r_1x$ and $b = r_2x$. Thus, $a - b = (r_1 - r_2)x \in S$. Also for any $a \in S$ and any $y \in R$, we have $ya = y(r_1x) = (y r_1)x \in S$. Because $y \in R$ and $r_1 \in R \Rightarrow (y r_1) \in R$. Thus, S is an ideal of R.

Finally, let T be any ideal of R such that T contains x. Let $a = r_1x$ be any element of S, where r_1 is an element of R, then $r_1x \in T$. This implies that $S \subseteq T$.

Hence, it is proved that S is a principal ideal of R generated by the element x.

Proved.

Theorem 6 Let S be an ideal of a commutative ring R. Let x be an element of S such that

$$z \in S \Rightarrow z = yx \text{ for some } y \in R$$

Then show that S is a principal ideal of R generated by x.

Proof: let T be any ideal of the given ring R such that T contains the element x. Let z be any element of S, thus by definition $\exists$ some element y in R such that $z = yx$. Since $y \in R$, $x \in T$ and T is an ideal of R $yx \in T$ i.e. $z \in T$. Thus $S \subseteq T$. Hence S is a principal ideal of R.

Proved.

9.3 Field

The gradual evolution of mathematical structures does not stop with division ring or skew field. If the second binary operation is commutative in a division ring (skew field) then the structure is called a ***field***. In other words, a mathematical structure (F, +, *) is called a field if

- F contains at least two elements,
- set F forms an abelian group with respect to +,

- non-zero elements of F form an abelian group with respect to *, and
- the binary operation of multiplication (*) distributes over the binary operation of addition (+) in F.

For example, (Q, +, *) is a field of rational numbers, (R, +, *) is a field of real numbers and (C, +, *) is a field of complex numbers. However set of integers, I, does not form a field with respect to integer addition and integer multiplication. More examples are discussed below.

Example 1 Show that set of numbers of the form $a + \sqrt{2}b$ where a and b are rational numbers is a field.

Solution: Let us denote the given set by F. To prove that (F, +, *) is a field, we shall prove that (F, +) and (F – {0}, *) are abelian groups and * distributes over + in F. Let x, y and z be any three elements in F such that $x = a + \sqrt{2}b$, $y = c + \sqrt{2}d$ and $z = e + \sqrt{2}f$.

Closure of +: For any x, y ∈ F, we have $x + y = (a + c) + \sqrt{2}(b + d)$, where both a + c and b + d are rational numbers. Thus x + y ∈ F.

Associativity of +: This is implied by the associativity of + on the set of rational numbers.

Existence of Zero (Identity of +): Since 0 is a rational number $0 = 0 + \sqrt{2}\,0 \in F$ and for any x ∈ F, we have $x + 0 = a + \sqrt{2}b + 0 + \sqrt{2}\,0 = a + \sqrt{2}b = 0 + x$. Hence zero element exists in F.

Existence of Inverse w. r. t. +: For any rational numbers a and b, there exists rational –a and –b respectively and hence $-a + \sqrt{2}(-b) = -x \in F$ whenever $a + \sqrt{2}b = x \in F$. And

$$x + (-x) = 0 = (-x) + x.$$

Thus, additive inverse for every element x ∈ F exists in F.

Commutativity of +: This is implied by the commutativity of + on the set of rational numbers.

Hence (F, +) is an abelian group. Now let us prove that non-zero element of F forms an abelian group w. r. t. *.

Closure of *: For any x, y ∈ F, we have

$$x * y = [(a * c) + 2(b * d)] + \sqrt{2}[(b * c) + (a * d)],$$

Where both [(a * c) + 2 (b * d)] and [(b * c) + (a * d)] are rational numbers. Thus x * y ∈ F.

Associativity of *: This is implied by the associativity of * on the set of rational numbers.

Existence of Unity (Identity of *): Since 1 & 0 are rational numbers, $1 = 1 + \sqrt{2}\,0 \in F$ and for any $x \in F$ and $x \neq \mathbf{0}$, we have $x * 1 = (a + \sqrt{2}\,b) * (1 + \sqrt{2}\,0) = a + \sqrt{2}\,b = 1 * x$. Hence unit element exists in F.

Existence of Inverse w. r. t. *: Let $x = a + \sqrt{2}\,b$ be any element of F such that $x \neq 0$. The multiplicative inverse of x is $\frac{1}{x}$. Therefor, we can write

$$\frac{1}{x} = \frac{1}{a+\sqrt{2}b} = \frac{1}{a+\sqrt{2}b} \times \frac{a-\sqrt{2}b}{a-\sqrt{2}b} = \frac{a-\sqrt{2}b}{a^2-2b^2} = \frac{a}{a^2-2b^2} + \frac{\sqrt{2}b}{a^2-2b^2} \in F$$

Thus, multiplicative inverse for every non-zero element $x \in F$ exists in F.

Commutativity of *: This is implied by the commutativity of * on the set of rational numbers.

Similarly, reader may verify that the distributive properties are satisfied in F.

Ans.

Example 2 If p is a prime number, then ring I_p of integers modulo p is a field under the binary operations of addition modulo p and multiplication modulo p.

Solution: Here $I_p = \{[0], [1], [2], \ldots, [p-1]\}$. It is easy to prove that $(I_p, +_p)$ is an abelian group. Many problems like this are solved in the chapter five. Let $N_p = I_p - \{[0]\}$. Now we have to prove that $(N_p, *_p)$ is an abelian group. Out of the five postulates, closure, associativity, commutativity and existence of unit element are obvious and can be easily proved. Thus, it is to be proved that every element of N_p possesses a multiplicative inverse in N_p. Let [x] be any elements of N_p. Then x is relatively prime to p as p is prime number. Thus there exists two integers a and b such that $ax + bp = 1$. Now if we divide ax by p, we get the remainder 1 i.e. $ax \equiv_p 1 \Rightarrow [a] *_p [x] \equiv_p [1] \Rightarrow [a]$ is multiplicative inverse of [x] under $*_p$. (However if p is not a prime, then we may get two elements [x] and [y] such that $[x] *_p [y] \equiv_p [0]$. For example $[2] *_6 [3] = [0]$. Hence in such a case I_p will not form even an integral domain.) Therefore $(I_p, *_p)$ is an abelian group for a prime number p. The distributive properties can be proved as in the example 5 of 9.2.

Ans.

Theorem 1 Every field is an integral domain.

Proof: Let (F, +, *) be a field. F is obviously a commutative ring with unity. In order to prove that F is an integral domain, it is sufficient to show that F has no zero divisors.

Let x and y be any two elements of F such that $x \neq 0$ and $x*y = 0$. Since x is non-zero element of F, its multiplicative inverse x^{-1} exists in F. Thus, we have

$$x*y = 0 \Rightarrow x^{-1}*(x*y) = x^{-1}*0$$
$$\Rightarrow (x^{-1}*x)*y = 0$$
$$\Rightarrow 1*y = 0 \Rightarrow y = 0$$

Similarly, we have $y \neq 0$ and $x*y = 0 \Rightarrow x = 0$. Thus $x*y = 0 \Rightarrow$ either $x = 0$ or $y = 0$. Therefore F has no zero divisors and hence F is an integral domain.

Proved.

Note: It implies from the above theorem that a skew field does not possess divisors of zero.

Theorem 2 Every finite integral domain is a field.

Proof: Let (D, +, *) be a finite integral domain. By the definition of D, it is obvious that D is a commutative ring with unity. Thus inorder to prove that D is a field, it is enough to show that every non-zero element of D possesses a multiplicative inverse in D.

Let x be any non-zero element of D. Let $|D| = n$ and $a_1, a_2, a_3, \ldots, a_n$ be the n distinct elements of D. Since D is a finite set its elements can be designated as $a_1, a_2, a_3, \ldots, a_n$ without loss of any generality. Now $x*a_1, x*a_2, x*a_3, \ldots, x*a_n$ are all distinct elements of D as $x*a_i = x*a_j \Rightarrow x*a_i - x*a_j = 0$

$\Rightarrow x*(a_i - a_j) = 0$ [By Left distribution]

$\Rightarrow a_i - a_j = 0$ [$x \neq 0$ and D had no zero divisors]

$\Rightarrow a_i = a_j$

Thus one the $x*a_1, x*a_2, x*a_3, \ldots, x*a_n$ equals the unit element of D. Let $x*a_k$ equals 1, the unit element of D. Since D is a commutative ring, $x*a_k = 1 = a_k*x$. This implies that a_k is multiplicative inverse of x. Hence every non-zero element of D possesses a multiplicative inverse. Therefore (D, +, *) is a field.

Example 3 If (R, +, *) is a ring such that $a^2 = a \; \forall a \in R$ where $a^2 = a*a$ then prove that

(i) $a + a = 0 \; \forall a \in R$

(ii) $a + b = 0 \Rightarrow a = b$

(iii) R is a commutative ring.

Solution: (i) $a \in R \Rightarrow a + a \in R$

$\Rightarrow (a + a) * (a + a) = a + a$ [By the definition]
$\Rightarrow (a + a) * a + (a + a) * a = a + a$ [By right distribution]
$\Rightarrow a*a + a*a + a*a + a*a = a + a$ [By right distribution]
$\Rightarrow a + a + a + a = a + a$ [By the definition]
$\Rightarrow a + a = 0$ [By the cancellation law]

(ii) $a + b = 0$ & $a + a = 0$ together $\Rightarrow a + b = a + a$
$\Rightarrow b = a$ [By left cancellation law]

(iii) Let a and b be any two elements of R then $a + b \in R$

$\Rightarrow (a + b) * (a + b) = a + b$ [By the definition]
$\Rightarrow a^2 + b*a + a*b + b^2 = a + b$ [By distribution]
$\Rightarrow a + b*a + a*b + b = a + b$
$\Rightarrow b*a + a*b = 0 \Rightarrow b*a = a*b$ [From (ii)]

Thus R is a commutative ring.

Ans.

Note: An element $a \in R$ is said to be ***idempotent*** if $a^2 = a$. A ring is said to be a ***Boolean ring*** if all its elements are idempotent. (Reader should note that $1*1 = 1$ and $0*0 = 0$).

A substructure called subfield can be defined in the same fashion as we have defined subgroup in the chapter 5 and subring in the previous section of this chapter. Let (F, +, *) be any field. A nonempty subset S of F is said to be a ***subfield*** of F if (S, +, *) is a field, where + and * are two binary operations of addition and multiplication of (F, +, *). In order to show whether (S, +, *) is a subfield of (F, +, *), it is sufficient to show that

- $a, b \in S \Rightarrow a - b \in R\ \forall a, b \in S$ and
- $a, b \in S \Rightarrow a * b^{-1} \in R\ \forall a, b \in S$ and

The notion of proper and improper subfield and simple field is the same as that defined for group and ring. A simple field is also called a ***prime*** field i.e. a field F is said to be prime if it has no subfield other than itself. Note that {0} cannot form a subfield and hence there is no concept like zero field. A field (Q, +, *) of rational numbers is a prime field whereas field (R, +, *) of real numbers is not a prime field. Another example of a prime field is $(I_p, +_p, *_p)$ for any prime p.

Since every field is an integral domain, the characteristic of a field F is 0 or $n > 0$ according as any non-zero element x of F is of order 0 or n when regarded as an element of additive group (F, +) of F. The field F is said to be an ***ordered field*** if it is ordered as an integral domain. For example, (C, +, *) is an unordered field.

Example 4 (Q, +, *) is a subfield of (R, +, *). ♦

Example 5 (R, +, *) is a subfield of (C, +, *). ♦

Example 6 Show that (I_p, $+_p$, $*_p$), where p is a prime number is not an ordered field.

Solution: Here I_p = {[0], [1], [2], ..., [p – 1]}. Let I_p^+ be the set of positive elements of the given field. The zero element of the field is [0]. The element [1] ≠ [0] has the additive inverse [p – 1]. Thus either [1] ∈ I_p^+ or [p – 1] ∈ I_p^+. Let [1] ∈ I_p^+. Since I_p^+ is closed with respect to $+_p$, we have 1 $+_p$ 1 $+_p$ 1 $+_p$ ...(p – 1) times = (p – 1) ∈ I_p^+.

Since both 1 and its additive inverse p – 1 belong to I_p^+, the principle of trichotomy is not satisfied and hence (I_p, $+_p$, $*_p$) is not an ordered field.

Proved.

Theorem 3 A commutative ring without zero divisors can be embedded in a field.

Solution: Let (D, +, *) be a commutative ring without zero divisors. We can always add unit element to D, if it is not already in D, to make D a commutative ring with unity. Since D has no divisors of zero the product of any non-zero element cannot be equal to zero. Thus we can always enhance the set D to include multiplicative inverse of all non-zero elements of D maintaining the closure property with respect to addition and multiplication of the enhanced set. Let x, y be any two non-zero elements of D then x^{-1} and y^{-1} are either already their in D or have been included by construction. Let us call this enhanced set F.

By this construction, closure property of F with respect to multiplication and addition is still preserved as $(x^{-1} * y^{-1}) = (y^{-1} * x^{-1}) = (x * y)^{-1}$. Since x * y ∈ D, either $(x*y)^{-1}$ is in D or has been included in F. Similarly, we can show for addition.

Now the mathematical structure (F, +, *) form a field that contains a subset D' such that (D, +, *) is isomorphic to (D', +, *). Therefore D can be embedded in a field F.

Proved.

The field F in which an integral domain D is embedded is called a ***quotient field*** of D. We discussed ideal in the previous section. So far as a field is concerned, it has no proper ideal i.e. a field has only improper ideal {0} and F itself. To prove this let us suppose that S is any non-zero ideal of a field F. Let x ∈ D such that x ≠ 0. Now

$$x \in S \Rightarrow x \in F \Rightarrow x^{-1} \in F$$

Since S is an ideal, it is closed with respect to multiplication, thus $x*x^{-1} \in S \Rightarrow 1 \in S$. For any $y \in F$, $1*y = y \in S$. Thus $F \subseteq S$. Since S is an ideal of F so $S \subseteq F$. Therefore $F = S$. This proves that only ideal of F is $\{0\}$ or F itself.

Proved.

Theorem 4 A commutative ring with unity is a field if it has no proper ideals.

Proof: Let (R, +, *) be a commutative ring with unity having no proper ideals. In order to prove that R is a field, it is enough to prove that every non-zero element of R possesses its multiplicative inverse.

Let $r \neq 0$ be any element of R. Then $Rx = \{rx \mid x \in R\}$ is an ideal of R. It is given that R has no proper ideal & $Rx \neq \{0\}$. Thus $Rx = R$. It means $\exists\, y \in R$ such that

$$y*x = 1 = x*y \Rightarrow x^{-1} = y.$$

Thus every non-zero element x has an multiplicative inverse in R. Therefore (R, +, *) is a field.

Proved.

We have discussed homomorphism of groups in the chapter 5. Similarly we can define homomorphism of ring and field. Let $(R_1, +_1, *_1)$ and $(R_2, +_2, *_2)$ be two rings. A function $f: R_1 \rightarrow R_2$ is said to be ***homomorphism onto*** from R_1 onto R_2 if

- $f(x +_1 y) = f(x) +_2 f(y) \ \forall\, x, y \in R_1$ and
- $f(x *_1 y) = f(x) *_2 f(y) \ \forall\, x, y \in R_1$

Similarly, a function $f: R_1 \rightarrow R_2$ is said to be ***homomorphism into*** from R_1 into R_2 if

- $f(x +_1 y) = f(x) +_2 f(y) \ \forall\, x, y \in R_1$ and
- $f(x *_1 y) = f(x) *_2 f(y) \ \forall\, x, y \in R_1$

Theorem 5 If f is a homomorphism of a ring R_1 into R_2 then

(i) $f(0_1) = 0_2$ where 0_1 and 0_1 are the zeros of R_1 and R_2 respectively.

(ii) $f(-x) = -f(x) \ \forall\, x \in R_1$

Proof: (i) Let $x \in R_1$ then $f(x) \in R_2$. Thus

$$f(x) + 0_2 = f(x) = f(x + 0_1) = f(x) + f(0_1)$$

$$\Rightarrow f(x) + 0_2 = f(x) + f(0_1)$$

$$\Rightarrow 0_2 = f(0_1) \qquad \text{[By left cancellation law]}$$

(ii) Let $x \in R_1$ be any element then

$$0_2 = f(0_1) \Rightarrow f(x + (-x)) = f(x) + f(-x)$$
$$\Rightarrow f(-x)) = -f(x)$$

Proved.

If f is a homomorphism of a ring R_1 into R_2 then the set S of all those elements of R_1 which are mapped to the zero element of R_2 is called ***kernel of ring homomorphism*** of f. Similarly we can define ***kernel of group homomorphism***. Let f is a homomorphism of a group G_1 into G_2 then the set H of all those elements of G_1 which are mapped to the identity element of G_2 is called ***kernel of group homomorphism***. Let us see the following two theorems.

Theorem 6 If f is a homomorphism of a ring R_1 into R_2 with kernel S, then S is an ideal of R_1.

Proof: Let 0_1 into 0_2 be the zeros of the rings R_1 and R_2 respectively. Thus

$$S = \{x \mid f(x) = 0_2\}$$

Since $f(0_1) = f(0_2)$, thus at least one element 0_1 is in S and S is not empty. Next let us suppose that x and y be any two elements in S then $f(x) = 0_2$ and $f(y) = 0_2$. Now,

$$f(x - y) = f(x +_1 (-y)) = f(x) +_2 f(-y) = f(x) +_2 (-f(y)) = 0_2 +_2 0_2 = 0_2$$

This implies that x – y is in S. Now if r be any element of R then, we have

$$f(r *_1 x) = f(r) *_2 f(x) = f(r) *_2 0_2 = 0_2.$$

Thus, $r *_1 x \in S \ \forall \ r \in R_1$ and $\forall \ x \in S$.
Therefore, S is an ideal of R_1.

Proved.

Theorem 7 If f is a homomorphism of a group $(G_1, *_1)$ into a group $(G_2, *_2)$ with kernel K, then K is a normal subgroup of G_1.

Proof: Let e_1 and e_2 be the identities of groups G_1 and G_2 respectively. Since K is kernel of f, $K = \{x \mid f(x) = e_2\}$. Obviously K is not empty as $f(e_1) = e_2$. Now let x and y be any two elements of K. Then $f(x *_1 y^{-1}) = f(x) *_2 f(y^{-1}) = f(x) *_2 [f(y)]^{-1} = e_2 *_2 e_2^{-1} = e_2$. Thus $x *_1 y^{-1} \in K$. This shows that K is a subgroup of G_1. Now we have to show that K is a normal subgroup. Let g be any element of G_1 and x be any element of K then $f(x) = e_1$. Then, we have

$$f(g *_1 x *_1 g^{-1}) = f(g) *_2 f(x) *_2 [f(g)]^{-1} = f(g) *_2 e_2 *_2 [f(g)]^{-1} = e_2.$$

Thus K is a normal subgroup of G_1.

Proved.

9.4 Vector Space

This is a type of algebraic structure that deals with dimension i.e. 2-dimensional (2D), 3-dimensional (3D), n-dimensional (nD) space. Up till now we have dealt with set of data of a type called scalars (having magnitude and no direction). Now we shall discuss an algebraic structure that deals with data that is vector (having magnitude as well as direction). A ***vector*** **v** in the xy-plane is an ordered pair of real numbers (a, b). The numbers a and b are called the components of the vector **v** or the x and y co-ordinate respectively. A ***zero vector*** in 2D is the vector (0, 0). $\mathbf{V_2}$ denotes a set of all vectors in a plane. A vector **v** in space contains at least three components. Thus an ordered triple of real numbers (a, b, c) is a vector in 3D space. Similarly an ordered n-tuple of real numbers $(a_1, a_2, a_3, \ldots, a_n)$ is a vector v in n-dimensional space. The vector v has n components in this case. In general, $\mathbf{V_n}$ denotes a set of all vectors in a n-dimensional space.

A vector can be combined with other vector using binary operations of addition, subtraction, scalar or **dot** product, vector or **cross** product and simple multiple of vector with some scalar. So far as vector space as an algebraic structure is concerned, we have to deal with addition, subtraction and scalar multiple of a vector. Let $x = (a_1, a_2)$ and $y = (b_1, b_2)$ be any two vectors then they are represented as $\vec{x} = a_1\vec{i} + a_2\vec{j}$ and $\vec{y} = b_1\vec{i} + b_2\vec{j}$ respectively. The arrow mark shows the direction of the vector. The addition of these two vectors is given as

$$\vec{x} + \vec{y} = (a_1 + b_1)\vec{i} + (a_2 + b_2)\vec{j}$$

And subtraction of these two vectors is given as

$$\vec{x} - \vec{y} = (a_1 - b_1)\vec{i} + (a_2 - b_2)\vec{j}$$

And if α be any scalar then α multiple of vector $\vec{x}$ is given as

$$\alpha\vec{x} = \alpha a_1\vec{i} + \alpha a_2\vec{j}$$

Based on these definitions, it can easily be proved that set of all vectors in a 2D xy- plane forms a group with respect to the binary operation of addition of vectors. See the following example.

Example 1 Let V_2 be the set of all vectors in 2D plane and + be the binary operation of addition of vectors, then show that $(V_2, +)$ is an abelian group.

Solution: Let us prove that the mathematical structure $(V_2, +)$ satisfies all the postulates of an abelian group.

Closure property: Let $\vec{x} = a_1\vec{i} + a_2\vec{j}$ and $\vec{y} = b_1\vec{i} + b_2\vec{j}$ be any two vectors in V_2, where a_1, a_2, b_1 and b_2 are real numbers. Then $\vec{x}+\vec{y} = (a_1+b_1)\vec{i} + (a_2+b_2)\vec{j}$ is again a vector in xy-pane as (a_1+b_1) and (a_2+b_2) are real numbers. Thus $\vec{x}+\vec{y} \in V_2 \forall\ \vec{x}, \vec{y} \in V_2$;

Associative property: Let $\vec{x} = a_1\vec{i} + a_2\vec{j}$, $\vec{y} = b_1\vec{i} + b_2\vec{j}$ and $\vec{z} = c_1\vec{i} + c_2\vec{j}$ be any three vectors in V_2. Then, we have

$$\begin{aligned}\vec{x}+(\vec{y}+\vec{z}) &= [a_1+(b_1+c_1)]\vec{i} + [a_2+(b_2+c_2)]\vec{j} \\ &= [(a_1+b_1)+c_1]\vec{i} + [(a_2+b_2)+c_2]\vec{j} \\ &= (\vec{x}+\vec{y})+\vec{z}\end{aligned}$$

Therefore, vector addition is associative in V_2.

Existence of Identity: Since 0 is a real number, the zero vector (0, 0) i.e. $\vec{0} = 0\vec{i} + 0\vec{j}$ is a vector in V_2. And for any vector $\vec{x} = a_1\vec{i} + a_2\vec{j}$, we have

$$\vec{x}+\vec{0} = [a_1+0]\vec{i} + [a_2+0]\vec{j} = [0+a_1]\vec{i} + [0+a_2]\vec{j} = a_1\vec{i} + a_2\vec{j} = \vec{x} = \vec{x}+\vec{0}$$

Thus $\vec{0} = 0\vec{i} + 0\vec{j}$ is the identity element of V_2.

Existence of Inverse: Let $\vec{x} = a_1\vec{i} + a_2\vec{j}$ be any vector in V_2. Then $-\vec{x} = -a_1\vec{i} - a_2\vec{j}$ is also a vector in V_2 because for any real number a, –a is also a real number. Then, we have

$$\vec{x}+(-\vec{x}) = [a_1+(-a_1)]\vec{i} + [a_2+(-a_2)]\vec{j} = 0\vec{i} + 0\vec{j} = \vec{0} \text{ and}$$

$$(-\vec{x})+\vec{x} = [(-a_1)+a_1]\vec{i} + [(-a_2)+a_2]\vec{j} = 0\vec{i} + 0\vec{j} = \vec{0}$$

Thus every vector in V_2 possesses its additive inverse in V_2.

Commutative property: Let $\vec{x} = a_1\vec{i} + a_2\vec{j}$ and $\vec{y} = b_1\vec{i} + b_2\vec{j}$ be any two vectors in V_2. Then, we have

$$\begin{aligned}\vec{x}+\vec{y} &= (a_1+b_1)\vec{i} + (a_2+b_2)\vec{j} \\ &= (b_1+a_1)\vec{i} + (b_2+a_2)\vec{j} \\ &= \vec{y}+\vec{x}\end{aligned}$$

Therefore, vector addition is commutative in V_2. And hence $(V_2, +)$ is an abelian group.

Ans.

Similarly we can, in general, prove that $(V_n, +)$ forms an abelian group. It is right time now to define, formally, a ***vector space.*** This algebraic structure uses the data and operations of earlier defined algebraic structure –field. A vector space is always defined over a field, usually called a scalar field. In the discussions above, we have used the field of real numbers to give basic notions about a vector.

Let (F, +, *) be a field and V be a set of all n-dimensional vector $v = (a_1, a_2, a_3, \ldots a_n)$, where $a_1, a_2, a_3, \ldots, a_n$ are components of v and each $a_i \in F$. Let $+_v$ be the binary operation of vector addition. The algebraic structure $(V, +_v)$ is said to be a ***vector space*** over field F if

1. $(V, +_v)$ is an abelian group.
2. $\alpha*(v +_v w) = \alpha*v +_v \alpha*w;\ \forall\ \alpha \in F$ and $\forall$ v, $w \in V$
3. $(\alpha + \beta)*v = \alpha*v +_v \beta*v;\ \forall\ \alpha, \beta \in F$ and $\forall\ v \in V$
4. $(\alpha * \beta)*v = \alpha * (\beta*v)\ \forall\ \alpha, \beta \in F$ and $\forall\ v \in V$
5. $1*v = v\ \forall\ v \in V$, where 1 is the identity element of F.

A vector space V over a field F is usually denoted as V(F).

Example 2 Show that $V_n(F)$ is a vector space, where F is any arbitrary field, n is any positive integer and V_n is the set of all n-tuples $(a_1, a_2, a_3, \ldots a_n)$. Here $a_1, a_2, a_3, \ldots, a_n$ are components of v and each $a_i \in F$.

Solution: It is already shown in the example 1 that $(V_2, +_v)$ is an abelian group. Similarly, reader can easily prove that $(V_n, +_v)$ is an abelian group. Next suppose that $v = (a_1, a_2, a_3, \ldots a_n)$ and $w = (b_1, b_2, b_3, \ldots b_n)$ be any two vectors in V_n and α, β be any scalars from F. Then

$$v +_v w = (a_1 + b_1, a_2 + b_2, a_3 + b_3, \ldots a_n + b_n) \text{ and}$$

$$\alpha*(v +_v w) = [\alpha*(a_1 + b_1), \alpha* (a_2 + b_2), \alpha* (a_3 + b_3), \ldots \alpha* (a_n + b_n)]$$

Also $\alpha*v +_v \alpha*w = [\alpha*(a_1 + b_1), \alpha* (a_2 + b_2), \alpha* (a_3 + b_3), \ldots \alpha* (a_n + b_n)]$

Hence,

$$\alpha*(v +_v w) = \alpha*v +_v \alpha*w;\ \forall\ \alpha \in F \text{ and } \forall\ v, w \in V$$

Next,

$$\begin{aligned}\alpha*v +_v \beta*v &= (\alpha*a_1 + \beta*a_1, \alpha*a_2 + \beta*a_2, \alpha*a_3 + \beta*a_3, \ldots \alpha*a_n + \beta*a_n)\\ &= [(\alpha + \beta)*a_1, (\alpha + \beta)*a_2, (\alpha + \beta)*a_3, \ldots (\alpha + \beta)*a_n]\\ &= (\alpha + \beta)*v\end{aligned}$$

Hence,

$$(\alpha + \beta)*v = \alpha*v +_v \beta*v;\ \forall\ \alpha, \beta \in F \text{ and } \forall\ v \in V$$

Next,

$$(\alpha * \beta)*v = [(\alpha*\beta)*a_1, (\alpha* \beta)*a_2, (\alpha* \beta)*a_3, \ldots(\alpha *\beta)*a_n]$$
$$= [\alpha*(\beta*a_1), \alpha* (\beta*a_2), \alpha* (\beta*a_3), \ldots\alpha *(\beta*a_n)]$$
$$=\alpha * (\beta*v) \ \forall\ \alpha, \beta\in F \text{ and } \forall\ v \in V$$

Finally, we have

$$1*v = [1*a_1, 1*a_2, 1*a_3, \ldots 1*a_n] = v\ \forall\ v \in V$$

Therefore V(F) is a vector space.

Proved.

Example 3 Let F[x] be the set of all polynomials in variable x over field F. Then show that F[x] is a vector space over the field F with respect to addition of polynomials.

Solution: Let $+_p$ be the binary operation of addition of polynomials. Then it can easily be shown that (F[x], $+_p$) is an abelian group. See example 16 of the section 9.2 of this chapter. Similarly reader may verify that the other four postulated are also satisfied by F[x]. ♦

Example 4 A field (F, +, *) can be regarded as a vector space over any subfield (K, +, *) of F.

Solution: Here F is a field so obviously (F, +) is an abelian group. The 2nd and 3rd postulates are just left and right distributive law in the field F and hence satisfied by default. The 4th one is the associative property in F and also satisfied. The last one is the existence of unity element in F. Hence all these postulates for F to be a vector space is implied from the definition of field itself.

Ans.

General Properties of vector Space

Let V(F) be a vector space and $\vec{0}$ be the zero vector of V, then

(i) $\alpha*\vec{0} = \vec{0}\ \forall\ \alpha\in F$

(ii) $0*\vec{v} = \vec{0}\ \forall \vec{v} \in V$

(iii) $\alpha*(-\vec{v}) = -(\alpha*\vec{v})\ \forall\alpha\in F\ \&\ \forall\ \vec{v} \in V$

(iv) $(-\alpha)*\vec{v} = -(\alpha*\vec{v})\ \forall\alpha\in F\ \&\ \forall\ \vec{v} \in V$

(v) $\alpha*(\vec{v} - \vec{w}) = \alpha*\vec{v} - \alpha*\vec{w}\ \forall\alpha\in F\ \&\ \forall\ \vec{v}, \vec{w} \in V$

(vi) $\alpha*\vec{v} = \vec{0} \Rightarrow \alpha = 0 \text{ or } \vec{v} = \vec{0}$

(vii) If $\alpha, \beta \in F$ and $\vec{v} \in V$ s. t. $\vec{v} \neq \vec{0}$ then, we have $\alpha*\vec{v} = \beta*\vec{v} \Rightarrow \alpha = \beta$

(viii) If $\vec{v}, \vec{w} \in V$ and $\alpha\in F$ s. t. $\alpha \neq 0$ then $\alpha*\vec{v} = \alpha*\vec{w} \Rightarrow \vec{v} = \vec{w}$

Proof: Here, the proof of (i), (v) and (viii) are given and reader is asked to prove the other properties.

(i) We have $\alpha * \vec{0} = \alpha * (\vec{0} + \vec{0}) = \alpha * \vec{0} + \alpha * \vec{0}$

i.e. $\vec{0} + \alpha * \vec{0} = \alpha * \vec{0} + \alpha * \vec{0} \Rightarrow \vec{0} = \alpha * \vec{0}$ [By right cancellation law in V]

(v) Here, we have

$\alpha * (\vec{v} - \vec{w}) = \alpha * (\vec{v} + (-\vec{w})) = \alpha * \vec{v} + \alpha * (-\vec{w})$ [By postulates 2 of vector space]

$= \alpha * \vec{v} - \alpha * \vec{w}$ [By property (iii) above]

(viii) Here, we have

$\alpha * \vec{v} = \alpha * \vec{w} \Rightarrow \alpha * \vec{v} - \alpha * \vec{w} = \vec{0} \Rightarrow \alpha * (\vec{v} - \vec{w}) = \vec{0} \Rightarrow \alpha = 0$ or $\vec{v} - \vec{w} = \vec{0}$

Since it given that $\alpha \neq 0$, we have $\vec{v} - \vec{w} = \vec{0}$ and hence $\vec{v} = \vec{w}$

Proved.

A non-empty subset W of a vector space V over a field F is called a ***vector subspace*** of V if W itself is vector space over F under the same binary operations of V and that of F. For any non-empty subset W of V(F) to be a vector subspace of V the following conditions are necessary and sufficient.

- $\forall\ \vec{v}, \vec{w} \in W \Rightarrow \vec{v} - \vec{w} \in W$
- $\forall \alpha \in F\ \&\ \forall\ \vec{v} \in W \Rightarrow \alpha * \vec{v} \in W$

Suppose V(F) is any vector space. Then V itself is a vector subspace of V. The other trivial vector subspace of V is the subset consisting of only zero vector $\vec{0}$. As usual these two vector subspaces are called ***improper vector subspaces.*** If V has any other subspace, then it is called ***proper subspace***. The subspace of V consisting of zero vector only is called the ***zero subspace.*** Intersection of two subspaces is again a subspace of the vector space. In general, an arbitrary intersection of vector subspaces is again a subspace. However this is not true in the case of union or difference of two subspaces. See the following examples and theorems.

Example 5 Suppose R is the field of real numbers. Which of the following are subspaces of $V_3(R)$?

(i) $\{(x, 2y, 3z) \mid x, y, z \in R\}$
(ii) $\{(x, x, x) \mid x \in R\}$
(iii) $\{(x, y, z) \mid x, y, z \in Q$, the set of rational numbers$\}$

Solution: (i) Let W = {(x, 2y, 3z) | x, y, z ∈R}. Let $\vec{v}, \vec{w} \in$ W, where $\vec{v} = (x_1, 2y_1, 3z_1)$ and $\vec{w} = (x_2, 2y_2, 3z_2)$. Obviously, x_1, y_1, z_1 x_2, y_2 and z_2 are all real numbers. Now

$$\vec{v} - \vec{w} = ((x_1 - x_2), 2(y_1 - y_2), 3(z_1 - z_2))$$

Obviously, $(x_1 - x_2)$, $2(y_1 - y_2)$, $3(z_1 - z_2)$ are real numbers and hence $\vec{v} - \vec{w} \in$ W.

Next let m be any real number, then $m\vec{v} = m(x_1, 2y_1, 3z_1) = (mx_1, 2my_1, 3mz_1)$ is a vector in W. Hence W is vector subspace of $V_3(R)$.

(ii) Let W = {(x, x, x) | x ∈R}. Let $\vec{v}, \vec{w} \in$ W, where $\vec{v} = (x_1, x_1, x_1)$ and $\vec{w} = (x_2, x_2, x_2)$. Clearly, x_1 and x_2 are real numbers. Now

$$\vec{v} - \vec{w} = ((x_1 - x_2), (x_1 - x_2), (x_1 - x_2))$$

Since $(x_1 - x_2)$ is a real numbers, $\vec{v} - \vec{w} \in$ W.

Next let m be any real number, then $m\vec{v} = m(x_1, x_1, x_1) = (mx_1, mx_1, mx_1)$ is a vector in W. Hence W is vector subspace of $V_3(R)$.

(iii) Let W = {(x, y, z) | x, y, z are rational number}. Let $\vec{v}, \vec{w} \in$ W, where $\vec{v} = (x_1, y_1, z_1)$ and $\vec{w} = (x_2, y_2, z_2)$. Obviously, x_1, y_1, z_1 x_2, y_2 and z_2 are all rational numbers. Now

$$\vec{v} - \vec{w} = ((x_1 - x_2), (y_1 - y_2), (z_1 - z_2))$$

Since $(x_1 - x_2)$, $(y_1 - y_2)$, $(z_1 - z_2)$ are rational numbers, $\vec{v} - \vec{w} \in$ W.

Next let m be any real (rational or irrational) number, then $m\vec{v} = m(x_1, y_1, z_1) = (mx_1, my_1, mz_1)$ ***may not*** be a vector in W. For example, let $m = \sqrt{2}$. Then $\sqrt{2}\, x_1$ is no longer a rational number. Hence W is not a vector subspace of $V_3(R)$.

Theorem 1 The intersection of any two vector subspaces W_1 and W_2 of a vector space V(F) is also a vector subspace of V.

Proof: Since zero vector $\vec{0}$ is in both the subsets W_1 and W_2, $W_1 \cap W_2$ is not empty. Let $\vec{v}, \vec{w} \in W_1 \cap W_2$, where $\vec{v} = (x_1, y_1, z_1)$ and $\vec{w} = (x_2, y_2, z_2)$. It is clear that x_1, y_1, z_1 x_2, y_2 and z_2 are all elements of F. Now

$$\vec{v} - \vec{w} = ((x_1 - x_2), (y_1 - y_2), (z_1 - z_2))$$

Here $(x_1 - x_2)$, $(y_1 - y_2)$, $(z_1 - z_2)$ are all in F because for x, y in F (a field), x and –y are also in F and hence x – y is in F. Thus $\vec{v} - \vec{w} \in W_1 \cap W_2$.

Next let m be any element of F, then $m\vec{v} = m(x_1, y_1, z_1) = (mx_1, my_1, mz_1)$. Since F is closed with respect to the multiplication, mx_1, my_1, mz_1 are all in F. Hence $m\vec{v}$ is in $W_1 \cap W_2$. Therefore, $W_1 \cap W_2$ is a vector subspace of V(F).

Proved.

Theorem 2 The union of two vector subspaces is a subspace if and only if one is contained in other.

Proof: Let W_1 and W_2 are any two subspaces of a vector space V(F). In the if part, let us suppose that one of the subspace is contained in the other i.e. either $W_1 \subseteq W_2$ or $W_2 \subseteq W_1$. If $W_1 \subseteq W_2$ then $W_1 \cup W_2 = W_2$. And if $W_2 \subseteq W_1$ then $W_1 \cup W_2 = W_1$. In either of the case $W_1 \cup W_2$ is a vector subspace of V(F).

Only if part: Let $W_1 \cup W_2$ is a vector subspace of V(F). Suppose that neither $W_1 \subseteq W_2$ nor $W_2 \subseteq W_1$. This implies that $\exists\ \vec{v} \in W_1$ such that $\vec{v} \notin W_2$ and $\exists\ \vec{w} \in W_2$ such that $\vec{w} \notin W_1$. Since $\vec{v} \in W_1$ and $\vec{w} \in W_2$, we have $\vec{v} + \vec{w} \in W_1 \cup W_2$. And

$$\begin{aligned} \vec{v} + \vec{w} \in W_1 \cup W_2 &\Rightarrow \vec{v} + \vec{w} \in W_1 \text{ or } \vec{v} + \vec{w} \in W_2 \\ &\Rightarrow \vec{v} + \vec{w} \in W_1 \text{ and } \vec{v} \in W_1 \\ &\Rightarrow (\vec{v} + \vec{w}) - \vec{v} \in W_1 \\ &\Rightarrow \vec{w} \in W_1, \end{aligned}$$

Which is a contradiction. Hence either $W_1 \subseteq W_2$ or $W_2 \subseteq W_1$.

Proved.

Let V(F) be a vector space. If $\vec{v}_1, \vec{v}_2, \vec{v}_3, \cdots \vec{v}_n \in V$ then any vector

$$\alpha_1\vec{v}_1 + \alpha_2\vec{v}_2 + \alpha_3\vec{v}_3 + \cdots \alpha_n\vec{v}_n$$

where $\alpha_1, \alpha_2, \alpha_3, \ldots \alpha_n \in F$, is called ***linear combination of vectors*** $\vec{v}_1, \vec{v}_2, \vec{v}_3, \cdots \vec{v}_n$. If S be any non-empty subset of V then set of all linear combinations of vectors of S is called ***linear span*** of S and is denoted by L(S). Thus if $S = \{\vec{v}_1, \vec{v}_2, \vec{v}_3, \cdots \vec{v}_n\}$ Then

$$L(S) = \{v \mid v = \alpha_1\vec{v}_1 + \alpha_2\vec{v}_2 + \alpha_3\vec{v}_3 + \cdots \alpha_n\vec{v}_n \text{ for } \alpha_1, \alpha_2, \alpha_3, \ldots \alpha_n \in F\}$$

A finite set of vectors $\vec{v}_1, \vec{v}_2, \vec{v}_3, \cdots \vec{v}_n$ are said to be ***linearly dependent*** if $\exists$ scalars $\alpha_1, \alpha_2, \alpha_3, \ldots \alpha_n \in F$ such that

$$\alpha_1\vec{v}_1 + \alpha_2\vec{v}_2 + \alpha_3\vec{v}_3 + \cdots \alpha_n\vec{v}_n = \vec{0}$$

where at least one (may be more or all) α_i is not equal to zero. If all α_i's are zero then the vectors $\vec{v}_1, \vec{v}_2, \vec{v}_3, \cdots \vec{v}_n$ are called ***linearly independent.*** Any infinite set V of vectors is called linearly independent if every finite subset of V is linearly independent otherwise it is called linearly dependent. Any non-empty subset S of V is said to be a ***basis*** of V if S contains linearly independent vectors that spans V i.e. L(S) = V. A vector space is said to be ***finite dimensional*** if it has a finite basis.

Theorem 3 The linear span L(S) of any subset S of a vector space V(F) is a subspace of V generated by S i.e. L(S) = {S}.

Proof: Suppose that S = $\{\vec{v}_1, \vec{v}_2, \vec{v}_3, \cdots \vec{v}_n\}$. If $\vec{v}, \vec{w}$ be any two vectors in L(S) then $\vec{v}$ *and* $\vec{w}$ are linear combinations of vectors in S and can be written as

$$\vec{v} = \alpha_1\vec{v}_1 + \alpha_2\vec{v}_2 + \alpha_3\vec{v}_3 + \cdots \alpha_n\vec{v}_n$$

$$\vec{w} = \beta_1\vec{v}_1 + \beta_2\vec{v}_2 + \beta_3\vec{v}_3 + \cdots \beta_n\vec{v}_n$$

where α_i's and β_i's $\in$ F. And

$$\vec{v} - \vec{w} = (\alpha_1 - \beta_1)\vec{v}_1 + (\alpha_2 - \beta_2)\vec{v}_2 + (\alpha_3 - \beta_3)\vec{v}_3 + \cdots (\alpha_n - \beta_n)\vec{v}_n$$

Since $\alpha_i - \beta_i \in$ F, $\vec{v} - \vec{w}$ is a linear combination of vectors in S thus $\vec{v} - \vec{w} \in$ L(S).

Next, let α be any scalar from F and $\vec{v}$ be any vector in L(S) then

$$\alpha\vec{v} = \alpha\alpha_1\vec{v}_1 + \alpha\alpha_2\vec{v}_2 + \alpha\alpha_3\vec{v}_3 + \cdots \alpha\alpha_n\vec{v}_n$$

Since $\alpha, \alpha_i \in F \Rightarrow \alpha\alpha_i \in F$, $\alpha\vec{v}$ is again a linear combination of vectors in S. Thus $\alpha\vec{v} \in$ L(S). Therefore L(S) is a vector subspace of V. Now we have to prove that L(S) is generated by S i.e. if there is any vector subspace W of V that contains S, then it must contain L(S) also.

Because 1 $\in$ F, for every vector $\vec{v}_i \in$ S, $1 * \vec{v}_i \in$ L(S). Thus S is contained in L(S) Let W be any vector subspace containing S. Then we shall show that W contains L(S) also. Since W is closed with respect to vectors addition and scalar multiplication, it must contain all possible linear combinations of vectors in S and S $\subseteq$ W. L(S) is the set of vectors spanned by S and hence L(S) $\subseteq$ W. This shows that any vector subspace that contains S also contains L(S).

The vector subspace L(S) spanned by S is the smallest vector subspace of V containing S i.e. L(S) is the vector subspace closure of a subset S. A vector subspace W is said to be ***vector subspace closure*** of a subset S of V if W contains S and is contained in every other subspace containing S.

Proved.

Theorem 4 In any vector space V(F) if $\vec{v}$ is a linear combination of vectors $\vec{v}_1, \vec{v}_2, \vec{v}_3, \cdots \vec{v}_n$, then the set of vectors $\vec{v}, \vec{v}_1, \vec{v}_2, \vec{v}_3, \cdots \vec{v}_n$ are linearly dependent.

Proof: Since $\vec{v}$ is a linear combination of $\vec{v}_1, \vec{v}_2, \vec{v}_3, \cdots \vec{v}_n$, $\exists$ scalars $\alpha_1, \alpha_2, \alpha_3, \ldots \alpha_n \in F$ such that

$$\vec{v} = \alpha_1\vec{v}_1 + \alpha_2\vec{v}_2 + \alpha_3\vec{v}_3 + \cdots \alpha_n\vec{v}_n$$

$$\Rightarrow 1 * \vec{v} - \alpha_1\vec{v}_1 + \alpha_2\vec{v}_2 + \alpha_3\vec{v}_3 + \cdots \alpha_n\vec{v}_n = \vec{0}$$

Where 1 is a scalar (multiplicative identity in F). Since there is at least one scalar (1) that is non zero, $\vec{v}, \vec{v}_1, \vec{v}_2, \vec{v}_3, \cdots \vec{v}_n$ are linearly dependent.

Proved.

Example 6 The subset {(1, 0, 0), (0, 1, 0)} of $V_3(R)$ generates the subspace W = {(a, b, 0) | a, b ∈ R}.

Solution: Let S = {(1, 0, 0), (0, 1, 0)}. Here we have to show that L(S) = W. Let (x, y, 0) be any element of W, then it can be written as

$$(x, y, 0) = x(1, 0, 0) + y(0, 1, 0) \Rightarrow (x, y, 0) \in L(S) \Rightarrow W \subseteq L(S)$$

Because (x, y, 0) is a linear combination of vectors in S.

Now let $\vec{v}$ be any vector in L(S). Then $\exists$ scalars α and β in R such that

$$\vec{v} = \alpha\,(1, 0, 0) + \beta\,(0, 1, 0) = (\alpha, \beta, 0)$$

And (α, β, 0) is an element of W, thus $\vec{v} \in W \Rightarrow L(S) \subseteq W$. Combining these two results, we have W = L(S).

Ans.

Note: 1 The subset S = {(1, 0, 0), (0, 1, 0), (0, 0, 1)} generates the vector space $V_3(R)$.

2. In general, S ={(1, 0, ...0), (0, 1, ...0), ...(0, 0,... 1)}, where each vector is ordered n-tuple, generates $V_n(R)$.

Example 7 If two vectors $\vec{v}$ & $\vec{w}$ are linearly dependent then show that one of them is a scalar multiple of other.

Solution: Let $\vec{v}$ & $\vec{w}$ be any two vectors of a vector space V(F). Since $\vec{v}$ & $\vec{w}$ are linearly dependent $\exists$ two scalars α and β ∈ F not both zero such that

$$\alpha\vec{v} + \beta\vec{w} = \vec{0}$$

If $\alpha \neq 0$ then α^{-1} exists in F and so we have

$$\alpha\vec{v} + \beta\vec{w} = \vec{0} \Rightarrow \alpha\vec{v} = -\beta\vec{w}$$

$$\Rightarrow \alpha^{-1}(\alpha\vec{v}) = \alpha^{-1}(-\beta\vec{w})$$

$$\Rightarrow (\alpha^{-1}\alpha)\vec{v} = -(\alpha^{-1}\beta)\vec{w}$$

$$\Rightarrow \vec{v} = -\alpha^{-1}\beta\vec{w}$$

And α^{-1} and $\beta \in F \Rightarrow \alpha^{-1}\beta \in F$. Thus $\vec{v}$ is a scalar multiple of $\vec{w}$. Similarly we can show that if $\beta \neq 0$ then $\vec{w}$ is a scalar multiple of $\vec{v}$

Ans.

Example 8 Let S = $\{\vec{v}_1, \vec{v}_2, \vec{v}_3, \cdots \vec{v}_n\}$ be a subset of V(F). If vectors in S are linearly independent then none of them can be a zero-vector.

Solution: Let $\vec{v}_m$ be a zero-vector where m lies between 1 and n i.e. $1 \leq m \leq n$. Thus for any $\alpha \in F$ and $\alpha \neq 0$, w ve

$$\vec{v}_1 + 0\vec{v}_2 + 0\vec{v}_3 + \cdots + \alpha\vec{v}_m + \cdots + 0\vec{v}_n = \vec{0}$$

This implies that S contains set of linearly dependent vectors. This is a contradiction with the given fact that S is a set of linearly independent vectors. Hence none of the vectors $\vec{v}_m$ can be zero-vector.

Ans.

Example 9 Let $V_3(R)$ be a 3-dimensional vector space over the field of real numbers. Test whether the following subsets of $V_3(R)$ is linearly dependent or independent:

(i) {(2, 1, 2), (8, 4, 8)}
(ii) {(1, 2, 0), (0, 3, 1), (–1, 0, 1)}
(iii) {(–1, 2, 1), (3, 0, –1), (–5, 4, 3)}
(iv) {(2, 3, 5), (4, 9, 25)}

Solution: (i) Let α and β be any two scalars such that

$$\alpha(2, 1, 2) + \beta(8, 4, 8) = \vec{0} = (0, 0, 0)$$
$$\Rightarrow (2\alpha + 8\beta, \alpha + 4\beta, 2\alpha + 8\beta) = (0, 0, 0)$$
$$\Rightarrow 2\alpha + 8\beta = 0, \alpha + 4\beta = 0 \text{ and } 2\alpha + 8\beta = 0$$

One of the obvious solution to the above set of equations is $\alpha = 4$ and $\beta = -1$. Thus the given vectors are linearly dependent.

(ii) Let α, β and γ be any three scalars such that

$$\alpha(1, 2, 0) + \beta(0, 3, 1) + \gamma(-1, 0, 1) = \vec{0} = (0, 0, 0)$$
$$\Rightarrow (1\alpha + 0\beta - 1\gamma, 2\alpha + 3\beta + 0\gamma, 0\alpha + 1\beta + 1\gamma) = (0, 0, 0)$$
$$\Rightarrow 1\alpha + 0\beta - 1\gamma = 0, 2\alpha + 3\beta + 0\gamma = 0 \text{ and } 0\alpha + 1\beta + 1\gamma = 0$$

We can represent the above set of equations in matrix form as AX = O, where A, X and O are as below:

$$A = \begin{bmatrix} 1 & 0 & -1 \\ 2 & 3 & 0 \\ 0 & 1 & 1 \end{bmatrix}, X = \begin{bmatrix} \alpha \\ \beta \\ \gamma \end{bmatrix} \text{ and } O = \begin{bmatrix} 0 \\ 0 \\ 0 \end{bmatrix}$$

Here $|A| = 1(3 - 0) - 2(0 + 1) + 0(0 + 3) = 1 \neq 0$. Thus A^{-1} exists and so $X = A^{-1}O = O$. This implies that $\alpha = \beta = \gamma = 0$. Therefore the given vectors are linearly independent.

(iii) Let α, β and γ be any three scalars such that

$$\alpha(-1, 2, 1) + \beta(3, 0, -1) + \gamma(-5, 4, 3) = \vec{0} = (0, 0, 0)$$
$$\Rightarrow (-1\alpha + 3\beta - 5\gamma, 2\alpha + 0\beta + 4\gamma, 1\alpha - 1\beta + 3\gamma) = (0, 0, 0)$$
$$\Rightarrow -1\alpha + 3\beta - 5\gamma = 0, 2\alpha + 0\beta + 4\gamma = 0 \text{ and } 1\alpha - 1\beta + 3\gamma = 0$$

We can represent the above set of equations in matrix form as AX = O, where A, X and O are as below:

$$A = \begin{bmatrix} -1 & 3 & -5 \\ 2 & 0 & 4 \\ 1 & -1 & 3 \end{bmatrix}, X = \begin{bmatrix} \alpha \\ \beta \\ \gamma \end{bmatrix} \text{ and } O = \begin{bmatrix} 0 \\ 0 \\ 0 \end{bmatrix}$$

Here $|A| = -1(0 + 4) - 3(6 - 4) - 5(-2 - 0) = -4 - 6 + 10 = 0$. Thus A is a singular matrix and so A^{-1} does not exist. However there exists non-zero cofactor of A of order 2. Thus A is of rank 2 < number of unknowns to be determined. This implies that ∃ non-zero solutions to the set of equations AX = O. Therefore the given vectors are linearly dependent.

(iv) Let α and β be any two scalars such that

$$\alpha(2, 3, 5) + \beta(4, 9, 25) = \vec{0} = (0, 0, 0)$$
$$\Rightarrow (2\alpha + 4\beta, 3\alpha + 9\beta, 5\alpha + 25\beta) = (0, 0, 0)$$
$$\Rightarrow 2\alpha + 4\beta = 0, 3\alpha + 9\beta = 0 \text{ and } 5\alpha + 25\beta = 0$$

Solving them we get only possible solution satisfying all the three equations is $\alpha = \beta = 0$. Thus the given vectors are linearly independent.

Ans.

Example 10 Show that the infinite set $S = \{1, x, x^2, x^3, \ldots x^n, \ldots\ldots\}$ is a basis of the vector space F[x] of all polynomials over the field F.

Solution: Let $T = \{x^{m_1}, x^{m_2}, x^{m_3}, \ldots x^{m_n}\}$ be any finite subset of S having n vectors. Let $\alpha_1, \alpha_2, \alpha_3, \ldots \alpha_n$ be n scalars from F such that

$$\alpha_1 x^{m_1} + \alpha_2 x^{m_2} + \alpha_3 x^{m_3} + \ldots + \alpha_3 x^{m_n} = 0$$

Then by the equality of two polynomials, we have $\alpha_1 = \alpha_2 = \alpha_3 = \ldots = \alpha_n = 0$. This implies that T contains linearly independent vectors. And since every finite subset T of S is linearly independent, S is linearly independent set of vectors. Now we have to show that L(S) = F[x].

Let f(x) be any polynomial of L(S). Then it is linear combination of vectors in S and so f(x) is a polynomial over the field F. Thus $f(x) \in F[x]$ i.e.

$$f(x) \in L(S) \Rightarrow f(x) \in F[x] \Rightarrow L(S) \subseteq F[x]$$

Now let f(x) be any polynomial in F[x]. Then f(x) is obviously of the form $\sum_{n=1}^{m} \alpha_n x^n$, where $\alpha_n \in F$. Obviously f(x) is a linear combination of vectors in S and so $f(x) \in L(S)$. Thus $F[x] \subseteq L(S)$. Combining these two results, we have L(S) = F[x]. Therefore S is the basis of F[x].

Ans.

The gradual evolution of different algebraic structures from a set of data and binary operation defined on them gives an insight into the way of object oriented approach. Though these topics are under the purview of pure mathematics, the concept is used tremendously in the information technology for design and development of software tools and bundling them in object oriented modules. It is not required to buy entire package of software, if you do not need them in entirety. Instead select only those parts that meet your requirement and pay for that only. This is possible only if the software is developed using object-oriented concept. *I need a group, one binary operation is enough, why to go for second operation?* The practical approach to solve the problems given in the exercise below will certainly help the reader in understanding various topics of object-oriented software engineering.

Exercise

1. Prove that the set R of numbers of the form $a + b\sqrt{3}$, where a and b are integers, is a ring w. r. t. ordinary addition and multiplication.
2. Prove that the set {km} where m is a fixed integer and k = 0, ±1, ±2, ±3, ... is a ring w. r. t. ordinary addition and multiplication of integers. Does it posses the unity element?
3. Verify that the set of residue classes modulo 6 is a ring. Give two non-zero elements in the ring such that their product is zero.

4. Let A be a set of all real numbers of the form x + ay + bz, where x, y, z are any rational number, **a** is *cube root of 2* and **b** is *cube root of 4*. Prove that A forms a commutative ring with unity w. r. t. the ordinary addition and multiplication.

5. Let S be a set and let M be the collection of all the subsets of S. Define 'addition' and 'multiplication' as follows:

$$A + B = \{x \mid x \in A \cup B, x \notin A \cap B\}$$
$$A * B = \{x \mid x \in A \cap B\}$$

6. (a) Prove that the ring on integers modulo 4 is not an integral domain.
   (b) Prove that the ring of integers modulo 5 is an integral domain.
   (c) Prove that the ring of integers modulo 6 is not an integral domain.
   (d) Prove that the ring of integers modulo 15 is not an integral dom

7. What condition must be true for n in order that the ring of integers modulo n be an integral domain.

8. Prove that the following sets of real numbers forms ring w. r. t. the binary operations of ordinary addition and multiplication.

$$(a)\ A = \{x \mid x = a + b\sqrt{2} + c\sqrt{3} + d\sqrt{6} + e\sqrt{12}\ \}$$
$$(b)\ B = \{x \mid x = a + b\sqrt{2} + c\sqrt{5} + d\sqrt{10} + e\sqrt{20}\ \}$$

Where a, b, c, d and e are rational numbers.

9. If p is a prime number, then the ring of integers modulo p is a field. Prove it.

10. Prove that a field is necessarily an integral domain.

11. Let (F, +, *) be a field. Let a, b $\in$ F with a $\neq$0. Prove that there exists a unique element x $\in$ F such that ax = b.

12. Prove that an integral domain with a finite number of elements is a field.

13. List the elements of the vector spaces $V_2(I/(2))$, $V_3(I/(2))$ and $V_3(I/(3))$,

14. Determine whether the following set forms a vector space over the indicated field under the binary operation of ordinary addition and multiplication.
    (a) The set of all real numbers of the form a + b$\sqrt{2}$, where a and b are elements of the field of rational numbers.
    (b) The set of all real functions f such that f(x+1) = f(x), where x is an element of the field of real numbers.

15. Let V be a vector space over a field F. Prove the following:
    (a) If a, b $\in$ F and v $\neq$ 0 be an element of V such that av = bv then a = b.
    (b) If u, w $\in$ V and a $\neq$ 0 be an element of F such that au = aw then u = w.

16. Determine whether the following vectors are linearly dependent or independent.
    (a) (1, -1, 2, 1), (2, 1, 1, 2)
    (b) (1, 2, 1, 2), (0, 1, 1, 0), (1, 4, 3, 2)
    (c) (2, 1, 1, 1), (1, 3, 1, -2), (1, 2, -1, 3)
    (d) (0, 1, 0, 1), (1, 0, 1, 0), (0, 3, 2, 0), (1, 2, 3, -1)
    (e) (1, 1, 0, 0), (0, 1, -1, 2), (1, 2, -1, 1), (2, 1, 0, 3)

17. A ***Gaussian integer*** is a complex number a +ib, where a and b are integers. Show that the set J[i] of Gaussian integers forms a ring under ordinary addition and multiplication of complex numbers. Is it an integral domain? Is it a field?

18. Prove that the totality R of all ordered pairs (a, b) of real numbers is a commutative ring with zero divisors under the addition and multiplication of ordered pairs defined as

$$(a, b) + (c, d) = (a+c, b+d)$$
$$(a, b) * (c, d) = (ac, bd)$$

19. Show that the set R of all real valued continuous functions defined in the closed interval [0, 1] is a commutative ring with unity with respect to the addition and multiplication of functions defined point-wise as follows:

$$(f+g)(x) = f(x) + g(x)$$
$$(fg)(x)=f(x)g(x)$$

20. Give an example of a skew field that is not a field.

21. Let p be a prime number. Prove that the set of integers I_p =(0, 1, 2, 3,......, p–1} forms a field with respect to addition and multiplication modulo p.

22. Define a filed. Prove that every filed is an integral domain, but there exist some integral domains which are not fields.

23. If two operations * and o on the set I of integers are defined as follows:

$$A * b = a + b - 1,\ a \text{ o } b = a + b - ab$$

24. Let A be the set of all real valued functions on (–∝, ∝). Define (f + g) (x) = f(x) + g(x) and (f×g)(x) = f(g(x)) for every x in (–∝, ∝). Is A a ring with respect to these two operations?

25. If R is a commutative ring, prove by induction that

$$(a + b)^n = a^n + {}^nC_1 a^{n-1} b^1 + {}^nC_2 a^{n-2} b^2 + {}^nC_3 a^{n-3} b^3 + ...+ b^n$$

for every positive integer n; here a and b are elements of R.

26. Show that the set of all 2 × 2 matrices over a field F forms a non-commutative ring under matrix addition and multiplication. Show also that this ring contains divisors of zero.

27. Let R be a non-zero ring such that for all a ∈ R, a^2 = a. Prove that R is a commutative ring of characteristic 2.

28. Show that every finite integral domain is of finite characteristic.

29. Define the characteristic of a ring and prove that if R is a finite ring then the characteristic of R is finite and is not equal to zero.

30. If U, V are ideals of a ring R, let U + V = {u + v | u ∈ U, v ∈ V}. Prove that U + V is also an ideal of R.

31. Verify the following for being true or false:
    (a) The set of all positive rational numbers is a sub ring of the ring of all rational numbers.
    (b) A sub ring of any field is a field.
    (c) Any sub ring of the ring of integers Z is an ideal of Z.

32. A field has no proper ideals i.e., if F is a field then its only ideal are (0) and F itself.

33. If R is a commutative ring and $a \in R$, then $Ra = \{ra \mid r \in R\}$ is an ideal of R.

34. A commutative ring with unity is a field if it has no proper ideals.

35. Let R be a ring with unit element, R not necessarily commutative, such that the only right ideals of R are (0) and R. Prove that R is a division ring.

36. Let R be a ring with unity element such that the only left ideals of R are (0) and R. Show that R is a division ring.

37. If D is an integral domain, then the polynomial ring D[x] is also an integral domain.

38. Show that if a ring R has no zero divisors, then the ring R[x] has also no zero divisors

39. If p is a prime integer, show that it need not be a prime Gaussian integer.

40. Show that the polynomial $x^3 - 9$ is reducible over the ring of integers modulo 11.

41. Show that a field F may be considered as a vector space over F if scalar multiplication is identified with field multiplication.

42. Show that the complex field C is a vector space over the real field R.

43. Let V be the set of all pairs (a, b) of real numbers, and let F be the field of real numbers. Define

$$(a, b) + (c, d) = (a + c, b + d)$$
$$c(a, b) = (ca, b)$$

Show that with these operations V is not a vector space over the field of real numbers.

44. Let R be the field of real numbers and let P_n be the set of all polynomials (of degree at the most n) over the field R. Prove that P_n is a vector space over the filed R.

45. How many elements are there in the vector space of polynomials of degree at the most n in which the coefficients are the elements of the field I(p) over the filed I(p), where p is a prime number?

46. Show that the set W of the elements of the vector space $V_3(R)$ of the form $(x + 2y, y, -x + 3y)$, where x, $y \in R$ is a subspace of $V_3(R)$.

47. Which of the following sets of vectors $\alpha = (a_1, a_2, a_3, \ldots, a_n)$ in R^n are subspaces of R^n $(n \geq 3)$?
   (a) all α such that $a_1 \leq 0$.
   (b) all α such that a_2 is an integer.
   (c) all α such that $a_2 + 4a_3 = 0$.

48. Find whether the vectors $2x^3 + x^2 + x + 1$, $x^3 + 3x^2 + x - 2$, and $x^3 + 2x^2 - x + 3$ of R[x], the vector space of all polynomials over the real number field, are linearly independent or not?

49. Show that a set of vectors which contains the zero vector is linearly dependent.

50. In the vector space R^3, let $\alpha = (1, 2, 1)$, $\beta = (3, 1, 5)$, $\gamma = (3, -4, 7)$. Show that the subspaces spanned by $\{\alpha, \beta\}$ and $\{\alpha, \beta, \gamma\}$ are the same.

51. Find a linearly independent subset T of the set S = {a, b, c, d}, where a = (1, 2, –1), b = (–3, –6, 3), c = (2, 1, 3) and d = (8, 7, 7) which spans the same space as S.

52. Let (D, +, *) be an ordered integral domain and D_+ be the set of positive elements of D. Then we define 'less than (<)', 'greater than (>)' relation in D as follows: for $\forall$ x, y $\in$ D, we have $x > y$ when $x - y \in D_+$ and $x < y$ when $y - x \in D_+$. Show that the order relation in an ordered integral domain is transitive.

53. If U and V are ideals of a ring R and UV be the set of all those elements of R which can be written as finite sums of elements of the form u*v where u $\in$ U and v $\in$ V. Prove that UV is an ideal of R.

54. The set of non-zero vectors $\vec{v}_1, \vec{v}_2, \vec{v}_3, \cdots \vec{v}_n$ of V(F) is linearly dependent if some $\vec{v}_k$ $2 \le k \le n$, is a linear combination of the preceding ones.

Bibliography & References

The list provided below is in no way complete and exhaustive. However, it is an honest attempt to cite all possible sources that have been referred while compiling this book. The list provided may be helpful in finding sources for further reading. In the era of Internet and World Wide Web, contents related to many topics are available online. It would not be fare to ignore these ready resources available for references. Keeping in mind the current trend and habit of web browsing, I have provided a number of web sites, where reader may visit to quench theirs thirst for knowledge.

1. *Aaron M. Tanenbaum & Moshe J. Augenstein: "Data Structure using Pascal", Prentice Hall, Englewood Cliffs, New Jersey, 07632.*
2. *Abbott, J. C.: "Sets, Lattices and Boolean Algebras", Allyn and Bacon, 1969.*
3. *Aho, A. V. and J. D. Ullman: "Principles of Compiler Design", Addison-Wesley Publishing Company, 1977.*
4. *Aho A. V.: "Indexed grammars-an extension of context-free grammars", ACM.*
5. *Aldeman L. and K. Manders: "Diophantine Complexity", Proc. Seventeenth Annual IEEE symposium on foundation of computer Science, page 81-88.*
6. *Aldeman L. and K. Manders: "Reducibility, randomness and intractability", Proc. Ninth Annual ACM symposium on the theory of computing, page 151-163.*
7. *Andrew S. Tanenbaum: "Computer Network", Prentice Hall of India, New Delhi, 1988.*
8. *Arbib M. A.: "Theories of Abstract Automata", Prentice Hall, Englewood Cliffs, New Jersey.*
9. *Arden D. N.: "Delayed logic and finite state machine: Theory of computing machine design", University of Michigan Press, Ann Arbor, Mich.*
10. *Augenstein M. A. Tanenbaum: "A lesson in Recursion and Structured Programming", SIGCSE Bulletin, 8(1), Feb, 1976.*
11. *Auslander M. A. and H. R. Strong: "Systematic Recursion Removal", Communication ACM, 21(2), Feb. 1978.*
12. *Batcher, K. E.: "Sorting Networks and Their Applications", AFIPS Proc. Of the 1968 SJCS, 32:307-314(1968).*
13. *Bauer M., D. Brand, M. J. Fischer, A. R. Meyer and M. S. Peterson: " A note on disjunctive form of tautologies".*

14. *Bayer R. and C. McCreight: "Organization and Maintenance of large ordered Indexes", Acta Informatica, 1(3), 1972.*

15. *Bellman R.: "Dynamic Programming", Princeton University Press, Princeton,, New Jersey, 1957.*

16. *Bentley J. L.: "Multidimensional Divide and Conquer", Communication ACM, 23(4), Apr. 1980.*

17. *Berge C.: "Graphs and Hypergraphs", North Holland, Amsterdam, 1973.*

18. *Berge C.: "Theory of Graphs and its Application", Methuen Press, 1962.*

19. *Berlekamp, E. R.: "Algebraic Coding Theory", McGraw-Hill Book Company, 1968.*

20. *Berman, G. and K.D. Fryer: "Introduction to combinatorics", Academic Press, 1972*

21. *Bird R. S.: "Improving Programs by the Introduction of Recursion", Communication ACM, 20(1), 1977.*

22. *Bird R. S.: "Notes on Recursion Elimination", Communication ACM, 20(6), June 1977.*

23. *Bitner J. R. and E. M. Reingold: "Backtrack Programming Techniques", Communication ACM, 18, 1975.*

24. *Blum M.: "A Machine independent theory of the complexity of recursive functions",J. ACM, 1967*

25. *Blum M.: "A machine-independent theory of the complexity of recursive functions", ACM 14:2.*

26. *Boasson L.: "Two iteration theorems for some families of languages", J. Computer and Systems Sciences 7(6), 1973.*

27. *Bondy, J. A. and U. S. R. Murthy: "Graph Theory with Applications", 1976.*

28. *Bose, R.C. and R. J. nelson: "A Sorting Problem", Journal of ACM, 9:282-296(1962).*

29. *Brainerd W. S. and L. H. Landweber: "Theory of computation", John Wiley and Sons, NY.*

30. *Brzozowski J. A.: "Derivatives of Regular Expressions", J, ACM, 11(4), 1964.*

31. *Burstall R. M. and J. Darlington: "A transformation System for developing Recursive Programs", Journal of the ACM,, 24(1), Jan. 1977.*

32. *Caldwell, S. H. : " Switching Circuits and Logical Design", John-Wiley & Sons, NY, 1958.*

33. *Cesarini F. and G. Soda: "An Algorithm to construct a Compact B-Tree in case of ordered Keys", Information Proceeding Letters, 17, 1983.*

34. *Chandler W. J.: "Abstract families of deterministic languages", Proc. First Annual ACM Symposium on the Theory of Computing, 1969.*

35. *Chartrand, G. Introductory Graph Theory. New York: Dover, p. 116, 1985.*

36. *Cheriton D. and R. E. Tarjan: " Finding Minimum Spanning Trees", SIAM J. Computing, 5, 1976.*

37. *Chomsky N. and G. A. Miller: "Finite State Languages", Information and Control, 1(2), 1958.*

38. *Chomsky N. and M. P. Schytzenberger: "The algebraic theory of Context Free Languages: Computer Programming and formal Systems", North Holland, Amsterdam, 1963.*

39. *Christo H. Papadimitriou,: "Elements of Theory of Computation", 2nd Ed, Printice Hall, 1997.*

40. *Chung, K. L., "Elementary Probability Theory with Stochastic Process", Springer-Verlag, 1974.*

41. *Codd, E. F.: "A Relational Model of Data for large Shared Data Banks", Communication of ACM, 13:377-387(1970).*

42. *Coffman, E. G.: "Computer and Job Shop Scheduling Theory", John Wiley & Sons, 1976.*

43. *Cohen, I. A. C.:"Basic Techniques of Combinational Theory", John Woley & Sons, New York, 1978.*

44. *Cohn, P. M.: " Universal Algebra", Harper and Row, NY, 1965.*

45. *Conway J. H.: "Regular Algebra and Finite Machines", Chapman and Hall, London, 1971.*

46. *Cook, S. A.: " The Complexity of Theorem-proving Procedures", 3rd ACM symposium on the theory of Computing", 1977.*

47. *Cooper C. D.: "Theorem proving in arithmetic without multiplication", Machine Intelligence 7, John Wiley and Sons, NY, 1972.*

48. *D. Loeb: "The world of Generating Functions and Umbral Calculus", Nov 16, 1995.*

49. *Dantzig G. B. and D. R. Fulkerson: " On the Max-flow Min-cut Theorem of Networks in Linear Inequalities and related systems", Annals of Math Study 38, Princeton University Press, Princeton, New Jersey, 1956.*

50. *Date, C. J.:"An Introduction to Data Base Systems", 3rd Ed, Addison-Wesley Publishing Company, 1981.*

51. *Dijkstra, E. W.:"A note on Two problems in connection with graphs", Nemerische Mathematik, 1:269 –271 (1951): Xeroxed notes referred.*

52. *Driscoll J. R. and Y. E. lien: "A selective Traversal Algorithm for Binary Search Trees", Cpmm, ACM, 21(6), jun, 1978.*

53. *Edmonds J. and R. M. Karp: "Theoretical Improvements in Algorithmic Efficiency for Network Flow Problem", Journal of ACM, 19, 1972.*

54. *Edwin H. Connel, "Elements of Abstract and Linear Algebra", From WWW.*

55. *Eilenberg S. and C. C. Elgot: "Recursiveness", Academic Press, NY, 1970.*

56. *Eitan Gurari, Ohio State University: "An Introduction to the Theory of Computation", Computer Science Press, 1989.*

57. *Even, S.: "Graph Algorithms", Computer Science Press, 1979.*

58. *Even S. and R. E. Tarjan: "Network Flow and Testing Graph Connectivity", SIAM J. of computing, 4(4), Dec, 1975.*

59. *Even S.: "Graph Algorithm", Computer Science Press, Potomac, Matrid, 1978.*

60. *Feller, W.: "An Introduction to Probability Theory and Its Applications", Vol. 1, 2nd Ed, John Wiley & Sons, 1957.*

61. *Feller W.:"An Introduction to probability theory and its Applications", 2nd Ed, John Wiley & Sons, 1950.*

62. *Fillmore J. P. and S. G. Williamson: "On Backtracking: A Combinatorial Description of the Algorithm", SIAM j. of computing, 3(1), Mar, 1974.*

63. *Floyd R. W.: "New proofs and old theorems in logic and formal linguistics", Computer Associates Inc, Wakefield, Mass, 1964.*

64. *Floyd R. W.: "On ambiguity in phrase structure languages", Comm. ACM, 5(10), 1962.*

65. *Ford L. R. and D. R. Fulkerson: "Flow in Networks", Princeton University Press, Princeton, New jersey, 1972.*

66. *Friedman A.: " Logical Design of Digital Systems", Computer Science Press, Potomac, Matrid, 1975.*

67. *Fulkerson D. R.: " Flow Networks and Combinational Operation Research", Amer. Math. Montbly, 73, 1966.*

68. Garey M. R., D. S. Johnson and R. E. Tarjan: "The planar Hamilton circuit problem is NP-complete", SIAM J. Computing 5(4), 1976.

69. Garey M. R. and D. S. Johnson: "The complexity of near-optimal graph coloring", J. ACM, 23(1), 1976.

70. Garey M. R.: "Optimal Binary Search Trees With Restricted Maximal Depth", SIAM J. Comp., 2: 1974.

71. Garsia A. M. and M. L. Wachs: " A new Algorithm for Minimum Cost Binary Trees", SIAM J. Comp., 6(4), Dec, 1977.

72. Gill, A.: "Introduction to the theory of Finite State Machines", McGraw-Hill Book Company, 1962.

73. Golomb S. W. and L. D. Baumert: "Backtrack Programming", J, ACM, 12: 516, 1965.

74. Halmos, P.: "Naïve Set Theory", D. Van Nostrand Company, Princeton, N. J., 1960.

75. Harary F.: "Graph Theory", Addison Wesley, Boston, 1969.

76. Hardy G. H. and E. M. Wright: "An Introduction to the Theory of Numbers", Oxford University press, London.

77. Harrison, M. A.: "Introduction to Formal Language Theory", Addison-Wesley Publishing Company, 1978.

78. Hohn, F. E.: "Applied Boolean Algebra", Macmillan Company, 1960.

79. Hopcroft, J. E. and J. D. Ullman: "Introduction to Automata Theory, Language and Computation", Addison-Wesley Publishing Company, 1979.

80. Hopcroft J. E. and J. D. Ullman: "Formal Languages and Their Relation to Automata", Addison-Wesley, Mass, 1969.

81. Horowitz, E., and S. Sahni: "Fundamentals of Computer Algorithms", Computer Science Press, 1978.

82. Hu, T. C.: "Integer Programming and Network Flows", Addison-Wesley Publishing Company, 1982.

83. Ital A and Y. Shiloach: "Maximal Flow in Planar Networks", SIAM J. Comp, May 1979.

84. John A Beachy and William D. Blair, "Abstract Algebra", 2nd Ed, Waveland Press, Illinois, 1996.

85. John E. Hopcroft & Jeffrey D. Ulalman: "Introduction to Automata Theory, Languages and Computation", Narosa Publishing House, 1989

86. *Jonassen A. and O. Dahl: "Analysis of an Algorithm for priority Queue Administration", BIT, 1975.*

87. *Jones N. D.: "Computability Theory : An Introduction", Academic Press, NY, 1973.*

88. *Kenneth H. Rosen, "Discrete Mathematics and its application", Tata McGraw-Hill, Fourth Ed, 2001.*

89. *Kleen S. C.: "General Recursive Functions of natural Numbers", Mathematische Annalen 112, 1936.*

90. *Knuth, D. E.: "The art of Computer Programming, vol 1, Fundamental algorithm", 2nd Ed, Addison-Wesley Publishing Company, 1973.*

91. *Kohavi, Z.: "Switching and Finite Automata Theory", 2nd Ed, McGraw-Hill Book Company, NY, 1978.*

92. *Kohavi Z.: "Switching and Finite Automata Theory", McGraw-Hill, NY, 1970.*

93. *Kolman, B., and R.C. Busby: "Discrete Mathematical Structures for Computer Science", PHI, 1984.*

94. *Korfhage, R. R.: "Discrete Computational Structures", Academic Press, New York, 1974.*

95. Korsh J. F.: "Greedy Binary Search Tree and Nearly Optimal", Information Proceeding Letters, 13(1), Oct 1981.

96. Lewis, H. R. and C. H. Papadimitriou: "Elements of the theory of Computation", Prentice-Hall, 1981.

97. *Lewis J. M.: "On the complexity of the maximum subgraph problem", Proc. 10th Annual ACM Symposium on the Theory of Computing", 1978.*

98. *Maibaum T. S. E.: "A generalized Approach to Formal Languages", J. Computer and System Sciences, 8(3), 1974.*

99. Manna Z. and A. Shamir: "The optimal Approach to Recursive programs", Comm. ACM, 20(11), Nov. 1977.

100. *Minsky M. L.: "Computation: Finite and Infinite Machines", Prentice Hall, Englewood Cliffs, New Jersey, 1967.*

101. *Mirsky L.: "Transversal Theory", Academic Press, New York, 1971.*

102. *Moore E. F. (ed): "Sequential Machines", Selected Papers, Addison-Wesley, reading, Mass, 1964.*

103. *Munro I.: " Efficient Determination of the Transitive Closure of a Directed Graph", Information Proceeding Letters, 1:56, 1971-72.*

104. *Myhill J.: "Finite Automata and the Representation of Events", WADD TR-57-624, Wright Patterson AFB, Ohio, 1957.*

105. *Myhill J.: "Linear Bounded automata", WADD TR-60-165, Wright Patterson AFB, Ohio, 1960.*

106. *Nievergelt J. J., J. C. Farrar and E. M. Reingold: " Computer Approaches to Mathematical Problems", Prentice Hall, Englewood Cliffs, New Jersey, 1974.*

107. *Nijenhuis A. and H. S. Wilf: "Combinatorial Algorithms", Academic Press, New York, 1975.*

108. *O'Neil P. E. and P. J. O'Neil: "A fast Expected Time Algorithm for Boolean Matrix Multiplication and Transitive Closure", Information and Control, 22:132-38, 1973.*

109. *Ore O.: "Graphs and their Uses", Random House and the L. W. Singer Co., New York, 1963.*

110. *Ore O.: "Theory of Graph", 38, American Mathematical Society, Providence, R. I., 1963.*

111. *Paley, H. and P. M. Weichsel: " A First Course in Abstract Algebra", Holt, Rinehart and Winston, NY, 1966.*

112. *Post E.: " A variant of recursively unsolvable problem", Bull, AMS, 52, 1946.*

113. *Reingold E.M., J. Nievergelt and N. Deo: "Combinatorial Algorithms: Theory and Practice", Prentice Hall, Englewood Cliffs, New Jersey, 1977.*

114. *Richard Johnsonbaugh, "Discrete Mathematics", Fifth Ed, Pearson Education Asia, 2002.*

115. *Robert W. Floyd: ``Algorithm 97 (SHORTEST PATH)", Communications of the ACM, 5(6): 345, **1962**.*

116. *Rogers H. Jr.: "The Theory of Recursive Functions and Effective Computability" McGraw-Hill, NY.*

117. *Roman, S. and Rota, G.-C. ``The Umbral Calculus." Adv. Math. 27, 95-188, 1978.*

118. *Roman, S. The Umbral Calculus. New York: Academic Press, 1984.*

119. *Rosenberg A. L. and L. Snyder: "Time and Space Optimality in B-Trees", ACM trans. Database System, march 1981.*

120. *Rounds W. C.: "Mappings and Grammars on trees", Math Systems Theory, 4(3), 1970.*

121. *Rutherford, D. E.: "Introduction to Lattice Theory", Oliver and Boyd, London, 1965.*

122. *Saaty, T. L. and Kainen, P. C. The Four-Color Problem: Assaults and Conquest. New York: Dover, p. 12, 1986.*

123. *Sahni, S.: "Concepts in Discrete Mathematics", Camelot Press, Fridley, Minn., 1981.*

124. *Sedgewick R.: "Permutation generation Method", ACM Computing Surveys, 9(2):137,june 1977.*

125. *Seiferas J. I.: "A note on prefixes of regular Languages", SIGACT News 6(1), 1974.*

126. *Shannon C. E. and J. McCarthy: " Automata Studies", Princeton University Press, Princeton, New Jersey.*

127. *Stanley R. J.: "Finite state representations of context free languages", Quarterly Prog. Rept. No. 76, MIT Research Lab. Elect., Cambridge Mass, 1965.*

128. *Stephan Warshall: ``A theorem on boolean matrices." Journal of the ACM,* 9(1):11-12, ***1962***.

129. *Stockmeyer L. J.: "The complexity of decision problems in automata theory and logic", MAC TR-133, project MAC, MIT, Cambridge, Mass, 1974.*

130. *Stoll, R. R.: "Set theory and Logic", W.H. Freeman and Company, San Francisco, 1963.*

131. *Strassen, V.: "Gaussian Elimination Is Not Optimal", Nemerische Mathematik, 13:354 –356 (1969): Xeroxed notes referred.*

132. *Tarjan R. E.: "Data Structures and Network Algorithms", Society for Industrial and applied Mathematics, Philadelphia, PA, 1983.*

133. *Thomas H. Cormen, Charles E. Leiserson and Ronald L. Rivest:* Introduction to Algorithms. *Cambridge: MIT Press,* ***1990***. *ISBN 0-262-03141-8.*

134. *Trembley, J. P., and R. P. Manohar,: "Discrete Mathematical Structures with application to Computer Science", McGraw-Hill Book Company, 1975.*

135. *Valiant, L. G. and M.S. Peterson: "Deterministic one-counter Automata", J. Computer and System sciences 10(3), 1975.*

136. *Vashistha A. R.: "Modern Algebra", 11th Ed, Krishna Prakashan Mandir, Meerut, India, 1981.*

137. *Walker W. A. and C. C. Gotlieb: "A Top Down Algorithm for Constructing Nearly Optimal Lexicographic Trees" in Graph Theory and computing, R read(ed) Academic Press, New York, 1972.*

138. *Warshall S.: "A theorem on Boolean Matrices", J. ACM, 9(1),11, 1962.*

139. *Weide B.: "A survey of analysis Techniques for Discrete Algorithm", ACM computing Survey, Dec 1977.*

140. *Whitworth W. A.: "DCC Exercises in Choice and chance", Hafner Publishing Company, 1965.*

141. *Wilson, R. J.: " Introduction to Graph Theory", Academic Press, 1972.*

142. *Yao A. C. and F. F. Yao: "The complexity of Searching an ordered Random Table", Proc. Sum on Foundation of Comp Science, 1976.*

143. *Yasuhara A.: "Recursive Function Theory and Logic", Academic Press, New York, 1971.*

144. *Zohar Manna: "mathematical theorey of Computation": McGraw-Hill, KogaKusha, 1974.*

Web Sites

1. http:// nms.lcs.mit.edu
2. http://childsupportguidelines.com
3. http://ciips.ee.uwa.edu.au/~morris/Year2/PLDS210/search_trees.html
4. http://datastructures.itgo.com/trees/traversal.htm
5. http://distance-ed.math.tamu.edu
6. http://eda.sci.univr.it/Publications/publications.html
7. http://kilby.stanford.edu/~rvg/352/duality.ps
8. http://math.skku.ac.kr
9. http://mathcircle.berkeley.edu
10. http://mathforum.org
11. http://matholymp.com
12. http://mathworld.wolfram.com
13. http://s13a.math.aca.mmu.ac.uk/PostAlevelMathsQs/Proof/Proof5.html
14. http://sampl.eng.ohio-state.edu
15. http://sweb.uky.edu/~jrbail01
16. http://www.ai.mit.edu
17. http://www.bit.umkc.edu
18. http://www.cbu.edu
19. http://www.cee.hw.ac.uk
20. http://www.cogs.susx.ac.uk
21. http://www.cs.byu.edu
22. http://www.cs.dartmouth.edu
23. http://www.cs.tufts.edu
24. http://www.cs.uleth.ca

25. http://www.cs.usm.maine.edu
26. http://www.cs.wpi.edu/~cs2223/d98
27. http://www.cse.iitd.ernet.in
28. http://www.cut-the-knot.com/
29. http://www.earlham.edu
30. http://www.foad.org
31. http://www.geocities.com
32. http://www.halcyon.com
33. http://www.math.caltech.edu/courses/01Ma6cHW8.ps
34. http://www.math.hawaii.edu
35. http://www.math.sc.edu
36. http://www.math.toronto.edu
37. http://www.mathematicshelpcentral.com
38. http://www.owlnet.rice.edu
39. http://www.physics.udel.edu
40. http://www.pms.informatik.uni-muenchen.de
41. http://www.seas.smu.edu
42. http://www.theomatics.com/theomatics
43. http://www.umr.edu
44. http://www.unc.edu/~rowlett/MATH148
45. http://www.visorloop.com
46. http://www.whatis.com
47. http://www.wikipedia.com
48. http://www2.edc.org
49. http://www-cad.eecs.berkeley.edu
50. http://www-cad.eecs.berkeley.edu
51. http://www-hkn.eecs.berkeley.edu
52. http://zimmer.csufresno.edu/~larryc/proofs/proofs.contradict.html

Index

A

Abelian group, 172
Absorption law, 245,
Abstractly Identical, 197, 435
Acyclic graph, 343
Addition of Matrix, 423
Addition Principle of Counting, 140
Adjacency matrix, 297, 298
Adjacent nodes, 295
Alternating group, 217
Alternating set, 215
Anti-chain, 227
Anti-parallel arc, 299
Anti-symmetric matrix, 428
Anti-symmetric relation, 19
Arborescence, 344
Arc disjoint path, 311
Arc, 292
Articulation point, 314
Associative law, property, 12, 171, 245
Asymmetric relation, 18
Atomic proposition, 276

B

Backtracking, 84
Backus Naur Form, 380
Basis, 458
Bijection, 51, 206
Binary operation, 171
Binary tree, 345
Binomial generating function, 68-75
Bipartite graph, 305
Block, 26
Boolean expression, 255
Boolean Function, 254, 255
Boolean matrix, 33
Boolean Polynomial, 255
Boolean ring, 447
Bounded lattice, 247
Bounded sequence, 63
Bounding elements, 233
Branches, 292
Breadth First Traversal, 351
Bridge, 316

C

Cancellation Law, 178
Cardinality, 4, 8
Cartesian Product, 11
Cell, 26
Central groupoid, 224
Chain, 227
Change of variable, 106
Characteristic equation, 88, 89-90
Characteristic Function, 66
Characteristics of ring, 437
Chomsky Grammar, 375
Chromatic number, 334
Chromatic polynomial, 335
Circuit, cycle, 309
Circular Permutation, 143
Circular relation, 45
Classification of recurrence relation, 82
Closed polygon, 333
Closure of relation, 32
Closure property, 171
Cofactor of matrix, 428
Coloring of Graph, 333
Column, 422
Column matrix, 425
Combination, 149
Commutative ring, 430
Commutative law, 12, 172
Comparable elements, 227
Compatible relation, 45
Complement of Graph, 304
Complement of relation, 16
Complement of set, 10
Complemented lattice, 250
Complete binary tree, 345
Complete bipartite graph, 306
Complete graph, 301
Complete lattice, 240
Complete n-ary tree, 345
Complete order, 227
Complete Solution, 95
Complete tripartite graph, 306
Complex, 185

Composition of function, 49
Composition of Permutations, 207
Composition Table, 180
Compound Proposition, 276
Congruence Modulo, 22
Congruence relation, 401
Conjunctive Normal Form, CNF, 256
Connected graph, 309
Constant function, 55
Context free grammar, 375
Context sensitive grammar, 375
Contingency, 278
Contradiction, 278
Convolution, 74
Coset, 191
Cotree, 352
Counter Example, 135
Cover of a set, 45
Cross product, 451
Cubic graph, 302
Cut point, 314
Cyclic Group, 201
Cyclic permutation, 210
Cyclomatic number, 352

D

Dangle node, 295
Degree of permutation, 206
Degree spectrum, 308
Degree, valance, 294
Degree of graph, 294
DeMorgan's law, 12, 245
Depth first Traversal, 352
Depth of a tree, 344
Derivation tree, 374
Descriptive Phrase Method, 4
Determinant, 428
Deterministic FSM, 391
Diagonal element, 425
Diagonal matrix, 425
Diagonal Relation, 32
Diagrammatic Way, 4
Diameter of graph, 308
Dictionary order, 232
Digraph and relation, 29
Dijsktra algorithm, 323-329
Diophantine Equation, 167
Direct proof, 125
Directed graph, 294
Directed tree, 344
Disconnected graph, 309
Discrete graph, 300
Discrete Numeric Function, 62
Disjoint cycle, 211
Disjoint set, 5
Disjunctive normal Form, DNF, 256
Distributive lattice, 247
Distributive law, 12, 280
Division ring, 433
Domain set, 23, 48
Don't Care Condition, 260
Dot product, 451
Dual Lattice, 241
Dual Poset, 226
Duet, Quad, Octet, 263

E

Edge, 292
Edge connectivity, 310
Elementary path, circuit, 308,309
Embedded ring, 438
Empty relation, 17
Empty set, 5
End points, 292
Endomorphism, 195
Enumerated way, 3
Equal matrix, 427
Equal set, 6
Equivalence relation, 21
Equivalent set, 7
Euler path, circuit, 314-318
Even permutation, 215
Everywhere Defined, 47
Existence of identity, 171
Existence of inverse, 171
Existential proof, 136
Explicit Formula, 64
Exponential generating function, 68-75
Expression tree, 349

F

Fermat's Theorem, see theorem 6, 138
Fibonacci, 69
Field, 443
Final State, 390

Finite automata, see FSM
Finite Boolean Algebra, 251
Finite dimensional , 458
Finite graph, 296
Finite group, 173
Finite lattice, 240
Finite poset, 225
Finite sequence, 62
Finite set, 4
Finite state Machine, 389
First element, see least upper bound
Fleury's Algorithm, 316-318
Floyd Warshall's Algorithm, 329-332
Forest, 347
Forward Chaining, 84
Four color theorem, 340
Free grammar, 376
Free Semigroup, 175
Function, 47

G

Gaussian elimination, 387
Gaussian integer, 464
General term, 62
Generating function, 62, 68-75, 147, 156
Generator, 201
Geodesic, 308
Graph, 293
Greatest element, see least upper bound
Greatest Lower bound, glb, Infimum, 236
Group, 171
Group of Permutations, see permutation group
Groupoid, 172

H

Hamiltonian path, circuit, 318-322
Hasse diagram, 228
Height of a tree, 344
Homogeneous recurrence equation, 82
Homomorphism, 195
Homomorphism of ring, 449
Homomorphism of field, 449
Hyper-graph, 308

I

Ideal, 438
Ideal closure, 439
Idempotent law, 12, 245, 280
Identity function, permutation, 49,
Identity Law, 12,
Identity matrix, 426
If and only if, 129
Image, 47
Improper Normal subgroup, 193
Improper Subgroup, 186
Improper subring, 436
Improper vector subspace, 455
Inaccessible state, 396, 400
Incidence Matrix, 297,298
Incident, 292
Indegree, 294
Induced Operation, 185
Induced subgraph, 303
Inductive step, 130
Infinite graph, 296
Infinite group, 173
Infinite lattice, 240
Infinite poset, 225
Infinite sequence, 62, 63
Infinite set, 4
Initial condition, 65
Initial State, start state, 390
Initial step, 130
Injection, 51
Inorder traversal, 350
Integral domain, 433
Internal node, 344
Intersection, 9
Into function, 52
Intransitive relation, 19
Inverse of complexes, 187
Inverse matrix, 428
Inverse relation, 17
Invertible function, 50
Invertible matrix, 428
Irreflexive relation, 18
Isograph, 294
Isolated node, 295
Isomorphic graph, 307
Isomorphic lattice, 250
Isomorphic Poset, 237
Isomorphism, 197

J

Join, 240

K

Karnaugh Map, 261-269
K-connected, 314
Kernel of group homomorphism, 450
Kernel of ring homomorphism, 450
Kleen, 394
K-regular, 294
Kruskal's algorithm, 356
Kuratowski's graph, theorem, 341

L

Labeled tree, 348
Language, 370
Last element, see least upper bound
Lattice, 240
Laws of Complement, 12,
Laws of inclusion, 12,
Lead element, 331
Leading element, 425
Leaf, 344
Least Upper Bound, lub, Supremum, 236
Least, Smallest, see least upper bound
Left Coset, 191
Left ideal, 438
Length of path, cycle, 308, 309
Lexicography Order, 232
Linear combination, 457
Linear graph, 300
Linear order relation, 227
Linear permutation, 143
Linear recurrence equation, 83
Linear span, 457
Linearly dependent, 457
Linearly independent, 457
Links, 292
List representation of graph, 296
Loop, 295
Lower Bound, 235
Lower triangular matrix, 428

M

Machine congruence, 401
Many to one, 53
Mapping, 48
Master diagram, 382
Master row, 331
Mathematical Induction, 130
Matrix, 422
Matrix and relation, 29
Matrix multiplication, 424
Matrix representation of graph, 297
Maximal element, 233
Maxterm, 256
Meet, 240
Member, 1
Membership of a set, 2
Method of Contradiction, see proof by contradiction
Method of proof, 124
Minimal element, 233
Minimum Hamiltonian path, circuit, 322
Minimum spanning tree, 352
Minor of matrix, 428
Minterm, 256
Mixed graph, 294
Modular lattice, 288
Monoid, 172
Monotonic decreasing, 63
Monotonic increasing, 63
Monotonic sequence, 63
Mother of determinants, 422
Multi graph, 299
Multi set, 8
Multiplication Principle of counting, 139
Multiplicity, 8

N

N Cube graph, 302
N-ary tree, 345
n-cycle graph, 301
Node disjoint path, 311
Nodes, 292
Non- Linear recurrence equation, relation, 82
Non-deterministic FSM, 391
Non-distributive lattice, 247
Non-homogeneous recurrence equation, 82
Non-planar graph, 340
Non-reflexive relation, 18
Non-singular matrix, 428
Non-symmetric relation, 20
Non-transitive relation, 19
Normal subgroup, 193
Null ideal, 438
Null relation, 17

Null ring, 430
Null set, 5
Numeric function, 62

O

Odd permutation, 215
One to one function, 51
Onto function, 51
Open polygon, 333
Open Selection, 152
Optimization of Boolean expression, 260
Order of element, 187
Order of function, 55
Order of group, 173
Order of recurrence relation, 83
Order of ring, 437
Ordered field, 447
Ordered integral domain, 433
Ordered pair, 11
Orientable graph, 363
Outclassed set, 363
Outdegree, 294
Overlapping set, 5

P

Parallel arc, 299
Partial Order relation, 225
Partial sum, 66, 283
Particular part, 93
Particular solution, 96
Partition, equivalence, 25, 27
Path, 308
Permutation, 139, 206
Permutation group, 206
Phrase Structure Grammar, 375
Pigeon Hole Principle, 159
Planar graph, 340
Platonic graph, 360
Point Set, 4
Polish infix, 281-285
Polish postfix, 281-285
Polish prefix, 281-285
Polygon, 333
Polynomial, 439
Polynomial ring, 443
Poset, 225
Positional tree, 349
Postorder traversal, 350
Power Set, 7
Predicate Calculus, 276
Pre-image, 47
Preorder Traversal, 350
Prim' algorithm, 353
Prime field, 447
Prime implicant, 260
Principal ideal, 439
Principal ideal ring, 439
Principle of Duality, 241
Principle of inclusion and exclusion, 13
Principle of optimality, 324
Principle of uniqueness, 93
Product Lattice, 243
Product of complexes, 187
Product Partial Order, 231
Product Poset, 232
Proof by Case by Case, 135
Proof by construction, 137
Proof by contradiction, 126
Proof by contra-positive, 128
Proof by Exhaustion, 135
Proper coloring, 333
Proper ideal, 438
Proper subgroup, 186
Proper subring, 436
Proper subset, 6
Proper subspace, 455
Properties of ring, 430
Properties of vector space, 454
Proposition, 276
Propositional Calculus, 276
Pseudo graph, 300
Pythagorean Triplet, 137

Q

Quasi Order relation, 232
Quaternion group, 185
Quine McClusky, 269
Quotient FSM, 402
Quotient field, 448
Quotient graph, 305
Quotient group, see residue class modulo the subgroup, 191
Quotient set, 26, 27

R

Range set, 23, 48
Reachability matrix, 35
Real Valued function, 55
Recognizer, 389
Rectangular matrix, 425
Recurrence equation, relation, 64, 83
Recurrence Formula, 64
Redundant state, 396
Reflexive closure, 32
Reflexive relation, 17
Region, 333
Regular expression, 384
Regular expression equation, 387
Regular grammar, 375, 412
Regular graph, 294
Regular set, 384
Relation, 11, 15
Relation of Isomorphism, 200
Restricted Permutation, 145
Right coset, 191
Right ideal, 438
Right linear grammar, 375, 412
Ring, 429
Ring isomorphism, 434
Ring with unity, 430
Ring with zero divisors, 432
Ring without zero divisors, 432
Rolling quad, 267
Root, 344
Rooted tree, 344
Row, 422
Row matrix, 425
R-relative Set, 23
Rule of reversal, 177

S

Scalar matrix, 425
Self complementary graph, 361
Semigroup, 172
Sentence, 370
Sentential form, 369-370
Separable, 314
Sequence, 61, 62
Series, 66
Set, 1
Set Difference, 9
Set Former Method, 3
Set Representation, 3
Set Theoretic Method, 4
Shortest path, 308, 322-332
Simple graph, 299
Simple group, 193
Simple order, 227
Simple path, circuit, 308, 309
Singleton set, See unit set
Singular matrix, 428
Sink node, 294
Skew field, 433
Skew-symmetric matrix, 428
Source node, 294
Spanning subgraph, 302
Spanning tree, 352
Square matrix, 425
Stable complex, 185
Standard sets, 2
State transition table, 390
Strong Order, 232
Strongly connected graph, 310
Subfield, 7
Subgraph, 302
Subgroup, 185
Sub-lattice, 240
Sub-matrix, 426
Subring, 436
Subring closure, 436
Subset, 6
Subtraction of Matrix, 424
Summation Method, 86
Superset, 6
Symbolic way, 3
Symmetric Closure, 33
Symmetric Difference, 9
Symmetric group, 210
Symmetric matrix, 428
Symmetric relation, 18
Symmetric set, 206
Syntax diagram, 380

T

Tabulation Method, 3
Tautology, 12, 278
Terminating Condition, 65
Topological Sorting, 238
Total order relation, 227

Total Solution, 95
Total vocabulary, 369
Transference of group Properties, 199
Transformation, 48
Transitive closure, 33
Transitive relation, 19
Transpose, 423, 427
Transposition, 211
Trapping state, 396
Traversal of graph, 348
Tree, 343
Tripartite graph, 306
Trivial graph, 300
Trivial subring, 436
Trivial subgroup, 186
Truth table, 254
Turing machine, 368
Type of lattice, 247
Type 0 grammar, 375
Type 1 Grammar, 375
Type 2 Grammar, 375
Type 3 grammar, 375

U

Unambiguous grammar, 412
Unbounded Sequence, 63
Undirected graph, 294
Unilaterally connected graph, 310
Union, 9
Uniqueness of identity, 176
Uniqueness of inverse, 176
Unit ideal, 438
Unit matrix, 426
Unit set, 5
Universal relation, 16
Universal set, 5
Unrestricted grammar, 375
Unstable complex, 185
Upper bound, 235
Upper triangular matrix, 428
Usual partial order, 225

V

Valence, degree, see degree
Vector, 451
Vector subspace, 455
Vector subspace closure, 458
Vector space, 452, 453
Venn Diagram, 4
Vertex connectivity, 310
Vertices, 292
Void relation, 17
Void set, 5

W

W-Compatible state, 403
Warshall's algorithm, 35
Weekly connected graph, 310
Well formed formula, 281
Well order relation, 240
Well Ordered Set, 240

Z

Zero matrix, 426
Zero ring, 430
Zero divisors, 432
Zero Polynomial, 439
Zero subspace, 455
Zero vector, 451